Advances in Intelligent Systems and Computing

188

Editor-in-Chief

Prof. Janusz Kacprzyk
Systems Research Institute
Polish Academy of Sciences
ul. Newelska 6
01-447 Warsaw
Poland
E-mail: kacprzyk@ibspan.waw.pl

T0135004

For further volumes:
http://www.springer.com/series/11156

Václav Snášel, Ajith Abraham,
and Emilio S. Corchado (Eds.)

Soft Computing Models in Industrial and Environmental Applications

7th International Conference, SOCO'12,
Ostrava, Czech Republic,
September 5th–7th, 2012

 Springer

Editors
Prof. Václav Snášel
VŠB-TU Ostrava
Ostrava
Czech Republic

Prof. Emilio S. Corchado
Universidad de Salamanca
Salamanca
Spain

Prof. Ajith Abraham
Machine Intelligence Research Labs
(MIR Labs)
Scientific Network for Innovation
and Research Excellence
Auburn, Washington
USA

ISSN 2194-5357
ISBN 978-3-642-32921-0
DOI 10.1007/978-3-642-32922-7
Springer Heidelberg New York Dordrecht London

e-ISSN 2194-5365
e-ISBN 978-3-642-32922-7

Library of Congress Control Number: 2012945408

Preface

This volume of Advances in Intelligent and Soft Computing contains accepted papers presented at SOCO 2012, held in the beautiful and historic city of Ostrava (Czech Republic), in September 2012.

Soft Computing represents a collection or set of computational techniques in machine learning, computer science and some engineering disciplines, which investigate, simulate, and analyze very complex issues and phenomena.

After a through peer-review process, the SOCO 2012 International Program Committee selected 76 papers which are published in these conference proceedings, and represents an acceptance rate of 38%. In this relevant edition a special emphasis was put on the organization of special sessions. Three special sessions were organized related to relevant topics as: Soft computing models for Control Theory & Applications in Electrical Engineering, Soft Computing Models for Biomedical Signals and Data Processing and Advanced Soft Computing Methods in Computer Vision and Data Processing.

The selection of papers was extremely rigorous in order to maintain the high quality of the conference and we would like to thank the members of the Program Committees for their hard work in the reviewing process. This is a crucial process to the creation of a high standard conference and the SOCO conference would not exist without their help.

SOCO 2012 enjoyed outstanding keynote speeches by distinguished guest speakers: Prof. Ponnuthurai Nagaratnam Suganthan, Prof. Jeng-Shyang Pan, Prof. Marios M. Polycarpou, Prof. Fanny Klett and Mr. Milan Kladnicek.

For this special edition, as a follow-up of the conference, we anticipate further publication of selected papers in special issues of prestigious international journal as Neurocomputing (ELSEVIER), Journal of Applied Logic (ELSEVIER) and Neural Network World (Institute of Computer Science CAS in cooperation with the Czech Technical University, Prague, Faculty of Transportation Sciences).

Particular thanks go as well to the Conference main Sponsors, IT4Innovations, VŠB-Technical University of Ostrava, IEEE.- Systems, Man and Cybernetics Society CzechoSlovakia, IEEE.- Systems, Man and Cybernetics Society Spain, MIR labs, Spanish Association for Artificial Intelligence, IFCOLOG and MIDAS project supported by

Spanish Ministry of Science and Innovation TIN2010-21272-C02-01 (funded by the European Regional Development Fund).

We would like to thank all the special session organizers, contributing authors, as well as the members of the Program Committees and the Local Organizing Committee for their hard and highly valuable work. Their work has helped to contribute to the success of the SOCO 2012 event.

September 2012

Václav Snášel
Ajith Abraham
Emilio S. Corchado

Organization

General Chair

Emilio Corchado   University of Salamanca, Spain

International Advisory Committee

Ashraf Saad	Armstrong Atlantic State University, USA
Amy Neustein	Linguistic Technology Systems, USA
Ajith Abraham	Machine Intelligence Research Labs - MIR Labs, Europe
Jon G. Hall	The Open University, UK
Paulo Novais	Universidade do Minho, Portugal
Antonio Bahamonde	President of the Spanish Association for Artificial Intelligence, AEPIA
Michael Gabbay	Kings College London, UK
Isidro Laso-Ballesteros	European Commission Scientific Officer, Europe
Aditya Ghose	University of Wollongong, Australia
Saeid Nahavandi	Deakin University, Australia
Henri Pierreval	LIMOS UMR CNRS 6158 IFMA, France

Industrial Advisory Committee

Rajkumar Roy	The EPSRC Centre for Innovative Manufacturing in Through-life Engineering Services, UK
Amy Neustein	Linguistic Technology Systems, USA
Jaydip Sen	Innovation Lab, Tata Consultancy Services Ltd., India

Program Committee Chair

Emilio Corchado	University of Salamanca, Spain
Václav Snášel	VSB-Technical University of Ostrava, Czech Republic
Ajith Abraham	VSB-Technical University of Ostrava, Czech Republic

Program Committee

Abdelhamid Bouchachia	Alps-Adriatic University of Klagenfurt, Austria
Aboul Ella Hassanien	Cairo University, Egypt
Abraham Duarte	University King Juan Carlos, Spain
Adil Baykasoglu	University of Gaziantep, Turkey
Alberto Freitas	University of Porto, Portugal
Alexander Gegov	University of Portsmouth, UK
Alexis Marcano-Cedeño	Polytechnic University of Madrid, Spain
Álvaro Herrero	University of Burgos, Spain
Amy Neustein	Linguistic Technology Systems, USA
Ana Almeida	Polytechnic of Porto, Portugal
Ana Carolina Lorena	Universidade Federal do ABC, Brazil
Ana Cristina Bicharra	Universidad Federal Fluminense, Brazil
Ana Gil	University of Salamanca, Spain
André CPLF de Carvalho	University of São Paulo, Brazil
Andrea Schaerf	University of Udine, Italy
Andrés Piñón Pazos	University of A Coruna, Spain
Ángel Arroyo	University of Burgos, Spain
Anna Bartkowiak	University of Wroclaw, Inst of Computer Science, Poland
Antonio Araúzo Azofra	University of Cordoba, Spain
Antonio Berlanga	University Carlos III of Madrid, Spain
Antonio Peregrín	University of Huelva, Spain
Ashish Umre	University of Sussex, UK
Aureli Soria-Frisch	Starlab Barcelona S.L., Spain
Ayeley Tchangani	University of Toulouse III, France
Belén Vaquerizo	University of Burgos, Spain
Benjamín Ojeda-Magaña	University of Guadalajara, Spain
Benoît Otjacques	Public Research Centre - Gabriel Lippmann, Luxembourg
Bernadetta Kwintiana Ane	University of Stuttgart, Germany
Bhavya Alankar	Hamdard University
Bogdan Gabrys	Bournemouth University, UK
Borja Sanz Urquijo	University of Deusto, Spain
Bruno Apolloni	University of Milan, Italy
Bruno Baruque	University of Burgos, Spain
Camelia Chira	Babes-Bolyai University, Romania
Carlos Laorden	University of Deusto, Spain
Carlos Pereira	Polytechnic Institute of Coimbra, Portugal
Carlos Redondo Gil	University of León, Spain
Cesar Analide	Universidade do Minho, Portugal
César Hervás	University of Cordoba, Spain
Chia-Chen Lin	Providence University, Taiwan

Crina Grosan   Babes-Bolyai University, Romania
Daniel Escorza   Infranor SAS, Spain
Daniela Zaharie   West University of Timisoara, Romania
Daryl Hepting   University of Regina, Canada
David Griol   University Carlos III of Madrid, Spain
David Meehan   Dublin Institute of Technology, Ireland
David Oro Garcia   Herta Security, Spain
Diego Andina   Universidad Politécnica de Madrid, Spain
Donald Davendra   VSB - Technical University of Ostrava,
                                      Czech Republic
Dragan Simic   University of Novi Sad, Serbia
Dusan Husek   Institute of Computer Science Academy of
                                      Sciences of the Czech Republic, Czech Republic
Eduardo Solteiro Pires   University of Trás-os-Montes and Alto Douro,
                                      Portugal
Eleni Mangina   University College Dublin, Ireland
Enrique Herrera-Viedma   University of Granada, Spain
Ernesto Damiani   University of Milan, Italy
Eva Volna   University of Ostrava, Czech Republic
Fanny Klett   German Workforce ADL Partnership Lab,
                                      Germany
Fatos Xhafa   Universtat Politècnica de Catalunya, Spain
Fernando Gomide   University of Campinas, Brazil
Florentino Fernández Riverola   University of Vigo, Spain
Francesco Marcelloni   University of Pisa, Italy
Francesco Masulli   University of Genova, Italy
Francisco Herrera   University of Granada, Spain
Francisco Martínez   Inmotia, Spain
Frank Klawonn   Ostfalia University of Applied Sciences, Denmark
Frederico G. Guimaraes   Universidade Federal de Minas Gerais, Brazil
Georgios Ch. Sirakoulis   Democritus University of Thrace, Greece
Gerald Schaefer   Loughborough University, UK
Gregg Vesonder   University of Pennsylvania, USA
Gregorio Sainz   CARTIF Technological Centre, Spain
Haibin Duan   Beihang University, China
Harleen Kaur   United Nations University-IIGH, Malaysia
Héctor Quintián   University of Salamanca, Spain
Horia Nicolai Teodorescu   "Gheorghe Asachi" Technical University of Iasi,
                                      Romania
Humberto Bustince   Public University of Navarra, Spain
Hussein Hiyassat   Al-Bayt University, Jordan
Ignacio Rojas   University of Granada, Spain
Igor Santos   University of Deusto, Spain
Irina Perfilieva   University of Ostrava, Czech Republic

Ivan Zelinka	VSB - Technical University of Ostrava, Czech Republic
Jan Platoš	VSB-Technical University of Ostrava, Czech Republic
Janez Brest	University of Maribor, Slovenia
Javier Carbó	University Carlos III of Madrid, Spain
Javier Nieves	University of Deusto, Spain
Javier Sedano	Instituto tecnológico de Castilla y León, Spain
Jean Caelen	CNRS, laboratoire LIG
Jerzy Grzymala-Busse	University of Kansas, USA
Jesús García-Herrero	University Carlos III of Madrid, Spain
Jesús Luna	Barcelona Digital Technology Centre, Spain
Jiří Dvorský	VSB-Technical University of Ostrava, Czech Republic
Jiří Pospíchal	Slovak University of Technology, Slovakia
Jonathan Lee	National Central University, Taiwan
Jorge Díez Peláez	University of Oviedo, Spain
Jorge Lopes	Brisa/IST
José Alfredo F. Costa	Federal University of Rio Grande de Norte, Portugal
José Antonio Gómez	University of Castilla la Mancha, Spain
José Antonio Lozano	University of País Vasco, Spain
José Fco. Martínez Trinidad	National Institute for Astrophysics, Optics and Electronics, Spain
José Luis Calvo Rolle	University of A Coruña, Spain
José Manuel Benítez Sánchez	University of Granada, Spain
José Manuel Molina	University Carlos III of Madrid, Spain
José María Peña	Polytechnic University of Madrid, Spain
José Ramón Villar	University of Oviedo, Spain
José Riquelme	University of Sevilla, Spain
José Valente de Oliveira	University of Algarve, Portugal
Josef Tvrdík	University of Ostrava, Czech Republic
Jouni Lampinen	University of Vaasa, Finland
Juan Álvaro Muñoz Naranjo	Universidad de Almería, Spain
Juan Gómez Romero	University Carlos III of Madrid, Spain
Juan José del Coz Velasco	University of Oviedo, Spain
Juan Manuel Corchado	University of Salamanca, Spain
Kai Qin	INRIA Grenoble Rhone-Alpes, France
Kai Xiao	Shanghai Jiao Tong University, China
Kalyamoy Deb	Indian Institute of Technology Kanpur, India
Lahcene MITICHE	University of Djelfa, Algeria
Laura García Hernández	University of Cordoba, Spain
Leocadio González Casado	Universidad de Almería
Leticia Curiel	University of Burgos, Spain

Luciano Sanchez Ramos   University of Oviedo, Spain
Luciano Stefanini   University of Urbino "Carlo BO", Italy
Luis Correia   Lisbon University, Portugal
Luis Nunes   ISCTE, Portugal
Luis Paulo Reis   University of Porto, Portugal
M. Chadli   UPJV Amiens France, France
Mª Dolores Muñoz   University of Salamanca, Spain
Maciej Grzenda   Warsaw University of Technology, Poland
Manuel Graña   University of País Vasco, Spain
Manuel J. Martínez   COIT, Spain
Marcin Paprzycki   Polish Academy of Science, Poland
Marco Cococcioni   University of Pisa, Italy
Marco Mora   Universidad Católica del Maule, Chile
María João Viamonte   Polytechnic of Porto, Portugal
María José del Jesus   University of Jaen, Spain
María N. Moreno   University of Salamanca, Spain
María Pia Fanti   Politecnico di Bari, Italy
Mario G.C.A. Cimino   University of Pisa, Italy
Mario Köppen   Kyushu Institue of Technology, Japan
Marius Balas   "Aurel Vlaicu" University of Arad, Romania
Martin Macaš   Czech Technical University in Prague,
                                  Czech Republic
Martin Štěpnička   University of Ostrava, Czech Republic
Mazdak Zamani   Universiti Teknologi Malaysia, Malaysia
Mehmet Aydin   University of Bedfordshire, UK
Michael N. Vrahatis   University of Patras, Greece
Michał Woźniak   Wroclaw University of Technology, Poland
Miguel Ángel Patricio   University Carlos III of Madrid, Spain
Milos Kudelka   VSB-Technical University of Ostrava,
                                  Czech Republic
Miroslav Burša   Czech Technical University, Czech Republic
Mohamed Elwakil   Cairo University, Egypt
Mohammed Eltaweel   Arab Academy for Science, Technology, and
                                  Maritime Transport, Egypt
Nabil Belacel   National Research Council of Canada, Canada
Nashwa El-Bendary   Arab Academy for Science, Technology, and
                                  Maritime Transport, Egypt
Nedhal A. Al-Saiyd   Applied Science University, Jordan
Neill Parkinson   Valence, UK
Neveen I. Ghali   Al-Azhar University, Egypt
Noelia Sánchez Maroño   University of A Coruña, Spain

Óscar Castillo	Tijuana Institute of Technology, Mexico
Óscar Fontenla Romero	University of Coruña, Spain
Óscar Luaces	University of Oviedo, Spain
Ovidio Salvetti	ISTI-CNR, Italy
Paulo Moura Oliveira	University of Trás-os-Montes and Alto Douro, Portugal
Paulo Novais	Universidade do Minho, Portugal
Pavel Kordík	Czech Technical University, Czech Republic
Pavel Krömer	VSB-Technical University of Ostrava, Czech Republic
Pedro Antonio Gutierrez	University of Córdoba, Spain
Pedro M. Caballero Lozano	CARTIF Technological Center, Spain
Petr Gajdoš	VSB-Technical University of Ostrava, Czech Republic
Petr Musilek	University of of Alberta, Canada
Petr Pošík	Czech Technical University in Prague, Czech Republic
Petrica Pop	North University of Baia Mare, Romannia
Petro Gopych	Universal Power Systems USA-Ukraine LLC, Ukraine
Pierre-François Marteau	Université de Bretagne-Sud, France
Rabie Ramadan	Cairo University, Egypt
Rafael Bello	Central University Marta Abreu, Cuba
Ramón Ferreiro García	University of Coruña, Spain
Raquel Redondo	University of Burgos, Spain
Richard Duro	University of Coruña, Spain
Robert Burduk	Wroclaw University of Technology, Poland
Roman Neruda	ASCR, Czech Republic
Roman Senkerik	Tomas Bata University in Zlin, Czech Republic
Rosa Basagoiti	Mondragon University, Spain
Rosario Girardi	Federal Universty of Maranhão, Brazil
Rui Sousa	Universidade of Minho, Portugal
Santiago Porras	University of Burgos, Spain
Sara Rodríguez	University of Salamanca, Spain
Sara Silva	INESC-ID, Lisboa, Portugal
Sebastián Ventura Soto	University of Córdoba, Spain
Shampa Chakraverty	NSIT, India
Shyue-Liang Wang	National University of Kaohsiung, Taiwan
Soumya Banerjee	Birla Institute of Technology, India
Stefano Pizzuti	Energy New technology and Environment Agency
Sung-Bae Cho	Yonsei University, Korea
Susana Ferreiro Del Río	TEKNIKER, Spain
Syed Aljunid	United Nations University-IIGH, Malaysia

Teresa B. Ludermir	Federal University of Pernambuco, Brazil
Turkay Dereli	University of Gaziantep, Turkey
Tzung Pei Hong	National University of Kaohsiung, Taiwan
Urko Zurutuza Ortega	Mondragon University, Spain
Valentina Casola	Università degli Studi di Napoli Federico II, Italy
Valentina E. Balas	"Aurel Vlaicu" University of Arad, Romania
Verónica Tricio	University of Burgos, Spain
Vicente Martín	Universidad Politécnica de Madrid, Spain
Vilém Novák	University of Ostrava, Czech Republic
Vivian F. López	University of Salamanca, Spain
Wei-Chiang Hong	Oriental Institute of Technology, Taiwan
Witold Pedrycz	University of Alberta, Canada
Xiao-Zhi Gao	Aalto University, Finland
Yin Hujun	University of Manchester, UK
Yuehui Chen	University of Jinan, China
Zhihua Cui	Taiyuan University of Science and Technology, China
Zita Vale	Polytechnic of Porto, Portugal
Zuzana Oplatkova	Tomas Bata University in Zlin, Czech Republic

Special Sessions

Soft Computing Models for Control Theory & Applications in Electrical Engineering

Pavel Brandstetter	VSB - Technical University of Ostrava, Czech Republic
Emilio Corchado	University of Salamanca, Spain
Daniela Perdukova	Technical University of Kosice, Slovak Republic
Jaroslav Timko	Technical University of Kosice, Slovak Republic
Jan Vittek	University of Zilina, Slovak Republic
Jaroslava Zilkova	Technical University of Kosice, Slovak Republic
Jiri Koziorek	VSB - Technical University of Ostrava, Czech Republic
Libor Stepanec	UniControls a.s., Czech Republic
Martin Kuchar	UniControls a.s., Czech Republic
Milan Zalman	Slovak University of Technology, Slovak Republic
Pavel Brandstetter	VSB - Technical University of Ostrava, Czech Republic
Pavol Fedor	Technical University of Kosice, Slovak Republic
Petr Palacky	VSB - Technical University of Ostrava, Czech Republic
Stefan Kozak	Slovak University of Technology, Slovak Republic

Soft Computing Models for Biomedical Signals and Data Processing

Lenka Lhotská	Czech Technical University, Czech Republic
Martin Macaš	Czech Technical University, Czech Republic
Miroslav Burša	Czech Technical University, Czech Republic
Chrysostomos Stylios	TEI of Epirus, Greece
Dania Gutiérrez Ruiz	Cinvestav, Mexico
Daniel Novak	Czech Technical University, Czech Republic
George Georgoulas	TEI of Epirus, Greece
Michal Huptych	Czech Technical University, Czech Republic
Petr Posik	Czech Technical University, Czech Republic
Vladimir Krajca	Czech Technical University, Czech Republic

Advanced Soft Computing Methods in Computer Vision and Data Processing

Irina Perfilieva	University of Ostrava, Czech Republic
Vilém Novák	University of Ostrava, Czech Republic
Antonín Dvořák	University of Ostrava, Czech Republic
Marek Vajgl	University of Ostrava, Czech Republic
Martin Štěpnička	University of Ostrava, Czech Republic
Michal Holcapek	University of Ostrava, Czech Republic
Miroslav Pokorný	University of Ostrava, Czech Republic
Pavel Vlašanek	University of Ostrava, Czech Republic
Petr Hurtik	University of Ostrava, Czech Republic
Petra Hodáková	University of Ostrava, Czech Republic
Petra Murinová	University of Ostrava, Czech Republic
Radek Valášek	University of Ostrava, Czech Republic
Viktor Pavliska	University of Ostrava, Czech Republic

Organising Committee

Václav Snášel - Chair	VSB-Technical University of Ostrava, Czech Republic (Chair)
Jan Platoš	VSB-Technical University of Ostrava, Czech Republic (Co-chair)
Pavel Krömer	VSB-Technical University of Ostrava, Czech Republic (Co-chair)
Katerina Kasparova	VSB-Technical University of Ostrava, Czech Republic
Hussein Soori	VSB-Technical University of Ostrava, Czech Republic
Petr Berek	VSB-Technical University of Ostrava, Czech Republic

Contents

Evolutionary Computation and Optimization

Intelligent Systems

Classification and Clustering Methods

Networks and Communication

Applications

A Hybrid Discrete Differential Evolution Algorithm for Economic Lot Scheduling Problem with Time Variant Lot Sizing

Srinjoy Ganguly[1], Arkabandhu Chowdhury[1], Swahum Mukherjee[1], P.N. Suganthan[2], Swagatam Das[3], and Tay Jin Chua[4]

[1] Dept. of Electronic & Telecommunication Engineering, Jadavpur University, Kolkata, India
[2] School of Electrical and Electronic Engineering, Nanyang Technological University, Singapore 639798
epnsugan@ntu.edu.sg
[3] Electronics and Communication Sciences Unit, Indian Statistical Institute, Kolkata, India
swagatam.das@isical.ac.in
[4] Singapore Institute of Manufacturing Technology (SIMTech), 71 Nanyang Drive, Singapore 638075
tjchua@SIMTech.a-star.edu.sg

Abstract. This article presents an efficient Hybrid Discrete Differential Evolution (HDDE) model to solve the Economic Lot Scheduling Problem (ELSP) using a time variant lot sizing approach. This proposed method introduces a novel Greedy Reordering Local Search (GRLS) operator as well as a novel Discrete DE scheme for solving the problem. The economic lot-scheduling problem (ELSP) is an important production scheduling problem that has been intensively studied. In this problem, several products compete for the use of a single machine, which is very similar to the real-life industrial scenario, in particular in the field of remanufacturing. The experimental results indicate that the proposed algorithm outperforms several previously used heuristic algorithms under the time-varying lot sizing approach.

Keywords: Lot scheduling, time-varying lot-sizes approach, discrete differential evolution, cyclic crossover, simple inversion mutation, greedy reordering local search, remanufacturing.

1 Introduction

It is a common practice in industries to produce several products on a single machine due to economic considerations. Typically, these facilities may produce only a single product at a time and have to be set-up (stopped and prepared) at the cost of time and money, before the start of the production run of a new product. A production scheduling problem arises due to the need to co-ordinate the set-up and production of a large number of different items. The main aim of the Economic Lot Scheduling Problem (ELSP) [1] is to find the best lot sizes and production schedule that does not allow any shortages for the items to be produced in the above described environment. Typical examples of such problems are:

V. Snasel et al. (Eds.): SOCO Models in Industrial & Environmental Appl., AISC 188, pp. 1–12.
springerlink.com   © Springer-Verlag Berlin Heidelberg 2013

- Metal forming and plastics production lines (press lines, and plastic and metal extrusion machines), where each product requires a different die that needs to be set up on the concerned machinery.
- Assembly lines which produce several products and different product models (electric goods, vehicles, etc.).
- Blending and mixing facilities (for paints, beverages, etc.), in which different products are poured into different containers for processing.

Typically, in an industrial scenario, a single machine of very high efficiency is purchased instead of several machines of lesser efficiency. This situation leads to the question of how one should schedule production on this high speed machine. The issue is one of selecting both a sequence, in which the products will be manufactured, and a batch size for each item run. The issue of batching arises because the system usually incurs a set-up cost and/or a set-up time when the machine switches from one product to a different product. Set-up times imply an idle-time during which the machine does nothing, which, in turn, implies a need to carry a large scale production facility. This problem has attracted the attention of many researchers over 40 years, partly because it is a representation of many frequently encountered scheduling problems, and simply because it appears to be unconquerable. In fact, many researchers also take into consideration another interesting variant of the problem wherein they allow the phenomenon of re-manufacture [2] to occur, i.e. items that have been returned by the consumers are re-manufactured and made as fresh as new. Ouyang and Zhu extended the classical ELSP to schedule the manufacturing and remanufacturing on the same single product line [19]. They assumed that the demand rate and return rate are constant and the product line has limited capacity of manufacturing and remanufacturing.

Typically, we may assume that the demand rates are known before-hand and are product-dependent while the set-up cost and set-up time are product-dependent but sequence-independent. Also, the majority of the research in ELSP literature focuses on cyclic production schedules, i.e. the schedule is repeated periodically. As the ELSP is an NP-hard problem, many heuristics have been devised that may solve this problem to near optimality. The three types of approaches generally taken are:

I.   **Common cycle approach:** This restricts all the products' cycle times to equal time (an item's cycle time is the duration between the starts of two consecutive runs of that item). This approach has the advantage of always generating a feasible schedule despite the use of a very simple procedure. This procedure, however, gives solutions far from the lower bound in some cases [3].

II. **Basic period approach:** This allows different cycle times for different products, but restricts each product's cycle time to be an integer multiple k of a time period called a basic period. This approach, in general, gives better solutions than the common cycle approach. However, its main drawback is the difficulty of ensuring that the production sequence is feasible [4].

III. **Time-varying lot size approach:** This allows different lot sizes for any given product during a cyclic schedule. It explicitly handles the difficulties caused by set-up times and always gives a feasible schedule as proved by Dobson [5]. It has been found to give fitter solutions in comparison to the previous two approaches.

The research on ELSP under the different policies discussed above mainly comprises of different algorithmic solutions since the restricted versions of the problem are also very difficult. Till date, most researchers have relied on genetic algorithms to solve the problem. There are only two studies in the literature that consider exact algorithms: Grznar and Riggle [6] for BP policy, and Sun et al. [7] for EBP policy. However, the exact algorithms are not very time-efficient especially when the utilization factor is high. The purpose of the current research is to develop a Hybrid Discrete Differential Evolution (HDDE) algorithm to solve the ELSP. Our HDDE is based on the time-varying lot sizes approach. In this paper, we present the ELSP formulation and proposed HDDE in Section 2 and the results and discussions in Section 3.

2 ELSP Problem Formulation and Algorithm

The following assumptions are normally used in the formulation of the ELSP:

- Several items compete for the use of a single production facility.
- Demand-rates, production-rates, set-up times and set-up costs are known before-hand and are constant.
- Backorders are not allowed.
- Inventory costs are directly proportional to inventory levels.

The following notations have been adopted:

- Item Index: $i = 1,2,...,M$
- Position Index: $j = 1,2,...,N$
- Constant Production Rate(Units per day): p_i ; $i = 1,2,...,M$
- Constant Demand Rate(Units per day): d_i ; $i = 1,2,...,M$
- Inventory Holding Cost(Price per unit per day): h_i ; $i = 1,2,...,M$
- Set-up Cost(Currency): a_i ; $i = 1,2,...,M$
- Set-up Time(days): s_i ; $i = 1,2,...,M$
- Item produced at position j: I^j ; $j = 1,2,...,N$
- Production-time for item produced at position j: t^j ; $j = 1,2,...,N$
- Idle-time for the item produced at position j: x^j ; $j = 1,2,...,N$
- Cycle Length(days): T

The ELSP in a nutshell is the situation where there is a single facility on which different products have to be produced. We try to find a cycle length and a production sequence $I=(I^1, I^2,..., I^N)$ where $I^j \in (1,2,...,M)$, production time durations $t=(t^1, t^2,..., t^N)$ and idle-times $u=(u^1, u^2,..., u^N)$ so that the given production cycle can be finished within the given cycle. This cycle has to be repeated over and over again, while at the same time, inventory and set-up costs have to be minimized. We may define μ as:

$$\mu = 1 - \sum_{i=1}^{m} \frac{d_i}{p_i}$$

We must note that μ represents the long-run proportion of time available for set-ups. For infinite horizon problems, $\mu > 0$ is absolutely necessary for the existence of a feasible solution. It can be shown that any production sequence can be converted into a feasible one if $\mu > 0$.

Let F represent the set of all the feasible finite sequences of the products and J_i denote the positions in the schedule where the product having the index i is produced. Let Y_k denote all the jobs in a given sequence starting from k up till that position in the given sequence where the item I^k is produced again. Then, the complete formulation of the ELSP is:

$$\inf_{j \in F} \min_{t \geq 0; x \geq 0; T \geq 0;} \frac{1}{T} (\sum_{j=1}^{N} \frac{1}{2} h^j (p^j - d^j)(\frac{p^j}{d^j})(t^j)^2 + \sum_{j=1}^{N} a^j) \tag{1}$$

Subject to the following boundary conditions:

- $$\sum_{j \in J^i} t^j p_j = d_i T \, ; i = 1, 2, ..., M \tag{2}$$

- $$\sum_{j \in Y_k} t^j + s^j + x^j = \frac{p^k}{d^k} t^k \, ; k = 1, 2, ..., N \tag{3}$$

- $$\sum_{j=1}^{N} (t^j + s^j + x^j) = T \tag{4}$$

The condition (2) ensures that enough space is allocated to each product so that it may meet its own demand during one complete cycle. Condition (3) ensures that we produce that much quantity of each product so that its demand is met till the time it is produced again. Condition(4) ensures that the total time taken for the complete cycle is numerically equal to the sum of the production time, setup time and idle time of the various items.

2.1 The Proposed HDDE Algorithm

The HDDE may be categorized into the following steps:

Step 1 The production frequencies are obtained by solving the lower bound (LB) model [8] as stated below. This lower bound is tighter than that obtained by using the so-called independent solution in which each product is taken in isolation by calculating its economic production quantity. The constraint here is that enough time must be available for set-ups. As stated previously, μ represents the average proportion of time available for set-up. T_i refers to the cycle length for the product i. However, the synchronization constant that states that no two items may be produced simultaneously is ignored. Hence, this results in a lower total daily cost for the ELSP. This scheme was initially proposed by Bomberger [9].

LB model:

$$\min_{T_1,...,T_m} \sum_{i=1}^{M} (\frac{a_i}{T_i}) + (\frac{h_i d_i T_i}{2})(1 - \frac{d_i}{p_i})$$

Given that, $\sum_{i=1}^{m} \frac{s_i}{T_i} \le \mu$; $T_i \ge 0; i = 1,2,...,M$ (5)

The objective function and the convex set in the above constraint model are convex in $T_i's$. Therefore, the optimal points of the LB model are those points that satisfy the KKT conditions as follows:

$$\sqrt{\frac{a_i + Rs_i}{H_i}} = T_i \text{ for all } i .$$ (6)

$R \ge 0$ with complementary slackness $\sum_{i=1}^{m} \frac{s_i}{T_i} \le \mu$, where $H_i = \dfrac{1 - \dfrac{d_i}{p_i}}{2}$. The

procedure that is adopted to find the optimal T_i's is described below.

Algorithm for the Lower Bound

1. The condition $R=0$ is checked for an optimal solution. The T_i's are found as per the formula $\sqrt{\frac{a_i}{H_i}}$ for all i. If $\sum_{i=1}^{m} \frac{s_i}{T_i} < \mu$, then the T_i's are proven to be an optimal solution. Else, the algorithm explained in step 2 is adopted.

2. R is taken at an arbitrary value greater than 0. The T_i's are computed as per $\sqrt{\frac{a_i + Rs_i}{H_i}} = T_i$. If $\sum_{i=1}^{m} \frac{s_i}{T_i} < \mu$, R is reduced and the afore-mentioned process is repeated. If $\sum_{i=1}^{m} \frac{s_i}{T_i} > \mu$, R is increased and the afore-mentioned process is repeated. If $\sum_{i=1}^{m} \frac{s_i}{T_i} = \mu$, the iterations stop and the set of T_i's obtained are optimal.

If the optimal cycle length for an item is T_i' , then the production frequency (f_i) of that item may be obtained as per the following formula: $f_i = \dfrac{\max(T_i')}{T_i'}$.

<u>**Step 2**</u> The production frequencies obtained in Step 1, are rounded off to the nearest integers.

Step 3 Using the frequencies obtained in the previous step, an efficient production sequence is obtained using the DE algorithm. This algorithm is discussed later in details.

Step 4 If we assume the approximation that there is no idle time (which works well for highly loaded facilities), we can easily solve for the set of production times using equation (3). This method is referred to as the Quick and Dirty heuristic [10]. Otherwise, we may adopt Zipkin's [11] parametric algorithm.

2.2 Discrete Differential Evolution

The key components of the proposed DDE are:

1)   Representation of the Schedule (Chromosome): The proper representation of a solution plays an important role in any evolutionary algorithm. In our paper, a string of positive integers (chromosome) is used to represent a solution. The length of the chromosome is the sum of the production frequencies obtained at the end of step 2. As standard operations such as cross-over and mutation are difficult using such a representation, we use an additional chromosome that possesses the absolute locations of each of the genes in the afore-mentioned chromosome. Suppose the problem at hand is a 4-product ELSP and it is observed at the end of step 2 that their frequencies are 1,2,2,1 respectively. We may represent this situation using two chromosomes A and B. The representation is as follows:

$$
\begin{array}{llllllll}
\text{Chromosome A} & \longrightarrow & (1 & 2 & 2 & 3 & 3 & 4) \\
\text{Chromosome B} & \longrightarrow & (1 & 2 & 3 & 4 & 5 & 6)
\end{array}
$$

Here, Chromosome A represents the item numbers simply, while Chromosome B represents their respective locations. All the operations shall be performed using chromosomes of the form of Chromosome B.

2)   The objective and fitness function. The fitness value for any chromosome may be computed as the inverse of Equation (1), which serves as the fitness function. Our objective is to minimize this function (Eq. 1) and therefore maximize the fitness, i.e. our aim is to find the string with the maximum fitness. For any chromosome (representing the production schedule) the production times of the different items may be computed as per step 4 of the main algorithm. After this, we obtain the value of the fitness function using Equation No. (1). The fitness function is represented by $f_{elsp}(V)$ for chromosome V.

3)   The Cross-Over operator. The Cyclic Crossover Operator [12] is of immense importance in the proposed HDDE because this operation is carried out at several points in the algorithm. This operator is unique one since it preserves characteristics of both the parents in the offspring. The crossover mechanism may be envisaged via the following example: Let us consider two flow sequences A and B, where A= (1 3 5 6 4 2) and B= (5 6 1 2 3 4). Let the first offspring begin with 1(the starting operation of parent A). Selection of '1' from A implies that '5' should be selected from B, because we want each operation to be derived from one of the two parents. Hence, C= 1 _ 5 _ _ _. This process continues on till after the selection and subsequent insertion of an operation from one of the two parents, the operation in the corresponding position in the other parent is already present in the offspring. After this, the remaining operations are filled in, as per their orders respective to one another in the other string. As can be seen in case of C, insertion of '5' implies that '1' should be inserted in the list, but as '1' is already present in the list (i.e. the starting operation), the cycle stops (hence, the name Cyclic Crossover) and the remaining operations are filled in from B. Hence, C = (1 6 5 2 3 4). Similarly, considering '5' as the starting operation, another offspring D can be obtained in a similar fashion. Hence, D= (5 3 1 6 4   2) .As our algorithm imposes a restriction that the result of each crossover operation results in only a single offspring, the fitter progeny is selected.

4)   Perturbation Operator. In our paper, we have adopted the Simple Inversion Mutation [13] as our perturbation operator. Two random cut-points are selected in the chromosome and that part of the chromosome between these two cut-points is inverted and placed in the original chromosome. For example, let us consider the chromosome represented by (**ABDEGCFJIH**). Suppose the first cut-point is between D and E and the second cut-point is between F and J .

Then the part of the chromosome between the two cut-points is reversed and then placed in the original chromosome to obtain (**ABDFCGEJIH**).

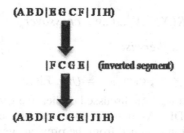

(ABD|EG CF|JIH)

|F C G E| (inverted segment)

(ABD|FCGE|JIH)

Fig. 1. Simple Inversion Mutation

DE [14] is one of the recent evolutionary optimization algorithms. Like other algorithms of the same class, DE is population-based and serves as a stochastic global optimizer. In the classic DE, candidate solutions are represented by a vector of floating point solutions. The mutation process entails the addition of the weighted difference of two members of the population to another third member of the same population to generate a mutated solution. Then, a cross-over operation occurs so as to generate a trial solution that is in turn compared with the target solution using a selection operator, which determines which of the two shall survive for the next generation. It is obvious that traditional DE equations shall not hold in this regard as they had been developed for continuous numerical optimization. As per the novel DDE algorithm proposed by Pan et. al. [15], the target individual is represented by a chromosome of type B (permutation of numbers). The mutant individual is obtained by perturbing the previous generation best solution in the target population. Hence, we achieve the difference variations by applying a perturbation (mutation) operator on the best solution present in the previous generation of the target population. These perturbations are achieved by applying the Simple Inversion Mutation operator on the best solution of the previous generation. Since these perturbations are stochastically different, each result is expected to be distinct from the other. To obtain the mutant, the following equation may be used:

$$V_i^t = \begin{cases} F(X_g^{t-1})..if..r \le mutation_{prob} \\ X_g^{t-1}..otherwise \end{cases} \quad (7)$$

where, X_g^{t-1} represents the best solution of the previous generation of the target population. $F()$ denotes the perturbation operator(Simple Inversion Mutation). Hence, a random number $r \in (0,1)$ is generated and if it is less than $mutation_{prob}$, then the trial solution is obtained by applying the perturbation operator on X_g^{t-1} else, the mutant is set to X_g^{t-1}. After this, we enter the recombination (cross-over) phase. In this, V_i^t is crossed over with X_i^t to obtain the trial solution U_i^t if $v \in (0,1)$ [generated randomly] is greater than $crossover_{prob}$. Else, U_i^t is taken to be equal to V_i^t. The pseudo-code representation for the afore-said is:

$$U_i^t = \begin{cases} CR(X_i^{t-1}, V_i^t..if..v > crossover_{prob} \\ V_i^t..otherwise \end{cases} \quad (8)$$

Hence, if U_i^t is fitter than X_i^{t-1}, then $X_i^t = U_i^t$. Else, $X_i^t = X_i^{t-1}$. Over here, $CR()$ refers to the cross-over operator. As discussed earlier, the Cyclic Cross-Over operator has been used in the HDDE. As it is pretty ostensible, our basic aim is to take advantage of the best solution obtained from the previous generation during the entire search process. Unlike its continuous counterpart, the differential evolution is achieved by the stochastic re-ordering of the previous generation's best fit solution in the target population.

Greedy Reordering Local Search (GRLS). In this paper, we propose a novel local search operator which restructures a chromosome iteratively on a probabilistic basis to search for fitter solutions. This re-ordering process goes on till the solutions which are obtained from reordering is better in comparison to the parent solution. The GRLS algorithm for a chromosome p is given in Fig. 2.

3 Results and Discussion

In our tests on ELSP, the proposed HDDE algorithm is coded in MATLAB and the experiments are executed on a Pentium P-IV 3.0 GHz PC with 512MB memory. We have compared the results that our algorithm provided when tested on Mallya's 5-item problem [17] and the famous Bomberger's 10-item problem (for the $\mu=0.01$ case) [8] with the hybrid GA proposed by Moon et. al.[10] , Khouza's GA [17], Dobson's heuristic [18] and the Common Cycle Solution [3] approach. In each of the runs, the number of iterations (generations) have been taken to be 80 and the population size has been taken as 100, the value of $mutation_{prob}$ has been taken to be 0.77, and the value of $crossover_{prob}$ has been taken to be 0.96 in all the runs. The value of $mutation_{prob}$ has been decided empirically by taking several runs of the proposed algorithm. The value of $crossover_{prob}$ has been intentionally kept high to ensure that most members of the target population undergo recombination because this not only facilitates increased convergence to the optimal solution but also ensures that diversity is improved. The results are as follows:

FOR $i=1; i \le n$

 Search for i in the *parent chromosome*;

 $current\_gene = i;$

 FOR $j=1; j \le n;$

  Search for $x_{right}(j);$ // $x_{right}(j)$ is the first element to the right of j which has not occurred in the list.

  Search for $x_{left}(j);$ // $x_{left}(j)$ is the first element to the left of j which has not occurred in the list.

  $p_R (\in [0,1])$ and $p_L (\in [0,1])$ are generated randomly.

  IF $p_R > p_L$

   $x_{right}(j)$ is inserted immediately after j ;

   $current\_gene = x_{right}(j);$

  ELSE

   $x_{left}(j)$ is inserted immediately after j ;

   $current\_gene = x_{left}(j);$

  END IF

 END FOR

END FOR

Out of the n chromosomes generated, let p' represent the fittest one.

IF $f(p') \ge f(p)$

 p is replaced by p'.

 CONTINUE GRLS

ELSE

 ABORT GRLS

END IF

Fig. 2. The pseudo-code for the GRLS

```
procedure Hybrid Discrete Differential Evolution(HDDE)
initialize parameters
initialize target population
evaluate target population
apply local search(Greedy Reordering Operator)
generation=0;
/* G is defined by the user */
while (generation<G)
obtain the mutant population
obtain the trial population
evaluate the trial population
Select the chromosomes that may progress to the next generation
Apply Local Search(GRLS operator)
generation=generation+1;
end while
Return the best found solution
end HDDE
```

Fig. 3. The pseudo-code for the Hybrid Discrete Differential Evolution(HDDE)

Table 1. Mallya's 5-item problem (Average Daily Cost)

HDDE	Hybrid GA proposed by Moon et al.[10]	Dobson's Heuristic[18]
59.87	60.91	61.83

Optimal Production sequence for Mallya's 5-item problem (as per the HDDE):

(4 , 5 , 3 , 1 , 3 , 2 , 3 , 4 , 3 , 1 , 2)

Table 2. Bomberger's 10-item Problem (Average Daily Cost)

HDDE	Hybrid GA proposed by Moon et al.[10]	Dobson's Heuristic[18]	Common Cycle Solution[3]	Khouza's GA [17]
125.28	126.12	128.43	231.44	196.14

Optimal Production sequence for Bomberger's 10-item problem (as per the HDDE):

(8 , 9 , 6 , 8 , 3 , 8 , 4 , 2 , 4 , 8 , 9 , 3 , 8 , 10 , 5 , 10 , 8 , 4 , 3 , 2 , 8 , 5 , 9 , 8 , 5 , 8 , 4 , 1 , 8 , 6 , 2 , 8 , 5 , 2 , 7 , 8 , 9 , 4 , 3 , 4 , 5 , 4)

In this paper, we have compared the HDDE that we proposed, with other algorithms that belong to the time varying lot sizes approach in the literature. We have used the data for Mallya's 5-item problem and Bomberger's 10-item problem (for the μ =0.01 case which represents a highly loaded facility). As it can be clearly seen, our algorithm clearly outperforms the hybrid GA proposed by Moon et al. by 1.7% and 0.66%, in case of Mallya's 5-item and Bomberger's 10-item problem, respectively.

4 Conclusions and Further Work

This paper has proposed a new type of Hybrid Discrete Differential Evolution algorithm that employs a novel Greedy Reordering operator for local search. We have compared our algorithm with those proposed by Dobson, Khouza and Moon on the Mallya's 5-item and Bomberger's 10-item problems (for the $\mu=0.01$ case) and it can be clearly seen that our algorithms clearly outperforms the rest. Since the ELSP is a very complex combinatorial optimization algorithm, most researchers have developed heuristics that may solve this algorithm efficiently. We have used a heuristics to generate the frequencies and a HDDE (coupled with a local search operator) to find the fittest ELSP sequence which is feasible. Hence, using this algorithm we have increased the speed and accuracy of the search process. Furthermore, this algorithm may be modified accordingly to solve the ELSP with remanufacturing model.

References

1. Hsu, W.: On the general feasibility test of scheduling lot sizes for several products on one machine. Management Science 29, 93–105 (1983)
2. Zanoni, S., Segerstedt, A., Tang, O., Mazzoldi, L.: Multi-product economic lot scheduling problem with manufacturing and remanufacturing using a basic policy period. Computers and Industrial Engineering 62, 1025–1033 (2012)
3. Hanssmann, F.: Operations Research in Production and Inventory. Wiley, New York (1962)
4. Tasgeterin, M.F., Bulut, O., Fadiloglu, M.M.: A discrete artificial bee colony for the economic lot scheduling problem. In: IEEE Congress on Evolutionary Computing (CEC), New Orleans, USA, pp. 347–353 (2012)
5. Dobson, G.: The economic lot scheduling problem: achieving feasibility using time-varying lot sizes. Operations Research 35, 764–771 (1987)
6. Grznar, J., Riggle, C.: An optimal algorithm for the basic period approach to the economic lot schedule problem. Omega 25, 355–364 (1997)
7. Sun, H., Huang, H., Jaruphongsa, W.: The economic lot scheduling problem under extended basic period and power-of-two policy. Optimization Letters 4, 157–172 (2010)
8. Maxwell, W.: The scheduling of economic lot sizes. Naval Research Logistics Quarterly 11, 89–124 (1964)
9. Bomberger, E.: A dynamic programming approach to a lot size scheduling problem. Management Science 12, 778–784 (1966)
10. Moon, I., Silver, E.A., Choi, S.: Hybrid Genetic Algorithm for the Economic Lot Scheduling Problem. International Journal of Production Research 40(4), 809–824 (2002)
11. Zipkin, P.H.: Computing optimal lot sizes in the economic lot scheduling problem. Operations Research 39(1), 56–63 (1991)
12. Dagli, C., Sittisathanchai, S.: Genetic neuro-scheduler for job shop scheduling. International Journal of Production Economics 41(1-3), 135–145 (1993)
13. Grefenstette, J.J.: Incorporating problem specific knowledge into genetic algorithms. In: Davis, L. (ed.) Genetic Algorithms and Simulated Annealing, pp. 42–60. Morgan Kaufmann, Los Altos (1987)
14. Das, S., Suganthan, P.N.: Differential evolution: a strategy of the state of the art. IEEE Transactions on Evolutionary Computation 15(1), 4–31 (2011)

15. Pan, Q.K., Tasgetiren, M.F., Liang, Y.: A discrete differential evolution algorithm for the discrete flow-shop scheduling problem. Computers and Industrial Engineering 55, 795–816 (2007)
16. Mallya, R.: Multi-product scheduling on a single machine: a case study. Omega 20, 529–534 (1992)
17. Khouza, M., Michalewicz, Z., Wilmot, M.: The use of genetic algorithms to solve the economic lot size scheduling problem. European Journal of Operational Research 110, 509–524 (1998)
18. Dobson, G.: The cyclic lot scheduling problem with sequence-dependent setups. Operations Research 40, 736–749 (1992)
19. Ouyang, H., Zhu, X.: An economic lot scheduling problem for manufacturing and remanufacturing. In: IEEE Conference on Cybernetics and Intelligent Systems, Chengdu, pp. 1171–1175 (2008)

An Ordinal Regression Approach for the Unequal Area Facility Layout Problem

M. Pérez-Ortiz, L. García-Hernández, L. Salas-Morera,
A. Arauzo-Azofra, and C. Hervás-Martínez

Department of Computer Science and Numerical Analysis, University of Córdoba,
Córdoba, Spain
{i82perom,ir1gahel,lsalas,chervas,arauzo}@uco.es

Abstract. This paper proposes the use of ordinal regression for helping the evaluation of Unequal Area Facility Layouts generated by an interactive genetic algorithm. Using this approach, a model obtained taking into account some objective factors and the subjective evaluation of the experts is constructed. Ordinal regression is used in this case because of the ordinal ranking between the different possible evaluations of the facility layouts made by the experts: {*very deficient, deficient, intermediate, good, very good*}. To do so, we will also make an approximation to some of the most successful ordinal classification methods in the machine learning literature. The best model obtained will be used in order to guide the searching of a genetic algorithm for generating new facility layouts.

1 Introduction

The efficiency of industrial production is widely influenced by the design of plant layouts [16]. In fact, it is estimated that between 20% and 50% of production costs are due to materials handling, and that these costs can be reduced at least by 10% and 30% through efficient design [25]. There are many kinds of layout problems; an example of a classification of them is in [17]. We will focus on the Unequal Area Facility Layout Problem (UA-FLP) as formulated by Armour and Buffa [2]. Briefly, UA-FLP has into account a rectangular plant layout that is made up of unequal rectangular facilities that have to be placed in the plant layout in the most effective way.

Most authors have solved this problem using quantitative criteria. Unfortunately, the approaches may not adequately consider all of the essential qualitative information that affects a human expert involved in design (e.g. engineers, stakeholders, regulators, etc.) [4]. So that, qualitative features are also important to be taken into account, as for example, location preferences of certain facilities, the way that remaining space is distributed in the layout, or any other subjective consideration that can be judged as relevant for the Decision Maker (DM). Besides, these features can be subjective, unknown a priori and changing during the procedure evolution. So that, it is difficult to take into account both quantitative and qualitative aspects at the same time, because they can not be easily formulated as an objective function. Consequently, including the expert knowledge is vital to incorporate qualitative considerations in the design. This fact also gives the following benefits: finding a solution that is satisfactory to the DM

V. Snasel et al. (Eds.): SOCO Models in Industrial & Environmental Appl., AISC 188, pp. 13–21.
springerlink.com   © Springer-Verlag Berlin Heidelberg 2013

(but that is not an optimal solution necessarily) [3,19], choosing the best trade-off solution when a conflict among objectives or constraints exists [15], assisting the algorithm in achieving the search process towards the DM preferences [20,10], excluding the need to give all the required preference information a priori, offering the DM the possibility to know about his/her own preferences [15]; stimulating the creativity of the DM [23], and getting original and feasible designs.

Many Evolutionary Computation (EC) approaches have been dealt with UA-FLP. Among these, the Genetic Algorithms (GAs) [14] are frequently applied. Brintup et al. [7] have emphasized that Interactive Evolutionary Computation (IEC) can greatly help to improving optimized design by involving experts in searching for a satisfactory solution [6].

Focus on the UA-FLP, Garcia-Hernandez et al. [13] proposed an Interactive Genetic Algorithm (IGA) that captures such aspects (quantitative and qualitative) that the DM would like in the final design. But, unfortunately, due to the fact that many features of the solutions should be considered at the same time, the DM can end up distracted and overloaded. In order to improve the reduction of DM fatigue, a classification of the data has been applied with the aim of learning the DM preferences. Because of each DM have a different expertise or preferences about the UA-FLP, it is necessary to diversify the data having into account the level of knowledge of each DM prototype.

So, in this case three experts have evaluated (in a subjective way) the different facility layouts using a Likert scale (where the possible evaluations for a facility layout are {*very deficient, deficient, intermediate, good, very good*}. This problem is then addressed by an ordinal point of view because of the nature of the categories, which (as can be seen) are ordered.

Ordinal regression has a wide range of applications in areas where the human evaluation plays an important role as psychology, medicine, information retrieval, etc. This type of classification shows similarities both with regression and classification, but also differences: regarding regression the number of classes is finite and the distances between them is not well-defined. Unlike classification the relationship between the classes is not nominal, but ordinal.

In ordinal classification, the variable to predict presents a natural order, so the main problem with this classification is that there is not a precise notion of the distance between classes. For example, on a scale of ranges we know that 3 is closer to 2 than 5, but on an ordinal scale as it could be {very good, good, medium, bad, very bad}, how can we know that medium is closer to good than very bad? Traditionally, ordinal classification problems have been solved ignoring the order among categories but this is not a good simplification, since the error obtained by confusing the very bad class with the medium one is equal to that of confusing the very bad class with the very good one (when in an ordinal scale, the latter should be more penalized).

At this point it is important to highlight the contribution of this paper both to the facility layout literature as well to the ordinal regression one. First, this paper introduces the use of subjective evaluations of different experts for the resolution of the area facility layout problem, which usually have been solved only taking into account objective information (as material flow or adjacency requirements, among others). This paper also presents a summary of some of the most used and successful methods in ordinal regression [21,12,18,24] and applies them to a problem of evaluation using a Likert scale.

This paper is organized as follows: a brief analysis of the ordinal used algorithms is given in Section 2. The UA-FLP database and its features are presented in Section 3. Finally, we present the results of the analysed algorithms comparing with other state-of-the art methods in Section 4 and Section 5 summarises the conclusions and future work.

2   Methodology

In this section, some of the main methodologies in ordinal regression [21,12,18,24] are slightly introduced for a better comprehension.

To clarify, a set of training samples will be denoted by $(\mathbf{x}_i, y_i) \in R^l \times R, i = \{1, ..., N\}$, where $\mathbf{x}_i \in R^l$ correspond to inputs and $y_i \in \{1, 2, ..., K\}$ are the ordered class labels, K is the number of classes for the problem, N is the sample size and finally, N_k is the number of samples for the k-th class.

2.1   Kernel Discriminant Learning for Ordinal Regression

This algorithm (Kernel Discriminant Learning Ordinal Regression) [24] is a modification of the well-known Kernel Discriminant Analysis (KDA) for dealing with ordinal regression. The main goal of the method is to find the optimal projection for classifying the patterns (i.e. the projection from which classes can be well separated), but in this case also preserving the ordinal information of the data, i.e. the classes should be ordered according to their rank in the projection.

For this, the algorithm analyses three different objectives: maximize the inter-class distance, minimize the intra-class distance and ensuring the ordinal information of the classes. The standard approach for that (LDA) uses a covariance matrix for measuring the between-class matrix ($\mathbf{S}_b$) and also a covariance matrix for the within-class distance ($\mathbf{S}_w$).

A kernel learning algorithm is used for solving non linear classification problems by mapping the original training data into a higher dimensional feature space using a function ϕ ($\phi : \mathbf{x} \to \phi(\mathbf{x})$). At this point, the data will be linearly separable and, as a result, the idea of discriminant learning can be applied. Thus, applying the kernel trick, the covariance matrixes would be:

$$\mathbf{S}_w = \frac{1}{N} \sum_{k=1}^{K} \sum_{i=1}^{N_k} (\phi(\mathbf{x}_i) - \mathbf{M}_k)(\phi(\mathbf{x}_i) - \mathbf{M}_k)^T,$$

$$\mathbf{S}_b = \frac{1}{N} \sum_{k=1}^{K} N_k (\mathbf{M}_k - \mathbf{M})(\mathbf{M}_k - \mathbf{M})^T,$$

where $\mathbf{M}_k = \frac{1}{N_k} \sum_{i=1}^{N_k} \phi(\mathbf{x}_i)$, and $\mathbf{M} = \frac{1}{N} \sum_{i=1}^{N} \phi(\mathbf{x}_i)$.

The optimization problem can be solved by the maximization of the so called Rayleigh coefficient:

$$J(\mathbf{w}) = \frac{\mathbf{w}^T \mathbf{S}_b \mathbf{w}}{\mathbf{w}^T \mathbf{S}_w \mathbf{w}}, \tag{1}$$

where $\mathbf{w}$ is the projection we are looking for.

In order to solve the previously presented problem, the method is reformulated applying Lagrange multipliers and Quadratic Programming. For further information, the procedure can be seen in [22].

2.2   Ensemble for Ordinal Regression by Frank & Hall

This algorithm [12] is a simple ensemble method which can be applied to any probabilistic nominal model. Given k ordered classes, the classification problem is transformed to $k - 1$ problems of binary classification. This is made by making a binarization of the categorical variables in such a way that given an ordinal attribute $\mathbf{A}$ with k possible values $(y_1, y_2, \ldots, y_k)$, $k - 1$ binary variables are obtained so the i-th attribute represents $A > y_i$. Then, $k - 1$ models (of the base classifier) are trained and the predicted probabilities are calculated too.

For predicting new unseen instances, the procedure consists of calculating the probabilities of the k categorical variables, using for that the $k - 1$ different models, and then, the predicted target for the new instance would be the class with maximum probability for this case.

Generally, the probability of belonging to one class is calculated as follows:

$$Pr(V_1) = 1 - Pr(Target > V_1)$$

$$Pr(V_i) = Pr(Target > V_{i-1}) - Pr(Target > V_i)$$

$$Pr(V_k) = Pr(Target > V_{k-1})$$

2.3   Extended Binary Classification

This method [18] is also based on binary classification but the procedure is different to the proposal of Frank & Hall [12] previously explained. In this case, the problems of binary classification are solved using only one model (although a decision rule is needed for the prediction).

The algorithm transforms the training data $(\mathbf{x}_i, y_i)$ to extended data $(\mathbf{x}_i^{(k)}, y_i^{(k)}), 1 \le k \le K - 1$:

$$\mathbf{x}_i^{(k)} = (\mathbf{x}_i, k), \ y_i^{(k)} = 2[\![k < y_i]\!] - 1, \tag{2}$$

but using the specific weights:

$$w_{y_i,k} = |C_{y_i,k} - C_{y_i,k+1}|, \tag{3}$$

where C is a cost matrix, with $C_{y_i,k-1} \ge C_{y_i,k}$ if $k \le y_i$ and $C_{y_i,k} \le C_{y_i,k+1}$ if $k \ge y_i$. Then, a binary classifier f will be used with all the extended data and it will generate probabilistic outputs (which will be used by the decision rule for predicting).

2.4 Proportional Odd Model

It is considered as one of the pioneer methods for ordinal regression since it was proposed in 1980 [21]. It applies a threshold method based on logistic regression, also making an assumption of stochastic ordering of the space X: for each pair x_1 and x_2 is satisfied that $P(y \leq C_i|x_1) \geq P(y \leq C_i|x_2)$ or $P(y \leq C_i|x_1) \leq P(y \leq C_i|x_2)$.

3 Experiments

3.1 Dataset Description

The dataset presented in this paper is obtained using a genetic interactive algorithm [13] which evolves taking into account the opinion of the experts and also, applies clustering for not overload the user. Three experts have evaluated 500 different area layouts and the three different evaluations have been joint in a unique one, using for that the median of the marks, in order to obtain a more robust evaluation for each facility layout. The choice of the median at this point is not arbitrary, it is justified because when using a $L1$ norm (as it could be the *MAE* metric, which is used for cross-validation and for estimating the performance of the different algorithms), the best centralized estimator is the median.

The plant layouts have been created using the Flexible Bay Structure (FBS) proposed by Tong [26] and the variables considered in the dataset are: the material flow between facilities, the adjacency relationship, the distance requests, the aspect ratio contraints, and the coordinates of each facility in the plant layout. These variables are calculated using the formulation given by Aiello et al. [1].

The class distribution of the patterns in the dataset is: $\{10\%, 26\%, 47\%, 11\%, 6\%\}$. If the dataset were absolutely balanced, the percentage per class would be 20% (as there are 5 classes), so in this case, we can consider that there are 3 unbalanced classes (class 1, 4 and 5).

3.2 Methods Compared

The methods compared are then:

- LDAOR and KDLOR: These algorithms are explained in Section 2. They were implemented and run using Matlab.
- SVM: Support Vector Machine is a well-known baseline nominal classifier which considers the standard C-SVC with soft margin and Gaussian kernel [11].
- Ensemble F&H: SVM method is also used as the base of this ensemble. It was implemented using Matlab.
- EBC(SVM): SVM is used as the base binary classifier for this method. SVM, EBC(SVM) and F&H(SVM) have been configured and run with the libsvm [9].
- POM: Logistic regression method for ordinal classification [21]. It was also implemented and run using Matlab.

3.3 Evaluated Measures

The performance of classification machine learning algorithms is typically evaluated using the accuracy metric. However, this is not appropriate when the data is unbalanced and/or the costs of different errors vary markedly (as in ordinal regression). For this reason, we also introduce the use of other metrics, which performs better for unbalanced and ordinal data. The metrics used are: CCR (threshold metric), R_s (ordinal metric invariant to class targets), MAE (usually used in ordinal classification) and $AMAE$ (for ordinal and unbalanced classification). A more detailed description is now given:

- Correct Classification Rate (CCR):

$$CCR = \left(\frac{1}{N}\right) \sum_{i=1}^{N} (I(\hat{y}_i = y_i)),$$

  where $I(\cdot)$ is the zero-one loss function, y_i is the desired output for pattern i, $\hat{y}_i$ is the prediction of the model and N is the total number of patterns in the dataset.
- Spearman's rank correlation coefficient (R_s): This metric is used in order to avoid the influence of the numbers chosen to represent the classes on the performance assessment [8]:

$$R_s = \frac{\sum (y_i - \mu(\mathbf{y})) \cdot (\hat{y}_i - \mu(\hat{\mathbf{y}}))}{\sqrt{\sum (y_i - \mu(\mathbf{y}))^2 \sum (\hat{y}_i - \mu(\hat{\mathbf{y}}))^2}},$$

  where $\mu(\mathbf{x})$ represents the mean for the vector $\mathbf{x}$.
- Mean Absolute Error (MAE): it is defined as the average deviation in absolute value of the predicted class from the true class. It is a commonly used for ordinal regression:

$$MAE = \frac{1}{N} \sum_{i=1}^{N} |y_i - \hat{y}_i|,$$

  where y_i is the true rank for the $i-th$ pattern, and $\hat{y}_i$ corresponds to the predicted rank for the same pattern. N is the number of test patterns.
- Average Mean Absolute Error ($AMAE$): it measures the mean of the classification errors across classes:

$$AMAE = \frac{1}{K} \sum_{j=1}^{K} MAE_j,$$

  where MAE_j is the Mean Absolute Error (MAE) taking into account only patterns from class j. $AMAE$ is preferred to MAE when dealing with unbalanced data [5].

3.4 Evaluation and Model Selection

For the evaluation of the results, a holdout stratified technique to divide the data has been applied 30 times, with the 75% of the patterns to train the model, and the remaining 25% to test it. The results are taken as the mean and standard deviation of the measures over the 30 test sets. The parameters of each algorithm are chosen using a k-fold method with $k = 5$ with each of the 30 training sets. The final combination chosen is the one which obtains in mean the best average performance for the validation sets, where the metric used is the *MAE* [5]. The kernel selected for all the algorithms (Kernel Discriminant Learning methods and SVM) is the Gaussian one, $K(\mathbf{x}, \mathbf{y}) = \exp\left(-\frac{\|\mathbf{x}-\mathbf{y}\|^2}{\sigma^2}\right)$ where σ is the standard deviation. The kernel width was selected within these values $\{10^{-3}, 10^{-2}, \ldots, 10^3\}$, and so was the cost parameter associated to the SVM methods. The parameter u for avoiding singularity (for the methods based on discriminant analysis) was selected within $\{10^{-4}, 10^{-3}, 10^{-2}, 10^{-1}\}$, and the C parameter for the KDLOR was selected within $\{10^{-1}, 10^0, 10^1\}$.

4 Results

The results of all the methods are included in Table 1. From these results, some conclusions can be drawn. First of all, good results are obtained by several methods, such as KDLOR or SVMRank. These results can be considered as good enough for the proposed objective in the application since most of the methods here used could work well in the genetic algorithm and guide the searching instead of using clustering. On the other hand the results obtained with the LDAOR (the non-kernel discriminant analysis), demonstrate the necessity of applying the kernel trick. Also, by comparing the results of the standard SVM, the SVMRank and the F&H(SVM), the application of an ordinal classification algorithm is justified, as it can significantly improve the results.

Another important issue to take into consideration is the execution complexity of the used algorithm, since it should not be forgotten that the model would be included in a genetic algorithm, for that, the KDLOR algorithm would be preferred rather than the SVMRank as this one can be classified as an ensemble method.

Table 1. Mean and Standard Deviation (SD) of *CCR*, R_s, *MAE* and *AMAE* from the 30 models obtained by the different methods for the test sets

Algorithm	Mean±SD			
	CCR	R_s	*MAE*	*AMAE*
KDLOR	**75.42±3.72**	**0.843±0.032**	**0.268±0.047**	*0.421±0.080*
LDAOR	46.43±3.89	0.679±0.047	0.619±0.053	0.555±0.078
POM	60.87±4.27	0.713±0.066	0.434±0.052	0.528±0.104
SVM	46.87±0.27	0.000±0.000	0.689±0.006	1.200±0.000
SVMRANK	*75.36±4.59*	*0.818±0.047*	*0.272±0.053*	**0.357±0.100**
F&H(SVM)	73.91±3.59	0.737±0.050	0.322±0.045	0.524±0.098

The best method is in bold face and the second best one in italics

A parametric test has been performed, comparing the algorithms which obtain the best results: KDLOR and LibSVMRank, in order to study the significance of the means obtained. Firstly of all, we checked that the conditions for the parametric test were satisfied: independence of the sample, same variances (using the Levene test), and normal distribution of the sample (Kolmogorov-Smirnov test). The performed test (Student's t test) shown no significant differences taking into account *CCR* and *MAE* (the associated p-values were respectively 0.954, and 0.762). However, the results shown significant differences for *AMAE* and R_s (the associated p-values were 0.09 and 0.028) respectly favouring SVMRank and KDLOR.

Summarising, we consider the KDLOR algorithm as the more appropiate for the proposed problem, as it obtained a better performance in three of the four considered metrics and a smaller standard deviations (which indicates that the results are more robust). The worst results are achieved taking into account the *AMAE* metric, which is more suitable for unbalanced classification. Due to this, the authors are working on a possible modification of this algorithm for dealing with unbanlaced datasets in order to obtain better results. This modification of the KDLOR would change the thresholds of the model according to the inverse of the number of patterns per class (so it would give more significance to the minority classes).

5   Conclusion and Future Work

This paper introduces the use of ordinal regression for solving the Unequal Area Facility Layout Problem, which takes into account subjective evaluations from different experts. Concerning the application, the results obtained have demonstrated to be good enough in such a way that the best model could be integrated in a genetic algorithm in order to relieve the arduous evaluating task of the experts and could guide a searching process so that new solutions can be discovered. Also, the application of ordinal regression seems a good approximation since (as demonstrated) it can significantly improve the results.

As future work, the two best obtained models will be integrated in the genetic algorithm, and the area layouts evaluated with the higher rank (i.e. very good) will be analysed by the experts in order to validate the methodology. Also, as the Kernel Discrimination Learning for Ordinal Regression algorithm is the one that shows the best results taking into account general metrics (not specialized in measuring unbalanced classification), a modification of this algorithm is being studied, changing in this case the thresholds of the model in order to deal and better perform with unbalanced problems. This "threshold moving" modification for unbalanced problems will be tested with other datasets, and applied to other ordinal classification algorithms (as SVM methods or the Frank & Hall approach previously explained).

Acknowledgement. This work has been partially subsidized by the TIN2011-22794 project of the Spanish Ministerial Commission of Science and Technology (MICYT), FEDER funds and the P08-TIC-3745 project of the "Junta de Andaluca" (Spain).

References

1. Aiello, G., Enea, M.: Fuzzy approach to the robust facility layout in uncertain production environments. International Journal of Production Research 39(18), 4089–4101 (2001)
2. Armour, G.C., Buffa, E.S.: A heuristic algorithm and simulation approach to relative location of facilities. Management Science 9, 294–309 (1963)
3. Avigad, G., Moshaiov, A.: Interactive evolutionary multiobjective search and optimization of set-based concepts. Trans. Sys. Man Cyber. Part B 39(4), 1013–1027 (2009)

4. Babbar-Sebens, M., Minsker, B.S.: Interactive genetic algorithm with mixed initiative inter-action for multi-criteria ground water monitoring design. Ap. Soft Comp. 12(1), 182 (2012)
5. Baccianella, S., Esuli, A., Sebastiani, F.: Evaluation measures for ordinal regression. In: Proc. of the Ninth Int. Conf. on Intelligent Systems Design and App. (ISDA), Pisa, Italy (2009)
6. Brintup, A.M., Ramsden, J., Tiwari, A.: An interactive genetic algorithm-based framework for handling qualitative criteria in design optimization. Computers in Ind. 58, 279 (2007)
7. Brintup, A.M., Takagi, H., Tiwari, A., Ramsden, J.: Evaluation of sequential, multi-objective, and parallel interactive genetic algorithms for multi-objective optimization problems. Journal of Biological Physics and Chemistry 6, 137–146 (2006)
8. Cardoso, J.S., Sousa, R.: Measuring the performance of ordinal classification. International Journal of Pattern Recognition and Artificial Intelligence 25(8), 1173–1195 (2011)
9. Chang, C., Lin, C.: Libsvm: a library for support vector machines (2001), http://www.csie.ntu.edu.tw/cjlin/libsvm
10. Chaudhuri, S., Deb, K.: An interactive evolutionary multi-objective optimization and deci-sion making procedure. Applied Soft Computing 10(2), 496–511 (2010)
11. Cortes, C., Vapnik, V.: Support vector networks. Maching Learning 20, 273–297 (1995)
12. Frank, E., Hall, M.: A simple approach to ordinal classification. In: Proceedings of the 12th European Conference on Machine Learning, pp. 145–156 (2001)
13. García-Hernández, L., Salas-Morera, L., Arauzo-Azofra, A.: An interactive genetic algo-rithm for the unequal area facility layout problem. In: SOCO, pp. 253–262 (2011)
14. Holland, J.H.: Adaptation in natural and artificial systems. MIT Press, Cambridge (1992)
15. Jeong, I., Kim, K.: An interactive desirability function method to multiresponse optimization. European Journal of Operational Research 195(2), 412–426 (2009)
16. Kouvelis, P., Kurawarwala, A.A., Gutierrez, G.J.: Algorithms for robust single and multi-ple period layout planning for manufacturing systems. European Journal of Operational Re-search 63(2), 287–303 (1992)
17. Kusiak, A., Heragu, S.S.: The facility layout problem. European Journal of Operational Re-search 29(3), 229–251 (1987)
18. Li, L., Lin, H.T.: Ordinal Regression by Extended Binary Classification. In: Advances in Neural Information Processing Systems, vol. 19 (2007)
19. Liu, F., Geng, H., Zhang, Y.Q.: Interactive fuzzy interval reasoning for smart web shopping. Applied Soft Computing 5(4), 433–439 (2005)
20. Luque, M., Miettinen, K., Eskelinen, P., Ruiz, F.: Incorporating preference information in interactive reference point methods for multiobjective optimation. Omega 37(2), 450 (2009)
21. McCullagh, P.: Regression models for ordinal data. Journal of the Royal Statistical Soci-ety 42(2), 109–142 (1980)
22. Pérez-Ortiz, M., Gutiérrez, P.A., García-Alonso, C., Carulla, L.S., Pérez, J.S., Hervás-Martínez, C.: Ordinal classification of depression spatial hot-spots of prevalence. In: Proc. of the 11th Int. Conf. on Intelligent Systems Design and App (ISDA), p. 1170 (2011)
23. Sato, T., Hagiwara, M.: Idset: Interactive design system using evolutionary techniques. Computer-Aided Design 33(5), 367–377 (2001)
24. Sun, B.Y., Li, J., Wu, D.D., Zhang, X.M., Li, W.B.: Kernel discriminant learning for ordinal regression. IEEE Transactions on Knowledge and Data Engineering 22, 906–910 (2010)
25. Tompkins, J., White, J., Bozer, Y., Tanchoco, J.: Facilities Planning, 4th edn. Wiley, New York (2010)
26. Tong, X.: SECOT: A Sequential Construction Technique For Facility Design. Doctoral Dis-sertation, University of Pittsburg (1991)

A Soft Computing Approach to Knowledge Flow Synthesis and Optimization

Tomas Rehorek and Pavel Kordik

Faculty of Information Technology, Czech Technical University in Prague, Thakurova 9, Prague, 16000, Czech Republic
{tomas.rehorek,kordikp}@fit.cvut.cz

Abstract. In the areas of Data Mining (DM) and Knowledge Discovery (KD), large variety of algorithms has been developed in the past decades, and the research is still ongoing. Data mining expertise is usually needed to deploy the algorithms available. Specifically, a process of interconnected actions referred to as knowledge flow (KF) needs to be assembled when the algorithms are to be applied to given data. In this paper, we propose an innovative evolutionary approach to automated KF synthesis and optimization. We demonstrate the evolutionary KF synthesis on the problem of classifier construction. Both preprocessing and machine learning actions are selected and configured by means of evolution to produce a model that fits very well for a given dataset.

1 Introduction

Several research attempts have been made in order to automate the construction of the KD process. In [1], an ontology of knowledge is modeled and the knowledge flow is constructed through the use of automated planning. In our opinion, an exhaustive enumeration is too limiting. There are thousands of different knowledge flows available online, some of which are highly specialized for a given problem. The KD ontology should be constructed in manner rather inductive than deductive.

A framework for an automated DM system using autonomous intelligent agents is proposed in [2]. In the FAKE GAME project [3], **Evolutionary Computation (EC)** was used to automate certain steps of the KD process. In FAKE GAME, genetic algorithm (GA) is used to evolve neural networks. In this paper, we further extend this work to automate the knowledge flow synthesis, this time by means of genetic programming (GP).

There are several motivations for process-oriented view of DM and KD. Wide range of algorithms for automated extraction of information from data has been developed in the last decades. These algorithms, however, are of specialized roles in the context of the whole KD procedure. While some of these algorithms focus on data preprocessing, others are designed to learning and model building, and yet others are able to visualize data and models in various forms. Only when orchestriated, these algorithms are able to deliver useful results. We believe that viewing KD tasks as problems of finding appropriate knowledge flows is the right approach to KD automation.

V. Snasel et al. (Eds.): SOCO Models in Industrial & Environmental Appl., AISC 188, pp. 23–32.

2   Knowledge Flows

The composition of KD processes can be expressed in form of oriented graphs called the knowledge flows (or workflows in general). These knowledge flows became standard way of representing interconnected actions that need to be executed in order to obtain useful knowledge [1].

In our context, **knowledge flow (KF)** is understood as directed acyclic graph (DAG) consisting of labeled nodes and edges. The nodes are called **actions** and can be thought as general transformations. The edges serve as transition channels between the actions. A general process is shown in Fig. 1.

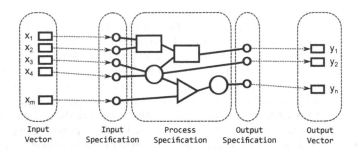

Fig. 1. A dynamic view of a general knowledge flow

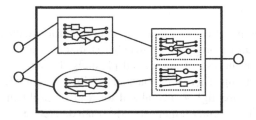

Fig. 2. A general nesting of knowledge flows

There is a special kind of actions that require not only connection of input ports, but also a specification of some **inner processes**. This is typical for meta-learning approaches that combine, for example, multiple machine learning methods to produce ensembles of models [4]. This is why we further extend our definition of KF with **nesting**, i.e. allowing a node to contain an inner process and thus creating DAG hierarchy. Hierarchical DAGs seem to be approach suitable enough to capture vast majority of thinkable KD processes. A general nesting is shown in Fig. 2.

3   Related Evolutionary Computation Methods

Evolutionary Computation (EC) is an area of optimization using population of candidate solutions that are optimized in parallel. These solutions interact with each other by means of principles inspired by biological evolution, such as selection, crossover, reproduction, and mutation [5].

There are many branches in EC, some of which arose independently on each other. The most prominent are the Genetic Algorithm (GA), Genetic Programming (GP), Evolutionary Programming (EP), and Evolution Strategies (ES). Several attempts have been done to adapt EC techniques for evolving DAGs as summarized below.

3.1   Approaches to Evolution of Graphs

In 1995, *Parallel Algorithm Discovery and Orchestration* (PADO) was introduced, aiming at automating the process of signal processing [6,?]. In its early version [6], it used GP to evolve tree-structured programs for classification task. In later version [7], trees were replaced by general graph structures.

A more systematic and general approach to DAG evolution is the *Cartesian Genetic Programming* (CGP) [8]. Nodes are placed into grid of fixed width and height, and a genome of inter-node connections and functions to be placed into the nodes is evolved. Although CGP may be used for various domains, it especially targets evolution of Boolean circuits.

In [9], evolution of analog electrical circuits using Strongly Typed GP (STGP) is introduced by J.R. Koza. The principle of cellular encoding proposed in [10] for evolution of neural networks is further extended and generalized. The algorithm starts with minimal *embryonic circuit*, changes to which are expressed as a GP tree. Non-terminals of the tree code topology operations such as splitting a wire into two parallel wires, whilst terminal code placing electrical components onto the wire. The principle is very general and allows us to design the GP grammar that suits best for a given problem.

Besides the aforementioned general approaches to graph evolution, a lot of attention has been paid to evolution of Artificial Neural Networks (ANNs). Research in this area, which is also known as *neuroevolution*, produced large variety of algorithms for both the optimization of synaptic weights and the topology. From our point of view, the only interesting are topology-evolving algorithms, sometimes called as Topology and Weight Evolving Neural Networks (TWEANNs) [11]. *Generalized Acquisition of Recurrent Links* (GNARL) evolves recurrent ANNs of arbitrary shape and size through the use of Evolutionary programming. *Symbiotic, Adaptive Neuro-Evolution* (SANE) evolves feed-forward networks of single hidden layer. There are two populations: the first one evolves neurons, and the second one evolves final networks connecting neurons together. In 2002, one of the most successful neuroevolution algorithms, *NeuroEvolution of Augmenting Topologies* (NEAT) [11] was introduced. Similarly to STGP embryonic evolution of circuits, it starts with minimal topology which grows during the evolution. It uses GA-like genome to evolve population of edges, and employs several sophisticated mechanisms such as niching and informed crossover.

From all the mentioned algorithms, the most suitable for evolution of KFs seem to be CGP, NEAT, and Embryonic STGP. CGP has two limitations: it bounds the maximal size of the graph and does not use strong typing as the actions in KF do. NEAT puts no limits on the size of the graph, but does not respect arities and I/O types of the actions. The most promising seems to be the Embryonic STGP proposed in [9] as it allows us to evolve KFs of arbitrary sizes and is able to respect both the I/O arities and types of the actions. That is why we decided to further investigate Embryonic STGP.

3.2 Embryonic Strongly Typed Genetic Programming (STGP)

Based on Frédéric Gruau's *cellular encoding* (CE) for neuroevolution [10], J. R. Koza introduced GP-based algorithm for automated design for both the topology and sizing of analog electrical circuits [9]. In CE, the program tree codes a plan for developing a complete neural network from some simple, initial topology. Similarly to CE, Koza's algorithm starts from trivial circuit, which is called the *embryonic* circuit.

In the embryonic circuit, there are several components (such as capacitors, inductors, etc.) fixed, and there are two wires that *writing heads* are pointing to. Accordingly, the root of the GP tree being evolved contains exactly two subtrees, each subtree expressing the development of one of these wires. The nodes are **strongly typed**, allowing us to evolve different aspects of the circuit. For the purposes of GP tree construction, there are several categories of nodes available:

1. **Circuit-constructing** functions and terminals that create the topology of the circuit. These include parallel and serial split of a wire, and polarity flip.
2. **Component-setting** functions that convert wires within the circuit into components, such as capacitors, resistors, and inductors.
3. **Arithmetic-performing** functions and numerical terminals that together specify the numerical value (sizing) of each component in the circuit.

This allows the resulting circuit to be fully specified within single GP tree. The results of the algorithm seems very promising. In [12], the algorithm is applied to other types of analog electrical circuits, and is reported as human-competitive as it was able to find circuits very similar to those that have been previously patented.

4   Evolution of Knowledge Flows

We will demonstrate our approach on particular subproblems of the knowledge discovery process: *data preprocessing* and *algorithm selection* tasks for **supervised classification problems**.

Generally, if the task is well-defined, there is usually some template for the KF that can be used to solve the task. In such a template, some of the blocks are fixed, while others may vary. The fixed parts of the template reflect the semantics of the task. The variable parts may be modified in order to adapt the template so it maximizes the performance on specific dataset. This exactly matches the presumptions of STGP approach to circuit evolution proposed by J. R. Koza in [9]. Given such an *embryonic* template and a dataset, our algorithm will evolve the variable parts to fit the final KF to the dataset as well as possible.

Specifically, for our exemplary problem of classifier construction, we are given a dataset, and a set of preprocessing and learning actions that can be integrated into the classifier template shown on Fig. 3. The algorithm then automatically constructs a classifier KF based on this template.

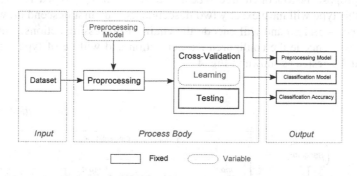

Fig. 3. A template of the Classifier selection process: Variable parts are to be evolved

The actions available may be used to complete/fine-tune the process. As shown in Fig. 3, our task is to find a combination of **preprocessing** steps (e.g. preprocessing model) and a learning algorithm that is able to **learn** from preprocessed data with high classification accuracy. The accuracy is estimated using 20-fold cross-validation.

Traditionally, there is human assistance needed in order to complete such a template. Furthermore, even the human expert often selects appropriate methods in trial-and-error manner. There is a large number of degrees of freedom, making it a difficult optimization problem. Moreover, dozens of parameters often need to be configured to fine-tune the classifier. Our task is to design an evolutionary algorithm that would, for a given dataset, construct the classification process automatically.

4.1 Our Approach

Based on problem analysis, we propose to adapt embryonic STGP according to [9] to construct the KF from a template. The method of encoding the topology of the graph, along with the parameters, into a single STGP program tree, is very flexible. In fact, the KF construction problem can be addressed easily by only slight modification of the algorithm. All the actions and their parameters (both numerical and categorical) can easily be embodied into the genotype through the use of type-specific subtrees. Moreover, nested processes can easily be encoded through the use of subtrees, as well.

In order to adapt the embryonic STGP algorithm to the evolution of KD processes, a problem-specific grammar needs to be designed. Hence the grammar is further investigated in this section.

4.2 Cross-Validation Process Grammar

In accordance with Fig. 3, there are two top-level points where the expansion of the process should be started. Basically, two crucial substructures are to be evolved: the sequence of configured preprocessing actions, and the configured learning action. For that reason, we propose the root of GP tree to be non-terminal strongly typed as `Workflow`. A node of this type will have exactly two descendants. The first descendant will be of type `Preprocessing` and will encode the chain of processing actions. The second descendant will encode the configured learner action and will be of type `Learner`. Learners can be nested as shown in Fig. 4.

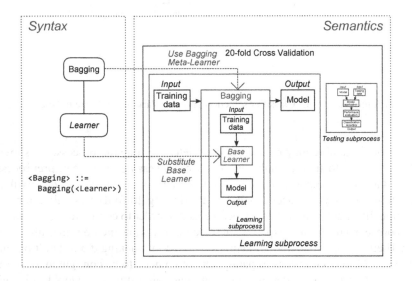

Fig. 4. Bagging Meta-Learner Grammar and Its Semantics

This can be expressed using Backus-Naur Form as:

```
     <Workflow> ::= KF(<Preprocessing,Learner>)
<Preprocessing> ::= <AttributeSelection> | <PCA> |
                    PreprocessTerminal
      <Learner> ::= <kNN> | <NaiveBayes> |
                    <DecisionTree> |
                    <MajorityVote> | <Bagging>
```

As can be seen, the structure can be constructed in hierarchical manner, making GP approach very flexible. Because our hierarchy of non-terminals is very large, we will further demonstrate the principle only briefly, showing the power and universality of GP approach. Further decomposition of `<AttributeSelection>` moving down until the level of integers follows:

```
<AttributeSelection> ::= AS(attrs=<SetOfIntegers>,
                           invert=<BooleanConstant>,
                           next=<Preprocessing>)
<SetOfIntegers> ::= SetMember(<SetOfIntegers>,<Integer>) |
                    SetTerminal
<Integer> ::= IntegerPlus(<Integer>,<Integer>) |
              IntegerMultiply(<Integer>,<Integer>) |
              IntegerDivide(<Integer>,<Integer>) |
              1 | 2 | ... | 10
```

Design of the rest of the grammar is completely analogous and can be designed easily for various knowledge flows solving different tasks. Very complex information can be encoded into the tree, allowing us to move downwards in the hierarchy and design actions that are appropriate for given layers of abstraction. We further investigate only the case study of classifier KF.

5 Experiments

The algorithm proposed in the previous section has been tested on several datasets with very promising results. Fig. 5 shows average accuracy over first 30 generations based on 10 runs of the algorithm on Ecoli dataset from the UCI Machine Learning Repository [13]. Many interesting solutions have been found for different datasets. An example GP tree, found for Ecoli dataset, is shown in Fig. 6.

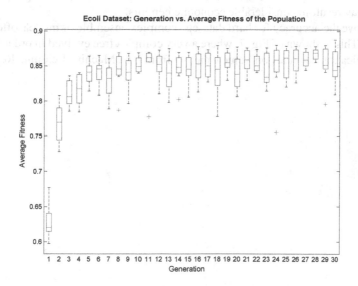

Fig. 5. Evaluation of the Evolutionary Classifier Process Construction on the Ecoli dataset

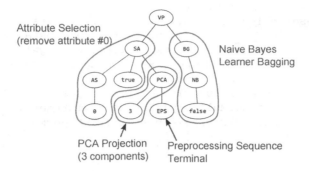

Attribute Selection
(remove attribute #0)

Naive Bayes
Learner Bagging

PCA Projection
(3 components)

Preprocessing Sequence
Terminal

Fig. 6. A sample tree evolved on Ecoli dataset

Despite the very limited number of available preprocessing and learning actions, the algorithm managed to find a satisfactory solution for most of the datasets. In most cases, a good solution was in the first generation within the randomly generated individuals. Indeed, this is because the limited set of learners. In the RapidMiner learning environment, for example, there are dozens of preprocessing and learning operators available. If included into the set actions used by our algorithm, the results could be further improved.

Even though the classification accuracy is improved only by units of percents during the evolution, in the field of DM, this is not insignificant. In fact, it is a very difficult task to improve the results at the point where the learning algorithms reach their limits. However, in many areas such as e-commerce, improvement of a single percent in accuracy may result in considerable economic benefits.

To improve the accuracy for hard datasets, complex ensemble of models often needs to be built. This is shown of Fig. 7, where a very complex tree evolved in order to maximize classification accuracy on Vehicle dataset from UCI Machine Learning Repository

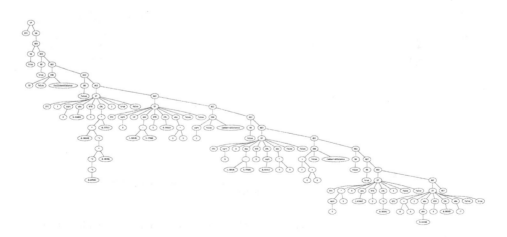

Fig. 7. A very complex tree evolved on Vehicle dataset

[13] is shown. As can be seen, it uses a majority vote on the top of 14 inner learners. This is consistent with the results obtained by a different strategy [14], where complex ensemble of models was also obtained for the Vehicle data.

6 Conclusion

In this paper an efficient algorithm for automated construction of KD process was introduced. Given list of appropriate preprocessing, modeling and parameter-tuning node types, the algorithm constructs a classifier that minimizes classification error on a given dataset. As it is based on universal concepts of GP and embryonic graph construction, it is reasonable to expect the algorithm to be suitable for many similar problems. Given a well-defined task and a set of appropriate actions, the algorithm constructs process of near-optimal behavior.

We have also presented promising preliminary results of the KD process evolution. Our approach futher extends capabilities of the SpecGEN algorithm [4] by incorporating data preprocessing operators. As a future work, we plan to benchmark the presented approach on bigger collection of problems and evaluate the effect of each improvement.

References

1. Žáková, M., Křemen, P., Železný, F., Lavrač, N.: Automatic knowledge discovery workflow composition through ontology-based planning. IEEE Transactions on Automation Science and Engineering 8(2), 253–264 (2011)
2. Rajan, J., Saravanan, V.: A framework of an automated data mining system using autonomous intelligent agents. In: ICCSIT 2008, pp. 700–704. IEEE Computer Society, Washington, DC (2008)
3. Kordík, P.: Fully Automated Knowledge Extraction using Group of Adaptive Models Evolution. PhD thesis, Czech Technical University in Prague, FEE, Dep. of Comp. Sci. and Computers, FEE, CTU Prague, Czech Republic (September 2006)
4. Černý, J.: Methods for combining models and classifiers. Master's thesis, FEE, CTU Prague, Czech Republic (2010)
5. Luke, S.: Essentials of Metaheuristics. Lulu (2009),
   http://cs.gmu.edu/~sean/book/metaheuristics/
6. Teller, A., Veloso, M.: PADO: Learning tree structured algorithms for orchestration into an object recognition system. Technical Report CMU-CS-95-101, Department of Computer Science, Carnegie Mellon University, Pittsburgh, PA, USA (1995)
7. Teller, A., Veloso, M.M.: Program evolution for data mining. International Journal of Expert Systems 8(3), 213–236 (1995)
8. Miller, J.F., Thomson, P.: Cartesian Genetic Programming. In: Poli, R., Banzhaf, W., Langdon, W.B., Miller, J., Nordin, P., Fogarty, T.C. (eds.) EuroGP 2000. LNCS, vol. 1802, pp. 121–132. Springer, Heidelberg (2000)
9. Koza, J.R., Bennett, F.H., Andre, D., Keane, M.A., Dunlap, F.: Automated synthesis of analog electrical circuits by means of genetic programming. IEEE Transactions on Evolutionary Computation 1(2), 109–128 (1997)
10. Gruau, F.: Cellular encoding of genetic neural networks. Technical Report RR-92-21, Ecole Normale Superieure de Lyon, Institut IMAG, Lyon, France (1992)

11. Stanley, K.O., Miikkulainen, R.: Evolving neural networks through augmenting topologies. Evolutionary Computation 10(2), 99–127 (2002)
12. Koza, J.R., Bennett III, F.H., Andre, D., Keane, M.A.: Four problems for which a computer program evolved by genetic programming is competitive with human performance. In: International Conference on Evolutionary Computation, pp. 1–10 (1996)
13. Frank, A., Asuncion, A.: UCI machine learning repository (2010)
14. Kordík, P., Černý, J.: Self-organization of Supervised Models. In: Jankowski, N., Duch, W., Grąbczewski, K. (eds.) Meta-Learning in Computational Intelligence. SCI, vol. 358, pp. 179–223. Springer, Heidelberg (2011)

The Combination of Bisection Method
and Artificial Bee Colony Algorithm
for Solving Hard Fix Point Problems

P. Mansouri[1,2,*], B. Asady[1], and N. Gupta[2]

[1] Department of Mathematics, Arak Branch, Islamic Azad University, Arak-Iran
[2] Department of Computer Science, Delhi University, Delhi, India
pmansouri393@yahoo.com, p-mansouri@iau-arak.ac.ir

Abstract. In this work, with combination Bisection method and Artificial bee colony algorithm together(BIS-ABC), We introduce the novel iteration method to solve the real-valued fix point problem $f(x) = x, x \in [a, b] \subseteq R$. Let $f(a).f(b) < 0$ and there exist $\alpha \in [a, b], f(\alpha) = \alpha$. In this way, without computing derivative of function f, real-roots determined with this method that is faster than ABC algorithm. In numerical analysis, NewtonRaphson method is a method for finding successively better approximations to the simple real roots, if $f'(x_i) - 1 = 0$,in i^{th}iteration, then Newton's method will terminate and we need to change initial value of a root and do algorithm again to obtain better approximate of α. But in proposed method, we reach to solution with direct search in $[a, b]$, that includesα(convergence speed maybe less than of Newton's method). We illustrate this method by offering some numerical examples and compare results with ABC algorithm.

Keywords: Bisection method, Root-finding method, Fix point problems, Artificial Bee Colony algorithm.

1 Introduction

Solving real-valued equation$f(x) = x, x \in [a, b] \subseteq R$(Finding simple roots)is one of the most important problem in engineering and science. The Bisection method in mathematics, is a root-finding method which repeatedly bisects an interval, then selects a subinterval in which a root must lie for further processing. It is a very simple and robust method, but it is also relatively slow. Because of this, it is often used to obtain a rough approximation to a solution which is then used as a starting point for more rapidly converging methods[1]. The method is also called the binary search method[1] and is similar to the computer science binary search, where the range of possible solutions is halved each iteration. Artificial bee colony (ABC) is one of the most recently defined algorithms by Dervis Karaboga in 2005, motivated by the intelligent behavior of honey bees [4],[8]. It is as simple as Particle Swarm Optimization (PSO) and Differential evolution

---

[*] Corresponding author.

V. Snasel et al. (Eds.): SOCO Models in Industrial & Environmental Appl., AISC 188, pp. 33–41.
springerlink.com

(DE) algorithms, Genetic algorithm (GA)[2], biogeography based optimization (BBO), and uses only common control parameters such as colony size and maximum cycle number. ABC as an optimization tool, provides a population-based search procedure in which individuals called foods positions are modified by the artificial bees with time and the bee's aim is to discover the places of food sources with high nectar amount and finally the one with the highest nectar. In ABC system, artificial bees fly around in a multidimensional search space and some (employed and onlooker bees) choose food. Development of an ABC algorithm for solving generalized assignment problem which is known as NP-hard problem is presented in detail along with some comparisons[7]. Sources depending on the experience of themselves and their nest mates, and adjust their positions. A novel iteration method for solve hard problems directly with ABC algorithm is presented with P. Mansouri et.all in 2011[9]. In this paper, we introduce a combination iteration method to solve equation $f(x) = x$, $x \in [a, b] \subseteq R$. At the first step, the ABC generates a randomly distributed initial population P of n solutions $x_i, i = 1, 2, ..., n$(approximate values for $\alpha \in [a, b]$), where n denotes the size of population. Let $f(a).f(b) < 0$, Bisection method repeatedly bisects an interval, then selects a subinterval in which a root must lie for further processing. At the kth iteration($k \geq 1$), interval width is $0.5^k.(b - a)$ and,

$$\Delta x_k = |x_k - x_{k1}| = (0.5)^k(b - a)$$

and the new midpoint is

$$x_k = \frac{a_{k-1} + b_{k-1}}{2}, \quad k = 1, 2, ...$$

In Section 1.1, the ABC algorithm, Bisection method and Newton method are described. In section 2, we explain and discussed about our method. In section 3, we compare accuracy and complexity of proposed method and ABC algorithm with some examples. In section 4, we discuss about ability of the proposed method.

1.1   Artificial Bee Colony Algorithm

Artificial bee colony algorithm (ABC) is an algorithm based on the intelligent foraging behavior of honey bee swarm, purposed by Karaboga in 2005 [3]. In ABC model, the colony consists of three groups of bees: employed bees, onlookers and scouts. It is assumed that there is only one artificial employed bee for each food source. In other words, the number of employed bees in the colony is equal to the number of food sources around the hive. Employed bees go to their food source and come back to hive and dance on this area. The employed bee whose food source has been abandoned becomes a scout and starts to search for a new food source. Onlookers watch the dances of employed bees and choose food sources depending on dances. The pseudo-code of the ABC algorithm is as following:

Algorithm 1:(artificial bee colony algorithm).

01. Initialize population with random solutions.
02. Evaluate fitness of the population.
03. **While** (stopping criterion not met) Forming new population.
04. Select sites for neighborhood search.
05. Recruit bees for selected sites (more bees for best sites) and evaluate fitnesses.
06. Select the fittest bee from each patch.
07. Assign remaining bees to search randomly and evaluate their fitnesses.
08. **End While**.

In ABC which is a population based algorithm, the position of a food source represents a possible solution to the optimization problem and the nectar amount of a food source corresponds to the quality (fitness) of the associated solution. The number of the employed bees is equal to the number of solutions in the population. At the first step, a randomly distributed initial population (food source positions) is generated. After initialization, the population is subjected to repeat the cycles of the search processes of the employed,onlooker, and scout bees,respectively. An employed bee produces a modification on the source position in her memory and discovers a new food source position. Provided that the nectar amount of the new one is higher than that of the previous source, the bee memorizes the new source position and forgets the old one. Otherwise she keeps the position of the one in her memory. After all employed bees complete the search process, they share the position information of the sources with the onlookers on the dance area. Each onlooker evaluates the nectar information taken from all employed bees and then chooses a food source depending on the nectar amounts of sources. As in the case of the employed bee, she produces a modification on the source position in her memory and checks its nectar amount. Providing that its nectar is higher than that of the previous one, the bee memorizes the new position and forgets the old one. The sources abandoned are determined and new sources are randomly produced to be replaced with the abandoned ones by artificial scouts.

Bisection Method. The Bisection Method is a numerical method for estimating the roots of the real-valued problem $f(x) = 0$. It is one of the simplest and most reliable, but it is not the fastest method. Assume that $f(x)$ is continuous.

Algorithm 2:(Bisection method algorithm).
   Given the continuous function $f(x)$. Find points a and b such that $a < b$ and $f(a).f(b) < 0$. Then we obtain midpoint x_1 of interval $[a, b]$ as follows:

$$\Delta x_i = |x_i - x_{i1}| = (0.5)^i (b - a)$$

and the new midpoint is

$$x_i = (a_{i1} + b_{i-1})/2$$

2   The Combination of Bisection Method and Artificial Bee Colony Algorithm

In this work, we introduced a novel iteration algorithm with combing the Bisection method and ABC algorithm(BIS-ABC)to obtained the solution of hard fix pint problem as $g(x) = x$, without need to compute derivative of $g(x)$. With this method, the sequence of closed intervals $I_i, i \in N$ constructed that I_i includes the solution. Let $f(x) = g(x) - x$, we obtain sequence of approximate values $x_i, x_i \in I_i = [a_i, b_i]$, $i \in N$. Initial value of solution approximated by ABC algorithm. Let $I_1 = [a_1, b_1]$, so $a_1 = a, b_1 = b$. At the first iteration, ABC algorithm randomly generates x_1. Bisection method computes $f(x_1).f(a)$, bisect interval to halve of that I_2. We assume that for each n, I_{n+1} is either $[a_n, (a_n + b_n)/2]$ or $[(a_n + b_n)/2, b_n]$. At the each step, we bisect the interval and choose one of the resulting half-intervals as the next interval and obtain randomly new value of root by ABC algorithm. It is clearly true that $I_{n+1} \subseteq I_n$, so the intersection of all the I_n's is nonempty, and there is at least one number c which belongs to all the intervals. We claim further that there is exactly one number c which is belong to all the intervals. Because of the bisection construction, we know that $b_n - a_n = 0.5^n(b - a)$. Suppose that $c \leq d$ are numbers each of which belongs to all of the intervals I_n. Clearly $0 \leq (d - c) \leq b_n - a_n$ for each n. Thus $0 \leq (d - c) \leq 0.5^n(b - a)$ for every n. But this means that $d - c = 0$, so $c = d$. Always the Bisection method is understood as a way of closing down on a uniquely determined number in $[a, b]$ with desired properties. We determine I_k as follows:

01. if $f(x_{k-1}).f(a_{k-1}) \leq 0$,
02. then $\alpha \in [a_{k-1}, x_{k-1}]$,
03. else $\alpha \in [x_{k-1}, b_{k-1}]$,

We focus to new interval $I_k = [a_k, b_k]$, that $a \leq a_k \leq b_k \leq b$ and $\alpha \in I_k$.

Algorithm 3:(Algorithm of Proposed method)
Source code for find I_k is as following:

01. **if** $(f(x_{k-1}).f(a_{k-1})) < 0$,
02. then $b_k = (x_{k-1} + b_{k-1})/2$, end,
03. $temp = (x_{k-1} + a_{k-1})/2$;
04. if $(f(temp).f(x_{k-1}) < 0$,
05. then $a_k = temp$,
06. else $temp = (temp + a_{k-1})/2$, end,
07. **else** $a_k = (x_{k-1} + a_{k-1})/2$,
08. $temp = (x_{k-1} + b_{k-1})/2$,
09. if $(f(temp).f(x_{k-1}) < 0$,
10. then $b_k = temp$,
11. else $temp = (temp + b_{k-1})/2$, end, **end,**
12. If $abs(b_k - a_k) \leq \varepsilon$,
13. then $\alpha \simeq (a_k + b_k)/2$.

We illustrate this algorithm with some examples and compare results of proposed method with Newton method and ABC algorithm.

3   Numerical Examples

In this section, BIS-ABC is applied to solution of some examples. Because the Newton method is one of best iterative method to solving hard problems, it's necessary to compare our result and Newton method's result.

3.1   Example1

Consider the problem $f(x) = x$ where,

$$x^2 - 8 = x \tag{1}$$

This example has two real-simple roots, we obtain one of them with our method, and compare accuracy of that with Newton method and then we solve this problem by using ABC algorithm and compare results. Assume $\alpha \in [a_0, b_0] = [-20, 2]$ and $f(x) = x^2 - 8 - x$.

By proposed algorithm at section3, after 50 iteration, solution achieved as follows,

$$\alpha = -2.37228, f(\alpha) = 7.10543e - 015,$$

Newton method with initial guess= -20 after 50 iteration reaches to solution as follows:

$$\alpha = -2.37228, f(\alpha) = -7.6016e - 006,$$

Result of solving this problem by ABC algorithm, after 50 iteration is as follows:

$$\alpha = -2.3728, \ f(\alpha) = 1.93489e - 03,$$

In Table1 we compare results of ABC algorithm and proposed method. As Table1 shows proposed method is too much faster than ABC algorithm and accuracy is good.

Table 1. Comparative results of example.1

Method	α	$f(\alpha)$	Iteration-No
ABC Method	-2.3728	1.93489e-03	50
Proposed Method	-2.37228	7.10543e-015	50

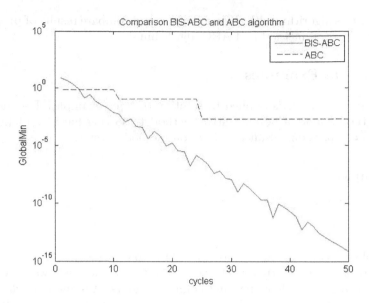

Fig. 1. The result of proposed method and ABC algorithms to solve the fix point problem of example.1

3.2   Example2

Consider the following equation:

$$\frac{x^2}{4000} - cos(x) + 1 = x; \tag{2}$$

Assumption that $f(x) = \frac{x^2}{4000} - cos(x) + 1 - x$, it is clear that $f(0) = 0$. By choosing initial interval $[a, b] = [-20, 20]$, proposed algorithm at section3, after 50 iteration achieved to solution as follows,

$$\alpha = -3.81139e - 015, \quad f(\alpha) = 3.81139e - 015,$$

By Newton method with initial guess $x_0 = 20$, after 50 iteration we achieve the solution:

$$\alpha = 0, \quad f(\alpha) = 0,$$

Then the proposed method's result is good. We obtained solution with ABC algorithm, after 50 iteration as follows:

$$\alpha = -0.0422, \quad f(\alpha) = 0.043048,$$

In Table2, we compare result of ABC algorithm and proposed method. Results show ABC algorithm is very slow, but convergence speed of proposed method is too much faster than that.

Table 2. Comparative results of proposed method and ABC algorithms on the example.2

Method	α	$f(\alpha)$	Iteration-No
ABC Method	-0.0422	0.043048	50
Proposed Method	-3.81139e-015	3.81139e-015	50

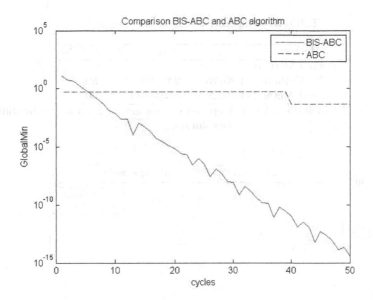

Fig. 2. The result of proposed method and ABC algorithms to solving of fix point problem of example.2

3.3 Example3

Consider the following equation:

$$x^3 - x + 2 = x; \tag{3}$$

With BIS-ABC algorithm, assumption that $\alpha \in [a, b] = [-10, 10]$, after 50 iteration, we obtained the solution as follows:

$$\alpha = -1.76929, f(\alpha) == -3.41949e - 014$$

By the Newton method with initial guess $x_0 = 0$, we achieved

$$x_1 = 1, f(x_1) = 1, x_2 = 0, f(x_2) = 2, x_3 = 1, x_4 = 0, ...,$$

With initial guess $x_0 = 0$, Newton method fall down to the loop and is fail to obtain solution. In this case we should change initial value or continue with another iterative methods as Bisection method to achieve good approximation value of the solution .

In Table3, we compare result of MABC and ABC algorithms. Results show convergence speed of BIS-ABC method is too much faster than ABC algorithm.

Table 3. Comparative results of example.3

Method	α	$f(\alpha)$	Iteration-No
*Newton Method	—	—	—
ABC Method	-1.80176	0.245589	50
Proposed Method	-1.76929	-3.41949e-014	50

*: Newton method fall down to the loop and cannot get an good approximation of the solution.

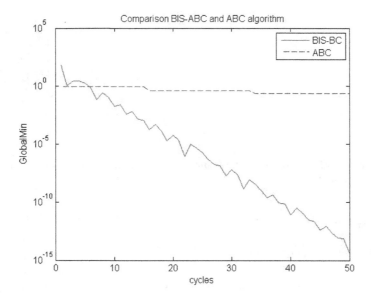

Fig. 3. The result of proposed method and ABC algorithms to solving of fix point problem of example.3

4   Conclusion

In this paper we study a novel iteration method for solving Hard fix point problems by using Artificial bee colony algorithm. If equation $g(x) = x$ to be hard, sometimes, it is difficult to determine suitable initial value close to the location of a root of given equation and compute derivative of the given function .

We solve these problems by using ABC algorithm, without need to choose initial close value to root and without compute derivative of the given function $f(x) = g(x) - x$. As we show that in example3. Hence, some times Newton method to solve hard problems may be fall to the loop and algorithm is stopping, but without any problem, proposed method directly search to approximate solution of given problem in suitable domain of optimal solution.. This algorithm is easy to use and reliable. As Comparison shows accuracy of proposed method also is good.

References

1. Burden, L.R., Faires, Douglas, J.: 2.1 The Bisection Algorithm, Numerical Analysis, 3rd edn (1985)
2. Holland, J.H.: Adaptation in Natural and Artificial Systems. University of Michigan Press, Ann Arbor (1975)
3. Karaboga, D.: An Idea Based on Honey Bee Swarm for Numerical Optimization. Technical Report-TR06, Erciyes University, Engineering Faculty, Computer Engineering Department (2005)
4. Karaboga, D., Basturk, B.: A powerful and efficient algorithm for numerical function optimization: Artificial Bee Colony (ABC) algorithm. Journal of Global Optimization 39, 459–471 (2007)
5. Karaboga, D., Basturk, B.: On the performance of Artificial Bee Colony (ABC) algorithm. Applied Soft Computing 8, 687–697 (2008)
6. Storn, R., Price, K.: Differential evolution-Asimple and efficient heuristic for global optimization over continuous spaces. Journal of Global Optimization 23, 689–694 (2010)
7. Baykaso, A., Özbakýr, L., Tapkan, P.: Artificial Bee Colony Algorithm and Its Application to Generalized Assignment Problem. University of Gaziantep, Department of Industrial Engineering Erciyes University, Department of Industrial Engineering, Turkey
8. Karaboga, D.: An Idea Based on Honey Bee Swarm for Numerical Optimization. Technical Report-TR06, Erciyes University, Engineering Faculty, Computer Engineering Department (2005)
9. Mansouri, P., Asady, B., Gupta, N.: A Novel Iteration Method for solve Hard Problems (Nonlinear Equations) with Artificial Bee Colony algorithm. World Academy of Science, Engineering and Technology 59, 594–596 (2011)
10. Abraham, A., Corchado, E., Corchado, J.: Hybrid learning machines. Neurocomputing 72, 13–15 (2009)
11. Corchado, E., Arroyo, A., Tricio, V.: Soft computing models to identify typical meteorological days. Logic Journal of the IGPL 19, 373–383 (2011)
12. Zhao, S.Z., Iruthayarajan, M.W., Baskar, S., Suganthan, P.N.: Multi-objective robust PID controller tuning using two lbests multi-objective particle swarm optimization. Inf. Sci. 181(16), 3323–3335 (2011)
13. Sedano, J., Curiel, Corchado, L.E., delaCal, E., Villar, J.R.: A soft computing method for detecting lifetime building thermal insulation failures. Integrated Computer-Aided Engineering 17, 103–115 (2010)

A System Learning User Preferences for Multiobjective Optimization of Facility Layouts

M. Pérez-Ortiz*, A. Arauzo-Azofra, C. Hervás-Martínez,
L. García-Hernández, and L. Salas-Morera

University of Córdoba, Spain

Abstract. A multiobjective optimization system based on both subjective and objective information for assisting facility layout design is proposed on this contribution. A data set is constructed based on the expert evaluation of some facility layouts generated by an interactive genetic algorithm. This dataset is used for training a classification algorithm which produces a model of user subjective preferences over the layout designs. The evaluation model obtained is integrated into a multi-objective optimization algorithm as an objective together with reducing material flow cost. In this way, the algorithm exploits the search space in order to obtain a satisfactory set of plant layouts. The proposal is applied on a design problem case where the classification algorithm demonstrated that it could fairly learn the user preferences, as the model obtained worked well guiding the search and finding good solutions, which are better in term of user evaluation with almost the same material flow cost.

1 Introduction

Facility Layout Design (FLD) determines the placement of facilities in a manufacturing plant with the aim of determining the most effective arrangement in accordance with some criteria or objectives, under certain constraints. In this respect, Kouvelis et al. (1992) [11] provided that FLD is known to be very important for production efficiency because it directly affects manufacturing costs, lead times, work in process and productivity. According to Tompkins et al. (2010) [20], well laid out facilities contribute to the overall efficiency of operations and can reduce between 20% and 50% of the total operating costs. There are many kinds of layout problems. This contribution focus on the Unequal Area Facility Layout Problem (UA-FLP) as formulated by Armour and Buffa (1963) [3]. In short, UA-FLP considers a rectangular plant layout that is made up of unequal rectangular facilities that have to be placed effectively in the plant layout.

Aiello et al. (2012) [2] stated that, generally speaking, the problem of designing a physical layout involves the minimization of the material handling cost as

---

* Partially subsidized by the TIN2011-22794 project of the spanish MICYT, FEDER funds and the P08-TIC-3745 project of the "Junta de Andalucía" (Spain).

the main objective. But, there are other authors that consider additional quantitative performance, as for example, Aiello et al. (2006) [1], who have addressed this problem taking into account criteria that can be quantified (e.g., material handling cost, closeness or distance relationships, adjacency requirements and aspect ratio), which are used in an optimization approach. However, Babbar-Sebens and Minsker (2012) [4] established that these approaches may not adequately represent all of the relevant qualitative information that affect a human expert involved in design (e.g. engineers). In this way, qualitative features sometimes also have to be taken into consideration. Brintup et al. (2007) [5] stipulated that such qualitative features are complicated to include with a classical heuristic or meta-heuristic optimization. Besides, according to Garcia-Hernandez et al. (2011) [10] these qualitative features can be subjective, not known at the beginning and can be changed during the process. As a consequence, the participation of the designer is essential to include qualitative considerations in the design. Moreover, involving the designer in the process provides additional advantages which have been detailed in its work.

The Interactive Genetic Algorithm (IGA) developed for FLD [10] consider user evaluation and handle subjective features. This algorithm uses a clustering mechanism to reduce the number of evaluations required from the user. However, running the IGA can be a tedious task for a designer, as many evaluations are still required. Fatigue is the main reason for an early stop of IGAs [14], thus reducing the possibilities of the system to find better designs. Moreover, user evaluation is some orders of magnitude slower than computed evaluation, leading necessarily to a much smaller search capacity. Learning user design preferences over a concrete layout problem would allow to simulate user responses. In this way, fatigue could be avoided and search could be performed much faster, which is specially useful in the context of the large search space of facility layouts. The goal of this contribution is to design a system that is able to learn these user layout preferences and perform a search considering both, the user preferences and other objective criteria.

For a layout design, the user evaluation considered in the IGA is of an absolute type deciding among five possible values for each design. From the user point of view, absolute evaluation is considered more practical than relative comparisons between layouts. Besides, absolute evaluation has shown better learning results when learning synthetic models of user evaluation [21].

Likert scales were firstly proposed in 1932 [13] as a way to produce attitude measures which could be interpreted as measurements on a proper metric scale. This technique is usually defined as a psychometric response scale, which is mainly used in questionnaires for obtaining the preference of different users or degree of agreement within a set of statements. In this paper, the most commonly likert scale is used in order to evaluate a set of facility layouts which have been synthetically created using an evolutionary algorithm. This rating technique can be seen as a 5-point (or granularity) scale ranging from "Strongly Disagree" on one and end to "Strongly Agree". Thus, the classes involved in the problem are:
{*Strongly disagree, Disagree, Neither agree or disagree, Agree, Strongly agree*}

where each class answers the question: *Could this plant layout be considered as a good solution for the Unequal Area Facility Layout Problem?*. Often, likert scale items are treated numerically in such a way that it is assumed that the distance between all points on the scale are equal, however, this assumption might be wrong as we are forgetting the underlying latent variable in the scale. Because of that, optimal scaling is a relevant issue to both ordinal predictor and outcome variables. Likert scales can also be addressed from an ordinal regression point of view, where there is a certain order among class labels. Ordinal regression (or classification) is a relatively new learning problem which is rapidly growing and enjoying a lot of attention from pattern recognition and machine learning communities [9,18]. In this case, the classification problem is quite different to the standard one, as the variable to predict is not numerical or nominal, but ordinal, so categories have a natural order (in the same way that the categories we aim to predict in this paper). The major problem with this kind of classification is that there is not a precise notion of the distance between categories and there are some misclassification errors which should be more penalized.

This paper is organized as follows: a brief analysis of the system constructed is given in Section 2. All the experimental design features and the way the system learn the FLD preferences from the expert are presented in Section 3. Finally, we present some of the results obtained in Section 4 and Section 5 summarizes the conclusions and future work.

2   Structure of the Proposed System

From a global perspective, the proposed system requires the designer to describe the problem and to evaluate several FLDs. At the end, the system must return a moderate number of designs according to the preferences found in the evaluation process and the optimization of the objectives factors. Thus, the main purpose would be combine both subjective and objective information in order to come to proper and fair decisions when evaluating these facility structures. To do so, a three-stage system has been developed as shown in Fig. 1:

1. Firstly, the IGA evolves towards FLDs that are preferred by the user. On each iteration, the user evaluates nine layouts which are stored for learning step. Although more layouts are generated and evaluated through clustering to better guide evolution, only user evaluated layouts are considered. Elicited evaluations from clustering may have been good for GA evolution but we have found that they may confuse the learning process.
2. After that, a machine learning classification algorithm will learn from the expert evaluations (using the dataset). In this stage, several nominal and ordinal algorithms are tested, in order to choose the one that achieved the best results. Once the more appropriated algorithm is selected, the model obtained is integrated in a multiobjective evolutionary algorithm, in such a way that this model will evaluate the facility layouts using the likert scale and this predicted target will be the first objective to maximize in the evolutionary learning process.

3. Finally, the second objective to maximize in the optimization will be an objective factor (in this case, the material flow between facilities). At this point, the multiobjective algorithm is able to search and evolve through all possible solutions, and will end with a pareto front containing a set of optimal plant layouts.

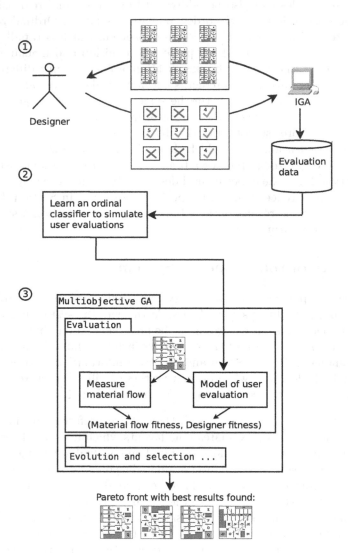

Fig. 1. Facility layout design system diagram

3   Learning Facility Layout Preferences from the Expert

Facility layouts from the IGA follow the Flexible Bay Structure (FBS) [19], where the facilities are placed in a series of rows of variable width. The data available to the learning algorithm is the number of bays and the geometrical coordinates of the rectangle assigned to each facility. Every of these layouts is tagged with an evaluation from a set of five ordered classes. Therefore, the user preferences will be learned with a supervised classification task. It must be noted that each problem case is different and usually will have different design preferences. For this reason, in principle, the knowledge and models created for one case are not applicable to others and a new model must be learned.

The facility layout problem case considered is composed of 20 facilities with different required areas that are arranged in a 61.7 × 61.7 meters plant. The IGA was run for 220 iterations using a population of 100 individuals with 0.5 probability of crossover and 0.05 probability of mutation. After removing some duplicated facility layouts, the IGA left us with a final database composed of 1969 patterns distributed in 5 classes and 86 attributes which contains information about the location and different characteristics of each facility distribution.

3.1   Selection of the Algorithm

Several state-of-the-art methods have been tested for this problem in order to choose the one which performs better taking into account metrics which measures different kind of errors: CCR, which is the standard classification metric (also known as accuracy), MAE which measures ordinal classification and finally, MS which measures the worst classified class so this measure will help us to detect trivial classifications and will be really useful in unbalanced problems (as the one treated in this paper).

The Mean Absolute Error (MAE) is defined as the average deviation in absolute value of the predicted class from the true class. It is a commonly used for ordinal regression:

$$MAE = \frac{1}{N} \sum_{i=1}^{N} |y_i - \hat{y}_i|,$$

where y_i is the true rank for the $i - th$ pattern, and $\hat{y}_i$ corresponds to the predicted rank for the same pattern. N is the number of test patterns.

The Minimum Sensitivity (MS) can be defined as the minimum value of the sensitivities for each class,

$$MS = \min\{S_i;\ i = 0, \ldots, K\},$$

where S_i is the sensitivity for the ith class. Sensitivity for class i corresponds to the correct classification rate for this specific class. In this way, MS reports the accuracy for the worst classified class.

In order to fairly compare the results obtained from different algorithms, a stratified 30-holdout and a nested 5-fold cross-validation have been performed. The algorithm tested for solving the given problem are the following ones:

- OCC (OrdinalClassClassifier): ensemble technique for ordinal regression [9] which applies C4.5 as base algorithm.
- KDLOR (Kernel Discriminant Learning for Ordinal Regression): method which combines discriminant analysis and kernel functions for ordinal regression [18].
- POM (Proportional Odd Model): one of the first models specifically designed for ordinal regression and arisen from a statistical background [15].
- EBC (Extended Binary Classification): ensemble method which performs multi-class classification with just a binary model [12]. Support Vector Machines are used as the base algorithm.
- SVC (Support Vector Classification): standard nominal classifier based on support vector machines which performs one-vs-one classification [7].
- C4.5: standard nominal decision tree [16] based on information entropy.

Table 1. CCR, MS, MAE obtained from the different methods

Algorithm	CCR	MAE	MS
OrdinalClassClassifier(C4.5)	88.69 ± 1.48	$\mathbf{0.169 \pm 0.025}$	37.50 ± 9.51
KDLOR	78.20 ± 2.19	0.231 ± 0.034	12.36 ± 1.87
POM	60.72 ± 1.24	0.468 ± 0.014	0.00 ± 0.00
EBC(SVM)	81.08 ± 2.76	0.249 ± 0.023	8.23 ± 2.50
SVC	80.15 ± 4.53	0.290 ± 0.047	13.81 ± 5.70
C4.5	$\mathbf{89.87 \pm 2.31}$	0.175 ± 0.037	$\mathbf{42.19 \pm 7.35}$

Taking into account these metrics, the best results in CCR and MS are achieved with the C4.5 algorithm [16], although it is not an ordinal method. But one can notice that best results in MAE are achieved using the OCC(C4.5). Nevertheless, as the differences are not significantly large and the algorithm selected will be integrated in a multiobjective algorithm, another important issue is the simplicity and one should take into account that the OCC is a ensemble model which makes use of probability functions for obtaining the final predicted targets. Because of that, the authors have considered the use of different ordinal cost matrices for training the C4.5 algorithm to see if the results could improve even more. This algorithm is also known as C4.5CS or cost-sensitive C4.5. C4.5CS [22] is a post-processor decision tree is used in conjunction with C4.5. This methodology implements cost-sensitive specialization by seeking to specialize leaves for high misclassification cost classes in favor of leaves for low misclassification cost classes. As said before, there are some misclassification errors which should be more penalized in the problem: confusing the "Strongly agree" with the "Strongly disagree" class should be considered by far a bigger mistake than confusing the "Neither agree or disagree" with the "Agree" class. Because of that, we have tested several approaches using cost matrices for solving this ordinal problem. The costs matrices used are shown in Table 2:

- Cost matrix #1: Usual cost matrix for the nominal classifiers, which assumes that all the misclassification errors are equal.
- Cost matrix #2: This matrix is the well-known standard ordinal one. It is widely used with nominal algorithms in order to weight the misclassification errors.
- Cost matrix #3: Quadratic ordinal cost matrix.
- Cost matrix #4: Cost matrix particularly proposed for the problem here addressed. Due to the fact that the best FLDs are a minority in the problem, authors have considered that it is critical not to miss any of them. Thus, errors when misclassifying an excellent plant layout are not the same as when misclassifying a bad one (because an expert will check over the final pareto front and will directly discard the non-proper ones).

Table 2. Different cost matrices considered for the problem

Cost matrix #1					Cost matrix #2					Cost matrix #3					Cost matrix #4				
0	1	1	1	1	0	1	2	3	4	0	1	4	9	16	0	1	2	3	4
1	0	1	1	1	1	0	1	2	3	1	0	1	4	9	1	0	1	2	3
1	1	0	1	1	2	1	0	1	2	4	1	0	1	4	4	1	0	1	2
1	1	1	0	1	3	2	1	0	1	9	4	1	0	1	9	4	1	0	1
1	1	1	1	0	4	3	2	1	0	16	9	4	1	0	16	9	4	1	0

The results obtained using these cost matrices with the same procedure as before (stratified 30-holdout) can be seen in Table 3 where one can notice that an important improvement of the results is produced by using these ordinal cost matrices. The cost matrix #4 is the one hybridized with the ordinal standard and the quadratic one, and it obtains the optimal results. Besides it is the one in concordance with the misclassifying errors to avoid in the problem, so it will be used for computing the final model to guide the evolutionary searching process.

Table 3. CCR, MS, MAE obtained with C4.5 and the different cost matrices

Cost matrix	CCR	MAE	MS
C4.5 (cost matrix #1)	88.69 ± 1.48	0.175 ± 0.037	42.19 ± 7.35
C4.5 (cost matrix #2)	89.82 ± 1.19	0.181 ± 0.014	51.56 ± 4.87
C4.5 (cost matrix #3)	*90.17 ± 1.24*	*0.173 ± 0.026*	*53.12 ± 6.56*
C4.5 (cost matrix #4)	**90.35 ± 1.56**	**0.163 ± 0.021**	**54.69 ± 3.42**

At this point, the procedure to train the final model have been established: use the C4.5 algorithm with a hybrid and ordinal cost matrix and the entire dataset (without any training and testing partitioning) to obtain a single final model.

4   Combining Subjective and Objective Criteria

Once a model of the evaluation according to design preferences is learned. It is also desired that the facility layout optimize other objective features. Material flow is considered a very important measure of a good FLD. It is calculated as the product of the distance and the material movement estimated between facilities. In order to find a good facility layout considering both objectives, a multi-objective genetic algorithm (MOGA) is applied in the last stage of the system.

NSGA2 [8] algorithm has been used because it applies well known important theory on multi-objective evolutionary algorithms [6] and achieves good results in facility layout problems [2]. The proposed system includes NSGA2 implemented using DEAP [17]. The encoding scheme uses a permutation of facilities and a binary vector with the points where bays are split. Crossover operators are PMX for facilities and two point crossover for split points, while mutation swaps the position of two facilities or toggles a bit in the split points vector.

Apart from optimizing material flow and designer preferences, facilities of a given area must also have an usable shape that allows allocation of machines or other resources. This is usually controlled with aspect ratio constraints. Rather than discarding infeasible solutions according to aspect ratio constraints, a penalty function is used. In this way, some infeasible solutions are preserved to allow convergence to solutions that lie in the boundary between feasible and infeasible solutions [6]. We considered including this penalty as an additional objective. However, we have achieved better results using the adaptive penalization over material flow by Tate et. al. [19].

Finally, the system returns the pareto front with at most five solutions (one for each user evaluation class considered). The designer can now choose the facility layout with the optimum equilibrium between material flow and the other subjective preferences.

4.1   Case Results

The described MOGA has been run 500 generations with a population of 200 individuals, using probabilities of 0.8 for crossover and 0.2 for mutation. Figure 2 shows the final results obtained by the system for this problem. There are no solutions with user evaluation values of 1 or 4 because all of them have been dominated by the ones shown. While material flow is pretty similar for all three layouts, the satisfaction of the user is much better in the third one. In this way, the proposed system improves the results of automatic facility layout because user preferences are included in the search without losing the fitting of the objective material flow measure.

In order to compare these results with those achieved by other algorithms not considering user preferences, a well known algorithm proposed by Aiello [1] has been run on the same problem with the same parameters. The best FLD found by this algorithm has a material flow of $189463u * m$ and it is necessary to increase the number of generations to reach FLDs with a similar material flow to

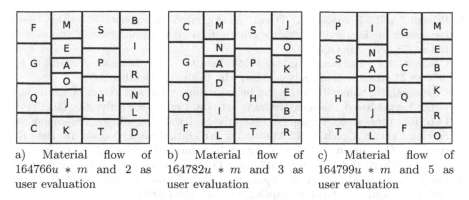

a) Material flow of $164766u * m$ and 2 as user evaluation

b) Material flow of $164782u * m$ and 3 as user evaluation

c) Material flow of $164799u * m$ and 5 as user evaluation

Fig. 2. Final pareto front with the FLDs found by the system

those obtained with the MOGA proposed. Intuitively, this may mean that user evaluation is helping to direct the search. This opens another research question requiring further experimentation with more data on diverse FLD problems.

5   Conclusions and Future Work

This paper proposes the construction of a system which combines user preferences and other objective factors. The proposed system allows considering subjective designer preferences without reducing search extension. In many problems, like the one used in experimentation, considering these user preferences does not necessarily mean getting worse layout in terms of the other objective measures. So, using the system proposed can have interesting results in terms of improving the quality of FLDs.

The use of ordinal regression in this case allowed us the integration of order information among labels in the model, and giving more importance to some kind of misclassification errors. Concerning future work, the system could be constructed by jointly taking into account the preferences of several users.

References

1. Aiello, G., Enea, M., Galante, G.: A multi-objective approach to facility layout problem by genetic search algorithm and electre method. Robotics and Computer-Integrated Manufacturing 22, 447–455 (2006)
2. Aiello, G., Scalia, G.L., Enea, M.: A multi objective genetic algorithm for the facility layout problem based upon slicing structure encoding. Expert Systems with Applications (2012)
3. Armour, G.C., Buffa, E.S.: A heuristic algorithm and simulation approach to relative location of facilities. Management Science 9, 294–309 (1963)
4. Babbar-Sebens, M., Minsker, B.S.: Interactive genetic algorithm with mixed initiative interaction for multi-criteria ground water monitoring design. Applied Soft Computing 12(1), 182–195 (2012)

5. Brintup, A.M., Ramsden, J., Tiwari, A.: An interactive genetic algorithm-based framework for handling qualitative criteria in design optimization. Computers in Indust. 58, 279–291 (2007)
6. Coello, C.A.C., Lamont, G.B., van Veldhuizen, D.A.: Evolutionary Algorithms for Solving Multi-Objective Problems, 2nd edn. Springer (October 2007)
7. Cristianini, N., Shawe-Taylor, J.: An Introduction to Support Vector Machines and Other Kernel-based Learning Methods, 1st edn. Cambridge University (2000)
8. Deb, K., Agrawal, S., Pratap, A., Meyarivan, T.: A Fast Elitist Non-dominated Sorting Genetic Algorithm for Multi-objective Optimisation: Nsga-ii. In: Deb, K., Rudolph, G., Lutton, E., Merelo, J.J., Schoenauer, M., Schwefel, H.-P., Yao, X. (eds.) PPSN 2000. LNCS, vol. 1917, pp. 849–858. Springer, Heidelberg (2000)
9. Frank, E., Hall, M.: A simple approach to ordinal classification. In: Proc. of the 12th Eur. Conf. on Machine Learning, pp. 145–156 (2001)
10. Hernandez, L.G., Morera, L.S., Azofra, A.A.: An Interactive Genetic Algorithm for the Unequal Area Facility Layout Problem. In: Corchado, E., Snášel, V., Sedano, J., Hassanien, A.E., Calvo, J.L., Ślęzak, D. (eds.) SOCO 2011. AISC, vol. 87, pp. 253–262. Springer, Heidelberg (2011)
11. Kouvelis, P., Kurawarwala, A.A., Gutierrez, G.J.: Algorithms for robust single and multiple period layout planning for manufacturing systems. European Journal of Operational Research 63(2), 287–303 (1992)
12. Li, L., Lin, H.T.: Ordinal Regression by Extended Binary Classification. In: Advances in Neural Information Processing Systems, vol. 19 (2007)
13. Likert, R.: A technique for the measurement of attitudes. Archives of Psychology 22(140) (1932)
14. Llor, X., Sastry, K., Goldberg, D.E., Gupta, A., Lakshmi, L.: Combating user fatigue in iGAs: partial ordering, support vector machines, and synthetic fitness. In: Proceedings of the 2005 Conference on Genetic and Evolutionary Computation, pp. 1363–1370 (2005)
15. McCullagh, P., Nelder, J.A.: Generalized Linear Models. Monographs on Statistics and Applied Probability, 2nd edn. Chapman & Hall/CRC (1989)
16. Quinlan, J.R.: C4.5: programs for machine learning. Morgan Kaufmann Publishers Inc., San Francisco (1993)
17. Rainville, F.M.D., Fortin, F.A., Gardner, M.A., Parizeau, M., Gagn, C.: Distributed evolutionary algorithms in python (deap) (2011), http://deap.googlecode.com
18. Sun, B.Y., Li, J., Wu, D.D., Zhang, X.M., Li, W.B.: Kernel discriminant learning for ordinal regression. IEEE Transactions on Knowledge and Data Engineering 22, 906–910 (2010)
19. Tate, D.M., Smith, A.E.: Unequal area facility layout using genetic search. IIE Transactions 27, 465–472 (1995)
20. Tompkins, J., White, J., Bozer, Y., Tanchoco, J.: Facilities Planning, 4th edn. Wiley, New York (2010)
21. Wang, S., Wang, X., Takagi, H.: User fatigue reduction by an absolute rating data-trained predictor in IEC. In: IEEE Conf. on Evolutionary Computation 2006, pp. 2195–2200 (2006)
22. Webb, G.I.: Cost Sensitive Specialisation. In: Foo, N.Y., Göbel, R. (eds.) PRICAI 1996. LNCS, vol. 1114, pp. 23–34. Springer, Heidelberg (1996)

Implementing Artificial Immune Systems for the Linear Ordering Problem

Pavel Krömer, Jan Platoš, and Václav Snášel

Department of Computer Science, FEECS, VŠB – Technical University of Ostrava,
IT4Innovations, Center of Excellence, VŠB – Technical University of Ostrava,
17. listopadu 15, 708 33 Ostrava-Poruba, Czech Republic
{pavel.kromer,jan.platos,vaclav.snasel}@vsb.cz

Abstract. Linear Ordering Problem (LOP) is a well know NP-hard combinatorial optimization problem attractive for its complexity, rich library of test data, and variety of real world applications. This study investigates the bio-inspired Artificial Immune Systems (AIS) as a pure metaheuristic soft computing solver of the LOP. The well known LOP library LOLIB was used to compare the results obtained by AIS and other pure soft computing metaheuristics.

Keywords: linear ordering problem, artificial immune systems, pure metaheuristics, soft computing.

1 Introduction

The linear ordering problem is a well-known NP-hard combinatorial optimization problem. It has been intensively studied and there are plenty of exact, heuristic and metaheuristic algorithms for LOP. With its large collection of well described testing data sets, the LOP represents an interesting testbed for metaheuristic and soft computing algorithms for combinatorial optimization [15,16].

The LOP can be formulated as a graph problem and as a matrix triangulation problem [15]. It can be seen as a search for simultaneous permutation of rows and columns of a matrix C such that the sum of the elements in the upper diagonal of the modified matrix is as large as possible. In this work we use the matrix formulation of the LOP.

2 LOP as a Matrix Triangulation Problem

Linear Ordering Problem (LOP) can be defined as a search for optimal column and row reordering of a weight matrix C [20,21,3,15]. Consider a matrix $C^{n \times n}$, permutation Π and a cost function f:

$$f(\Pi) = \sum_{i=1}^{n} \sum_{j=i+1}^{n} c_{\Pi(i)\Pi(j)} \tag{1}$$

V. Snasel et al. (Eds.): SOCO Models in Industrial & Environmental Appl., AISC 188, pp. 53–62.
springerlink.com

The LOP is defined as a search for permutation Π so that $f(\Pi)$ is maximized, i.e. the permutation restructures the matrix C so that the sum of its elements above main diagonal is maximized. The LOP is a NP-hard problem with a number of applications in scheduling (scheduling with constraints), graph theory, economy, sociology (paired comparison ranking), tournaments and archaeology among others.

In economics, LOP algorithms are deployed to triangularize input-output matrices. The resulting permutation provides an useful information about the stability of the investigated economy. In archaeology, LOP algorithms are used to process the Harris Matrix, a matrix describing most probable chronological ordering of samples found in different archaeological sites [21]. Other applications of the LOP include the equivalent graph problem, the related graph problem, the aggregation of individual preferences, ranking in sport tournaments and e.g. the minimization of crossing [15].

2.1 LOP Data Sets

There are several test libraries used for benchmarking LOP algorithms. They are well preprocessed, thoroughly described and the optimal (or so far best) solutions are available. Majority of investigated algorithms were tested against the LOLIB library. The original LOLIB library contains 49 instances of input-output matrices describing European economies in the 70s. Optimal solutions of LOLIB matrices are available. Although LOLIB features real world data, it is considered rather simple and easy to solve [19].

Mitchell and Bochers [17] published an artificial LOP data library and LOP instance generator to evaluate their algorithm for Linear Ordering Problem. The data (from now on addressed as MBLB) and code are available at Rensselaer Polytechnic Institute[1].

Schiavinotto and Stützle [20,21] showed that the LOLIB and MBLB instances are significantly different, having diverse high-level characteristics of the matrix entries such as sparsity or skewness. The search space analysis revealed that MBLB instances typically have higher correlation length and also a generally larger fitness-distance correlation than LOLIB instances. It suggests that MBLB instances should be easier to solve than LOLIB instances of the same dimension. Moreover, a new set of large artificial LOP instances (based on LOLIB) called XLOLIB was created and published. Another set of LOP instances is known as the Stanford Graph Base (SGB). The SGB is composed of larger input-output matrices describing US economies. In this study we use the widely used LOLIB library for initial evaluation of an AIS-based LOP solver.

2.2 LOP Algorithms

There are several exact and heuristic algorithms for the linear ordering problem. The exact algorithms are strongly limited by the fact that LOP is a NP-hard

---

[1] http://www.rpi.edu/ mitchj/generators/linord/

problem (i.e. there are no exact algorithms that could solve LOP in polynomial time). Among the exact algorithms, branch & bound approach based on LP-relaxation of the LOP for the lower bound, a branch & cut algorithm and interior point/cutting plane algorithm attracted attention [21]. Exact algorithms are able to solve rather small general instances of the LOP and bigger instances (with the dimension of few hundred rows and columns) of certain classes of LOP [21].

A number of heuristics and soft computing algorithms was used for solving LOP instances: greedy algorithms, local search algorithms, elite tabu search, scattered search and iterated local search [21,9,15].The metaheuristic algorithms used to solve LOP in the past include genetic algorithms [12,13], differential evolution [23], and ant colony based algorithms [4]. The investigation of meta-heuristics and soft computing algorithms is motivated by previous success of such methods in real world and industrial applications [1,5,29,22]

3   Artificial Immune Systems

Artificial immune systems (AIS) constitute a family of bio-inspired algorithms based on the models known from the studies of biological immune systems and immunology [7,26,25,8]. Informally, biological immune systems protect the body from dangerous substances presented in the form of pathogens. They combine the ability to protect from both, general pathogens and specific attackers (e.g. different types of viruses, bacteria and so on) that cannot be be eliminated by the static (innate) part of the immune system.

The AIS algorithms implement a variety of principles observed in the adaptive part of biological immune systems of the vertebrae [26,25] including the negative selection, clonal selection, and immune networks. The *clonal selection* is based on the clonal selection theory describing the basic response of the immune system to an antigen (a pathogen recognized by the immune system). In such a situation, the immune system cells (lymphocytes, B-cells and T-cells) act as immune agents that participate in the recognition and elimination of the alien substances [26,8]. The B-cells are being cloned and hypermutated in order to create an optimal response to intruding cells. The B-cells turn into plasma cells that eliminate the antigen and memory cells that store the information about the antigen for better (i.e. more rapid) response to this type of attack in the future. *Negative selection* is in the biological immune systems used to ensure that newly created lymphocytes will be able to adapt to new types of threats and remain tolerant to body's own cells at the same time. An *immune network* is a reinforced network of B-cells that uses the ability of some parts of the cell (paratopes) to match against another parts of the cell (idiotopes). Such an immune network contributes to the stable memory structure of the immune system that is able to retain the information about the antigens even without their presence. However, it should be noted that the immune network theory is no longer widely accepted by the immunologists [26].

The AIS algorithm used in this study (a modified B-cell algorithm [25]) based on clonal selection is shown in algorithm 1.

Algorithm 1. An outline of AIS used in this study.

1 Initialize an population of artificial lymphocytes (ALCs) P;
2 Compute affinity (fitness) for all $a_i \in P$;
3 **while** *Termination criteria not satisfied* **do**
4   Select $a_i \in P$ proportionally according to its affinity;
5   Create from a_i clone population C;
6   **for** $a_i' \in C$ **do**
7   Apply the hypermutation operator;
8   Evaluate a_i'. If a_i' is better solution than a_i, replace a_i by a_i' in P;
9   **end**
10 **end**

The AIS algorithms based on clonal selection are quite similar to the GAs. They evolve a population of candidate solutions by cloning, (hyper)mutation, and selection. Another type of AISs are e.g. the swarm-like AIS [26]. They have been used for anomaly detection, intrusion detection, function and combinatorial optimization, clustering, classification, and others [26]. An application of AIS for task scheduling was shown in [28].

In this study, we compare an AIS-based approach to LOP with two other pure metaheuristics, genetic algorithms (GA) [2] and differential evolution (DE) [18], because they have been used to solve LOP recently [10,12,13,23]. All three metaheuristics represent different populational approaches to global optimization based on various bio-inspired paradigms and essentially performing different kind of search in the solution space of the problem. The operations of each metaheuristic yield different efficiency for different problems and their experimental evaluation is needed [27].

4 Computational Evaluation

In order to solve LOP instances, the GA, DE, and AIS were implemented from scratch in C++. A steady state GA with elitism [2], a DE/rand/1 [18] variant of the DE, and the basic AIS as outlined in algorithm 1 were used to search for LOLIB solutions. The algorithms evolved a population of 64 candidate solutions encoded using the random keys encoding [24,11]. Other GA parameters were mutation probability $p_M = 0.04$ and crossover probability $p_C = 0.8$. The DE was executed with $F = 0.9$ and $C = 0.9$ and the AIS was executed with hypermutation with $p_{HM} = 0.04$ and clone population size 3. The parameters were selected after initial experiments.

Because of the different number of fitness function evaluations in single iteration of the algorithms, the execution time was set as the stopping criterion. Due to the stochastic nature of the metaheuristics, the evolution of each LOLIB instance by every algorithm was repeated 10 times. The results of the evolution of LOLIB solutions after 5, 10, and 20 seconds are shown in table 1, table 2,

Table 1. Solution error (in %) after 5 seconds

LOLIB Instance	Optimum	GA		DE		AIS	
		lowest	average	lowest	average	lowest	average
be75eec	264940	2.125E-03	5.031E-03	1.767E-02	2.707E-02	**1.574E-03**	**3.189E-03**
be75np	790966	**7.459E-05**	**8.559E-04**	1.169E-02	3.463E-02	5.752E-04	1.087E-03
be75oi	118159	**1.642E-03**	**2.852E-03**	1.311E-02	1.676E-02	2.082E-03	4.147E-03
be75tot	1127387	**6.706E-04**	**1.390E-03**	2.253E-02	4.436E-02	1.726E-03	3.164E-03
stabu1	422088	**2.732E-03**	**3.966E-03**	3.595E-02	4.713E-02	3.412E-03	6.233E-03
stabu2	627929	**2.986E-03**	**5.313E-03**	4.519E-02	5.183E-02	5.142E-03	6.877E-03
stabu3	642050	**3.539E-03**	**5.591E-03**	4.412E-02	5.385E-02	5.554E-03	8.464E-03
t59b11xx	245750	2.014E-03	9.656E-03	7.976E-03	1.262E-02	**1.750E-03**	**2.621E-03**
t59d11xx	163219	0	**3.357E-03**	9.349E-03	1.617E-02	3.860E-04	6.341E-03
t59f11xx	140678	**8.530E-05**	1.237E-03	8.018E-03	1.651E-02	**8.530E-05**	**4.265E-04**
t59i11xx	9182291	0	1.606E-03	9.096E-03	1.533E-02	1.761E-04	**1.279E-03**
t59n11xx	25225	**6.739E-04**	**3.100E-03**	1.296E-02	1.864E-02	2.498E-03	7.707E-03
t65b11xx	411733	1.780E-03	**3.080E-03**	1.817E-02	2.788E-02	**1.312E-03**	5.312E-03
t65d11xx	283971	0	**1.433E-03**	5.596E-03	2.081E-02	1.944E-03	2.775E-03
t65f11xx	254568	1.721E-03	2.675E-03	1.199E-02	2.008E-02	**1.430E-03**	**2.435E-03**
t65i11xx	16389651	0	**7.841E-04**	5.230E-03	1.170E-02	5.355E-04	1.260E-03
t65l11xx	18359	0	0	3.323E-03	4.586E-03	0	1.089E-04
t65n11xx	38814	**6.699E-04**	**2.365E-03**	5.848E-03	1.407E-02	2.783E-03	3.901E-03
t65w11xx	1.66E+08	**9.080E-04**	**1.896E-03**	8.241E-03	1.179E-02	1.495E-03	3.128E-03
t69r11xx	865650	**2.357E-04**	**3.977E-03**	1.589E-02	2.043E-02	3.884E-03	7.776E-03
t70b11xx	623411	**4.203E-04**	**1.195E-03**	8.648E-03	1.395E-02	1.006E-03	1.781E-03
t70d11xn	438235	2.966E-04	6.275E-04	1.281E-02	2.167E-02	**2.579E-04**	**2.284E-03**
t70d11xx	450774	0	**1.067E-03**	8.088E-03	1.786E-02	6.300E-04	2.405E-03
t70f11xx	413948	0	**2.788E-03**	1.243E-02	2.569E-02	3.080E-03	4.764E-03
t70i11xx	28267738	**1.555E-04**	**8.327E-04**	5.864E-03	1.107E-02	2.353E-04	1.275E-03
t70k11xx	69796200	**8.883E-05**	**1.142E-03**	8.946E-03	1.439E-02	5.101E-04	4.433E-03
t70l11xx	28108	0	1.857E-03	3.131E-03	8.055E-03	3.558E-03	**7.613E-04**
t70n11xx	63944	4.520E-03	4.835E-03	1.592E-02	2.479E-02	**1.048E-03**	**3.365E-03**
t70u11xx	27296800	2.198E-04	1.808E-03	2.069E-02	2.914E-02	**2.198E-04**	**1.267E-03**
t70w11xx	2.68E+08	**7.360E-06**	**1.057E-03**	8.541E-03	1.633E-02	3.883E-04	1.367E-03
t70x11xx	3.43E+08	**5.607E-06**	**6.354E-04**	9.151E-03	1.577E-02	1.714E-04	1.861E-03
t74d11xx	673346	8.762E-05	**1.031E-03**	1.041E-02	1.838E-02	**1.485E-05**	1.311E-03
t75d11xx	688601	**1.888E-05**	**2.507E-03**	1.474E-02	1.901E-02	5.097E-04	5.691E-03
t75e11xx	3095130	2.226E-04	9.596E-04	9.899E-03	1.788E-02	**6.979E-05**	**7.560E-04**
t75i11xx	72664466	3.694E-04	**1.118E-03**	1.654E-02	2.400E-02	**2.226E-04**	1.805E-03
t75k11xx	124887	2.963E-04	1.081E-03	7.239E-03	1.186E-02	0	**5.525E-04**
t75n11xx	113808	0	4.235E-03	1.652E-02	2.517E-02	4.657E-04	**2.074E-03**
t75u11xx	63278034	0	1.546E-03	8.045E-03	1.686E-02	7.893E-04	3.180E-03
tiw56n54	112767	**1.782E-03**	**3.769E-03**	3.260E-02	4.340E-02	2.075E-03	5.693E-03
tiw56n58	154440	1.930E-03	**2.739E-03**	2.740E-02	3.700E-02	**1.774E-03**	3.419E-03
tiw56n62	217499	**9.379E-04**	**2.400E-03**	2.659E-02	5.002E-02	2.929E-03	3.848E-03
tiw56n66	277593	**8.033E-04**	**1.780E-03**	4.276E-02	4.893E-02	1.733E-03	4.568E-03
tiw56n67	277962	**1.155E-03**	**2.493E-03**	2.886E-02	3.925E-02	6.443E-03	7.997E-03
tiw56n72	462991	**9.611E-04**	**3.067E-03**	4.007E-02	5.039E-02	2.024E-03	9.352E-03
tiw56r54	127390	**8.635E-04**	**3.148E-03**	4.895E-02	5.647E-02	1.578E-03	5.330E-03
tiw56r58	160776	2.053E-03	**3.234E-03**	3.406E-02	4.474E-02	**1.300E-03**	3.732E-03
tiw56r66	256326	**2.142E-03**	**4.799E-03**	4.037E-02	4.858E-02	2.294E-03	6.129E-03
tiw56r67	270497	**1.231E-03**	**3.172E-03**	3.048E-02	4.803E-02	1.763E-03	4.969E-03
tiw56r72	341623	3.220E-05	**3.194E-03**	3.176E-02	4.833E-02	**1.750E-03**	3.896E-03

Table 2. Solution error (in %) after 10 seconds

LOLIB Instance	Optimum	GA lowest	GA average	DE lowest	DE average	AIS lowest	AIS average
be75eec	264940	9.059E-04	**2.291E-03**	1.200E-02	1.473E-02	**2.491E-04**	4.069E-03
be75np	790966	**3.793E-06**	**7.978E-04**	5.888E-03	1.522E-02	1.656E-04	8.572E-04
be75oi	118159	7.871E-04	1.540E-03	8.649E-03	1.050E-02	0	**1.092E-03**
be75tot	1127387	**1.597E-04**	**1.878E-03**	8.801E-03	2.002E-02	5.322E-04	2.171E-03
stabu1	422088	**3.317E-04**	**2.369E-03**	2.001E-02	2.717E-02	1.694E-03	5.499E-03
stabu2	627929	1.674E-03	**3.566E-03**	2.094E-02	2.844E-02	**1.365E-03**	3.757E-03
stabu3	642050	**1.355E-04**	**3.137E-03**	1.712E-02	2.712E-02	2.394E-03	5.412E-03
t59b11xx	245750	**8.545E-04**	2.901E-03	6.022E-03	9.908E-03	1.974E-03	**2.629E-03**
t59d11xx	163219	0	3.523E-03	3.217E-03	8.118E-03	1.715E-03	**2.506E-03**
t59f11xx	140678	0	**2.986E-04**	4.947E-03	1.122E-02	7.108E-05	8.672E-04
t59i11xx	9182291	5.097E-05	1.101E-03	5.706E-03	9.520E-03	0	**9.650E-04**
t59n11xx	25225	0	**3.631E-03**	7.968E-03	1.162E-02	8.325E-04	4.963E-03
t65b11xx	411733	5.878E-04	2.757E-03	7.284E-03	1.930E-02	**5.198E-04**	**2.448E-03**
t65d11xx	283971	0	**9.473E-04**	3.972E-03	8.673E-03	3.028E-04	5.729E-03
t65f11xx	254568	1.371E-03	2.003E-03	7.633E-03	1.242E-02	**4.282E-04**	**1.603E-03**
t65i11xx	16389651	**1.355E-05**	7.292E-04	3.759E-03	4.609E-03	1.806E-05	**3.942E-04**
t65l11xx	18359	0	0	2.342E-03	4.390E-03	0	4.358E-05
t65n11xx	38814	**2.576E-05**	2.118E-03	4.792E-03	6.472E-03	1.288E-04	**1.999E-03**
t65w11xx	1.66E+08	0	**9.261E-04**	4.703E-03	1.044E-02	1.025E-03	1.221E-03
t69r11xx	865650	8.768E-04	5.249E-03	6.491E-03	1.310E-02	**6.977E-04**	3.049E-03
t70b11xx	623411	4.010E-04	1.407E-03	2.530E-03	3.930E-03	**8.181E-05**	**1.269E-03**
t70d11xn	438235	2.510E-04	**5.477E-04**	5.474E-03	1.379E-02	**8.215E-05**	1.919E-03
t70d11xx	450774	0	**4.304E-04**	4.242E-03	1.368E-02	5.812E-04	1.138E-03
t70f11xx	413948	7.199E-04	2.669E-03	1.177E-02	1.550E-02	**6.523E-05**	**2.063E-03**
t70i11xx	28267738	**4.889E-05**	**7.251E-04**	3.526E-03	7.572E-03	7.807E-04	1.073E-03
t70k11xx	69796200	0	**2.333E-04**	4.059E-03	1.156E-02	8.883E-05	2.171E-03
t70l11xx	28108	0	8.467E-04	1.423E-03	6.539E-03	0	**2.562E-04**
t70n11xx	63944	0	2.133E-03	6.162E-03	9.249E-03	1.564E-05	**1.436E-03**
t70u11xx	27296800	2.198E-04	1.264E-03	2.480E-03	1.099E-02	0	**6.382E-04**
t70w11xx	2.68E+08	0	**7.400E-04**	5.073E-03	8.847E-03	7.360E-06	**5.803E-04**
t70x11xx	3.43E+08	0	**4.224E-04**	5.998E-03	9.763E-03	0	6.412E-04
t74d11xx	673346	**8.911E-06**	6.609E-04	7.855E-03	1.239E-02	5.792E-05	**3.906E-04**
t75d11xx	688601	**1.452E-06**	2.236E-03	8.381E-03	1.495E-02	9.004E-05	**2.155E-03**
t75e11xx	3095130	**2.226E-04**	**7.560E-04**	2.629E-03	1.165E-02	3.321E-04	8.885E-04
t75i11xx	72664466	0	1.242E-03	4.259E-03	7.705E-03	0	**7.564E-04**
t75k11xx	124887	0	**4.404E-04**	5.365E-03	1.155E-02	0	4.564E-04
t75n11xx	113808	0	**1.019E-03**	5.571E-03	1.447E-02	1.845E-04	2.856E-03
t75u11xx	63278034	1.473E-03	3.071E-03	5.497E-03	1.694E-02	**6.766E-04**	**9.090E-04**
tiw56n54	112767	**4.700E-04**	**1.268E-03**	1.330E-02	2.127E-02	1.880E-03	2.953E-03
tiw56n58	154440	**2.202E-04**	**9.389E-04**	1.901E-02	2.692E-02	1.386E-03	2.202E-03
tiw56n62	217499	7.264E-04	**1.802E-03**	2.314E-02	2.884E-02	**6.943E-04**	2.212E-03
tiw56n66	277593	**1.264E-03**	**2.215E-03**	1.924E-02	2.797E-02	1.617E-03	2.958E-03
tiw56n67	277962	**5.396E-05**	**2.932E-03**	1.900E-02	2.197E-02	1.004E-03	4.501E-03
tiw56n72	462991	2.549E-04	7.821E-03	1.349E-02	2.563E-02	**6.480E-05**	**5.477E-03**
tiw56r54	127390	1.311E-03	**2.096E-03**	1.353E-02	2.337E-02	**8.949E-04**	3.438E-03
tiw56r58	160776	**9.143E-04**	**2.606E-03**	2.252E-02	2.777E-02	1.455E-03	2.824E-03
tiw56r66	256326	**7.061E-04**	**2.220E-03**	1.686E-02	3.006E-02	2.497E-03	3.191E-03
tiw56r67	270497	**4.067E-05**	**6.248E-04**	2.150E-02	2.849E-02	1.179E-03	1.930E-03
tiw56r72	341623	**4.976E-05**	**1.493E-03**	1.432E-02	2.286E-02	1.051E-03	2.409E-03

Table 3. Solution error (in %) after 20 seconds

LOLIB Instance	Optimum	GA		DE		AIS	
		lowest	average	lowest	average	lowest	average
be75eec	264940	**6.983E-04**	3.771E-03	8.583E-03	1.595E-02	1.242E-03	**3.457E-03**
be75np	790966	3.793E-06	**8.635E-04**	6.349E-03	1.235E-02	**2.529E-06**	3.108E-03
be75oi	118159	5.924E-05	1.921E-03	1.100E-03	8.759E-03	**8.463E-06**	**1.413E-03**
be75tot	1127387	1.091E-04	3.545E-03	1.072E-02	2.592E-02	**1.002E-04**	**6.537E-04**
stabu1	422088	1.796E-03	**2.860E-03**	8.534E-03	1.601E-02	**1.407E-03**	3.791E-03
stabu2	627929	2.511E-03	2.983E-03	1.868E-02	2.473E-02	**1.745E-03**	**2.629E-03**
stabu3	642050	**1.455E-03**	**2.314E-03**	1.499E-02	1.888E-02	1.716E-03	3.430E-03
t59b11xx	245750	**8.342E-04**	3.638E-03	5.717E-03	9.607E-03	**8.342E-04**	**2.580E-03**
t59d11xx	163219	4.172E-03	5.453E-03	4.822E-03	1.280E-02	0	2.922E-03
t59f11xx	140678	0	**2.772E-04**	3.981E-03	1.296E-02	0	5.473E-04
t59i11xx	9182291	0	0	7.634E-03	1.283E-02	0	1.430E-03
t59n11xx	25225	1.427E-03	2.387E-03	9.514E-03	1.782E-02	0	1.197E-03
t65b11xx	411733	9.934E-04	**2.370E-03**	1.989E-02	2.519E-02	**5.950E-04**	2.387E-03
t65d11xx	283971	2.081E-03	2.282E-03	7.395E-03	1.133E-02	**2.888E-04**	**1.535E-03**
t65f11xx	254568	1.371E-03	**1.520E-03**	6.918E-03	9.369E-03	**2.357E-04**	1.654E-03
t65i11xx	16389651	**1.355E-05**	**3.875E-04**	3.762E-03	9.485E-03	2.793E-04	9.122E-04
t65l11xx	18359	0	**5.447E-05**	2.615E-03	3.486E-03	0	2.832E-04
t65n11xx	38814	0	1.546E-03	2.267E-03	9.991E-03	2.576E-05	**1.536E-03**
t65w11xx	1.66E+08	1.052E-04	1.516E-03	5.593E-03	8.604E-03	0	**6.070E-04**
t69r11xx	865650	3.400E-03	4.282E-03	7.858E-03	1.348E-02	0	**3.944E-03**
t70b11xx	623411	4.010E-04	5.133E-04	3.588E-03	1.565E-02	0	**4.395E-04**
t70d11xn	438235	1.278E-04	**3.742E-04**	3.411E-03	1.223E-02	7.758E-05	3.948E-04
t70d11xx	450774	0	6.899E-04	7.738E-03	1.102E-02	0	**3.838E-04**
t70f11xx	413948	0	**1.971E-03**	8.107E-03	1.850E-02	2.307E-03	3.392E-03
t70i11xx	28267738	0	**5.886E-04**	5.307E-03	9.932E-03	1.507E-04	8.008E-04
t70k11xx	69796200	0	2.151E-03	6.956E-03	1.137E-02	0	**5.427E-04**
t70l11xx	28108	0	7.115E-04	2.668E-03	6.568E-03	0	**2.419E-04**
t70n11xx	63944	1.220E-03	3.156E-03	6.271E-03	1.007E-02	**4.692E-05**	**2.361E-03**
t70u11xx	27296800	0	**6.704E-04**	3.521E-03	1.391E-02	2.125E-04	8.668E-04
t70w11xx	2.68E+08	0	8.968E-04	5.211E-03	9.601E-03	0	**7.325E-04**
t70x11xx	3.43E+08	0	**3.239E-04**	3.778E-03	9.774E-03	2.699E-05	1.258E-03
t74d11xx	673346	2.970E-06	1.598E-03	9.864E-03	2.205E-02	0	**3.609E-04**
t75d11xx	688601	0	**1.689E-03**	9.012E-03	1.831E-02	1.699E-04	1.891E-03
t75e11xx	3095130	0	**9.919E-04**	2.045E-03	6.528E-03	2.455E-05	1.166E-03
t75i11xx	72664466	2.359E-04	1.263E-03	3.552E-03	7.728E-03	0	**6.918E-04**
t75k11xx	124887	0	6.886E-04	1.593E-03	6.198E-03	0	**4.804E-04**
t75n11xx	113808	0	1.678E-03	1.061E-02	1.630E-02	0	**1.177E-03**
t75u11xx	63278034	**6.766E-04**	1.840E-03	3.620E-03	2.546E-02	**6.766E-04**	1.891E-03
tiw56n54	112767	**2.926E-04**	**9.577E-04**	1.229E-02	1.846E-02	9.932E-04	2.368E-03
tiw56n58	154440	7.770E-04	2.350E-03	8.392E-03	1.353E-02	**3.238E-04**	**8.482E-04**
tiw56n62	217499	**8.736E-05**	**1.113E-03**	1.813E-02	2.306E-02	2.023E-04	1.301E-03
tiw56n66	277593	**5.043E-05**	**1.617E-03**	8.667E-03	1.833E-02	1.866E-03	4.204E-03
tiw56n67	277962	**3.598E-05**	**1.601E-03**	1.306E-02	2.110E-02	1.979E-04	3.069E-03
tiw56n72	462991	**1.060E-03**	5.043E-03	1.722E-02	2.389E-02	1.123E-03	**4.698E-03**
tiw56r54	127390	**1.256E-04**	**1.656E-03**	9.962E-03	2.275E-02	1.437E-03	2.159E-03
tiw56r58	160776	4.603E-04	**1.468E-03**	1.083E-02	1.617E-02	**4.292E-04**	2.364E-03
tiw56r66	256326	**2.731E-05**	**8.076E-04**	1.730E-02	2.636E-02	9.246E-04	3.043E-03
tiw56r67	270497	1.142E-03	4.577E-03	1.097E-02	2.070E-02	**1.885E-04**	**2.115E-03**
tiw56r72	341623	3.688E-04	**5.708E-04**	6.888E-03	1.667E-02	**2.547E-04**	1.885E-03

and table 3 respectively. The tables show for each algorithm the lowest error obtained by the best solution found in the 10 independent runs and the average error of all solutions found in the 10 independent runs. The best average error and the best lowest error is typed in bold for each LOP instance.

It can be seen that the GA and AIS have found better solutions than DE in all cases. The best solution was found after 5s by AIS for 16 LOLIB instances and by GA for 36 instances (the same solutions were found by both algorithms in 3 cases). The average error of the solutions evolved by AIS was better for 12 instances while the average error of solutions evolved by GA was better for 37 LOLIB instances.

AIS evolved after 10 seconds better best solutions for 20 LOLIB matrices and GA for 34 matrices (both algorithms have found the same best solution in 5 cases). The average error of the solutions found by AIS was better for 20 matrices and the average error of solutions evolved by GA was better for 29 matrices.

The 20-seconds long evolution of LOP solutions resulted in better best solutions found by AIS for 32 LOLIB matrices and by GA for 28 LOLIB instances. The average solution found by AIS was better for 24 matrices and the average solution found by GA was better for 25 instances.

5   Conclusions

The experiments have shown that both GA and AIS were able to evolve better LOLIB solutions than DE in the same time. Both GA and AIS used the same random keys based encoding and the same mutation (hypermutation) implementation consisting of randon changes in the real encoded chromosome. The crossover used in the GA was a simple 1-point crossover.

A comparison of GA and AIS has revealed that GA delivered better solutions than AIS when using shorter execution time but AIS has improved when the execution time was longer. It suggests that the GA has quickly found promising solutions while the AIS needed more time to converge to solutions of such quality. The fast convergence of the GA could be attributed to the implemented variant of GA (steady-state GA with elitism) that yields higher selection pressure on the account of diversity [14,6]. The AIS seems to be robust and prone to premature convergence as it becomes more successful with longer execution times. The results presented in this work show that AIS is a promising metaheuristic for linear ordering problem that delivers good results for the LOLIB library.

In the future, we will use AIS to find solutions of other known LOLIB instances from e.g. XLOLIB, SGB, or other suitable problem libraries. Moreover, the parameters of AIS for LOP will be fine-tuned and different encoding schemes will be used.

Acknowledgement. This work was supported by the European Regional Development Fund in the IT4Innovations Centre of Excellence project (CZ.1.05/ 1.1.00/02.0070) and by the Bio-Inspired Methods: research, development and knowledge transfer project, reg. no. CZ.1.07/2.3.00/20.0073 funded by Operational Programme Education for Competitiveness, co-financed by ESF and state budget of the Czech Republic. This work was also supported by the Ministry of Industry and Trade of the Czech Republic, under the grant no. FR-TI1/420 and by SGS, VSB – Technical University of Ostrava, Czech Republic, under the grant No. SP2012/58.

References

1. Abraham, A.: Editorial - hybrid soft computing and applications. International Journal of Computational Intelligence and Applications 8(1) (2009)
2. Affenzeller, M., Winkler, S., Wagner, S., Beham, A.: Genetic Algorithms and Genetic Programming: Modern Concepts and Practical Applications. Chapman & Hall/CRC (2009)
3. Campos, V., Glover, F., Laguna, M., Martí, R.: An experimental evaluation of a scatter search for the linear ordering problem. J. of Global Optimization 21(4), 397–414 (2001)
4. Chira, C., Pintea, C.M., Crisan, G.C., Dumitrescu, D.: Solving the linear ordering problem using ant models. In: Proceedings of the 11th Annual Conference on Genetic and Evolutionary Computation, GECCO 2009, pp. 1803–1804. ACM, New York (2009)
5. Corchado, E., Arroyo, A., Tricio, V.: Soft computing models to identify typical meteorological days. Logic Journal of the IGPL 19(2), 373–383 (2011)
6. Dyer, J.D., Hartfield, R.J., Dozier, G.V., Burkhalter, J.E.: Aerospace design optimization using a steady state real-coded genetic algorithm. Applied Mathematics and Computation 218(9), 4710–4730 (2012)
7. Engelbrecht, A.: Computational Intelligence: An Introduction, 2nd edn. Wiley, New York (2007)
8. Hart, E., Timmis, J.: Application areas of ais: The past, the present and the future. Applied Soft Computing 8(1), 191–201 (2008)
9. Huang, G., Lim, A.: Designing a hybrid genetic algorithm for the linear ordering problem. In: GECCO, pp. 1053–1064 (2003)
10. Krömer, P., Platos, J., Snasel, V.: Differential evolution for the linear ordering problem implemented on cuda. In: Smith, A.E. (ed.) Proceedings of the 2011 IEEE Congress on Evolutionary Computation, June 5-8, pp. 790–796. IEEE Computational Intelligence Society, IEEE Press, New Orleans (2011)
11. Krömer, P., Platoš, J., Snášel, V.: Modeling permutations for genetic algorithms. In: Proceedings of the International Conference of Soft Computing and Pattern Recognition (SoCPaR 2009), pp. 100–105. IEEE Computer Society (2009)
12. Krömer, P., Snášel, V., Platoš, J.: Evolving feasible linear ordering problem solutions. In: CSTST 2008: Proceedings of the 5th International Conference on Soft Computing as Transdisciplinary Science and Technology, pp. 337–342. ACM, New York (2008)
13. Krömer, P., Snášel, V., Platoš, J., Husek, D.: Genetic Algorithms for the Linear Ordering Problem. Neural Network World 19(1), 65–80 (2009)
14. Lozano, M., Herrera, F., Cano, J.: Replacement Strategies to Maintain Useful Diversity in Steady-State Genetic Algorithms, pp. 85–96 (2005)
15. Martí, R., Reinelt, G.: The Linear Ordering Problem - Exact and Heuristic Methods in Combinatorial Optimization. Applied Mathematical Sciences, vol. 175. Springer, Heidelberg (2011)

16. Martí, R., Reinelt, G., Duarte, A.: A benchmark library and a comparison of heuristic methods for the linear ordering problem. In: Computational Optimization and Applications, pp. 1–21 (2011)
17. Mitchell, J.E., Borchers, B.: Solving linear ordering problems with a combined interior point/simplex cutting plane algorithm. Tech. rep., Mathematical Sciences, Rensselaer Polytechnic Institute, Troy, NY 12180–3590 (September 1997), http://www.math.rpi.edu/~mitchj/papers/combined.ps; accepted for publication in Proceedings of HPOPT 1997, Rotterdam, The Netherlands
18. Price, K.V., Storn, R.M., Lampinen, J.A.: Differential Evolution A Practical Approach to Global Optimization. Natural Computing Series. Springer, Berlin (2005)
19. Reinelt, G.: The Linear Ordering Problem: Algorithms and Applications, Research and Exposition in Mathematics, vol. 8. Heldermann Verlag, Berlin (1985)
20. Schiavinotto, T., Stützle, T.: Search Space Analysis of the Linear Ordering Problem. In: Raidl, G.R., Cagnoni, S., Cardalda, J.J.R., Corne, D.W., Gottlieb, J., Guillot, A., Hart, E., Johnson, C.G., Marchiori, E., Meyer, J.-A., Middendorf, M. (eds.) EvoIASP 2003, EvoWorkshops 2003, EvoSTIM 2003, EvoROB/EvoRobot 2003, EvoCOP 2003, EvoBIO 2003, and EvoMUSART 2003. LNCS, vol. 2611, pp. 322–333. Springer, Heidelberg (2003)
21. Schiavinotto, T., Stützle, T.: The linear ordering problem: Instances, search space analysis and algorithms. Journal of Mathematical Modelling and Algorithms 3(4), 367–402 (2004)
22. Sedano, J., Curiel, L., Corchado, E., de la Cal, E., Villar, J.R.: A soft computing method for detecting lifetime building thermal insulation failures. Integr. Comput.-Aided Eng. 17(2), 103–115 (2010)
23. Snášel, V., Krömer, P., Platoš, J.: Differential Evolution and Genetic Algorithms for the Linear Ordering Problem. In: Velásquez, J.D., Ríos, S.A., Howlett, R.J., Jain, L.C. (eds.) KES 2009, Part I. LNCS, vol. 5711, pp. 139–146. Springer, Heidelberg (2009)
24. Snyder, L.V., Daskin, M.S.: A random-key genetic algorithm for the generalized traveling salesman problem. European Journal of Operational Research 174(1), 38–53 (2006)
25. Timmis, J., Hone, A., Stibor, T., Clark, E.: Theoretical advances in artificial immune systems. Theoretical Computer Science 403(1), 11–32 (2008)
26. Timmis, J., Andrews, P.S., Hart, E.: Special issue on artificial immune systems. Swarm Intelligence 4(4), 245–246 (2010)
27. Wolpert, D.H., Macready, W.G.: No free lunch theorems for optimization. IEEE Transactions on Evolutionary Computation 1(1), 67–82 (2002)
28. Yu, H.: Optimizing task schedules using an artificial immune system approach. In: Proceedings of the 10th Annual Conference on Genetic and Evolutionary Computation, GECCO 2008, pp. 151–158. ACM, New York (2008)
29. Zhao, S.Z., Iruthayarajan, M.W., Baskar, S., Suganthan, P.: Multi-objective robust pid controller tuning using two lbests multi-objective particle swarm optimization. Information Sciences 181(16), 3323–3335 (2011)

Genetic Programming of Augmenting Topologies for Hypercube-Based Indirect Encoding of Artificial Neural Networks

Jan Drchal and Miroslav Šnorek

Department of Computer Science and Engineering,
FEE CTU, Karlovo náměstí 13, 121 35, Praha 2, Czech Republic
{drchajan,snorek}@fel.cvut.cz

Abstract. In this paper we present a novel algorithm called GPAT (Genetic Programming of Augmenting Topologies) which evolves Genetic Programming (GP) trees in a similar way as a well-established neuro-evolutionary algorithm NEAT (NeuroEvolution of Augmenting Topologies) does. The evolution starts from a minimal form and gradually adds structure as needed. A niching evolutionary algorithm is used to protect individuals of a variable complexity in a single population. Although GPAT is a general approach we employ it mainly to evolve artificial neural networks by means of Hypercube-based indirect encoding which is an approach allowing for evolution of large-scale neural networks having theoretically unlimited size. We perform also experiments for directly encoded problems. The results show that GPAT outperforms both GP and NEAT taking the best of both.

1 Introduction

Recently, there has been a growing interest in the techniques which evolve large-scale artificial neural networks (ANNs). Classical training algorithms (such as back-propagation, second order optimization and evolutionary computation) suffer from poor convergence caused by the large dimension of the optimized synaptic weights (or other parameters). This problem was already successfully addressed by neuro-evolutionary approaches [1,2,3] which employ indirect encoding of ANNs. The idea of the indirect encoding is inspired by the Nature, where relatively short genomes encode highly complex structures. This immense compression of information is, among others, facilitated by regularities found at all scales of magnification.

This paper deals with a related state-of-the-art approach called HyperNEAT [4] which employs the so-called Hypercube-based indirect encoding and a well-established neuro-evolutionary algorithm NEAT. HyperNEAT works in a following way: at first, one have to decide for the structure of a *final network* (number and types of neurons and possible connections), also each neuron is assigned coordinates. Such template is called the *substrate*. Second, NEAT algorithm is used to evolve the CPPNs (Compositional and Pattern Producing Networks). The CPPN in HyperNEAT has a form of a common neural network, with an exception of using nodes with special transfer functions which are either symmetric, periodic or of a different type to reflect

V. Snasel et al. (Eds.): SOCO Models in Industrial & Environmental Appl., AISC 188, pp. 63–72.
springerlink.com

the above mentioned regularities found in living creatures. Third, the *final network* is constructed according to the *substrate*. The CPPN is then used to determine synaptic weights of all possible connections found in the *substrate* (coordinates of all pairs of neurons are fed to inputs). Fourth, the *final network* is evaluated on a problem domain to get the fitness value and provide a feedback to NEAT. HyperNEAT was successfully used to evolve networks having almost 8 millions of connections [4].

In [5] Buk et al. presented a modification of HyperNEAT, where NEAT was replaced by Genetic Programming (GP), as the *base algorithm* evolving CPPNs. It was shown to be superior in a speed of convergence to NEAT on a simulated robotic task. Here, we combine advantages of both GP and NEAT and propose a novel algorithm called Genetic Programming of Augmenting Topologies (GPAT). We compare all three *base algorithms* on problems having both directly and indirectly encoded genomes. GPAT in combination with Hypercube-based encoding will be denoted as HyperGPAT.

The paper[1] is organized as follows. Section 2 describes the theoretical background. Section 3 introduces GPAT algorithm. Section 4 describes the test problems and the experimental setup. Section 5 discusses the results. Final section concludes the paper.

2   Background

In this section, we briefly describe NEAT and GP base algorithms.

NEAT. NeuroEvolution of Augmenting Topologies (NEAT) [6] is an algorithm originally developed for evolution of both parameters (weights) and topology of artificial neural networks. It was later enhanced to produce the CPPNs with heterogenous nodes for the HyperNEAT algorithm instead of producing the neural networks directly. It works with genomes of variable size. NEAT introduced a concept of *innovation numbers*, which are gene labels allowing effective genome alignment in order to facilitate crossover-like operations. Moreover, innovation numbers are used for computation of a genotypical distance between two individuals. The distance measure is needed by niching evolutionary algorithm, which is a core of NEAT. Because NEAT evolves networks of different complexity (sizes) niching was found to be necessary for protection of new topology innovations. An important NEAT property is the *complexification* – it starts with simple networks and gradually adds new neurons and connections.

The niching algorithm used in NEAT is *Explicit Fitness Sharing* [7] (EFS) as it uses a genotypical distance measure. Fitness sharing reduces the fitness of an individual given the number of similar individuals (similar individuals form a niche) in the population. The version used in NEAT divides the population into mutually exclusive niche (species). When a new individual is to be assigned to a species it goes through a list of the already existing species and is always compared with the individual designated as the *species' representative*. The comparison is done via evaluation of a distance measure. If the two individuals are similar enough (their distance is below a predefined *speciation threshold* δ) the new individual is assigned to the species, otherwise, it tries the next one in the list. If no species is compatible enough, a new one is created and the individual becomes its *representative*.

---

[1] Note, the the source codes and detailed experimental settings can be found at:
http://neuron.felk.cvut.cz/~drchajl/.

GP. Genetic Programming [8] is a well-known evolutionary approach which evolves syntactic trees (or forests of trees). In this paper, we use a slightly simplified version of GP as a *base algorithm*, omitting a commonly used crossover operator as it was not found beneficial by Buk [5] for Hypercube-based domain. More specifically, to create a new generation, we employ a tournament selection (tournament of size 2) to select N (the population size) parents. Each selected parent then produces a single offspring. Algorithm continues by sorting all $2N$ individuals by their fitness and successively reduces their number back to N, keeping only the fittest.

The initial population is created using the *grow method* [8]. The depth of trees is limited to avoid bloating. We use two types of mutations: a *structural mutation* selects a random node in a tree and replaces it by a random subtree again using the *grow method* or replaces the node by a random node of the same arity. A *parametric mutation* selects a random constant node (if exists) and applies a Cauchy mutation [9] to it. A newly created random constant node is initialized by a random value from a selected interval.

3   Our Approach

In this section we describe GPAT algorithm. The algorithm works exactly as NEAT with an exception of genome representation, genetic operators and distance measure. Unlike in NEAT, where genomes encode neural networks, GPAT genomes represent trees (more specifically forests of trees, to facilitate multiple outputs). GPAT uses nodes of variable arity starting with zero children. Moreover, each children of a node is assigned a constant which has a similar function as a synaptic weight in ANN. The constants are therefore associated with links contrary to constants represented only by terminal nodes in GP.

For a detailed description of NEAT, especially speciation and selection, see [6]. In our implementation, we use an adaptive speciation threshold δ: it is doubled when a number of species exceeds the target number of species n_S, otherwise it is divided by 2.

GPAT Nodes. We use the following non-terminals for GPAT: $+$ $\left(\sum_{i=1}^{C} c_i x_i\right)$, $\star$ $\left(\prod_{i=1}^{C} x_i\right)$, A $\left(\arctan\left(\sum_{i=1}^{C} x_i\right)\right)$, S $\left(\sin\left(\sum_{i=1}^{C} x_i\right)\right)$ and G $\left(e^{\sum_{i=1}^{C} x_i}\right)$ for all experiments in this paper. The output of each node is given in parentheses, C stands for the node arity, c_i is the i-th child's constant and x_i is the i-th child's output. For $C = 0$ the output is defined as 0 for all node types. Note, that constants are used only by the $+$ node, they are stored but not used for other types of nodes. The given set is optimized for Hypercube-based indirect encoding, it contains symmetric, periodic and other functions. The set of terminals is composed of a problem-specific number of inputs and a fixed constant 1.

Similarly to NEAT, GPAT starts with a population of simplest forests possible. In our case this means a forests with all trees having a $+$ terminal as the root with zero children (such trees always output zero).

Genetic Operators. A new individual is created by the following structural and parametric mutations which can be applied with a given probability for each tree of the forest:

Add Link Mutation. With a probability p_{AL} generate a random terminal t and connect it to a random tree non-terminal node. This mutation closely resembles *add link mutation* as found in NEAT, except, links always connect terminals to non-terminals.

Add Node Mutation. With a probability p_{AN} choose a random link l and replace it by the structure $l_1 \rightarrow n \rightarrow l_2$, where l_1 and l_2 are new links and n a new non-terminal random node. This mutation was again inspired by its NEAT counterpart.

Insert Root Mutation. Create a new random root node and assign the original root as its only child with a probability p_{IR}.

Switch Node Mutation. Chose a random non-terminal or terminal node and change its type randomly with a probability p_{SN}.

Cauchy Parametric Mutation. Apply a Cauchy mutation as described in GP section to each constant of the tree with a probability p_{CM}.

Replace Constant Parametric Mutation. With a probability p_{RC} for each constant of the tree reset its value to a random number from a given interval $[-a_R, a_R]$.

Currently we do not employ any crossover-like operator, which would require to implement *innovation number* mechanism (see [6]). To limit bloating no structural mutation was allowed for genotypes large enough (here the limit was: depth > 5, number of constants > 10 and the number of nodes > 12).

Distance Measure. Unlike in NEAT we have decided to use a simpler distance measure not based on innovation numbers. This is facilitated by the fact that we employ trees instead of networks[2].

Our approach called the *generalized* distance measure, covers many already published approaches for GP trees, it is mostly inspired by [10], although we do not take constant values into account. It allows to change behavior according to parameters. At first, all node types are partitioned to sets $A_0, A_1, A_2, \ldots$, where $A_0 = \{NIL\}$ contains only a special *NIL* node and A_1 is reserved for constants. Then we define an auxiliary function to compute the distance between two node types $x \in A_i$ and $y \in A_j$:

$$d'(x,y) = \begin{cases} 0 \text{ if } x = y \text{ (note, this implies } i = j), \\ 1 \text{ otherwise.} \end{cases}$$

The *generalized* measure is defined as $d(p,q) = d'(p,q)$, if neither p, nor q have descendants, $d(p(s_1,\ldots,s_n), q(t_1,\ldots,t_m)) = d(q(t_1,\ldots,t_m), p(s_1,\ldots,s_n)))$, for $m < n$ and $d(p(s_1,\ldots,s_n), q(t_1,\ldots,t_m)) = d'(p,q) + \max(\alpha, \delta_{pq}) \frac{1}{K} (\sum_{i=1}^{n} d(s_i, t_i) + \beta \sum_{j=n+1}^{m} d(NIL, t_j))$ for $n \leq m$. The meaning of constants is as follows:

- $\alpha \in \{0, 1\}$ decides whether to descend into subtrees, when nodes have different types (descends when $\alpha = 1$),

---

[2] We have also experimented with NEAT-like distance measure, the preliminary results show, that both approaches are comparable.

- $\beta \in \{0,1\}$ decides whether to include information concerning the missing subtree in one of the trees (includes when $\beta = 1$),
- δ_{pq} is the Kronecker delta,
- $K > 0$ controls the influence of a node depth in the tree (for $K = 1$ the node depth does not matter).

Along with the *generalized* distance measure we have also used the *random* distance as a control treatment, where the distance between two trees is defined as a random number from interval $[0,1)$ using uniform distribution. GPAT with the *random* measure will be denoted as GPAT-R. Note, that although this section dealt with tree distance measures, in fact we evolve forests with GPAT to allow multiple outputs. The distance between two forests is computed as an average of distances between corresponding trees.

4   Experimental Setup

In this section, we present both directly (Symbolic Regression and Maze Navigation) and indirectly encoded problems used to compare the *base algorithms* (NEAT, GP, GPAT and GPAT-R). Such experiments will be helpful to show, whether there is a difference in a choice of the right distance measure for direct and indirect encoding domains.

Symbolic Regression. The first, most straightforward problem is Symbolic Regression, the test functions are in Tab. 1. The equations are divided into four groups according to their dimensionality (the number of inputs). They were selected from a larger set, of which they presented the most diverse behaviors. Note, that the equation 4D-V was obtained by solving a Visual Discrimination problem (see below). The 1D, 2D and 3D functions were sampled in a hypercube having minimum and maximum coordinates -10 and 10, 4D used boundaries of -1 and 1. We have used 20 equidistant samples in each dimension for 1D, 7 for 2D and 3D and 5 for 4D. The fitness is computed using *Mean Squared Error* as $1/(1+MSE)$ which lies in $(0,1]$. The problem was considered to be solved, when target fitness 0.95 was reached. Note, neither testing nor validation sets were used.

Table 1. Symbolic Regression functions

ID	Function	ID	Function
1D-F	$1.5x_1^3 + 2.3x_1^2 - 1.1x_1 + 3.7$	3D-K	$x_1x_2^2x_3 - x_1x_2 + x_2x_3$
1D-H	$0.1x_1^2 + 0.2\sin(x)$	4D-C	$1.5x_1x_2x_3x_4$
2D-I	$1.5x_1x_2^2 + 2.3x_1x_2 - 1.1x_2^2$	4D-F	$1.5x_1x_2x_3 + 2.3x_1 - 1.1x_2 - 1.1x_4$
2D-K	$x_1x_2^2 + x_1x_2 - x_2^2$	4D-G	$1.5x_1x_3x_4x_2^2 - 1.1x_2 + 2.3x_1$
3D-E	$1.5x_1x_2 + 2.3x_1 + x_2x_3 - 1.1x_3$	4D-V	$e^{-ax_2^2}x_1\left(be^{-x_3^2} - c\sin(1 - x_1)\right)$
3D-H	$1.5x_1x_2^2x_3 + 2.3x_1x_2x_3 - 1.1x_2 + 5.3$		$(a,b,c) = (1.36709, 2.43454, 0.393788)$

Maze Navigation. Maze navigation (see Fig. 1) presents a *deceptive task* to show abilities to overcome local extremes. The problem is a simple target approach, where a simulated robot tries to navigate through a maze from a starting position to a target one. Fitness is derived from the Manhattan distance between the robot and the target after 150 steps. We have used not only two different maps, but for each map also different types of sensors. For MAZE-1 the sensors were: distance to target, binary wall sensors (1 step ahead, ahead-left, ahead-right) and binary target detection (ahead, left, right and behind). For MAZE-2 we have used four wall range finders instead of three wall detection sensors. The controller has a single output: for output values less than -0.5 the robot turns left and advances, for values greater than 0.5 it turns right and advances, otherwise it makes only a step forward. The problem was considered to be solved if and only if the robot reaches the target. Both tasks are deceptive as there are local extremes in a proximity of the global one.

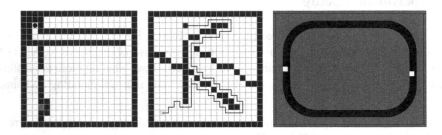

Fig. 1. MAZE-1 (left), MAZE-2 (middle) maps, Mobile Robot Navigation arena (right). MAZE-2 shows a possible solution. Mobile Robot Navigation map shows roads (fast) and grass (slow) surfaces. Two white squares determine starting positions.

Bit Permutation. Bit permutation is a set of three problems which are similar to the Bit Mirroring in [11]. All possible combinations of inputs (0s and 1s) were tested. In this paper we did experiments with $n = 6$ inputs/outputs (64 possibilities). The fitness is computed as a number of all correct output bits over all test patterns (in contrary to [11].) The problem was considered to be solved when all input patterns generated correct output patterns. The substrate is composed of two layers of neurons (input and output). Nodes in each layer are placed equidistant (with a distance 1). A bias neuron (having coordinate 0) is connected to all output neurons. The CPPN has therefore 2 outputs. The three problems are:

Bit Reverse. Bits $1, 2, \ldots n$ labeled from the LSB to the MSB: $o_j = i_{n-j+1}$, for $j \in \{1, 2, \ldots, n\}$.

Bit Shift. Logical shift to the right: $o_n = 0$, $o_j = i_{j+1}$, for $j \in \{1, \ldots, n-1\}$.

Bit Rotate. Circular shift to the right: $o_j = i_{j+1} \mod n$, for $j \in \{1, \ldots, n\}$.

Parity Problem. In this experiment we evolve a network computing odd parity. The function has n inputs (in our case 4) and a single output. The substrate is similar to the previous one, although we added a biased hidden layer (CPPN with 4 outputs). A *final network* must give correct answers for all 16 test cases.

Visual Discrimination. As presented in [4], the task is to detect the larger object of two objects, projected on a two-dimensional array of input neurons. Unlike in original experiment, we employed a smaller input resolution of 5×5 and the problem was considered to be solved only for success in all test cases.

Mobile Robot Navigation. In the last experiment, we have employed ViVAE (Visual Vector Agent Environment) [5,12]. The task was to evolve robotic controller to drive two robots on a map (See Figure 4) at a maximum possible speed. The best strategy is to drive on a side of a road to avoid collisions. The problem was considered to be solved when a fitness reached 0.797 (value found experimentally).

Parameter Settings. The size of population was set to 100, the maximum number of generations was 1000 (for Mobile Robot Navigation they were set to 100 and 50) for all algorithms (GP, GPEFS and NEAT). All experiments were repeated for 200 times with an exception of the Mobile Robot Navigation, which was repeated for 20 times, only.

We have used the same node types for all experiments. In the case of NEAT they were: Bipolar Sigmoid, Sin, Linear and Gaussian. For GP: Add, Multiply, ArcTan, Sin and Gaussian, see [5]. GPAT nodes were described above. Along with inputs, GP uses a fixed constant of -1. Note, that other parameter settings, e.g., mutation rates are available at the paper support page (see above).

To test the properties of the *generalized* distance measure, we did experiments for all 8 combinations of $K \in \{1,2\}$, $\alpha \in \{0,1\}$ and $\beta \in \{0,1\}$, we will write these configurations as tuples, e.g., $(2,0,1)$, in the following text.

5   Results

The results of all experiments are summarized in Fig. 2 showing success rates for both direct and indirect problems. One can see that NEAT performs the worst of all algorithms with an exception of Visual Discrimination where it is comparable with GPAT-R and MAZE-2 where it significantly[3] outperforms GP. Note, that NEAT never outperforms GP on indirect encoding problems: this supports the results in [5].

GP is a significant winner on three symbolic regression problems (2D-K, 3D-K and 4D-V) and a nonsignificant winner on Parity. Interestingly 2D-K and 3D-K are equations with a minimum number of constants (zero and one), while 4D-V is a solution of the Visual Discrimination evolved by GP (GP is most probably biased to evolve such solutions).

---

[3] We use Fisher's Exact test (two-tailed, significance level 5%) to compare the numbers of successful experiments.

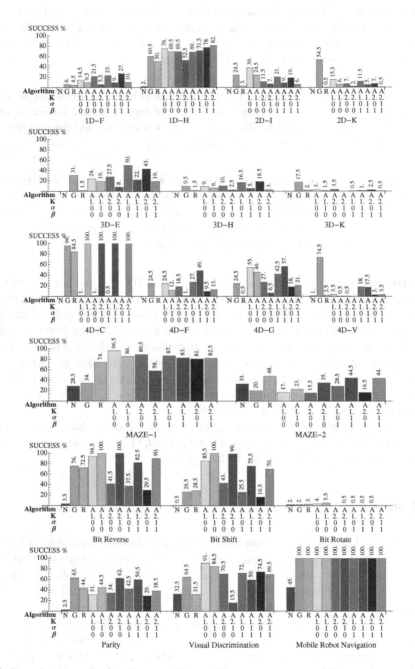

Fig. 2. Experiment results for all test problems. The success rate over 200 experiments (20 for Mobile Robot Navigation) is shown above bars. NEAT is labeled as N, GP as G, GPAT as A and GPAT-R as R. The values of distance measure constants K, α and β are shown.

GPAT-R, the control treatment, performed significantly better than NEAT in most cases. On the other hand it performed worse than GP on all with an exception of both maze tasks (significantly better) and Bit Shift (non-significantly better).

Finally, with a proper choice of the parameters K, α and β, GPAT is a significant winner on all problems with an exception of the before mentioned 2D-K, 3D-K, and 4D-V dominated by GP and MAZE-2 and Parity where GPAT is non-significantly worse than GPAT-R and GP. Even without selecting the best configuration for each problem and having them fixed either to $(1,0,0)$ or $(1,1,1)$, GPAT is a winner (at least non significantly worse than the best of NEAT, GP or GPAT-R) for 8 out of 13 direct problems and 5 out of 6 indirect problems.

As one can see, GPAT success rate shows a noticeable dependence on distance measure parameters. The cases where GPAT-R outperforms GPAT clearly indicates where inappropriate parameter settings were chosen. The choice of $K = 2$ (for $K > 1$ the deeper the node is in a tree the smaller influence it has on a distance) seems to rather harm the performance when compared to $K = 1$ (with few exceptions, e.g., Parity) which contradicts results in [13]. The choice of α (enables/disables a comparison of subtrees with different root node types) and β (take or not the size and shape of subtrees missing in one of the trees into account) parameters is highly problem-dependent. However, for all indirect encoding problems with an exception of Visual Discrimination, setting $\alpha = 1$ improves the performance. For 4D-C (direct encoding) the difference even makes up to 100%. Setting $\beta = 1$ most often leads to a reduced performance. On the other hand, when $\alpha = 1$, setting $\beta = 1$ led to improvement (see Visual Discrimination, Parity, 4D-F, 4D-G, 4D-V and maze tasks). This influence of α and β should be further examined.

6 Conclusions

In this paper, we have proposed a novel algorithm called GPAT which is inspired by the well-known NEAT and GP algorithms. It takes the complexification property and the niching evolutionary algorithm from NEAT while operating on tree-structures known from GP. The use of trees allowed for a simpler distance measure and one such the *generalized* measure was proposed and tested. Although the choice of node types for GPAT was optimized for Hypercube-based indirect encoding tasks, the algorithm is general and can be simply adapted to any task where GP is applicable.

GPAT outperformed both NEAT and GP on most benchmarks. For all indirect encoding problems GP was superior to NEAT which supports results in [5]. Our estimation is, that this is caused by a larger search-space for NEAT as it has to optimize more random constants than a comparable tree. Also, some of the tested problems demanded high accuracy (e.g., Symbolic Regression) which might be disadvantageous for NEAT. This has to be, however, further examined.

In the future we plan to experiment with *innovation number* based distance measure resembling the one used in NEAT and compare it to the proposed *generalized* measure. The use of innovation numbers will also allow to implement crossover (mating) operator efficiently. The *generalized* measure can be further extended to take values of constants into account similalrly as in [10]. Moreover, there is a possibility to employ other measures, i.e., edit-distance based measures [14].

Acknowledgement. This research has been supported by the Ministry of Education, Youth and Sports of the Czech Republic as a part of the institutional development support at the CTU in Prague.

References

1. Eggenberger-Hotz, P.: Creation of Neural Networks Based on Developmental and Evolutionary Principles. In: Gerstner, W., Hasler, M., Germond, A., Nicoud, J.-D. (eds.) ICANN 1997. LNCS, vol. 1327, pp. 337–342. Springer, Heidelberg (1997)
2. Gruau, F.: Neural Network Synthesis Using Cellular Encoding and the Genetic Algorithm. PhD thesis, Ecole Normale Supirieure de Lyon, France (1994)
3. Koutnik, J., Gomez, F., Schmidhuber, J.: Evolving Neural Networks in Compressed Weight Space. In: Proceedings of the 12th Annual Conference on Genetic and Evolutionary Computation - GECCO 2010, p. 619. ACM Press, New York (2010)
4. Gauci, J., Stanley, K.O.: Generating Large-Scale Neural Networks Through Discovering Geometric Regularities. In: Proceedings of the 9th Annual Conference on Genetic and Evolutionary Computation - GECCO 2007, pp. 997–1004. ACM Press, New York (2007)
5. Buk, Z., Koutník, J., Šnorek, M.: NEAT in HyperNEAT Substituted with Genetic Programming. In: Kolehmainen, M., Toivanen, P., Beliczynski, B. (eds.) ICANNGA 2009. LNCS, vol. 5495, pp. 243–252. Springer, Heidelberg (2009)
6. Stanley, K.O.: Efficient Evolution of Neural Networks through Complexification. PhD thesis, The University of Texas at Austin (2004)
7. Mahfoud, S.W.: A Comparison of Parallel and Sequential Niching Methods. In: Proceedings of the Sixth International Conference on Genetic Algorithms, pp. 136–143. Morgan Kaufmann (1995)
8. Poli, R., Langdon, W.B., Mcphee, N.F.: A Field Guide to Genetic Programming (March 2008), Published via http://lulu.com
9. Yao, X., Yong, L., Guangming, L.: Evolutionary Programming Made Faster. IEEE Transactions on Evolutionary Computation 3, 82–102 (1999)
10. Ekárt, A., Németh, S.Z.: A Metric for Genetic Programs and Fitness Sharing. In: Poli, R., Banzhaf, W., Langdon, W.B., Miller, J., Nordin, P., Fogarty, T.C. (eds.) EuroGP 2000. LNCS, vol. 1802, pp. 259–270. Springer, Heidelberg (2000)
11. Clune, J., Stanley, K.O., Pennock, R.T., Ofria, C.: On the Performance of Indirect Encoding Across the Continuum of Regularity. IEEE Transaction on Evolutionary Computation 15(3), 346–367 (2011)
12. Drchal, J., Koutnik, J., Snorek, M.: HyperNEAT Controlled Robots Learn How to Drive on Roads in Simulated Environment. In: CEC 2009 Proceedings of the Eleventh Conference on Congress on Evolutionary Computation, Trondheim, pp. 1087–1092. IEEE Press (2009)
13. Igel, C., Chellapilla, K.: Investigating the Influence of Depth and Degree of Genotypic Change on Fitness in Genetic Programming. In: Proceedings of the Genetic and Evolutionary Computation Conference, Orlando, FL, USA, pp. 1061–1068. Morgan Kaufmann (1999)
14. Nguyen, T.H., Nguyen, X.H.: A Brief Overview of Population Diversity Measures in Genetic Programming. In: Proceedings of the Third Asian Pacific Workshop on Genetic Programming, pp. 128–139 (2006)

Master Slave LMPM Position Control Using Genetic Algorithms

Tatiana Radičová and Milan Žalman

Slovak University of Technology in Bratislava
Faculty of Electrical Engineering and Information Technology, Ilkovičova 3, 812 19 Bratislava,
Slovakia
{tatiana.radicova,milan.zalman}@stuba.sk

Abstract. Recently, in the era of high speed computers, nanotechnology and intelligent control; genetic algorithms belong to the essential part of this high tech world. Therefore, this paper sticks two actual topics together - linear motor and genetic algorithm. It is generally known that linear motors are maintenance free and they are able to evolve high velocity and precision which is why we made closer look on this topic. To make the linear motor more precise, genetic algorithm was applied. The GA role was to design optimal parameters for PID regulator, lead compensator and Luenberger observer to ensure the most precise positioning. Eventually, some experiments were done to demonstrate the impact of Luenberger observer and it will be also shown responses of position, velocity, force, and position error, which were gained from the experiment using GA.

Keywords: linear motor, genetic algorithm, master-slave control, Luenberger observer.

1 Introduction

Linear motors with permanent magnets (LMPM) found their right place within all kinds of motors covering the applications where the standard rotary motors have no chance to succeed. Thanks belong to their matchless features like better positioning and precision, no maintenance and no bearing ware. Therefore these motors can be found in various sectors, whether in electro-technical or electronic production – drives used in the elevators, conveyors, pumps, compressors, paper machines, robots, etc. Above all it has to be mentioned that linear motors purchase the popularity mainly by their implementation in the velocity trains such as Maglev or Trans-rapid.

This paper will be focused on position servo-drive control design of LMPM comparing two methods: Genetic Algorithm (GA) and Pole Placement Method with 4D Master slave control and Luenberger observer. It will be unveiled the positive effect of Luenberger observer in one of the experiments, as well.

GA is one of the most famous and the most used representatives of evolutionary computing techniques with wide range of application [1]. Control performance possesses highly important function in servo-drives that is why we took advantage of GA to improve the overall performance. GA is able to design the parameters for PID controller, lead compensator and Luenberger observer at the same time, 9 parameters,

V. Snasel et al. (Eds.): SOCO Models in Industrial & Environmental Appl., AISC 188, pp. 73–82.
springerlink.com

which is the main reason GA was applied. There is no problem to design these parameters independently with Pole-placement method, but it is naturally time consuming and mathematical techniques have to be well-known.

The main idea of designing controller parameters using GA has been publicly adopted in the 1990's, but remains popular in the present as well, which is proven by number of papers in relevant journals [6][7]. Interesting is also attempt of PI position controller design of SMPM drive by Khater and others [5].

The goals of creating artificial intelligence and artificial life can be traced back to the very beginnings of the computer age. The earliest computer scientists - Alan Turing, John von Neumann, Norbert Wiener, and others were motivated in large part by visions of imbuing computer programs with intelligence, with the life-like ability to self-replicate and with the adaptive capability to learn and to control their environments. These early pioneers of computer science were as much interested in biology and psychology as in electronics, and they looked to natural systems as guiding metaphors for how to achieve their visions. It should be no surprise, then, that from the earliest days computers were applied not only to calculating missile trajectories and deciphering military codes, but also to modeling the brain, mimicking human learning and simulating biological evolution. These biologically motivated computing activities have waxed and waned over the years, but since the early 1980s they have all undergone resurgence in the computation research community. The first has grown into the field of neural networks, the second into machine learning, and the third into what is now called "evolutionary computation", of which genetic algorithms are the most prominent example [3].

2   LMPM Position Control

2.1   Position Servo-Drive, Implementation Block Scheme

Position servo-drive can be performed by various algorithms (PID, PIV, P+PI…). PID algorithm with lead compensator is applied, referring to the article [2]. However, the entire block diagram consists of Luenberger observer and 4D master generator in addition.

The entire position servo-drive structure may be seen in the Fig.1.

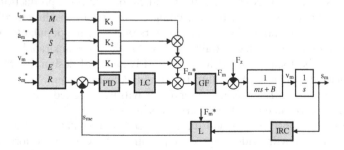

Fig. 1. Entire diagram of position servo-drive (PID-proportional–integral–derivative controller, LC-Lead Compensator, GF-Force generator, L-Luenberger observer, IRC-Incremental sensor)

2.2 Force Generator

Force generator GF is one of the most important blocks of linear servo-drive control structure and works on a principal of vector frequency-current control synchronous motor with PM (Fig.2). It contains blocks of Park's transformation, compensation block, IRC sensor, two current controllers (CC_d, CC_q) and block LMPM – particular servo-drive realized by following equations

$$u_d = R_s i_d + \frac{d\psi_d}{dt} - \omega_s \psi_q$$

$$u_q = R_s i_q + \frac{d\psi_q}{dt} + \omega_s \psi_d$$

$$\psi_f = L_{md} i_f \qquad (1)$$

$$\psi_d = L_d i_d + \psi_f$$

$$\psi_q = L_q i_q$$

$$F_m - F_z = m\frac{dv_m}{dt}$$

$$\omega_s = K_x v_m ; \qquad \upsilon = K_x s_m$$

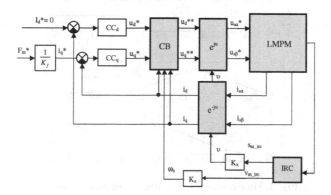

Fig. 2. Force generator LMPM structure

Equation (2) represents a relation between rotary and linear parameters issued from the physical interpretation.

$$v_m = 2\tau_p f_s \qquad (2)$$

τ_p – Pole spacing [m]
f_s – Power supply frequency [Hz]

Then generally holds the equation (3).

$$K_x = \frac{\omega_s}{v_m} = \frac{\pi}{\tau_p} \qquad (3)$$

2.3 Master-Slave Control

Master slave control can be assign to the status control or model control [4]. Its significant advantage is the inutility of knowing the exact mathematical model of controlled system. The quality model of regulating system is highly sufficient providing that you are familiar with the scale of the main parameters.

2.3.1 Master-Slave Generator

Master serves as a generator of control state variables and surprisingly the control vector can be greater than number of measured variables. Its task is to generate desired waveforms of state variables – control vectors which shape can be rectangular, trapezoidal or sinusoidal, either 3-dimensional or 4-dimensional. In this paper is used 4D master slave generator and it generates state variables of the position, velocity, acceleration and jerk (Fig.3).

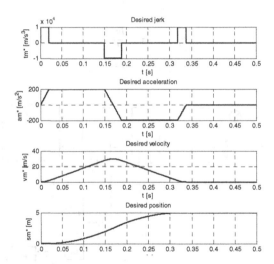

Fig. 3. Time responses of Master slave output values

2.3.2 Precorrection Constants

Among indispensable parts in Master-slave control belong precorrection constants. Their task is to enhance position accuracy and consequently lower the position error. Calculating precorrection coefficients (K_1, K_2, and K_3) starts from the condition for feed forward control and force generator dynamics is considered.

$$G_x(s) = \frac{1}{G_2(s)} \tag{4}$$

$$G_2(s) = \frac{1}{T_{gm}s+1} \frac{1}{ms+B} \frac{1}{s} = \frac{1}{T_{gm}ms^3 + \left(T_{gm}B+m\right)s^2 + Bs}$$

$$G_x(s) = \frac{1}{G_2(s)} = T_{gm}ms^3 + \left(T_{gm}B+m\right) + Bs$$

$$K_1 = B; \quad K_2 = T_{gm}B+m; \quad K_3 = T_{gm}m$$

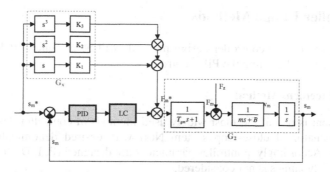

Fig. 4. Force Position servo-drive block diagram with PID structure and marked precorrection Master-3D (PID-proportional–integral–derivative controller, LC-Lead Compensator, GF-Force generator,)

2.4 Luenberger Observer

Observers are algorithms that combine sensed signals with other knowledge of the control system to produce observed signals [2]. They can enhance the accuracy and reduce the sensor-generated noise. Consequently, among various observers, Luenberger observer was chosen. Basically it is the observer of velocity and acceleration. In general, it may contain different algorithm structures. In this paper is chosen PID algorithm for controlling the third order system, though.

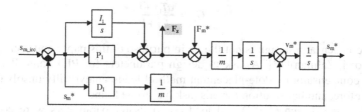

Fig. 5. Luenberger observer block diagram

Pole-placement method is applied. It compares denominator of close-loop system $N(s)$ with desired denominator $N_D(s)$ by equal power.

$$N(s) = s^3 + \left(\frac{D_1}{\tilde{m}}\right)s^2 + \frac{P_1}{\tilde{m}}s + \frac{I_1}{\tilde{m}}$$

$$N_D(s) = \left(s^2 + 2\xi_1\omega_{01}s + \omega_{01}^2\right)\left(s + k_1\omega_{01}\right)$$

$$N(s) = N_D(s)$$

$$P_1 = \tilde{m}\omega_{01}^2\left(2\xi_1 k_1 + 1\right)$$

$$I_1 = \tilde{m}k_1\omega_{01}^3$$

$$D_1 = \omega_{01}\left(2\xi_1 + k_1\right)\tilde{m}$$

(5)

Parameters setup variables ξ_1, k_1 and ω_{01} are further explained in the Table 3.

3 Controller Design Methods

The comparison of two controller design methods will be presented. Pole Placement method is used for designing the PID controller.

3.1 Pole Placement Method

Pole placement is one of the most widely used methods of controller design. It compares denominator of close-loop system N(s) with desired denominator $N_D(s)$ by equal power. Accordingly, controller parameters are designed (P, I, D), however force generator GF dynamics is not considered.

$$N(s) = N_D(s)$$

$$N(s) = s^3 + \left(\frac{D+B}{m}\right)s^2 + \frac{P}{m}s + \frac{I}{m} \tag{6}$$

$$N_D(s) = \left(s^2 + 2\xi\,\omega_0 s + \omega_0^2\right)(s + k\omega)$$

$$P = m\omega_0^2(2\xi k + 1)$$

$$I = mk\omega_0^3$$

$$D = \omega_0(2\xi + k)m - B$$

Parameters setup variables ξ, k and ω_0 are further explained in the Table 3.

Lead compensator coefficients are design by well-known method using relation (lead - lag).

$$G_{LC} = \frac{aT_1 s + 1}{T_1 s + 1} \tag{7}$$

The lead compensator design is not the main purpose of this paper and you can find it in (Radičová, Žalman)[12]. A task to design parameters for PID controller together with lead compensator by Pole-placement method led to analytically unsolvable problem. Therefore, another solution for this task had to be found.

However, the design of PID and lead compensator parameters were continuous, realization was discreet.

3.2 Genetic Algorithm

GA, one of the mostly used representatives of evolutionary computing, is based on finding optimal solution (optimal structure and controller parameters) for the given problem. Accordingly, the base rule for success is the precise fitness function design. Hence the fitness function represents minimization of position error using the following

$$Fitness = \sum|e| + a\sum|dy| \tag{8}$$

Genetic algorithm toolbox was used as a solving tool [10]. It is not the standard part of MATLAB distribution. The Toolbox can be used for solving of real-coded search and optimization problems. Toolbox functions minimize the objective function and maximizing problems can be solved as complementary tasks, as well.

The process of searching is adjusted very sophistically. First of all, a random population is generated with a predefined number of chromosomes in one population within prescribed limits for controller, lead compensator and Luenberger observer parameters (for relevant values see Table 1). To achieve preferable parameters, fitness function, which minimized position error from Luenberger observer, was applied.

$$Fitness_{Luenberger} = \sum |e| \qquad (9)$$

Then two best strings according the first fitness function were selected to the next generation. Bigger number of strings was selected to the next generation by tournament. Then number of crossovers and mutations are applied to the population to achieve bigger chances to reach the global optimum. This progress is the same for the fitness function used for Luenberger observer. Finally, the best parameters are chosen.

Table 1. Table of parameters extracted from GA

Number of generations	100
Number of chromosomes in one population	30
Number of genes in a string	3
Parameter "a" weight	0.7

This algorithm, using the method mentioned above, is able to design 9 parameters at once. Therefore, GA belongs to the very effective algorithms which employ easiness compared to the incredible mathematic severity. Eq. 10 represents transfer function of parameters obtained from GA for the lead compensator, the PID controller and the representation of Luenberger observer (Fig.5). Table 2 shows concrete parameters designed with GA and Pole Placement method according to Fig.1.

$$G_{LC}(z) = \frac{a_1 z + a_2}{z + b_2}$$

$$G_{PID}(z) = P + I \frac{Tz}{z-1} + D \frac{z-1}{Tz} \qquad (10)$$

$$G_{LO}(z) = P_1 + I_1 \frac{Tz}{z-1} + D_1 \frac{z-1}{Tz}$$

Table 2. Table of parameters extracted from GA

	P	I	D	a_1	a_2	b_2	P_1	I_1	D_1
GA	6193	61.55	138.4	8	-0.987	-0.293	3917	251080	129.25
Pole Placement	4737	99220	75.39	20	-19.95	-0.9454	4737	992200	301.59

4 Simulation Results

At the beginning, it has to be mentioned that following experiments were performed on the simulation model in Matlab Simulink environment using Luenberger observer and precorrection constants according to the Fig.1.

First experiment (Fig.6, Fig.7) compares the behavior of force and position error with/without Luenberger observer. Table 3 shows used simulation parameters.

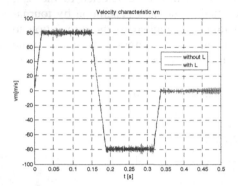

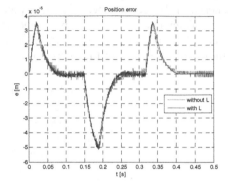

Fig. 6. The time response of force comparing the response with/without Luenberger observer

Fig. 7. The time responses of position error comparing the response with/without Luenberger observer

Table 3. Table of acronyms

Acronym	Meaning	Value
T	Sampling period	0.2 ms
T_{gm}	Time constant of GF	0.5 ms
Parameters for PID controller		
ξ	Damping index	1
k	Shift pole index	1
ω_0	Bandwidth	$2\pi f_0$
f_0	Frequency	10 Hz
Parameters for Luenberger observer		
ξ_l	Damping index	1
k_l	Shift pole index	10
ω_{0l}	Bandwidth	$2\pi f_0$
f_{0l}	Frequency	10 Hz
Parameters for precorrection		
K_1	Precorrection constant	B = 0.01 kg.s^{-1}
K_2	Precorrection constant	m = 0.4 kg
Parameters for IRC sensor		
N	Resolution	2 μm

Second experiment represents linear servo-drive behavior using parameters gained from GA and Pole Placement method with the parameters listed in the Table 3 and Table 4 (Fig.8, 9, 10, 11).

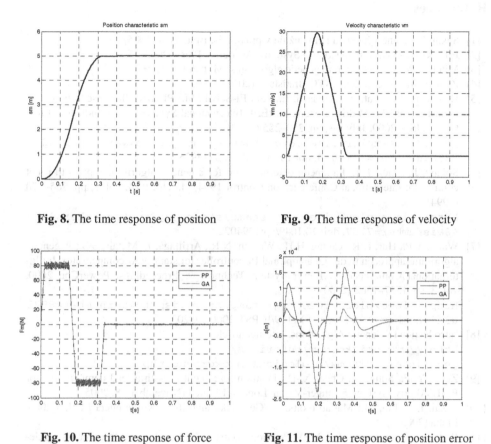

Fig. 8. The time response of position   **Fig. 9.** The time response of velocity

Fig. 10. The time response of force   **Fig. 11.** The time response of position error

5 Conclusion

It has been performed two experiments which confirm that genetic algorithm toolbox in connection with MATLAB is a very powerful tool for optimization and search problems. GA was able to design nine optimal parameters for linear servo drive that is the significant contribution to this area. In addition, as can be seen in the Fig. 11, the positioning accuracy and dynamics of the system is very high using GA. Eventually, it has to be mentioned that using Luenberger observer and precorrection constant led to the more precise positioning of LMPM.

Acknowledgment. This work was supported by the Slovak Research and Development Agency under the contract No. VMSP-II-0015-09.

References

[1] Sekaj: Evolučné výpočty a ich využitie v praxi. Iris, Bratislava (2005)
[2] Ellis, G.H.: Observers in Control Systems. Academic Press (2002)
[3] Mitchell, M.: Introduction to genetic algorithms. MIT Press (1998)
[4] Žalman, M.: Akčné členy, STU Bratislava (2003)
[5] Khater, F., Shaltout, A., Hendawi, E., Abu El-Sebah, M.: PI controller based on genetic algorithm for PMSM drive system. In: IEEE International Symposium on Industrial Electronics, ISIE 2009, July 5-8, pp. 250–255 (2009),
    http://ieeexplore.ieee.org/stamp/stamp.jsp?tp=&arnumber=5217
    925&isnumber=5213059, doi:10.1109/ISIE, 5217925
[6] Solano, J., Jones, D.I.: Parameter determination for a genetic algorithm applied to robot control. In: International Conference on Control 1994, March 21-24, vol. 1, pp. 765–770 (1994),
    http://ieeexplore.ieee.org/stamp/stamp.jsp?tp=&arnumber=3270
    48&isnumber=7757, doi: 10.1049/cp:19940229
[7] Wang, Y.P., Hur, D.R., Chung, H.H., Watson, N.R., Arrillaga, J., Matair, S.S.: A genetic algorithms approach to design an optimal PI controller for static VAr compensator. In: International Conference on Power System Technology, Proceedings, PowerCon 2000, vol. 3, pp. 1557–1562 (2000),
    http://ieeexplore.ieee.org/stamp/stamp.jsp?tp=&arnumber=8982
    03&isnumber=19429, doi:10.1109/ICPST.2000.898203
[8] Introduction to Genetic Algorithms, Internet article,
    http://www.obitko.com/tutorials/
    genetic-algorithms/biological-background.php
[9] Zhou, Y.-F., Song, B., Chen, X.-D.: Position/force control with a lead compensator for PMLSM drive system. Springer-Verlag London Limited (November 18, 2005)
[10] Sekaj, I., Foltin, M.: Matlab toolbox – Genetické algoritmy. Konferencia Matlab 2003. Praha (2003)
[11] Radičová, T., Žalman, M.: LMPM Position Control with Luenberger Observer Using Genetic Algorithms. In: ELEKTRO 2012, Rajecké Teplice (2012)
[12] Radičová, T., Žalman, M.: Master-slave position servo-drive design of aircore linear motor with permanent magnets. AT&P Journal Plus (January 2010)

An Approach of Genetic Algorithm to Model Supplier Assessment in Inbound Logistics

Dragan Simić[1], Vasa Svirčević[2], and Svetlana Simić[3]

[1] University of Novi Sad, Faculty of Technical Sciences
Trg Dositeja Dobradovića 6, 21000 Novi Sad, Serbia
dsimic@eunet.rs
[2] Lames Ltd., Jarački put bb., 22000 Sremska Mitrovica, Serbia
vasasvircevic@lames.rs
[3] University of Novi Sad, Faculty of Medicine Hajduk Veljkova 1-9,
21000 Novi Sad, Serbia
drdragansimic@gmail.com

Abstract. In times of economic crises and increasing market competition, business stability, quality, safety and supply chain flexibility and cost optimization play an increasing role in companies that strive to stay and survive in the market. A wise choice of suppliers, in such circumstances, becomes increasingly important prerequisite for the success of any company. This paper presents a novel model for supplier assessment. The proposed model considers the performance of suppliers classified into several different groups of questions related to all the relevant issues: finance, logistics, competitiveness, quality and level of supplier services. This model can be applied in a variety of companies and for different supplier categories based on their purchase categories and therefore achieve a realistic assessment.

1 Introduction

In times of economic crises and increasing market competition, business stability, quality, safety and supply chain flexibility and cost optimization play an increasing role in companies that strive to stay and survive in the market. Dynamic market changes demand selection of business partners who are logistically and otherwise able to follow changes in company requirements. Purchase activities in large part support a firm's inbound logistics and are vital to value creation [8]. Managing the purchasing task in the supply chain has been a challenge for many companies in the last decade.

In such circumstances a wise choice of suppliers becomes increasingly important prerequisite for the success of any company. A firm's sourcing strategy is characterized by three key decisions [2]: (a) criteria for establishing a supplier base; (b) criteria for the selection of the suppliers (a subset of the base) who will receive an order from the firm and (c) the selected quantity of goods to order from each supplier.

On the other hand, purchase analysis, in general, could be divided in the following way: supplier analysis, consumption analysis and savings, cost analysis

V. Snasel et al. (Eds.): SOCO Models in Industrial & Environmental Appl., AISC 188, pp. 83–92.
springerlink.com

and planning and analysis of supply of raw materials for production. First of all the criteria for establishing a supplier base is defined and afterwards, supplier usage by supplier analysis is assessed and then supplier base with all suppliers, where a small number of suppliers is evaluated as appropriate, is created.

Soft computing (SC) methods have been successfully applied to solve non-linear problems in business, engineering, medicine and sciences. These methods, which indicate a number of methodologies used to find approximate solutions for real-word problems which contain kinds of inaccuracies and uncertainties, can be alternative to different kind of statistical methods. The underlying paradigms of soft computing are neural computing, fuzzy logic computing and genetic computing [16].

This paper presents a novel model for supplier assessment based on genetic algorithm that is used in a multinational company with operations in more than a dozen countries in the world as well as in Serbia. Supplier assessment is a continuous process, and its results must be conveyed to the suppliers so they could create a corrective action plan to develop and improve their weaknesses. The main goal of assessment processes and supplier selection is to ensure a successful and long-term cooperation between all parties in a supply chain.

The rest of the paper is organized as follows. The following Section overviews similar available implementation while Section 3 elaborates a short part of a previous research. The Section 4 describes the genetic algorithm performance value constrained model. Section 5 presents experimental results while Section 6 concludes the paper and offers notes on future work.

2   Literature Review

It is proven that the application of SC has two main advantages. First, it made solving nonlinear problems in which mathematical models ate not available, poss-ible. Secondly, it introduced the human knowledge such as cognition, recog-nition, understanding, learning, and other skills into fields of computing. This resulted in the possibility of constructing intelligent systems such as autonomous self-tuning systems, and automated design systems.

2.1   Literature Review of Soft Computing

Soft computing models are capable of identifying patterns that can be charac-terized as 'typical day' in terms of its meteorological conditions. The application of a series of statistical and soft computing models to identify what may be called 'Typical Days' in terms of previously selected meteorological variables is discussed in [3].

The multidisciplinary study [9] presents a novel four-step soft computing knowledge identification model to perform thermal insulation failure detection. These are all commonly used pre-processing tools in soft computing that under-take pattern recognition based on dimensionality reduction issue. The aim is gene-rating a model which will estimate how indoor temperature in a build-ing of a specific configuration behaves. Once the model has been obtained, it is used as a reference model and soft computing model should be developed to automatically detect the insulation failure detection.

The design of multi-objective robust proportional-integral-derivative (PID) controller using various multi-objective optimization algorithms. In [17], two-lbest based multi-objective, a soft computing technique and particle swarm optimizer algorithm are applied for the design of robust PID controller by minimizing integral squared error.

Several computation intelligence paradigms have been established, and recently hybridization of computational intelligence techniques are becoming popular due to their capabilities in handling many real-word complex problems, involving imprecision, uncertainty, vagueness, and high dimensionality [1]. Hybrid methods are based on coupling of the procedures from different optimization methods to improve accuracy and effectiveness of basis optimization methods.

For example, in science, a novel hybrid method based on learning algorithm of fuzzy neural network for the solution of differential equation with fuzzy initial value is presented in [7]. In [11] an application of hybrid particle swarm optimizer is presented and it is applied to identify complex impedances of room walls based on the mechanism discovered in the nature during observations of the social behavior of animals. And finally, in business, the usage of hybrid genetic algorithms and support vector regression model for stock selection model is showed in [4]. A review, approach of fuzzy models and applications in logistics is in detail presented in [10].

2.2 Literature Review of Supply Chain

Some mathematical programming approaches have been used for supplier selection in the past. A multi-phase mathematical programming approach for effective supply chain design was presented in 2002 [12]. More specifically, a combination of multi-criteria efficiency models, based on game theory concepts, and linear and integer programming methods was developed and applied. A max-min productivity based approach was proposed in 2003 [13]. It derives variability measures of vendor performance, which are then utilized in a nonparametric statistical technique in identifying vendor groups for effective selection. Fuzzy goal programming approach was applied in 2004 to solve the vendor selection problem with multiple objectives [5].

According to recent research work conducted in 2009, the quantitative decision methods for solving the supplier selection problem can be classified into three categories: (1) multi-attribute decision-making, (2) mathematical programming models and (3) intelligent approaches [14]. Furthermore, in the latest literature survey from 2010, it can be seen that the mathematical programming models are grouped into the following five models: (1) linear programming, (2) integer linear programming, (3) integer non-linear programming, (4) goal programming and (5) multi-objective programming [6].

As was mentioned in previous section, criteria for establishing a supplier base and criteria for selecting suppliers as subset of the supplier base are discussed in some important surveys. In [15] 74 articles discussing supplier selection criteria were reviewed. It was also concluded that supplier selection is a multi-criteria problem and the priority of criteria depends on each purchasing situation.

Hundreds of criteria were proposed, and the most often criterion is quality, followed by delivery, price/cost, manufacturing capability, service, management, technology, research and development, finance, flexibility, reputation, relationship, risk, and safety and environment. Various quality related attributes have been found, such as: "compliance with quality", "continuous improvement program", "six sigma program or total quality management", "corrective and preventive action system", "ISO quality system installed". As mentioned before, delivery is second most popular criterion, as well as: appropriateness of the delivery date", delivery and location", delivery conditions"," delivery lead time"," delivery mistakes". The third most popular criterion is price/cost and related attributes including: "competitiveness of cost", "cost reduction capability", "cost reduction performance", "logistics cost" , "total shipment costs".

Based on the above mentioned findings, it was revealed that price/cost is not the most widely adopted criterion. The traditional single criterion approach based on lowest cost bidding is no longer supportive and robust enough in contemporary supply management.

3   The Previous Research

Supplier assessment and selection mapping as an essential component of supply chain management is usually a multi-criteria decision problem which, in actual business contexts, may have to be solved in the absence of precise information. Suppliers are evaluated continuously, at least once a year, and so the supplier base is a time variable category.

Procurement categories are: A) Raw Material; B) Industrial Products & Consumables; C) Industrial Services; D) Utilities; E) Transport Services; F) General Supplies & Services; G) Plants and Equipments; Z) Out of Scope. All of these segments are divided into sub-categories: A.1) Electrical components; A.2) Plastic components; A.3) Rubber parts; A.4) Metal castings; A.5) Small components; A.6) Other components. The parameters of the defined target levels of supplier performance for every segment of a purchase segments are defined separately.

Suppliers are grouped into categories, and every supplier is assessed in its particular category, so it can happen that one supplier gets represented in several categories. It is important to mention that a supplier can be well placed in one category, and not so well placed in another.

Supplier performances represent groups of questions that are to be assessed in the following business activities: 1) Finance; 2) Logistics; 3) Competitiveness; 4) Quality; 5) Service/Commercialism. All of performances which are assessed are ranked in three levels (1, 2 or 3) respectively. On the other hand, as has been mentioned before, target levels for supplier performances for every purchase segment are defined separately. It is necessary because the importance of one performance is not same for suppliers in different purchase segments. In other

Table 1. Performance groups and individual performances for supplier assessment

Financial	Debt	Ratio Debt/Equity
	Dependency	% Turnover with Lames/Supplier Global Turnover
	Profitability	Net result/Turnover
Logistics	Delivery	Respect of delivery time
	Distribution	Ability to manage the distribution chain
	Capacity	Capacity to supply required prod./service range
	Commitment/Stock	Commitment/Availability on supplier stock
	Coverage	Supplier's geographical coverage
	B to B Solutions	Ability to implement EDI or B to B solution
Competitiveness	Technical offer	Management of the technical part of its offer
	Technical know-how	Technical/Business know-how
	Proposition force	Proposition force (process improvement, ability to innovate)
	Price transparency	Transparency : on prices, organization
	Cost level/market	Cost Level compared to the market
	Cost level/competitors	Price level/competitors
Quality	Quality Certification	ISO certification, control process or various certification
	Quality compliance	Compliance with specifications
	Corrective actions	Ability to implement corrective actions
	Sustainable development	Safety & Environmental information, child labor
Service	K.A.M. approach	Is the Key Account Manager approach working adequately?
	Reactivity	Reactivity towards a demand
Commercialism	R&D	R&D department existence
	Export turnover	% of export turnover/Global turnover
	Reporting	Reporting

words, there is a difference between the importance of criteria for suppliers who deal with raw materials and those who supply office supplies or provide the service of cleaning office space. Table 1 shows all performance groups and all individual performances and the method of evaluation.

For every question concerning supplier performance a target level of performance is set on a scale form 1 to 3 so that the sum of the entire performance set should be 50. This should be done in order to apply the following formula for calculating the percentage of supplier performance on a scale from 0% to 100%:

$$\% \text{ supplier performance} = \Sigma(\text{target level} \times \text{current level})2/3. \qquad (1)$$

It should be noted that one-dimensional supplier assessment models consider only supplier performance. It is possible to extend the previous performance supplier system with some other relevant issues which represent a new observed supplier performance.

4   The GA Performance Value Constrained Model

In graphical representation of the results, as it can be seen, the questions are grouped by 3, 4, and 6, in this way graphical representation of the results corresponds to a triangle, or tetragon or pentagon or hexagon (Fig. 1).

As it can be seen individual performance can be lower, equal to, or higher than defined target values. If individual performance shows value lower than the target value, it means that the supplier has not satisfied company requirements yet. If individual performance shows value equal to the target value then the performance of the supplier is as expected.

If individual performance shows value higher than the target value, it means that assessed supplier develops in the direction which is not requisite to the company. This led to constraints on performances that exceed expected target values when assessing a supplier.

The proposed genetic algorithm performance value constrain model is based on performance value constraints during supplier assessment for one performance on maximum target values. This limits the possibility of a supplier performance which is not requisite to heighten supplier assessment value. Further on, the proposed model calculates the area enclosed by the performance values which are lower or equal to target values on one side, and on the other, by target values of those supplier performances which are higher than the target values. One such example, supplier performance and the enclosed area by the proposed methodology is shown in Fig 2. In accordance with what was previously mentioned, light grey color of enclosed area shows that the criteria was satisfied, grey enclosed area shows where supplier performances are higher than target performance, and black enclosed area where target performance values are grater than supplier performance. For this experimental research all of three areas are important for discussions. The Fig 2 shows only six target - supplier individual performances but in experiment all of 24 performances were used.

Considering that it is necessary to determine maximum bordered area general genetic algorithm and partially crossover and mutation GA effects are used. The proposed GA performance value constrain model maximizes the observed enclosed area. Also, the assessment of the minimum value of the observed area can be done using the GA performance value constrain model in the same way. In this way the estimation of the maximum and minimum affected area is completed thus eliminating the possibility of subjectivity in positioning performances or groups of performances. Our research can be classified as empirical multi-attribute decision making model and scoring models are typically used to evaluate suppliers for inclusion in the supplier base.

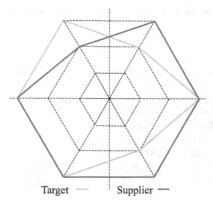

Fig. 1. Hexagonal graphical representation

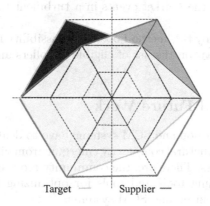

Fig. 2. Supplier individual performances, target level performances and the enclosed areas

5 Experimental Results

Experimental results show minimum and maximum supplier assessment and supplier performance values. Considering that it is necessary to determine maximum and minimum bordered area, general genetic algorithm and partially crossover and mutation GA effects are used. The proposed GA performance value constrain model maximizes the observed enclosed area.

Table 2 shows only some of the experimental results but nevertheless some vary important conclusions can be made. Experimental results obtained by GA constrain value model are shown and compared with regard to the empirical method used in our previous research. Presented comparison shows that the values obtained by supplier assessment GA model are somewhat higher than those obtained by the established empirical method. This fact implies that GA

Table 2. Experimental results: Empirical model, Surface without constrains, GA values constrain model

Company	Year	Empirical model	Surface without constrain	GA value constrain model	Distinction GA & Empirical
AB SOFT Ltd.	2006	70.00%	97.20%	75.70%	7.53%
Adam Šped Sys.	2010	74.67%	108.33%	88.89%	16.00%
Lames Ltd.	2010	82.00%	130.84%	87.85%	6.66%
Staklo Enterijer	2011	82.00%	130.84%	75.70%	-8.32%
Belem	2011	84.00%	151.46%	97.09%	13.48%
Elab	2012	78.67%	121.36%	72.82%	-8.03%
Ninagro	2012	77.33%	117.82%	74.26%	-4.13%
Average Error					3.21%

model is slightly less strict model when compared to the previous empirical model which probably reflects the Market events in a turbulent business environment in a much better way.

The results obtained by this method open the possibility for a greater number of the suppliers to appear on the list of eligible suppliers and get a chance as a potential bidder.

6   Conclusion and Future Work

Today's logistics distribution and SCM systems have to deal with ever-changing markets and intrinsic structural complexity emerging from virtually infinite number of interacting entities. Therefore, the community requires effective soft computing methods, modeling tools and tools for optimizing large scale complex inbound logistics distribution and SCM systems.

The GA performance value constrained model is presented in this paper. The experimental results of this method are obtained from the realistic data seta of the supplier performances of a multinational company operating in Serbia. The experimental results were later on compared to the results obtained by the empirical method used in our previous research. Presented comparison shows that the values obtained by supplier assessment GA model are somewhat higher than those obtained by the established empirical method. This fact implies that GA model is slightly less strict model when compared to the previous empirical model which probably reflects the Market events in a turbulent business environment in a much better way.

The results obtained by this method open the possibility for a greater number of the suppliers to appear on the list of eligible suppliers and get a chance as a potential bidder. Future development can be directed towards adding weight to every performance group; business constraints can be considered as well as technical, legal, geographical and commercial nature of every performance.

Supplier assessment is a continuous process, and its results must be conveyed to the suppliers so they could create a corrective action plan to develop and improve their weaknesses. This process is important for both sides; the buying company can secure and improve its supply chain, and the supplying company can develop and advance its competitiveness. The main goal of assessment processes and supplier selection is to ensure a successful and long-term cooperation between all parties in a supply chain.

Acknowledgment. The authors acknowledge the support for research project TR 36030, funded by the Ministry of Science and Technological Development of Serbia.

References

1. Abraham, A.: Hybrid Soft Computing and Applications. International Journal of Computational Intelligence and Applications 8(1), 1–2 (2009)
2. Burke, G.J., Carrillo, J.E., Vakharia, A.J.: Single versus multiple supplier sourcing strategies. European Journal of Operational Research 182(1), 95–112 (2007)
3. Corchado, E., Arroyo, A., Tricio, V.: Soft computing models to identify typical meteorological days. Logic Journal of the IGPL 19(2), 373–383 (2011)
4. Huang, C.F.: A hybrid stock selection model using genetic algorithms and support vector regression. Applied Soft Computing 12(2), 807–818 (2012)
5. Kumar, M., Vrat, P., Shankar, R.: A fuzzy goal programming approach for vendor selection problem in a supply chain. Computers and Industrial Engineering 46(1), 69–85 (2004)
6. Ho, W., Xu, X., Dey, P.K.: Multi-criteria decision making approaches for supplier evaluation and selection: a literature review. European Journal of Operational Research 202(1), 16–24 (2010)
7. Mosleh, M., Otadi, M.: Simulation and evaluation of fuzzy differential equations by fuzzy neural network. Applied Soft Computing (2012), doi:10.1016/j.asoc.2012.03.041
8. Porter, M.: Competitive Advantage, pp. 38–40. The Free Press, New York (1985)
9. Sedano, J., Curiel, L., Corchado, E., Cal, E., Villar, J.R.: A soft computing method for detecting lifetime building thermal insulation failures. Integrated Computer-Aided Engineering 17(2), 103–115 (2012)
10. Simić, D., Simić, S.: A Review: Approach of Fuzzy Models Applications in Logistics. In: Burduk, R., Kurzyński, M., Woźniak, M., Żołnierek, A. (eds.) CORES 2011. AISC, vol. 95, pp. 717–726. Springer, Heidelberg (2011)
11. Szczepanik, M., Poteralski, A., Ptaszny, J., Burczyński, T.: Hybrid Particle Swarm Optimizer and Its Application in Identification of Room Acoustic Properties. In: Rutkowski, L., Korytkowski, M., Scherer, R., Tadeusiewicz, R., Zadeh, L.A., Zurada, J.M. (eds.) EC 2012 and SIDE 2012. LNCS (LNAI), vol. 7269, pp. 386–394. Springer, Heidelberg (2012)
12. Talluri, S., Baker, R.C.: A multi-phase mathematical programming approach for effective supply chain design. European Journal of Operational Research 141(3), 544–558 (2002)
13. Talluri, S., Narasimhan, R.: Vendor evaluation with performance variability: A max-min approach. European Journal of Operational Research 146(3), 543–552 (2003)

14. Wang, T.Y., Yang, Y.H.: A fuzzy model for supplier selection in quantity discount environments. Expert Systems with Applications 36(10), 12179–12187 (2009)
15. Weber, C.A., Current, J.R., Benton, W.C.: Vendor selection criteria and methods. European Journal of Operational Research 50(1), 2–18 (1991)
16. Zadeh, L.: Soft computing and fuzzy logic. Computer Journal of IEEE Software 11(6), 48–56 (1994)
17. Zhao, S.Z., Iruthayarajan, M.W., Baskar, S., Suganthan, P.N.: Multi-objective robust PID controller tuning using two lbests multi-objective particle swarm optimization. Information Sciences 181(16), 3323–3335 (2011)

Optimization of the Batch Reactor by Means of Chaos Driven Differential Evolution

Roman Senkerik[1], Donald Davendra[2], Ivan Zelinka[2],
Zuzana Oplatkova[1], and Michal Pluhacek[1]

[1] Tomas Bata University in Zlin,
Faculty of Applied Informatics,
T.G. Masaryka 5555, 760 01 Zlin, Czech Republic
{senkerik,oplatkova,pluhacek}@fai.utb.cz
[2] VŠB-Technical University of Ostrava,
Faculty of Electrical Engineering and Computer Science,
Department of Computer Science, 17. listopadu 15,
708 33 Ostrava-Poruba, Czech Republic
{donald.davendra,ivan.zelinka}@vsb.cz

Abstract. In this paper, Differential Evolution (DE) is used in the task of optimization of a batch reactor. The novality of the approach is that a discrete chaotic dissipative standard map is used as the chaotic number generator to drive the mutation and crossover process in DE. The results obtained are compared with original reactor geometry and process parameters adjustment.

1 Introduction

These days the methods based on soft computing such as neural networks, evolutionary algorithms, fuzzy logic, and genetic programming are known as powerful tool for almost any difficult and complex optimization problem. They are often applied in engineering design problems [1] or for the purpose of optimization of processing plants or technological devices [2] - [5]. Chemical industry produces a whole range of products through chemical reactions. Generally, it may be stated that key technological points are chemical reactors. Designing optimal reactor parameters including control constitutes is one of the most complex tasks in process engineering.

This paper is aimed at investigating the chaos driven DE. Although a number of DE variants have been recently developed, the focus of this paper is the embedding of chaotic systems in the form of chaos number generator for DE and its application to optimization of batch reactor.

This research is an extension and continuation of the previous successful initial application based experiment with chaos driven DE [6].

The chaotic systems of interest are discrete dissipative systems. The Dissipative standard map was selected as the chaos number generator for DE.

Firstly, batch processes are explained. The next sections are focused on the description of the batch reactor, differential evolution, used chaotic systems and problem design. Results and conclusion follow afterwards.

V. Snasel et al. (Eds.): SOCO Models in Industrial & Environmental Appl., AISC 188, pp. 93–102.

2 Characteristics of the Batch Processes

The optimization of batch processes has attracted attention in recent years [7], [8]. Batch and semi-batch processes are of considerable importance in the fine chemicals industry. A wide variety of special chemicals, pharmaceutical products, and certain types of polymers are manufactured in batch operations. Batch processes are typically used when the production volumes are low, when isolation is required for reasons of sterility or safety, and when the materials involved are difficult to handle. In batch operations, all the reactants are charged in a tank initially and processed according to a pre-determined course of action during which no material is added or removed. From a process systems point of view, the key feature that differentiates continuous processes from batch and semi-batch processes is that continuous processes have a steady state, whereas batch and semi-batch processes do not [9], [10].

Lot of modern control methods for chemical reactors were developed embracing the approaches such as iterative learning model predictive control [11], iterative learning dual-mode control [12] or adaptive exact linearization by means of sigma-point kalman filter [13]. Also the fuzzy approach is relatively often used [14]. Finally the methods of artificial intelligence are very frequently discussed and used. Many papers present successful utilization of either neural networks [15] - [17] or genetic (evolutionary) algorithms [18] - [20].

This paper presents the static optimization of the batch reactor by means of chaos driven differential evolution.

3 Description of the Reactor

This work uses a mathematical model of the reactor shown in Fig. 1. The reactor has two physical inputs (one for chemical substances and one for cooling medium) and one output (cooling medium).

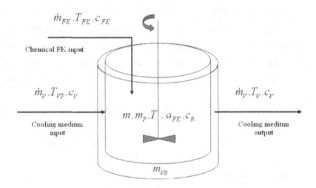

Fig. 1. Batch reactor

Chemical FK (Filter Cake) flows into the reactor through the input denoted "Chemical FK input", with parameters temperature-T_{FK}, mass flow rate- $\dot{m}_{FK}$ and specific heat-c_{FK}. The coolant flows into reactor through the second input denoted "Cooling medium", which is usually water of temperature T_{VP}, mass flow rate- $\dot{m}_V$ and specific heat-c_V.

Cooling medium flows through the jacket inner space of the reactor, with volume related to mass-m_{VR}, and flows out through the second output, with parameters mass flow rate m_V, temperature-T_V and specific heat-c_V.

At the beginning of the process there is an initial batch inside the reactor with parameter mass-m_P. The chemical FK is then added to this initial batch, so the reaction mixture inside the reactor has total mass-m, temperature-T and specific heat-c_R, and also contains partially unreacted portions of chemical FK described by parameter concentration a_{FK}.

This technique partially allows controlling the temperature of reaction mixture by the controlled feeding of the input chemical FK.

The main objective of optimization is to achieve the processing of large amount of chemical FK in a very short time. In general, this reaction is highly exothermal, thus the most important parameter is the temperature of the reaction mixture. **This temperature must not exceed 100°C** because of safety aspects and quality of the product. The original design of the reactor was based on standard chemical-technological methods and gives a proposal of reactor physical dimensions and parameters of chemical substances. These values are called within this paper **original parameters**.

Description of the reactor applies a system of four balance equations (1) and one equation (2) representing the term "k".

$$\dot{m}_{FK} = m'[t]$$

$$\dot{m}_{FK} = m[t]\, a'_{FK}[t] + k\, m[t] a_{FK}[t]$$

$$\dot{m}_{FK} c_{FK} T_{FK} + \Delta H_r k\, m[t] a_{FK}[t] = K\, S\, (T[t] - T_V[t]) + m[t] c_R T'[t]$$

$$\dot{m}_V c_V T_{VP} + K\, S\, (T[t] - T_V[t]) = \dot{m}_V c_V T_V[t] + m_{VR} c_V\, T'_V[t] \qquad (1)$$

$$k = A e^{-\frac{E}{RT[t]}} \qquad (2)$$

4 Differential Evolution

DE is a population-based optimization method that works on real-number-coded individuals [21]. A schematic is given in Fig. 2.

There are essentially five sections to the code. Section 1 describes the input to the heuristic. D is the size of the problem, G_{max} is the maximum number of generations, NP is the total number of solutions, F is the scaling factor of the solution and CR is the factor for crossover. F and CR together make the internal tuning parameters for the heuristic.

Section 2 outlines the initialization of the heuristic. Each solution $x_{i,j,G=0}$ is created randomly between the two bounds $x^{(lo)}$ and $x^{(hi)}$. The parameter j represents the index to the values within the solution and i indexes the solutions within the population. So, to illustrate, $x_{4,2,0}$ represents the fourth value of the second solution at the initial generation.

After initialization, the population is subjected to repeated iterations in section 3. Section 4 describes the conversion routines of DE. Initially, three random numbers r_1, r_2, r_3 are selected, unique to each other and to the current indexed solution i in the population in 4.1. Henceforth, a new index j_{rand} is selected in the solution. j_{rand} points to the value being modified in the solution as given in 4.2. In 4.3, two solutions, $x_{j,r1,G}$ and $x_{j,r2,G}$ are selected through the index r_1 and r_2 and their values subtracted. This value is then multiplied by F, the predefined scaling factor. This is added to the value indexed by r_3.

However, this solution is not arbitrarily accepted in the solution. A new random number is generated, and if this random number is less than the value of CR, then the new value replaces the old value in the current solution. The fitness of the resulting solution, referred to as a perturbed vector $u_{j,i,G.}$, is then compared with the fitness of $x_{j,i,G.}$. If the fitness of $u_{j,i,G}$ is greater than the fitness of $x_{j,i,G.}$, then $x_{j,i,G.}$ is replaced with $u_{j,i,G}$; otherwise, $x_{j,i,G.}$ remains in the population as $x_{j,i,G+1}$. Hence the competition is only between the new *child* solution and its *parent* solution.

Description of the used DERand1Bin strategy is presented in (3). Please refer to [21] and [22] for the detailed complete description of all other strategies.

$$u_{j,i,G+1} = x_{j,r1,G} + F \bullet \left(x_{j,r2,G} - x_{j,r3,G} \right) \tag{3}$$

1. Input: $D, G_{max}, NP \geq 4, F \in (0,1+), CR \in [0,1]$, and initial bounds: $\vec{x}^{(lo)}, \vec{x}^{(hi)}$.

2. Initialize:
$$\begin{cases} \forall i \leq NP \land \forall j \leq D : x_{i,j,G=0} = x_j^{(lo)} + rand_j[0,1] \bullet \left(x_j^{(hi)} - x_j^{(lo)} \right) \\ i = \{1,2,...,NP\}, \ j = \{1,2,...,D\}, \ G = 0, \ rand_j[0,1] \in [0,1] \end{cases}$$

3. While $G < G_{max}$

    4. Mutate and recombine:

        4.1 $r_1, r_2, r_3 \in \{1,2,.....,NP\}$, randomly selected, except: $r_1 \neq r_2 \neq r_3 \neq i$

        4.2 $j_{rand} \in \{1,2,...,D\}$, randomly selected once each i

$\forall i \leq NP$ 4.3 $\forall j \leq D, u_{j,i,G+1} = \begin{cases} x_{j,r_3,G} + F \cdot (x_{j,r_1,G} - x_{j,r_2,G}) \\ \quad \text{if } (rand_j[0,1] < CR \lor j = j_{rand}) \\ x_{j,i,G} \quad \text{otherwise} \end{cases}$

    5. Select

        $\vec{x}_{i,G+1} = \begin{cases} \vec{u}_{i,G+1} & \text{if } f(\vec{u}_{i,G+1}) \leq f(\vec{x}_{i,G}) \\ \vec{x}_{i,G} & \text{otherwise} \end{cases}$

$G = G + 1$

Fig. 2. DE Shematic

5 Chaotic Maps

This section contains the description discrete chaotic maps used as the random generator for DE. Iterations of the chaotic maps were used for the generation of real numbers in the process of crossover based on the user defined CR value and for the generation of the integer values used for selection of individuals [23].

5.1 Dissipative Standard Map

The Dissipative Standard map is a two-dimensional chaotic map. The parameters used in this work are $b = 0.1$ and $k = 8.8$ as suggested in [24]. For these values, the system exhibits typical chaotic behavior and with this parameter setting it is used in the most research papers and other literature sources. The Dissipative standard map is given in Fig. 3. The map equations are given in Eq. 3 and 4.

$$X_{n+1} = X_n + Y_{n+1} (\mathrm{mod}\, 2\pi) \tag{3}$$

$$Y_{n+1} = bY_n + k \sin X_n (\mathrm{mod}\, 2\pi) \tag{4}$$

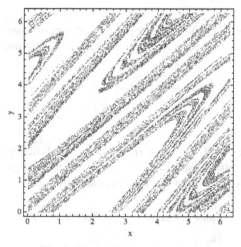

Fig. 3. Dissipative standard map

6 Problem Design

The cost function (CF) that was minimized is given in (5). It is divided into three time intervals, and has two penalizations. The first part ensures minimizing the area between required and actual temperature of the reaction mixture, and the second part ensures the rapid reaction of the whole batch of chemical FK, thus the very low value of concentration a_{FK} of partly unreacted portions of chemical FK in reaction mixture.

The first penalization helps to find solutions in which the temperature of reaction mixture cools down fast to its initial state, and the process duration is shortened. The second corresponds to the critical temperature.

$$f_{cost} = \sum_{t=0}^{t_1} |w - T[t]| + \sum_{t=0}^{t_1} a_{FK}[t] + pen.1 + pen.2 \tag{5}$$

$$pen.1 = \begin{cases} 0 & Max(T[\tau]) \leq 323,15 \\ 50000 & else \end{cases}$$

for $\tau \in \langle t_2, t_3 \rangle$

$$pen.2 = \begin{cases} 0 & Max(T[\tau]) \leq 373,15 \\ 50000 & else \end{cases}$$

for $\tau \in \langle 0, t_3 \rangle$

Where the time intervals were set for example as: $t_1 = 15000$ s; $t_2 = 20000$ s; $t_3 = 25000$ s.

The minimizing term, presented in (6), limits the maximum mass of one batch. Moreover, many parameters were interrelated due to the optimization of the reactor geometry.

$$m[t] \leq m_{max} \tag{6}$$

7 Results

The parameter settings for ChaosDE were obtained analytically based on numerous experiments and simulations (see Table 2). Experiments were performed in an environment of Wolfram Mathematica and were repeated 50 times.

Table 1. Parameter set up for Chaos DE

Parameter	Value
PopSize	50
F	0.8
CR	0.8
Generations	200
Max. CF Evaluations (CFE)	10000

The optimization proceeded with the parameters shown in Table 2, where the internal radius of reactor is expressed in parameter r and is related to cooling area S. Parameter d represents the distance between the outer and inner jackets and parameter h means the height of the reactor.

Table 2. Optimized reactor parameters, difference between original and the optimized reactor

Parameter	Range	Original value	Optimized value
$\dot{m}_{FK}$ [kg.s^{-1}]	0 – 10.0	0 - 3	0,0696
T_{VP} [K]	273.15 – 323.15	293.15	276.077
$\dot{m}_V$ [kg]	0 – 10.0	1	4,6355
r [m]	0.5 – 2.5	0.78	1.0
h [m]	0.5 – 2.5	1.11	1.0
d [m]	0.01 – 0.2	0.03	0.0749

The design of the reactor was based on standard chemical-technological methods and gives a proposal of reactor physical dimensions and parameters of chemical substances. These values are called within this paper **original parameters (values)**. The best results of the optimization are shown in Fig.4 and Table 2. From these results it is obvious that the temperature of reaction mixture did not exceed the critical value. The maximum temperature was 372.8 K (99.65°C). The required temperature w used in cost function was 370.0 K (96.85°C). Another fact not to be neglected is the shortened duration of the process (approx 20740 s compared to the original approx 25000 s). Fig 4 shows the time evolution of CF value for the best individual solution, whereas Fig 5 confirms the robustness of ChaosDE in finding the best solutions for all 50 runs.

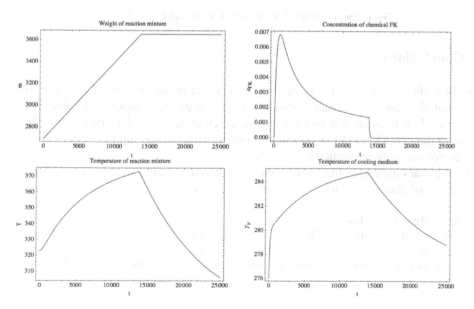

Fig. 4. Results of optimization, Simulation of the **best solution**, course of the weight of reaction mixture (upper left), concentration of chemical FK (upper right), temp. of reaction mixture (lower left), and temp. of cooling medium (lower right)

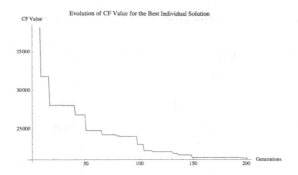

Fig. 5. Evolution of CF value for the best individual solution

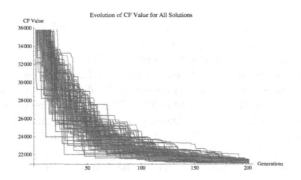

Fig. 6. Evolution of CF value for all 50 runs of Chaos-DE

8 Conclusions

Based on obtained results, it may be claimed, that the presented ChaosDE driven by means of the chaotic Dissipative standard map has given satisfactory results. The behavior of an uncontrolled original reactor gives quite unsatisfactory results. The reactor parameters found by optimization brought performance superior compared to the reactor set up by an original parameters.

Future plans include testing of different chaotic systems, comparison with different heuristics and obtaining a large number of results to perform statistical tests.

Acknowledgments. This work was supported by European Regional Development Fund under the project CEBIA-Tech No. CZ.1.05/2.1.00/03.0089 and project IT4Innovations Centre of Excellence No. CZ.1.05/1.1.00/02.0070, and by Internal Grant Agency of Tomas Bata University under the project No. IGA/FAI/2012/037.

References

[1] He, Q., Wang, L.: An effective co-evolutionary particle swarm optimization for constrained engineering design problems. Engineering Applications of Artificial Intelligence 20, 89–99 (2007)

[2] Silva, V.V.R., Khatib, W., Fleming, P.J.: Performance optimization of gas turbine engine. Engineering Applications of Artificial Intelligence 18, 575–583 (2005)

[3] Tan, W.W., Lu, F., Loh, A.P., Tan, K.C.: Modeling and control of a pilot pH plant using genetic algorithm. Engineering Applications of Artificial Intelligence 18, 485–494 (2005)

[4] Lepore, R., Wouwer, A.V., Remy, M., Findeisen, R., Nagy, Z., Allgöwer, F.: Optimization strategies for a MMA polymerization reactor. Computers and Chemical Engineering 31, 281–291 (2007)

[5] Rout, B.K., Mittal, R.K.: Optimal manipulator parameter tolerance selection using evolutionary optimization technique. Engineering Applications of Artificial Intelligence 21, 509–524 (2008)

[6] Davendra, D., Zelinka, I., Senkerik, R.: Chaos driven evolutionary algorithms for the task of PID control. Computers & Mathematics with Applications 60(4), 898–1221 (2010) ISSN 0898-1221

[7] Silva, C.M., Biscaia, E.C.: Genetic algorithm development for multi-objective optimization of batch free-radical polymerization reactors. Computers and Chemical Engineering 27, 1329–1344 (2003)

[8] Arpornwichanop, A., Kittisupakorn, P., Mujtaba, M.I.: On-line dynamic optimization and control strategy for improving the performance of batch reactors. Chemical Engineering and Processing 44(1), 101–114 (2005)

[9] Srinisavan, B., Palanki, S., Bonvin, D.: Dynamic optimization of batch processes I. Characterization of the nominal solution. Computers and Chemical Engineering 27, 1–26 (2002)

[10] Srinisavan, B., Palanki, S., Bonvin, D.: Dynamic optimization of batch processes II. Role of Measurement in Handling Uncertainly. Computers and Chemical Engineering 27, 27–44 (2002)

[11] Wang, Y., Zhou, D., Gao, F.: Iterative learning model predictive control for multi-phase batch processes. Journal of Process Control 18, 543–557 (2008)

[12] Cho, W., Edgar, T.F., Lee, J.: Iterative learning dual-mode control of exothermic batch reactors. Control Engineering Practice 16, 1244–1249 (2008)

[13] Beyer, M.A., Grote, W., Reinig, G.: Adaptive exact linearization control of batch polymerization reactors using a Sigma-Point Kalman Filter. Journal of Process Control 18, 663–675 (2008)

[14] Sarma, P.: Multivariable gain-scheduled fuzzy logic control of an exothermic reactor. Engineering Applications of Artificial Intelligence 14, 457–471 (2001)

[15] Sjöberg, J., Mukul, A.: Trajectory tracking in batch processes using neural controllers. Engineering Applications of Artificial Intelligence 15, 41–51 (2002)

[16] Mukherjee, A., Zhang, J.: A reliable multi-objective control strategy for batch processes based on bootstrap aggregated neural network models. Journal of Process Control 18, 720–734 (2008)

[17] Mujtaba, M., Aziz, N., Hussain, M.A.: Neural network based modelling and control in batch reactor. Chemical Engineering Research and Design 84(8), 635–644 (2006)

[18] Causa, J., Karer, G., Nunez, A., Saez, D., Skrjanc, I., Zupancic, B.: Hybrid fuzzy predictive control based on genetic algorithms for the temperature control of a batch reactor. Computers and Chemical Engineering (2008), doi:10.1016/j.compchemeng.2008.05.014

[19] Altinten, A., Ketevanlioglu, F., Erdogan, S., Hapoglu, H., Alpbaz, M.: Self-tuning PID control of jacketed batch polystyrene reactor using genetic algorithm. Chemical Engineering Journal 138, 490–497 (2008)

[20] Faber, R., Jockenhövel, T., Tsatsaronis, G.: Dynamic optimization with simulated annealing. Computers and Chemical Engineering 29, 273–290 (2005)

[21] Price, K.: An Introduction to Differential Evolution. In: Corne, D., Dorigo, M., Glover, F. (eds.) New Ideas in Optimization, pp. 79–108. McGraw-Hill, London (1999) ISBN 007-709506-5

[22] Price, K., Storn, R.: Differential evolution homepage (2001), http://www.icsi.berkeley.edu/~storn/code.html

[23] Caponetto, R., Fortuna, L., Fazzino, S., Xibilia, M.: Chaotic sequences to improve the performance of evolutionary algorithms. IEEE Trans. Evol. Comput. 7(3), 289–304 (2003)

[24] Sprott, J.C.: Chaos and Time-Series Analysis. Oxford University Press (2003)

Differential Evolution Classifier with Optimized Distance Measures for the Features in the Data Sets

David Koloseni[1], Jouni Lampinen[2,3], and Pasi Luukka[1]

[1] Laboratory of Applied Mathematics, Lappeenranta University of Technology,
P.O. Box 20, FIN-53851 Lappeenranta, Finland
{david.koloseni,pasi.luukka}@lut.fi
[2] Department of Computer Science,
University of Vaasa, P.O. Box 700, FI-65101 Vaasa, Finland
jouni.lampinen@uwasa.fi
[3] VSB-Technical University of Ostrava,17. listopadu 15,
70833 Ostrava-Poruba, Czech Republic

Abstract. In this paper we propose a further generalization of differential evolution based data classification method. The current work extends our earlier differential evolution based nearest prototype classifier that includes optimization of the applied distance measure for the particular data set at hand. Here we propose a further generalization of the approach so, that instead of optimizing only a single distance measure for the given data set, now multiple distance measures are optimized individually for each feature in the data set. Thereby, instead of applying a single distance measure for all data features, we determine optimal distance measures individually for each feature. After the optimal class prototype vectors and optimal distance measures for each feature has been first determined, together with the optimal parameters related with each distance measure, in actual classification phase we combine the individually measured distances from each feature to form an overall distance measure between the class prototype vectors and sample. Each sample is then classified to the class assigned with the nearest prototype vector using that overall distance measure. The proposed approach is demonstrated and initially evaluated with three different data sets.

1 Introduction

Differential evolution algorithm (DE) has lately gained increasing popularity in solving classification problems. The recent research in the field include *e.g.* bankruptcy prediction [1], classification rule discovery [2], nearest neighbor prototype search [3] and feature selection [4]. Since its introduction in 1995 DE have emerged among the most frequently applied evolutionary computing methods [5]. DE has also been used in many areas of pattern recognition, i.e. in remote sensing imagery [6] and hybrid evolutionary learning in pattern recognition systems [7], to mention a few examples.

The focus of this paper is in solving classification problems by extending the earlier DE based classifier [8],[9], [10] further on with an extension for optimizing the selection of applied distance measure individually for each feature in the data set to be classified. In our previous work [10] we generalized our original DE classifier version [8] by extending the optimization process to cover also the selection of the applied distance

V. Snasel et al. (Eds.): SOCO Models in Industrial & Environmental Appl., AISC 188, pp. 103–111.
springerlink.com

measure from a predefined pool of alternative distance measures. In [10], however, we applied the same distance measure for all features in the dataset, while the proposed approach are optimizing the distance measures individually for each feature of the classified data. The rationale behind this extension is that in classification, the data set to be classified is often in a form where each sample consists of several measurements, and it is not guaranteed that all the measured values from one particular sample obey the same optimal distance measure. It is clear that an optimal distance measure for a feature may not be optimal for another feature in the same data set. This is the underlying motivation to further generalize our earlier method so, that instead of optimizing a single vector based distance measure for all features, we concentrate on the problem at the feature level, and optimize the selection of applied distance measure individually for each particular feature.

Examples on situations where a considerably improved classification accuracy have been obtained by applying some other distance measure than a simple euclidean metric have been reported in several articles *i.e.* [11], [12], [13]. However, typically in these types of studies one has simply tested with a few different distance measures for classifying the data set at hand. So far none of them have concentrated on selecting distance measures optimally at feature level, but on data set level instead.

Thereby, in case of classifying a dataset containing T features using the proposed approach, we need to determine T different optimal distance measures. In addition we need to determine also the possible parameters related to each distance measure and the class prototype vectors representing each class. All these should be optimized so, that the classification accuracy over the current dataset will be maximized. We apply DE algorithm for solving the resulted global optimization problem in order to determine all mentioned values optimally. In particular, if the distance measure applied to a particular feature has any free parameters, also their values need to be optimized as well.

After the optimal class prototype vectors and distance measures with their related parameters have been determined by DE algorithm, the actual classification by applying the determined values takes place. After we have first computed the individual distances between a sample and the class prototype vectors individually for each feature, then we compute the overall distance value by normalizing the individual distances first, and then simply calculate the sum of all feature wisely computed and normalized distances. Finally, each sample is to be classified into the class represented by the nearest class prototype vector that is providing the lowest overall distance value.

2   Differential Evolution Based Classifier with Optimized Distances for the Features in the Data Sets

2.1   Differential Evolution Based Classification

The DE algorithm [15], [5] was introduced by Storn and Price in 1995 and it belongs to the family of Evolutionary Algorithms (EAs). As a typical EA, DE starts with a randomly generated initial population of candidate solutions for the optimization problem to be solved that is then improved using selection, mutation and crossover operations. Several ways exist to determine a stopping criterion for EAs but usually a predefined upper limit G_{max} for the number of generations to be computed provides an appropriate stopping condition. Other control parameters for DE are the crossover control parameter CR, the mutation factor F, and the population size NP.

In each generation G, DE goes through each D dimensional decision vector $v_{i,G}$ of the population and creates the corresponding trial vector $u_{i,G}$ as follows in the most common DE version, DE/rand/1/bin [16]:

$$r_1, r_2, r_3 \in \{1, 2, \ldots, NP\}, \text{(randomly selected,}$$
$$\text{except mutually different and different from } i)$$
$$j_{rand} = \text{floor}\,(rand_i[0,1) \cdot D) + 1$$
$$\text{for}(j = 1; j \leq D; j = j + 1)$$
$$\{$$
$$\quad \text{if}(rand_j[0,1) < CR \vee j = j_{rand})$$
$$\quad\quad u_{j,i,G} = v_{j,r_3,G} + F \cdot \left(v_{j,r_1,G} - v_{j,r_2,G}\right)$$
$$\quad \text{else}$$
$$\quad\quad u_{j,i,G} = v_{j,i,G}$$
$$\}$$

In this DE version, NP must be at least four and it remains fixed along CR and F during the whole execution of the algorithm. Parameter $CR \in [0,1]$, which controls the crossover operation, represents the probability that an element for the trial vector is chosen from a linear combination of three randomly chosen vectors and not from the old vector $v_{i,G}$. The condition "$j = j_{rand}$" is to make sure that at least one element is different compared to the elements of the old vector. The parameter F is a scaling factor for mutation and its value is typically $(0, 1+]$[1].

After the mutation and crossover operations, the trial vector $u_{i,G}$ is compared to the old vector $v_{i,G}$. If the trial vector has an equal or better objective value, then it replaces the old vector in the next generation. This can be presented as follows (in this paper minimization of objectives is assumed) [16]:

$$v_{i,G+1} = \begin{cases} u_{i,G} \text{ if } f(u_{i,G}) \leq f(v_{i,G}) \\ v_{i,G} \text{ otherwise} \end{cases}.$$

DE is an elitist method since the best population member is always preserved and the average objective value of the population will never get worse. As the objective function, f, to be minimized we applied the number of incorrectly classified learning set samples. Each population member, $v_{i,G}$, as well as each new trial solution, $u_{i,G}$, contains the class vectors for all classes and the power value p. In other words, DE is seeking the vector $(y(1), \ldots, y(T), p)$ that minimizes the objective function f. After the optimization process the final solution, defining the optimized classifier, is the best member of the last generation's, G_{max}, population, the individual $v_{i,G_{max}}$. The best individual is the one providing the lowest objective function value and therefore the best classification performance for the learning set. For control parameter values see [8], [9]. Next into the actual classification. We suppose that T is the number of different kinds of features that we can measure from objects. The key idea is to determine for each class the ideal vector $\mathbf{y}_i$, $\mathbf{y}_i = (y_{i1}, \ldots, y_{iT})$ that represents class i as well as possible. Later on we call

---

[1] Notation means that the upper limit is about 1 but not strictly defined.

these vectors as class vectors. When these class vectors have been determined we have to make the decision to which class the sample $\mathbf{x}$ belongs to according to some criteria. This can be done e.g. by computing the distances d_i between the class vectors and the sample which we want to classify. For computing the distance usual way is to use Minkowsky metric, $d(\mathbf{x}, \mathbf{y}) = \left(\sum_{j=1}^{T} |x_j - y_j|^p \right)^{1/p}$. After we have the distances between the samples and class vectors then we can make our classification decision according to the shortest distance. For $\mathbf{x}, \mathbf{y} \in R^n$. We decide that $\mathbf{x} \in C_m$ if

$$d\langle \mathbf{x}, \mathbf{y}_m \rangle = \min_{i=1,\ldots,N} d\langle \mathbf{x}, \mathbf{y}_i \rangle \tag{1}$$

In short the procedure for our algorithm is as follows:

1. Divide data into learning set and testing set
2. Create trial vectors to be optimized which consists of classes and parameter p, $v_{i,G}$
3. Divide $v_{i,G}$ into class vectors and parameter p.
4. Compute distance between samples in the learning set and class vectors
5. Classify samples according to their minimum distance by using (1)
6. Compute classification accuracy (accuracy = no. of correctly classified samples/total number of all samples in learning set)
7. Compute the fitness value for objective function using $f = 1 - accuracy$
8. Create new pool of vectors $v_{i,G+1}$ for the next population using selection, mutation and crossover operations of differential evolution algorithm, and goto 3. until stopping criteria is reached. (For example maximum number of iterations reached or 100% accuracy reached)
9. Divide optimal vector $v_{i,G_{max}}$ into class vectors and parameter p.
10. Repeat steps 4, 5 and 6, but now with optimal class vectors, p parameter and samples in the testing set.

For more thorough explanation we refer to [8]. The proposed extension to the earlier DE classifier will be described in detail next.

2.2 The Proposed Extension for Optimizing Distance Measures Individually for Each Feature in the Data Set

Basically the vector to be optimized consists now of following components:

$$v_{i,G} = \{\{class1, class2 \cdots classN\}, \{switch\}, \{parameters\}\}$$

where $\{class1, class2 \cdots classN\}$ are the class vectors which are to be optimized for the current data set, $\{switch\}$ is an Integer valued parameter pointing the particular distance measure to be applied from choices d_1 to d_8 (see Table 1). In the proposed approach an individual value of $\{switch\}$ is assigned for each data feature. In other words our $\{switch\}$ is T dimensional vector consisting of integer number within $[1,8]$. Since DE algorithm operates internally with floating point representations and $\{switch\}$ is a vector of Integer values, the corresponding adaptations are needed. Therefore we use in our

Algorithm 1. Pseudo code for classification process with optimal parameters from DE.

Require: $Data[1,\ldots,T]$, $classvec1[1,\ldots,T]$, $classvec2[1,\ldots,T]$,...,$classvecN[1,\ldots,T]$, $switch$, $p1[1,\ldots,T]$, $p2[1,\ldots,T]$,
$p3[1,\ldots,T]$,$p4[1,\ldots,T]$ $center=[classvec1;classvec2;...classvecN]$
  for $j=1$ to T **do**
    for $i=1$ to N **do**
      if $switch(j) == 1$ **then**
        $d(:,i,j) = dist1(data, repmat(center(i,j),T,1),p1(j))$
      else if $switch(j) == 2$ **then**
        $d(:,i,j) = dist2(data, repmat(center(i,j),T,1),p2(j))$
      else if $switch(j) == 3$ **then**
        $d(:,i,j) = dist3(data, repmat(center(i,j),T,1),p3(j))$
      else if $switch(j) == 4$ **then**
        $d(:,i,j) = dist4(data, repmat(center(i,j),T,1))$
      else if $switch(j) == 5$ **then**
        $d(:,i,j) = dist5(data, repmat(center(i,j),T,1))$
      else if $switch(j) == 6$ **then**
        $d(:,i,j) = dist6(data, repmat(center(i,j),T,1),p4(j))$
      else if $switch(j) == 7$ **then**
        $d(:,i,j) = dist7(data, repmat(center(i,j),T,1))$
      else
        $d(:,i,j) = dist8(data, repmat(center(i,j),T,1))$
      end if
    end for
  end for
  $D = scale(d)$
  for $i=1$ to T **do**
    $d_{total}(:,i) = sum(D(:,:,i),2);$
  end for
  for $i=1$ to $length(d_{total})$ **do**
    $class(:,i) = find(d_{total}(i,j) == min(d_{total}(i,:)));$
  end for

optimization real numbers that are boundary constrained $[0.5, \quad 8.499]$, and all DE operations are performed in real space. Only when we are applying the $\{switch\}$ vector to point the actual distance measures to be applied, we first round the values to the nearest integer. In addition we have the $\{parameters\}$ which is a vector of possible parameters from the distance measures. In this case $\{parameters = \{\mathbf{p_1}, \mathbf{p_2}, \mathbf{p_3}, \mathbf{p_4}\}\}$ where all $\{\mathbf{p_1}, \mathbf{p_2}, \mathbf{p_3}, \mathbf{p_4}\}$ are again vectors of T dimensions (i.e. $\mathbf{p_1} = \{\mathbf{p_{1,1}}, \mathbf{p_{1,2}}, \cdots, \mathbf{p_{1,T}}\}$).

Each component of vector $\{switch\}$ is referring to a distance measure in the pool given in Table 1. A goal of the optimization process is to select the distance measures optimally from this pool individually for each feature in the current data set. A collection of applicable distance measure is provided in[17], from where also the measures in Table 1 have been taken.

The pseudocode Algorithm1 is describing the actual classification process after the optimal class prototype vectors, distance measures and the possible free parameters of each distance measure have been first determined by DE algorithm.

Next we discuss a bit more in detail the main modifications done to our previous method. After we had created the pool of distances we needed to find a way to optimize the selection of distance from the pool of possible choices to the data set at hand. For this we used the $switch$ operator which was needed to optimize. For the eight distances in the pool now we had to add vector of parameters of length of T to be optimized in order to select the suitable distance. By optimizing the integer numbers within $[1,8]$ we performed the needed optimal selection. In addition to this, as can be noticed from the pool of distances and from the pseudo code there are different parameter values with different distances which needs to be optimized as well. For this we created additional parameters to be optimized for each of the different parameters in pool of distances. This part of the vector we call $\{parameters\}$ which is possible parameters from the distance measures. In this case $\{parameters = \{\mathbf{p_1}, \mathbf{p_2}, \mathbf{p_3}, \mathbf{p_4}\}\}$. At this point we are facing the situation where we can have i.e. situation where optimal distance for feature one can be d_1 and also i.e. optimal distance for feature two can be d_1 and clearly the optimal parameter value can be different. For this reason we made $\{\mathbf{p_1}, \mathbf{p_2}, \mathbf{p_3}, \mathbf{p_4}\}$ also vectors of length T. This results in having several optimized parameter values which are not used, making the optimization task more challenging, but also makes it possible that different features with same optimal distance measure can have different optimal parameter. After the vector $v_{i,G}$ is divided in its corresponding parts we can calculate the distances between the samples and the class vectors. This results in a vector of distances for one sample and one class vector. This vector is then normalized properly and after that we aggregate this vector simply by computing the sum of normalized distances. This is now stored in d_{total} in the pseudo code. This process is repeated for all the classes and all the samples. This way we end up having a distance matrix consisting of samples and distances to each particular class. After we have created this distance matrix the selection of to which class the particular sample belongs is made according to minimum distance.

Table 1. Distance measures in a pool of distances

$d_1(x,y) = (\lvert x-y\rvert^{P_1}); p_1 \in [1,\infty)$	$d_2(x,y) = \lvert x-y\rvert^{P_2}/max\{\lvert x\rvert, \lvert y\rvert\}; p_2 \in [1,\infty)$
$d_3(x,y) = \lvert x-y\rvert^{P_3}/min\{1+\lvert x\rvert, \lvert y\rvert\}; p_3 \in [1,\infty)$	$d_4(x,y) = \lvert x-y\rvert/max\{1+\lvert x\rvert, \lvert y\rvert\};$
$d_5(x,y) = \lvert x-y\rvert/[1+\lvert x\rvert+\lvert y\rvert];$	$d_6(x,y) = \lvert x/[1+\lvert x\rvert] - y/[1+\lvert y\rvert]\rvert$
$d_7(x,y) = p_4(x-y)^2/(x+y); p_4 \in (0,\infty)$	$d_8(x,y) = \lvert x-y\rvert/(1+\lvert x\rvert)(1+\lvert y\rvert)$

3 Classification Experiments and Comparisons of the Results

The data sets for experimentation with the proposed approach were taken from UCI machine learning data repository [14]. Chosen data sets were all such where optimal distance measure was not euclidean distance. The data sets were subjected to computation of 1000 generations ($G_{max} = 1000$) of DE algorithm and the data was divided 30 times into random splits of testing sets and learning sets, based on which mean accuracies and variances were then computed. The fundamental properties of the data sets are summarized in Table 2.

Table 2. Properties of the data sets

Name	Nb of classes	Nb of features	Nb of instances
Horse-colic	2	11	368
Hypothyroid	2	25	3772
Balance scale	3	5	625

The proposed differential evolution classifier with optimal distance measures for each feature was tested with Hypothyroid, Horsecolic and Balance scale data set. The mean classification accuracies were recorded as well as the corresponding optimal distance measure that was found optimal in each individual experiment. In each case 30 repetitions were performed dividing the data randomly into learning and testing tests. The applied crossvalidation technique was two fold crossvalidation, where samples are divided randomly into folds 30 times and required statistics was calculated from that basis. Folds were normalized to achieve efficient and precise numerical computations. Results from the experiments are reported in Table 3.

As can be observed from the Table 3 for Hypothyroid data set the mean accuracy of 99.57% was reached and using 99% confidence interval (by $\mu \pm t_{1-\alpha} S_\mu / \sqrt{n}$) accuracy was 99.57 ± 0.63. With Balance scale data we observed the accuracy of 91.30 ± 0.12, and with Horsecolic data the classification accuracy of 83.35 ± 2.23 were observed.

To enable initial comparisons with some of the most frequently applied classifiers and with the previous DE classifier, we calculated the corresponding classification results also by using k-NN classifier, Back Propagation Neural Network (BPNN) and DE classifier [8]. The results and their comparisons are provided in Tables 4.

In Table 4, results from Horsecolic, hypothyroid and Balancescale data are compared between four classifiers. For Hypothyroid data, the proposed method performed significantly better than k-NN. Also in comparison with BPNN the proposed method gave significantly higher mean accuracy in 0.999 confidence interval. The original DE classifier performed here slightly better than the proposed method, but the observed difference was not found to be statistically significant.

Table 3. Classification results for the three data sets using $N = 30$ and $G_{max} = 1000$. Mean classification accuracies, variances and optimal distance found are reported in columns 2 to 6. TS is referring to test set and LS to the learning set.

Data	Mean (TS)	Variance (TS)	Mean (LS)	Variance (LS)
Horsecolic	83.35	24.67	88.59	2.36
Balance scale	91.30	0.075	92.22	0.028
Hypothyroid	99.57	1.98	99.98	0.00092

Table 4. Comparison of the results from the proposed method to other classifiers with Horse-colic data, Hypothyroid data and Balancescale data.

Method	Horsecolic		Hypothyroid		Balancescale	
	Mean accuracy	Variance	Mean accuracy	Variance	Mean accuracy	Variance
KNN	68.53	5.44	98.37	0.006	88.06	1.42
BPNN	80.85	19.27	97.29	0.018	87.90	2.22
DE classifier	71.76	107.78	99.95	0.0061	88.66	4.71
Proposed method	83.35	24.67	99.57	1.98	91.30	0.075

The results from the comparisons with the Horsecolic data set the proposed method achieved a clearly higher mean classification accuracy when compared with KNN, BPNN and DE classifiers. Improvement with mean accuracy compared to original DE classifier was significant and remarkably high, more than 10%.

In Table 4 also the results with balance scale data set are reported. Here the proposed method significantly and rather clearly outperformed the other compared classifiers by reaching the mean accuracy of 91.30%.

4 Discussion

In this paper we proposed an extension for the differential evolution based nearest prototype classifier where the selection of the applied distance measure is optimized individually for each feature of the classified data set. Earlier a single distance measure was applied for all data features, and thereby we were able to optimize the selection of the distance measure in the data set level only. The proposed generalization extends the optimization of distance measures to the feature level.

To demonstrate the proposed approach, and to enable a preliminary evaluation of it, we carried out experimentation with three different data sets. In two cases the classification accuracy of the proposed method outperformed all compared classifiers significantly. In the remaining case no statistically significant difference to the earlier DE classifier version were observed, despite the difference to the other compared classifiers were again significant and rather clear. The results are suggesting that the proposed generalization is advantageous from the classification accuracy point of view. However, this conclusion should be interpreted as a preliminary one due to limited number of data sets investigated so far.

An important aspect is, that the proposed approach is not limited to DE classifiers only, and can be applied generally in connection with any other similar type of classification method that is based on global optimization. However, this is assuming that an effective enough global optimizer like differential evolution algorithm is applied to solve the resulting optimization problem.

Despite the current results are promising, they should be considered preliminary and need to be further confirmed by applying a broader selection of data sets. That includes into our further research plans. Concerning the possibilities for further developments of the proposed approach, so far we have applied a simple summation of feature wisely computed distances to calculate the final overall distance measure. However, in future also a suitable aggregation method can be used for the purpose. This is also one of our future directions for further investigations of this method.

Acknowledgment. The work of Jouni Lampinen was supported by the IT4Innovations Center of Excellence project. CZ.1.05/1.1.00/02.0070 funded by the Structural Funds of EU and Czech Republic ministry of education.

References

1. Chauhan, N., Ravi, V., Chandra, D.K.: Differential evolution trained wavelet neural networks: Application to bankruptcy prediction in banks. Expert Systems with Applications 36(4), 7659–7665 (2009)
2. Su, H., Yang, Y., Zhao, L.: Classification rule discovery with DE/QDE algorithm. Expert Systems with Applications 37(2), 1216–1222 (2010)
3. Triguero, I., Garcia, S., Herrera, F.: Differential evolution for optimizing the positioning of prototypes in nearest neighbor classification. Pattern Recognition 44, 901–916 (2011)
4. Khushaba, R.N., Al-Ani, A., Al-Jumaily, A.: Feature subset selection using differential evolution and a statistical repair mechanism. Expert Systems with Applications 38, 11515–11526 (2011)
5. Price, K., Storn, R., Lampinen, J.: Differential Evolution - A Practical Approach to Global Optimization. Springer (2005)
6. Ujjwal, M., Saha, I.: Modified differential evolution based fuzzy clustering for pixel classification in remote sensing imagery. Pattern Recognition 42, 2135–2149 (2009)
7. Zmudaa, M.A., Rizkib, M.M., Tamburinoc, L.A.: Hybrid evolutionary learning for synthesizing multi-class pattern recognition systems. Applied Soft Computing 2(4), 269–282 (2009)
8. Luukka, P., Lampinen, J.: Differential Evolution Classifier in Noisy Settings and with Interacting Variables. Applied Soft Computing 11, 891–899 (2011)
9. Luukka, P., Lampinen, J.: A Classification method based on principal component analysis differential evolution algorithm applied for predition diagnosis from clinical EMR heart data sets. In: Computational Intelligence in Optimization: Applications and Implementations. Springer (2010)
10. Koloseni, D., Lampinen, J., Luukka, P.: Optimized Distance Metrics for Differential Evolution based Nearest Prototype Classifier. Accepted to Expert Systems With Applications
11. Shahid, R., Bertazzon, S., Knudtson, M.L., Ghali, W.A.: Comparison of distance measures in spatial analytical modeling for health service planning. BMC Health Services Research 9, 200 (2009)
12. Yu, J., Yin, J., Zhang, J.: Comparison of Distance Measures in Evolutionary Time Series Segmentation. In: Third International Conference on Natural Computation, ICNC 2007, pp. 456–460 (2007)
13. Jenicka, S., Suruliandi, A.: Empirical evaluation of distance measures for supervised classification of remotely sensed image with Modified Multivariate Local Binary Pattern. In: International Conference on Emerging Trends in Electrical and Computer Technology (ICETECT), pp. 762–767 (2011)
14. Newman, D.J., Hettich, S., Blake, C.L., Merz, C.J.: UCI Repository of machine learning databases. University of California, Department of Information and Computer Science, CA (1998), http://www.ics.uci.edu/~mlearn/MLRepository.html
15. Storn, R., Price, K.V.: Differential Evolution - a Simple and Efficient Heuristic for Global Optimization over Continuous Space. Journal of Global Optimization 11(4), 341–359 (1997)
16. Price, K.V.: New Ideas in Optimization. An Introduction to Differential Evolution, ch. 6, pp. 79–108. McGraw-Hill, London (1999)
17. Bandemer, H., Näther, W.: Fuzzy Data Analysis. Kluwer Academic Publishers (1992)

Modifications of Differential Evolution with Composite Trial Vector Generation Strategies

Josef Tvrdík

Department of Computer Science, University of Ostrava,
30. dubna 22, 701 03 Ostrava, Czech Republic
`josef.tvrdik@osu.cz`

Abstract. Differential evolution algorithm with composite trial vector generation strategies and control parameters has been proposed recently. The performance of this algorithm is claimed to be better or competitive in comparison
with the state-of-the-art variants of differential evolution. When we attempted to
implement the algorithm according to the published description, several modified variants appear to follow the description of the algorithm. These variants
of the algorithm were compared experimentally in benchmark problems. One of
newly proposed variants outperforms the other variants significantly, including
the variant used by the authors of the algorithm in their published experimental
comparison.

1   Introduction

Differential evolution (DE) was proposed by Storn and Price [9] as a global optimizer
for unconstrained continuous optimization problems with a real-value objective function. The search space (domain) S is specified by lower (a_j) and upper (b_j) limits of
each component j, $S = \prod_{j=1}^{D}[a_j, b_j]$, $a_j < b_j$, $j = 1, 2, \ldots, D$, D is the dimension of the
problem. The global minimum point x^*, satisfying condition $f(x^*) \leq f(x)$ for $\forall x \in S$ is
the solution of the problem.

DE algorithm has become one of the evolutionary algorithms most frequently used
for solving the continuous global optimization problems in recent years [7]. Comprehensive summarizations of the up-to date results in DE are presented by Neri and Tirronen [5] and by Das and Suganthan [3].

Algorithm of DE works with a population of individuals (*NP* points in domain *S*)
that are considered as candidates of solution. Parameter *NP* is called the size of the
population. The population is developed iteratively by using evolutionary operators of
selection, mutation, and crossover. Each iteration corresponds to an evolutionary generation. Let us denote two subsequent generations by P and Q. Applications of evolutionary operators in the old generation P create a new generation Q. After completing
the new generation Q, the Q becomes the old generation for next iteration. The basic
structure of DE algorithm is shown in Algorithm 1.

The trial vector y is generated (line 5 in Algorithm 1) by crossover of two parent
vectors, the current (target) vector x_i and a mutant vector v. The mutant vector v is
obtained by a mutation. Several kinds of mutation have been proposed in last years.
Three kinds of mutation used in algorithms compared in this study are described below.

V. Snasel et al. (Eds.): SOCO Models in Industrial & Environmental Appl., AISC 188, pp. 113–122.
springerlink.com   © Springer-Verlag Berlin Heidelberg 2013

Algorithm 1. Differential evolution

1: generate an initial population $P = (x_1, x_2, \ldots, x_{NP})$, $x_i \in S$ distributed uniformly
2: evaluate $f(x_i)$, $i = 1, 2, \ldots, NP$
3: **while** stopping condition not reached **do**
4:     **for** $i = 1$ to NP **do**
5:         generate a trial vector y
6:         evaluate $f(y)$
7:         **if** $f(y) \leq f(x_i)$ **then**
8:             insert y into new generation Q
9:         **else**
10:            insert x_i into new generation Q
11:         **end if**
12:     **end for**
13:     $P := Q$
14: **end while**

Symbols r_1, r_2, r_3, r_4, and r_5 denote mutually distinct points taken randomly from the current generation P, not coinciding with the target point x_i, $F > 0$ is an input control parameter, and U is uniformly distributed random value between 0 and 1.

- *rand/1*

$$v = r_1 + F(r_2 - r_3). \tag{1}$$

- *rand/2.*

$$v = r_1 + F(r_2 - r_3) + F(r_4 - r_5). \tag{2}$$

- *current-to-rand/1*

$$y = x_i + U(r_1 - x_i) + F(r_2 - r_3). \tag{3}$$

The current-to-rand/1/ mutation generates directly a trial point y because it includes so called arithmetic crossover represented by the second term on the right side of (3).

The crossover operator constructs the trial vector y from current individual x_i and the mutant vector v. Two types of crossover were proposed in [9]. One of them is binomial crossover which generates a new trial vector y by using the following rule

$$y_j = \begin{cases} v_j & \text{if} \quad U_j \leq CR \quad \text{or} \quad j = l \\ x_{ij} & \text{if} \quad U_j > CR \quad \text{and} \quad j \neq l, \end{cases} \tag{4}$$

where l is a randomly chosen integer from $\{1, 2, \ldots, D\}$, and $U_1, U_2, \ldots, U_D$ are independent random variables uniformly distributed in $[0, 1)$. $CR \in [0, 1]$ is a control parameter influencing the number of elements to be exchanged by the crossover. Eq. (4) ensures that at least one element of x_i is changed, even if $CR = 0$.

Mutation according to (1), (2), or (3) could cause that a new trial point y moves out of the domain S. In such a case, the values of $y_j \notin [a_j, b_j]$ is turned over into S by using transformation $y_j = 2 \times a_j - y_j$ or $y_j = 2 \times b_j - y_j$ for the violated component.

2   DE with Composite Trial Vector Generation Strategies

DE algorithm with composite trial vector generation strategies and control parameters, in abbreviation *composite DE*, was presented by Wang et al. [12] and compared with the state-of-the-art DE variants considered best performing, namely *jDE* [1], *EPSDE* [4], *SaDE* [8], and *JADE* [13]. From the results In benchmark tests [10], *composite DE* outperformed all of these algorithms except *EPSDE*. In comparison with *EPSDE*, *composite DE* was competitive.

The *composite DE* combines three well-studied trial-vector strategies with three control parameter settings in a random way to generate trial vectors. The strategies are: rand/1/bin, rand/2/bin, and current-to-rand/1/. All the strategies are carried out with a pair of *F* and *CR* values randomly chosen from a parameter pool. It results in having three candidates to a trial vector in each iteration step (line 5 in Algoritmus 1) that compete by tournament. Thus, three function evaluations are needed in each step. The vector with the least function value of those three candidates is then used as a trial vector. The parameter pool used in [12] contains the following pairs of control parameters: $[F = 1.0, CR = 0.1]$, $[F = 1.0, CR = 0.9]$, and $[F = 0.8, CR = 0.2]$.

After the first reading of the paper [12], the *composite DE* algorithm was implemented as described above. This variant of the algorithm is labeled by *CoDE* hereafter.

However, the following part of the paper [12] induced doubts about the right form of the algorithm (the symbols in the text within quotation marks are changed to be compatible with the symbols used in this paper): "*After mutation, the current-to-rand/1 strategy uses the rotation-invariant arithmetic crossover rather than the binomial crossover, to generate the trial vector [2,6]. As a result, this strategy is rotation-invariant and suitable for rotated problems. The arithmetic crossover in this strategy linearly combines the mutant vector with the target vector as follows:*

$$y = x_i + U\,(v - x_i) \tag{5}$$

where U is a uniformly distributed random number between 0 and 1. Note that for the arithmetic crossover the crossover control parameter CR is not needed". It is not quite clear if this text is meant only as an explanation how the current-to-rand/1 works or if the current-to-rand/1 defined in (3) is used for generation of a mutant vector *v* and then followed by the arithmetic crossover according to (5). If we decide for the latter explanation and substitute *y* from (3) instead of *v* into (5), it results in generating the trial vector according to

$$y = x_i + U_1 U_2\,(r_1 - x_i) + U_2 F\,(r_2 - r_3), \tag{6}$$

where U_1 and U_2 are independent random variables uniformly distributed in $[0, 1]$, U_1 corresponding to (1) and U_2 corresponding to (5). However, the expected value of the multiplicative coefficient at $(r_1 - x_i)$ is $E(U_1 U_2) = E(U_1)E(U_2) = 1/4$ and the distribution of $U_1 U_2$ has positive skewness, which means that smaller values of the multiplicative coefficient are more frequent. It results in generating the trial point *y* preferably in smaller distance from x_i compared to (3). This version of *composite DE* is denoted by *CoDE0* hereafter.

The uncertainty about the correct form of *composite DE* algorithm was consulted via e-mail with Y. Wang, the first author of [12]. From his response follows that no other crossover in current-to-rand/1 strategy is applied and the (5) is just to explain why the current-to-rand/1 strategy is rotation-invariant. He also wrote, that the current-to-rand/1 strategy can be considered as two-step procedure:

1) generation of the mutant vector v by rand/1 mutation (1),
2) arithmetic crossover according to (5).

However, then we obtain the rule for generating the trial point as it follows:

$$y = x_i + U(r_1 - x_i) + UF(r_2 - r_3). \tag{7}$$

It is obvious that such explanation of the current-to-rand/1 strategy is not quite correct, (7) differs from (3) in the multiplicative coefficient at the second difference of the vectors. Meanwhile after a very careful reading of [12], a short passage hidden in the text was found in the paper [12]: *"In order to further improve the search ability of the rand/2/bin strategy, the first scaling factor F in the rand/2 mutation operator is randomly chosen from 0 to 1 in this paper"*. It means that version of *composite DE* algorithm described in the paper [12] uses the rand/2 mutation in the form as follows:

$$v = r_1 + U(r_2 - r_3) + F(r_4 - r_5), \tag{8}$$

and the current-to-rand/1 strategy generates the trial vector just according to (3). This "classic" version of composite DE which follows the description in [12] is labeled by *CoDE1* hereafter.

Through the inspection of Matlab source text of the *composite DE* algorithm and comments therein downloaded from Q. Zhang's home page[1] it was found that the current-to-rand/1 strategy is implemented in a slightly different way in the *composite DE* algorithm which is tested in the paper [12]. The points r_1, r_2, and r_3 for the mutation are chosen as a sample with replications and the current point x_i is not excluded from the sampling. They need not be mutually distinct. This approach seems to be strange, but the comment in the source code explains the reason: *"We found that using the following mechanism to choose the indices for mutation can improve the performance to certain degree"*. This variant using rand/2 according to (8) and modified current-to-rand/1 strategy with replications described above is marked by *CoDE2* hereafter.

The last variant of *composite DE* for experimental comparison. labeled by *CoDE3* hereafter, was implemented with the improved rand/2 strategy according to (8) and the current-to-rand/1 strategy according to

$$y = x_i + U_1(r_1 - x_i) + U_2 F(r_2 - r_3), \tag{9}$$

where U_1 and U_2 are independent random variables uniformly distributed in $[0, 1]$. Comparing with the current-to-rand/1 strategy (6) used in efficient *CoDE1* variant, the first scaling coefficient is again randomly distributed but uniformly with $E(U_1) = 1/2$ and the second scaling coefficient is randomized in the same way as in (6).

---

[1] http://dces.essex.ac.uk/staff/qzhang/

3 Experiments

Six well-known scalable test functions [7,9] were used as a benchmark, namely first De-Jong (sphere model), Ackley, Griewank, Rastrigin, Rosenbrock, and Schwefel function. The first four of the test functions in their non-shifted form have the global minimum point in the center of domain S, $x^* = (0,0,\ldots,0)$, which makes the search of the solution easier for many stochastic algorithms. That is why they were used in their shifted version. The shifted function is evaluated at the point $z = x - o$, $o \in S$, $o \neq (0,0,\ldots,0)$. The shift o is generated randomly from uniform D-dimensional distribution before each run. The solution of the global minimization problem is then $x^* = o$ for all the shifted functions. The definition of the test functions and the range of search domains are presented in [11]. The test functions names or their self-explaining abbreviations are used as labels when reporting the results.

The six functions with the problem dimension of $D = 30$ were used as a benchmark in experimental comparison of tested DE variants. One hundred of independent runs was executed for each test problem and each *composite DE* variant in comparison. Population size was set to $NP = 30$ for all the problems, i.e. the same as in [12].

The minimum function value (f_{min}) in the final population and the number of objective function evaluations (nfe) needed for the search was recorded in each run. The run is terminated if the difference of function values in the current population is very small or if the number of objective function evaluations was over the given limit. Small difference of function values in the current population indicates that the points of the population aggregated in a very small part of the search space and the population lost the ability to change its place considerably. The given maximum allowed number of objective function evaluations expresses our willingness to wait for the results. Such form of stopping condition is appropriate either for the benchmark tests or for the real-world applications. The run is finished if the following condition is reached:

$$f_{max} - f_{min} < 1 \times 10^{-6} \quad OR \quad nfe \geq 2 \times 10^4 \times D, \tag{10}$$

where f_{max} and f_{min} are maximum and minimum function values in the current generation, respectively. The same stopping condition is used in all the experiments.

The solution of the problem found by a DE variant was considered acceptable if $f_{min} - f(x^*) < 1 \times 10^{-4}$. If an acceptable solution is found in a run, the number of objective function evaluations needed for its finding was also returned from the run. This number of the function evaluations is denoted by nfe_near in results.

The reliability rate R of the search is assessed by the count of acceptable solutions obtained in 100 runs. The number of objective function evaluations (nfe) and the reliability rate R are fundamental experimental characteristics of the efficiency of the search.

4 Results

Two basic characteristics are needed to assess algorithm's performance. The first one is the number of function evaluations (nfe) needed for the termination of the search. The second one is the reliability rate of the search (R) expressed by the number of runs giving an acceptable solution out of total 100 runs. The values of these characteristics are shown in Table 1 for all the tested *composite DE* variants and benchmark problems. The algorithms except *CoDE3* perform with high reliability, $R \geq 98$. The reliability of *CoDE3* is much lower in three benchmark problems, even no acceptable solution was found in Rosenbrock problem. The least values of *nfe* for each problem are underlined, only the the variants with $R \geq 98$ are taken into account into this *nfe* comparison.

Table 1. Average number of function evaluations and values of reliability rate of *composite DE* variants for all the benchmark problems in experimental tests

	Ackley		DeJong1		Griewank		Rastrigin		Rosenbrock		Schwefel	
Alg	nfe	R	nfe	R	nfe	R	nfe	R	nfe	R	nfe	R
CoDE	189764	100	98703	100	152818	100	207422	100	507140	100	165572	100
CoDE0	46002	100	23920	100	35018	100	131766	100	259676	98	73691	100
CoDE1	162412	100	85365	100	132715	100	198209	100	467466	100	144816	100
CoDE2	122313	100	65035	100	100560	100	169936	100	322318	100	117908	100
CoDE3	27113	23	20296	100	26779	57	121764	99	594310	0	62238	100

Computational costs of the algorithms (expressed by *nfe*) are also compared visually using the boxplots in Figure 1. The same scale of vertical axis is used for all the problems in order to make inter-problem comparison easier. It is evident from the Figure 1 that the variants differ in their computational costs substantially. It was also confirmed statistically by one-way analysis of variance (ANOVA) carried out for the variants with almost full reliability of the search, i.e. *CoDE3* was excluded from ANOVA as the variant with low reliability in three out of six benchmark problems. ANOVA tests rejected the null hypotheses on the equivalence of expected *nfe* values in all the test problems. Tukey-Kramer multiple comparison reveals the significant differences among the variants. In all the problems, the increasing sequence of variants with respect to *nfe* is the same:

$$CoDE0 \prec CoDE2 \prec CoDE1 \prec CoDE$$

with significant difference between each consequent pairs of variants at 5 % significance level.

Another simple characteristic of the search process is the success rate. The success rate is defined as the percentage of iterations, when the trial point is better than current point of the population (condition in Line 7 in Algorithm 1 is satisfied) with respect to the total number of function evaluations *nfe*. The values of the success rate are shown in Table 2. Higher success rate means that the results of previous process are more exploited. However, too high success rate causes the suppression of exploration, which can result in premature convergence. High success rate of *CoDE3* likely causes its bad reliability in solving Rosenbrock problem.

The average number of the function evaluations necessary to find an acceptable solution gives us a view to the convergence of algorithms in the early and the last stage

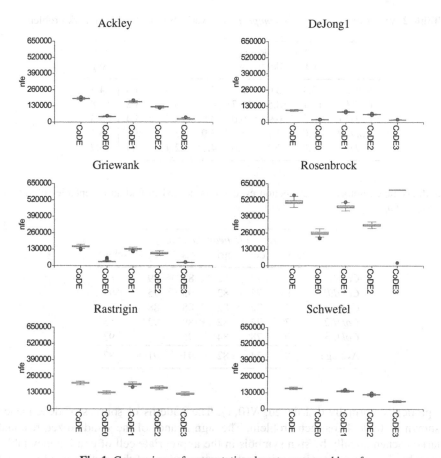

Fig. 1. Comparison of computational costs expressed by *nfe*

of the search. The average values given as the percentage of *nfe* are shown in Table 3. The number of function evaluations needed to find an acceptable solution is specific rather to the problem than variant.

Statistical comparison of the success rates of the DE strategies used in the search process is helpful for the explanation of the different performance of various *composite DE* variants in the experimental tests. The success of a given strategy means that the strategy generates the trial vector with the minimum function value among three strategies and simultaneously the function value of the trial point is less than $f(x_i)$. The counts of the success found by each strategy were recorded in the experiments, the values of counts in 5 by 3 contingency tables are evaluated cumulatively in 100 runs for each benchmark problem separately. The contingency tables were analyzed by the χ^2 tests. Null hypotheses (independence of *composite DE* variant and kind of strategy) were rejected in all the problems at significance level of $\alpha = 0.001$. The sources of dependence were assessed by standardized residuals. The standardized residuals are

Table 2. Values of success rate of *composite DE* variants in all the benchmark problems

	Success Rate (% of *nfe*)						
	ack	dej	gri	ras	ros	schw	Avg
CoDE	5.4	6.0	5.3	3.6	3.3	4.8	4.7
CoDE0	16.1	18.5	16.7	4.7	18.5	8.2	13.8
CoDE1	6.2	6.9	6.0	3.8	3.7	5.4	5.3
CoDE2	7.0	7.7	6.8	3.9	5.5	5.8	6.1
CoDE3	20.0	22.2	21.6	4.7	22.4	9.4	16.7

Table 3. Average number of the function evaluations needed to find an acceptable solution as percentage of *nfe*

	nfe_near (% of *nfe*)					
	ack	dej	gri	ras	ros	schw
CoDE	77	72	82	87	89	83
CoDE0	79	75	82	95	95	92
CoDE1	77	72	82	88	88	84
CoDE2	77	72	82	89	92	85
CoDE3	80	78	84	96	–	93
Average	78	74	82	91	91	87

asymptotically normally distributed, $N(0,1)$. The patterns of sources of dependence are shown in Table 4 for each problem. The significance of the standardized residual is marked schematically by sign symbols in the appropriate cell of contingency table. The symbol "$+++$" denotes significantly positive value of the standardized residual (i.e. this strategy is successful more frequently than expected under independence), the symbol "$---$" denotes significantly lower frequency of success at the level of significance $\alpha = 0.001$, the symbol "$++$" means significant positive value at $\alpha = 0.01$, If the value of the standardized residual is not significant, the cell is empty.

Table 4 shows almost the same patterns of strategy's success for *CoDE0* and *CoDE3* variants. In the both variants, rand/2 strategy has significantly higher success than the other strategies in all the test problems. Notice, the rand/2 strategy in *CoDE0* differs from the rand/2 strategy applied in *CoDE3*. *CoDE0* and *CoDE3* were the most efficient variants in all the test problems except Rosenbrock, see Figure 1. The patterns of *CoDE0* and *CoDE3* are different from the patterns of the other variants. Despite the similarity of success patterns, there is a big difference in the performance of *CoDE0* and *CoDE3* variants. While the reliability rate of *CoDE0* is almost 100, the reliability of *CoDE3* is much less in three test problems. It seems that *CoDE3* is too greedy, likely due to the using the version of current-to-rand/1 according to (9).

Table 4. Comparison of success rate of strategies – significance of standardized residuals

	rand1	rand2	currtorand	rand1	rand2	currtorand
		ackley			dejong1	
CoDE	+++	---	+++	+++	---	+++
CoDE0	---	+++	---	---	+++	---
CoDE1	+++	---	+++	+++	---	+++
CoDE2	++	---	+++	++	---	++
CoDE3	---	+++	---	---	+++	---
		griewank			rosenbrock	
CoDE	+++	---	++	+++	---	+++
CoDE0	---	+++	---	---	+++	---
CoDE1	+++	---	+++	+++	---	+++
CoDE2	+++	---	+++	+++	---	+++
CoDE3	---	+++	---	---	+++	---
		rastrigin			schwefel	
CoDE	+++	---	+++	+++	---	+++
CoDE0	---	+++	---	---	+++	---
CoDE1	+++	---	+++	+++	---	+++
CoDE2	---		++	++	---	++
CoDE3	---	+++	---	---	+++	---

5 Conclusion

The experimental comparison of *composite* DE variants verifies significantly different performance. The results impeaches the role of rationality in the design of well-performing stochastic global optimizers. Best performing *CoDE0* variant was proposed due to misunderstanding of *composite DE* description in [12]. Second best *CoDE2* variant uses a strange modification of the current-to-rand/1 strategy (random points are sampled with replications) described only in the source code of the algorithm. Any rational explanation of using the sampling with replications in this strategy can be hardly found. However, this variant is significantly more efficient than the classic *CoDE1* variant corresponding to the description of algorithm in [12]. The worst performing variant *CoDE3* was designed taking into account all the previous results with a rational effort to increase the algorithm's efficiency. This variant is very efficient in some problems but also appears unreliable in some other problems. However, the best performing *CoDE0* variant is a promising modification of composite DE algorithm and it should be tested in hard benchmark problems, e.g. those defined in [10] in order to verify its superiority in more thorough and convincing way. The current-to-rand/1 strategy and the choice of distribution in randomizing its multiplicative coefficients is another topic for next research.

Acknowledgement. This work was supported by the European Regional Development Fund in the IT4Innovations Centre of Excellence project (CZ.1.05/1.1.00/02.0070) and partly supported by University of Ostrava, the project SGS13/PrF/2012. The autor also thanks to Y. Wang for his e-mail discussion on the composite DE algorithm.

References

1. Brest, J., Greiner, S., Boškovič, B., Mernik, M., Žumer, V.: Self-adapting control parameters in differential evolution: A comparative study on numerical benchmark problems. IEEE Transactions on Evolutionary Computation 10, 646–657 (2006)
2. Das, S., Abraham, A., Chakraborty, U.K., Konar, A.: Differential evolution using a neighborhood-based mutation operator. IEEE Transactions on Evolutionary Computation 13, 526–553 (2009)
3. Das, S., Suganthan, P.N.: Differential evolution: A survey of the state-of-the-art. IEEE Transactions on Evolutionary Computation 15, 27–54 (2011)
4. Mallipeddi, R., Suganthan, P.N., Pan, Q.K., Tasgetiren, M.F.: Differential evolution algorithm with ensemble of parameters and mutation strategies. Applied Soft Computing 11, 1679–1696 (2011)
5. Neri, F., Tirronen, V.: Recent advances in differential evolution: a survey and experimental analysis. Artificial Intelligence Review 33, 61–106 (2010)
6. Price, K.V.: An introduction to differential evolution. In: New Ideas in Optimization, pp. 293–298. McGraw-Hill, London (1999)
7. Price, K.V., Storn, R., Lampinen, J.: Differential Evolution: A Practical Approach to Global Optimization. Springer (2005)
8. Qin, A., Huang, V., Suganthan, P.: Differential evolution algorithm with strategy adaptation for global numerical optimization. IEEE Transactions on Evolutionary Computation 13, 398–417 (2009)
9. Storn, R., Price, K.V.: Differential evolution - a simple and efficient heuristic for global optimization over continuous spaces. J. Global Optimization 11, 341–359 (1997)
10. Suganthan, P.N., Hansen, N., Liang, J.J., Deb, K., Chen, Y.P., Auger, A., Tiwari, S.: Problem definitions and evaluation criteria for the CEC 2005 special session on real-parameter optimization (2005), http://www.ntu.edu.sg/home/epnsugan/
11. Tvrdík, J.: A comparison of control-parameter-free algorithms for single-objective optimization. In: Matousek, R. (ed.) 16th International Conference on Soft Computing Mendel 2010, pp. 71–77 (2010)
12. Wang, Y., Cai, Z., Zhang, Q.: Differential evolution with composite trial vector generation strategies and control parameters. IEEE Transactions on Evolutionary Computation 15, 55–66 (2011)
13. Zhang, J., Sanderson, A.C.: JADE: Adaptive differential evolution with optional external archive. IEEE Transactions on Evolutionary Computation 13, 945–958 (2009)

Multi-Objective Differential Evolution on the GPU with C-CUDA

Fernando Bernardes de Oliveira[1], Donald Davendra[2],
and Frederico Gadelha Guimarães[3]

[1] Universidade Federal de Ouro Preto – Rua 36, 115, 35931-026, João Monlevade, MG, Brazil
Graduate Program in Electrical Engineering, Universidade Federal de Minas Gerais,
Av. Antônio Carlos 6627, 31270-901, Belo Horizonte, MG, Brazil
fbo.fernando@gmail.com
[2] Department of Computer Science, VSB Technical University of Ostrava,
17. listopadu 15, 708 33 Ostrava-Poruba, Czech Republic
donald.davendra@vsb.cz
[3] Department of Electrical Engineering, Universidade Federal de Minas Gerais,
Av. Antônio Carlos 6627, 31270-901, Belo Horizonte, MG, Brazil
fredericoguimaraes@ufmg.br

Abstract. In some applications, evolutionary algorithms may require high computational resources and high processing power, sometimes not producing a satisfactory solution after running for a considerable amount of time. One possible improvement is a parallel approach to reduce the response time. This work proposes to study a parallel multi-objective algorithm, the multi-objective version of Differential Evolution (DE). The generation of trial individuals can be done in parallel, greatly reducing the overall processing time of the algorithm. A novel approach to parallelize this algorithm is the implementation on the Graphic Processing Units (GPU). These units present high degree of parallelism and they were initially developed for image rendering. However, NVIDIA has released a framework, named CUDA, which allows developers to use GPU for general-purpose computing (GPGPU). This work studies the implementation of Multi-Objective DE (MODE) on the GPU with C-CUDA, evaluating the gain in processing time against the sequential version. Benchmark functions are used to validate the implementation and to confirm the efficiency of MODE on the GPU. The results show that the approach achieves an expressive speed up and a highly efficient processing power.

Keywords: Multi-Objective problem, Differential Evolution, GPU, C-CUDA.

1 Introduction

Evolutionary algorithms (EAs) periodically require significant computational resources to produce efficient solutions. Nonetheless, the use of a population-based approach allows the parallelization of the main operations, which can be applied independently on different individuals. As a result, the parallelization and more efficient search models may not only improve EAs performance, but also be more effective than a sequential approach, even when executed on a single processor [1].

V. Snasel et al. (Eds.): SOCO Models in Industrial & Environmental Appl., AISC 188, pp. 123–132.
springerlink.com
© Springer-Verlag Berlin Heidelberg 2013

DE [2] is a population-based optimizer that inherits the terminology and ideas of EA's in general. It begins by sampling the objective function with multiple and random initial points. These points (vectors in the search space) are generated within the bounds of the variables. New points are generated from the perturbations of existing points. In DE, perturbations are defined according to the scaled differences between pairs of candidate solutions in the population, in a operation called differential mutation. These differences depend on the spatial distribution of the population, self-adapting to the characteristics of the problem, which would explain the power of the algorithm as a general-purpose optimizer. A simplified form of DE is shown in Algorithm 1.

Algorithm 1. Pseudo-code for a simplified form of DE

---

1 Initialize(pop, Np);
2 Evaluate(pop);
3 **while** *(convergence criterion not met)* **do**
4   **foreach** *individual i in pop* **do**
5   select randomly $r1 \neq r2 \neq r3$;
6          $p_{i,g}$ = GeneratePerturbation(x_{r1}, x_{r2});
7          $v_{i,g}$ = Mutation(x_{r3}, F, $p_{i,g}$);
8          $u_{i,g}$ = Crossover($x_{i,g}$, $v_{i,g}$);
        // Selection
9   **if** $f(u_{i,g}) \leq f(x_{i,g})$ **then**
10         $x_{i,g+1} = u_{i,g}$;
11   **else**
12         $x_{i,g+1} = x_{i,g}$;
13 print(solution);

Multi-objective optimization problems (MOPs) have more than one objective function to be optimized. Usually, in this problem, some functions are in conflict with each other, representing competing trade-off between different designs. The result for this problem is a solution set, called the Pareto Front [3]. In such a situation, it is essential to define comparisons between solutions, such as Pareto dominance. The Nondominated Sorting Genetic Algorithm II (NSGAII) classifies and sorts the solutions using Fast Non-Dominated Sorting Procedure (FNDSP). This process defines dominant and dominated solutions. Xue et al. [4,5] proposed a multi-objective approach based on DE, named Multi-Objective Differential Evolution (MODE) by the authors, which extends the ideas from NSGAII. MODE uses Pareto-based dominance to implement the selection of the best individual for the mutation operation of an individual. Similarly, [6] proposed the Differential Evolution for Multiobjective Optimization (DEMO), which uses Pareto-based ranking and crowding distance sorting and three different methods to select an individual after these operations.

The existence of a population of candidate solutions makes EAs naturally parallel search methods. There is a vast literature on how to parallelize EAs [1]. A novel approach to parallelize EAs is their implementation on the Graphic Processing Units

(GPU). These units are highly parallel, multithreaded and having manycore processors. They were initially developed for image rendering, but the use of GPU for general-purpose computing (GPGPU) with C/C++ became possible after NVIDIA released its framework, a parallel computing architecture named Compute Unified Device Architecture (CUDA) [7]. Given this scenario, this work studies the implementation of Multi-Objective Differential Evolution on the GPU with C-CUDA (MODE+GPU), evaluating the gain in processing time against the sequential version.

There are some works in the literature concerning the implementation of DE in GPUs. For instance, in [8] the authors present an implemention of DE in a single objective context. They state that their work is the first implementation of DE in C-CUDA. Some problems with random numbers are reported by the authors. In this work we implement random numbers on the GPU. In [9], the authors report an implementation of a co-evolutionary DE in C-CUDA for solving min-max problems. The optimization process uses two independent populations: one for the optimization variables and another one for the Lagrange multiplier vector. DE and Markov Chain Monte Carlo (MCMC) are combined in [10] to solve multi-objective continuous optimization problems. A population of Markov chains evolves toward a diversified set of solutions at the Pareto optimal front. In our work the solutions are ranked with Fast Non-Dominated Sorting Procedure and Crowding Distance.

The paper is structured as follows. In Section 2, we show the parallelization of DE, together with considerations and some limitations. The experimental setups and the performance comparison between the algorithms are presented in Section 3. We conclude this article with some discussions in Section 4.

2   Multi-objective Differential Evolution on the GPU

The most important question is to define which part of an algorithm will be parallelized. The performance will be influenced by the appropriate choice for parallelism in the sequential code. This presents the greatest difficulty, because it involves many issues, such as shared memory, concurrency, synchronism amongst others. DE executes its evolution operations for each individual of the population with size Np. Then, there are Np operations for generation. (see Algorithm 1, lines 4–8). This part is a suitable choice to be parallelized, in which the evolution operation for each individual will run on GPU. As a result, each trial individual is created simultaneously. This approach, named MODE+GPU, is shown in Algorithm 2.

The function RunDEonGPU (lines 4) is a CUDA kernel, which is specified using an execution configuration ($<<< ... >>>$). This function is invoked with Nb blocks and Nt CUDA threads for each block and is executed on the GPU (called *device*). The operations (lines 6–9) are executed in parallel by Nt different CUDA threads. The others instructions (lines 1–2; 3; 10–14) are executed in sequence on the CPU (called *host*).

The random numbers generation on GPU is carried out by CURAND Library [11]. The pseudorandom numbers are created on the device and the results are stored in a global memory. These numbers are used to select individuals randomly (line 6) and to create a trial individual.

After creation of the trial population in parallel, it is merged with the current population (line 10). This process creates a new population with size $2 * Np$ [3,12]. The

---

Algorithm 2. Pseudo-code for MODE+GPU

---

1 Initialize(pop, Np);
2 Evaluate(pop);
3 **while** (*convergence criterion not met*) **do**

4   `// Each thread will run`
    trialPop = **RunDEonGPU** $<<< Nb, Nt >>>$(pop)
5   **begin**
6   select randomly $r1 \neq r2 \neq r3$;
7            $p_{i,g}$ = GeneratePerturbation(x_{r1}, x_{r2});
8            $v_{i,g}$ = Mutation($x_{r3}, F, p_{i,g}$);
9            $u_{i,g}$ = Crossover($x_{i,g}, v_{i,g}$);

    `// to generate the trial population`

10   newPop = merge(pop, trialPop); `// 2 * Np`
11   ranking(newPop, 2 * Np); `// Fast Nondominated Sorting`
12   crowdingDistance(newPop, 2 * Np);
13   pop = select(newPop, Np);

14 print(solution);

---

individuals are then classified and sorted using Fast Non-Dominated Sorting Procedure (FNDSP) (line 11). Crowding distance is calculated (line 12) to distinguish individuals belonging to the same front (same rank). The new population, for the next generation, is built using the dominance criterion and crowding distance to select the Np best individuals (line 13). The algorithm runs until some convergence criterion is met (line 3). The data are stored in RAM memory and must be copied to GPU memory to be processed. This process occurs before the function RunDEonGPU (lines 4) is invoked. After the execution of the CUDA threads and immediately before the merge operation (line 10), the result is copied again from GPU memory to RAM memory.

In the next section, the experimental setups are described and the performance comparison between the algorithms are presented. Moreover, the computational environment, the parameters and the benchmark functions are explained.

3 Experiments

In this section, the parallel approach proposed is compared with the sequential context (CPU only). Benchmark functions are used to validate and to confirm the efficiency of MODE+GPU and are shown in Section 3.1. The computational environment is presented in Section 3.2 and the parameters are explained in Section 3.3. At the end, the results are shown in Section 3.4.

3.1 Benchmark Functions

We have used four functions from the ZDT benchmark functions to test and validate the parallel approach. These functions are used by [3,13,14] and are shown in Table 1. All objective functions are to be minimized.

Table 1. Benchmark functions

Func.	Var.	Var. bounds	Comments	Objective functions
ZDT1	30	$[0,1]$	convex	$f_1(x) = 1 - \exp(-4x_1)\sin^6(6\pi x_1);$ $f_2(x) = p(x)\left[1 - \left(\frac{f_1(x)}{p(x)}\right)^2\right]$ $p(x) = 1 + 9\left[\frac{(\sum_{i=2}^n x_i)}{n-1}\right]^{0.25}$
ZDT2	30	$[0,1]$	nonconvex	$f_1(x) = x_1;$ $f_2(x) = p(x)\left[1 - \left(\frac{x_1}{p(x)}\right)^2\right]$ $p(x) = 1 + \frac{9(\sum_{i=2}^n x_i)}{n-1}$
ZDT3	30	$[0,1]$	convex, disconnected	$f_1(x) = x_1;$ $f_2(x) = p(x)\left[1 - \left(\sqrt{\frac{x_1}{p(x)}} - \frac{x_1}{p(x)}\sin(10\pi x_1)\right)\right]$ $p(x) = 1 + \frac{9(\sum_{i=2}^n x_i)}{n-1}$
ZDT6	10	$[0,1]$	nonconvex, nonuniformly spaced	$f_1(x) = 1 - \exp(-4x_1)\sin^6(6\pi x_1);$ $f_2(x) = p(x)\left[1 - \left(\frac{f_1(x)}{p(x)}\right)^2\right]$ $p(x) = 1 + 9\left[\frac{(\sum_{i=2}^n x_i)}{n-1}\right]^{0.25}$

3.2 Computational Environment

Both algorithms were executed on a MacBookPro, with a Intel Core 2 Duo (2,4 GHz) processor, 4 GB RAM and Mac OS X (update 10.6.8) operating system. The algorithms were implemented using NetBeans 7.0.1. The NVIDIA *Cuda Compiler* (NVCC) is version 4.0 and GCC/G++ Compiler is version 4.2.1.

The GPU unit is the *GeForce 320M* and is integrated on the main board. This GPU has 48 CUDA Cores, 254 MB of global memory, 512 CUDA *threads* per block, each block with dimension $512 \times 512 \times 64$ and the grids with dimension $65535 \times 65535 \times 1$. Furthermore, it supports CUDA *Capability* 1.2, with single point precision only (float) and the CUDA Driver Version is 4.0.

3.3 Parameters

The parameters of the algorithm are shown on Table 2. For the *F weight* and the *Cr probability*, we have configurations with fixed values and random values, making up 9 different algorithms. Each configuration set was executed 5 times on each problem, totaling 180 results. In this paper, we present the graphics related to some results chosen randomly. The stopping criteria is the maximum number of generations.

Table 2. Parameters

Parameter	Value
Population size (Np)	100 individuals
Max generations	1,000
Runs for configuration set	5
F weight	0.4
	0.7
	random (in each generation)
Cr probability	0.5
	0.9
	random (in each generation)

3.4   Results

At first, the average processing time is verified in 3.4, with some considerations and comments. In Section 3.4 we present the tests with benchmark functions (Section 3.1) to validate MODE+GPU.

Average Processing Time. This experiment aims at evaluating the performance of algorithms on the CPU only and on the GPU. This is a very important comparison to verify if the MODE+GPU can indeed generate efficient solutions with less processing time or at least better solutions in the same time. The results are shown on Table 3. The time corresponds to 100 individuals and 1,000 generations (Table 2).

Table 3. Result – Average processing time

Version	Average processing time (sec.)
CPU only	3,000
MODE+GPU	10
Improvement	30,000%

The difference between the two implementations is very expressive, representing a gain of 30,000%. MODE+GPU needs to transfer data between RAM memory and GPU memory, which represents a relevant overhead in the implementation. Nonetheless, its mean time was the lowest. This shows that MODE+GPU can obtain equivalent solutions with less processing time. As a result, MODE+GPU could run for approximately 300,000 generations or 30,000. If the population is increased other effects must be observed, such as function evaluation, memory, time to transfer data between CPU and GPU. Besides, the number of started CUDA threads cannot be supported by GPU, which results in a process queue. This situation could seriously damage the performance. The recommendation is to find a suitable balance between the number of generations and individuals.

Benchmark Functions. This section presents the graphical results for the Functions ZDT1, ZDT2, ZDT3 and ZDT6, which were explained in Section 3.1. The figures report MODE+GPU results. The first function, ZDT1, is a convex one and the results for that problem are shown in Figures 1(a) and 1(b). These results are very satisfactory, because the points have reached the true Pareto Front with good proximity and diversity. Overall, both parameter sets produced the similar results.

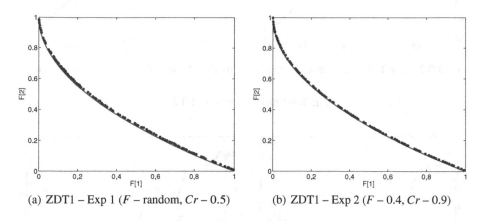

(a) ZDT1 – Exp 1 (F – random, Cr – 0.5)   (b) ZDT1 – Exp 2 (F – 0.4, Cr – 0.9)

Fig. 1. Results – Function ZDT1

Figures 2(a) and 2(b) show the results for Function ZDT2, which is a nonconvex function. The points in Figure 2(a) are closer to the true Pareto front than in the other figure. This can be explained by a random Cr probability, which might have hampered the convergence. Nevertheless, the results are also quite satisfactory.

The Function ZDT3 is convex and disconnected. The results are displayed in Figures 3(a) and 3(b). In the first figure, there are some points far from the Pareto Front, in the 4th segment. In this case, the random parameter is the F weight, which could have caused such a fluctuation of solutions. The Cr probability in the second experiment is also random based, however it has not affected the results. In conclusion, the results can be considered satisfactory.

Figures 4(a) and 4(b) show the results for Function ZDT6, which is a nonconvex function and nonuniformly spaced. The first experiment has used all parameters randomly. This situation has not implicated the result in this experiment. The points obtained were very close to the Pareto Front. The second experiment shows that a fixed F might slow down the convergence. Nevertheless, these results can be considered acceptable.

Generally, the results were very satisfactory. The Pareto Front was covered in the majority of cases. The random parameters had not implicated the results severely. On the contrary, this approach has produced good solutions, diversifying the search. Therefore, the MODE+GPU efficiency is illustrated with an expressive speed up and good results in the benchmark functions.

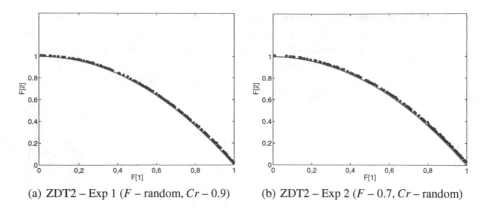

(a) ZDT2 – Exp 1 (F – random, Cr – 0.9)   (b) ZDT2 – Exp 2 (F – 0.7, Cr – random)

Fig. 2. Results – Function ZDT2

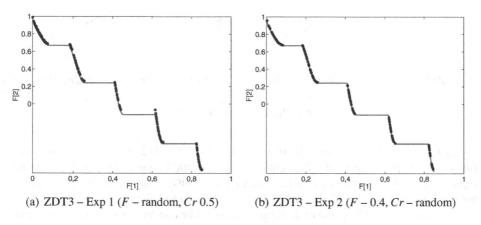

(a) ZDT3 – Exp 1 (F – random, Cr 0.5)   (b) ZDT3 – Exp 2 (F – 0.4, Cr – random)

Fig. 3. Results – Function ZDT3

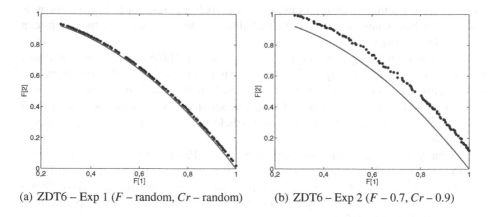

(a) ZDT6 – Exp 1 (F – random, Cr – random)   (b) ZDT6 – Exp 2 (F – 0.7, Cr – 0.9)

Fig. 4. Results – Function ZDT6

4   Conclusions

In this paper, a new approach to multi-objective optimization problems using Differential Evolution on the GPU with C-CUDA was presented and discussed. In the proposed approach, some parts of DE were parallelized and implemented to run on graphics processing unit with the NVIDIA's CUDA framework.

This work presented a significant application with Evolutionary algorithms. The proposed approach showed a high speed up when compared with the CPU version. The improvement in the processing time was very expressive. GPUs are fast, highly parallel processor units and may improve the response time of EAs significantly. The MODE+GPU efficiency was confirmed by the good results achieved in benchmark functions. The Pareto Front was covered in the majority of cases. The random parameters had not implicated the results severely.

Some questions remain to be investigated. First, identifying other parts of the algorithm MODE+GPU that can be parallelized. Moreover, new operators will be studied to improve the results and a coevolutionary model with subpopulations will be developed for the GPU. The variation of the speed up as the dimensionality of the problem increases is another important point to evaluate.

References

1. Nedjah, N., de Macedo Mourelle, L., Alba, E. (eds.): Parallel Evolutionary Computations. SCI, vol. 22. Springer (2006)
2. Price, K.V., Storn, R.M., Lampinen, J.A.: Differential evolution: a practical approach to global optimizations. Springer (2005)
3. Deb, K., Pratap, A., Agarwal, S., Meyarivan, T.: A Fast Elitist Multi-Objective Genetic Algorithm: NSGA-II. IEEE Transactions on Evolutionary Computation 6, 182–197 (2000)
4. Xue, F., Sanderson, A., Graves, R.: Pareto-based multi-objective differential evolution. In: The 2003 Congress on Evolutionary Computation, CEC 2003, vol. 2, pp. 862–869 (December 2003)
5. Xue, F.: Multi-objective Differential Evolution: Theory and Applications. PhD thesis, Rensselaer Polytechnic Institute, New York (September 2004)
6. Robič, T., Filipič, B.: DEMO: Differential Evolution for Multiobjective Optimization. In: Coello Coello, C.A., Hernández Aguirre, A., Zitzler, E. (eds.) EMO 2005. LNCS, vol. 3410, pp. 520–533. Springer, Heidelberg (2005)
7. NVIDIA: CUDA C Programming Guide. 4.0 edn. NVIDIA (June 2011)
8. de Veronese, L., Krohling, R.: Differential evolution algorithm on the GPU with C-CUDA. In: 2010 IEEE Congress on Evolutionary Computation (CEC), pp. 1–7 (July 2010)
9. Fabris, F., Krohling, R.A.: A co-evolutionary differential evolution algorithm for solving min-max optimization problems implemented on GPU using C-CUDA. Expert Systems with Applications (2011)
10. Zhu, W., Yaseen, A., Li, Y.: DEMCMC-GPU: An Efficient Multi-Objective Optimization Method with GPU Acceleration on the Fermi Architecture. New Generation Computing 29, 163–184 (2011), doi:10.1007/s00354-010-0103-y

11. NVIDIA: NVIDIA cuRAND (January 2012),
    http://developer.nvidia.com/curand
12. Coello Coello, C.A.: Evolutionary multi-objective optimization: a historical view of the field.
    IEEE Computational Intelligence Magazine 1(1), 28–36 (2006)
13. Zitzler, E., Deb, K., Thiele, L.: Comparison of Multiobjective Evolutionary Algorithms: Empirical Results. Evolutionary Computation 8, 173–195 (2000)
14. Batista, L.S., Campelo, F., Guimarães, F.G., Ramírez, J.A.: Pareto Cone ε-Dominance: Improving Convergence and Diversity in Multiobjective Evolutionary Algorithms. In: Takahashi, R.H.C., Deb, K., Wanner, E.F., Greco, S. (eds.) EMO 2011. LNCS, vol. 6576, pp. 76–90. Springer, Heidelberg (2011)

Competitive Differential Evolution Algorithm in Comparison with Other Adaptive Variants

Radka Poláková and Josef Tvrdík

Centre of Excellence IT4Innovations division of University of Ostrava
Institute for Research and Applications of Fuzzy Modeling
{radka.polakova,josef.tvrdik}@osu.cz

Abstract. The differential evolution algorithm using competitive adaptation was compared experimentally with the state-of-the-art adaptive versions of differential evolution on CEC2005 benchmark functions. The results of experiments show that the performance of the algorithm with competitive adaptation is comparable with the state-of-the-art algorithms, outperformed only by CoDE and JADE algorithms in this test. A modification of competitive differential evolution preferring successful strategy for a longer period of search was also investigated. Such modification brings no improvement and the standard setting of the competition recommended in previous papers is suitable for applications.

1 Introduction

Differential evolution (DE) is simple population-based algorithm for the global optimization introduced by Storn and Price [8]. DE has become one of the evolutionary algorithms most frequently used for solving the global optimization problems in recent years [6]. Compared with other evolutionary algorithms, DE has a very small number of control parameters. However, it is commonly known that the performance of DE algorithm is strongly dependent on the values of these parameters. Tuning the proper values of control parameters for solving a particular optimization problem by trial-and-error can take a lot of time. Because of this fact many adaptive approaches in DE have appeared in literature. Four of them, namely jDE [1], SADE [7], JADE [16], and EPSDE [4] are usually considered as the state-of-the-art adaptive variants of DE algorithm. DE algorithm with composite trial vector generation strategies and control parameters (CoDE) has appeared recently and its performance was found comparable with the state-of-the-art algorithms [13].

The main goal of the study is to compare the performance of recently proposed variant of competitive differential evolution (CDE) [12] with the other well-performing adaptive DE algorithms on the hard benchmark test functions [9]. Another goal is to study the influence of small changes in competitive mechanism on the performance.

2 Differential Evolution

DE works with two alternating generations of population, P and Q. The points of population are considered as candidates of solution. At the beginning, the generation P is

V. Snasel et al. (Eds.): SOCO Models in Industrial & Environmental Appl., AISC 188, pp. 133–142.
springerlink.com

initialized randomly in the search domain S, $S = \prod_{j=1}^{D}[a_j, b_j]$, $a_j < b_j$, $j = 1, 2, \ldots, D$. A new point y (trial point) is produced by mutation and crossover operations for each point $x_i \in P$, $i \in \{1, 2, \ldots, NP\}$, where NP is the size of population. Assuming minimization, the point y is inserted into new generation Q if $f(y) \leq f(x_i)$, otherwise the point x_i enters into Q. After completing the new generation, Q becomes the old generation P and the whole process continues until the stopping condition is satisfied. The basic scheme of DE is shown in a pseudo-code in Algorithm 1.

Algorithm 1. Differential evolution

---

generate an initial population $P = (x_1, x_2, \ldots, x_{NP})$, $x_i \in S$ distributed uniformly
while stopping condition not reached **do**
    for $i = 1$ to NP **do**
        generate a trial vector y
        if $f(y) \leq f(x_i)$ **then**
            insert y into new generation Q
        else
            insert x_i into new generation Q
        end if
    end for
    $P := Q$
end while

The trial vector y is generated by crossover of two parent vectors, the current (target) vector x_i and a mutant vector v. The mutant vector v is obtained by a kind of mutation. Many kinds of mutation have been proposed, see e.g. [2,5,6,8], we mention those used in algorithms compared in this study. Suppose that r_1, r_2, r_3, r_4, and r_5 are five mutually distinct points taken randomly from population P, not coinciding with the current x_i, $F > 0$ is an input control parameter, and $\text{rand}(0, 1)$ is a random number uniformly distributed between 0 and 1. The mutant vector v can be generated as follows:

- rand/1
$$v = r_1 + F(r_2 - r_3), \tag{1}$$
- rand/2
$$v = r_1 + F(r_2 - r_3) + F(r_4 - r_5), \tag{2}$$
- best/2
$$v = x_{\text{best}} + F(r_1 - r_2) + F(r_3 - r_4), \tag{3}$$

where x_{best} is the point with the minimum function value in the current population.
- rand-to-best/2
$$v = r_1 + F(x_{\text{best}} - r_1) + F(r_2 - r_3) + F(r_4 - r_5), \tag{4}$$
- current-to-rand/1
$$y = x_i + \text{rand}(0, 1) \times (r_1 - x_i) + F(r_2 - r_3). \tag{5}$$

Note that the current-to-rand/1 mutation generates a trial point y directly, because (5) includes so called arithmetic crossover.

- randrl/1

$$v = r_1^x + F(r_2 - r_3),\qquad(6)$$

where the point r_1^x is not chosen randomly like in rand/1, but tournament best among r_1, r_2, and r_3, i.e. $r_1^x = \arg\min_{i\in\{1,2,3\}} f(r_i)$, as proposed in [3].
- current-to-pbest/1 [16]

$$v = x_i + F(x_{\text{pbest}} - x_i) + F(r_1 - r_2),\qquad(7)$$

where x_{pbest} is randomly chosen from $100\,p\,\%$ best individuals with input parameter $p \in (0,1]$, value of $p \in [0.05, 0.20]$ is recommended. The vector $r_1 \neq x_i$ is randomly selected from P, r_2 is randomly selected from the union $P \cup A$ of the current population P and the archive A.

The crossover operator constructs the trial vector y from current individual x_i and the mutant vector v. Two types of crossover were proposed by Storn and Price in [8]. Binomial crossover replaces the elements of vector x_i using the following rule

$$y_j = \begin{cases} v_j & \text{if} \quad U_j \leq CR \quad \text{or} \quad j = l \\ x_{ij} & \text{if} \quad U_j > CR \quad \text{and} \quad j \neq l, \end{cases}\qquad(8)$$

where l is a randomly chosen integer from $\{1, 2, \ldots, D\}$, and $U_1, U_2, \ldots, U_D$ are independent random variables uniformly distributed in $[0, 1)$. $CR \in [0, 1]$ is a control parameter influencing the number of elements to be exchanged by crossover. Eq. (8) ensures that at least one element of x_i is changed, even if $CR = 0$.

In exponential crossover, the starting position of crossover is also chosen randomly from $1, \ldots, D$, but L consecutive elements (counted in circular manner) are taken from the mutant vector v. Probability of replacing the kth element in the sequence $1, 2, \ldots, L$, $L \leq D$, decreases exponentially with increasing k. L adjacent elements are changed in exponential variant, in binomial one the changed coordinates are dispersed randomly over the coordinates $1, 2, \ldots, D$. While the relation between the probability of mutation and the CR is linear in binomial crossover, in the exponential crossover this relation is nonlinear and the deviation from linearity enlarges with increasing dimension of problem. Probability of mutation (p_m) controls the number of exchanged elements in crossover, $p_m \times D$ is the expected number of mutant elements used in producing offsprings. Zaharie [14,15] derived the relation between p_m and CR for exponential crossover. Her result can be rewritten in the form of polynomial equation

$$CR^D - D\,p_m\,CR + D\,p_m - 1 = 0.\qquad(9)$$

The value of CR for given value of $p_m \in (1/D, 1)$ can be evaluated as the root of the equation (9).

The combination of mutation and crossover is called DE strategy, usually abbreviated by DE/m/n/c, where m stands for a kind of mutation, n for the number of differences of randomly selected points in mutation, and c for the type of crossover.

3   Competitive Differential Evolution

Adaptive mechanism for DE algorithm based on the competition of different strategies or different settings of $[F, CR]$ was introduced in [10]. Let us have H strategies or different $[F, CR]$ settings in a pool. For simplicity, we speak on H strategies in the pool. Any of H strategies can be chosen for the generation of a new trial point y. A strategy is selected randomly with probability q_h, $h = 1, 2, \ldots, H$. At the start the values of probability are set uniformly, $q_h = 1/H$, and they are modified according to their success rates in the preceding steps of the search process. The hth strategy is considered successful if it generates such a trial vector y satisfying $f(y) \le f(x_i)$. Probability q_h is evaluated as the relative frequency according to

$$q_h = \frac{n_h + n_0}{\sum_{j=1}^{H} (n_j + n_0)} \,, \tag{10}$$

where n_h is the current count of the hth strategy successes, and $n_0 > 0$ is an input parameter. The setting of $n_0 > 1$ prevents from a dramatic change in q_h by one random successful use of the hth strategy. To avoid degeneration of the search process, the current values of q_h are reset to their starting values if any probability q_h decreases below some given limit δ, $\delta > 0$.

Several variants of competitive DE differing both in the pool of DE strategies and in the set of control-parameters values were tested [11]. A variant of competitive DE appeared well-performing and robust in different benchmark tests in [12]. In this variant, denoted *b6e6rl* hereafter, 12 strategies are in competition ($H = 12$), six of them with the binomial crossover and six ones with the exponential crossover. The randrl/1 mutation (6) is applied in all the strategies. Two different values of control parameter F ($F = 0.5$ and $F = 0.8$) are combined with three values of CR, which gives six different setting for each crossover. The binomial crossover uses the values of $CR \in \{0, 0.5, 1\}$. The values of CR for exponential crossover are evaluated as the roots of the equation (9), corresponding three values of probability p_m are set up equidistantly in the interval $(1/D, 1)$. The input parameters controlling competition are standardly set up to $n_0 = 2$ and $\delta = 1/(5 \times H)$.

4   Experiments and Results

The adaptive DE algorithms experimentally compared in this study including the basic features of their adaptive mechanism are briefly summarized in Table 1.

Two modifications of b6e6rl algorithm are tested. In the first one, hereafter called b6e6rl60, the parameter δ controlling the competition was set to conventional value, i.e. $\delta = 1/(5*H) = 1/60$. In the second one, hereafter called b6e6rl480, the parameter δ is set to a much less value, $\delta = 1/(40*H) = 1/480$, in order to find out how the preference of successful strategies for a longer period influences the performance of the algorithm.

These two modifications of b6e6rl algorithm were applied to twenty five test functions defined for CEC2005 competition [9]. Tests were carried out for the $D = 30$ dimension of the problems under the experimental conditions specified in [13], 25

Table 1. Adaptive DE variants in experimental comparison

Algorithm	Strategy #	type	Adaptive mechanism
jDE [1]	1	rand/1/bin	evolutionary self-adaptation of F and CR with respect to their success
JADE [16]	1	curr-to-pbest/1/bin	adaptation of F and CR with respect to their success in the current generation
SADE [7]	4	rand/1/bin rand/2/bin rand-to-best/2/bin curr-to-rand/1	competition of strategies, F random without adaptation, CR – median of successful values in last LP generations
EPSDE [4]	3	rand/1/bin best/2/bin curr-to-rand/1	competition of strategies, evolutionary selection of successful strategy and control parameters, mutation by random selection from the pool of strategies and the pools of F and CR fixed values
CoDE [13]	3	rand/1/bin rand/2/bin curr-to-rand/1	F and CR selected at random from three pairs of fixed values, tournament selection among strategies
b6e6rl [12]	2	randrl/1/bin randrl/1/exp	competition of strategies with assigned fixed values of F and CR, probability of strategy selection proportional to its success, resetting the values of probability if any of probability values is too small

independent runs for each benchmark function were carried out, each run was stopped if the number of function evaluations $FES = 3 \times 10^5$ was achieved.

A comparison of b6e6rl algorithm modifications with other algorithms is presented in Tables 2 and 3. The average and standard deviation of the function error values are given here. The function error value in a run is computed as $f(x_{min}) - f(x^*)$, where x_{min} is the best solution found by the algorithm in a run and x^* is the global minimum of the function. The results of the other algorithms are taken from [13]. The minimum average error values for each function are printed in bold, all the error values less than 1×10^{-10} are considered equal. The numbers of wins are given in the last rows of the tables.

With respect to the number of wins, the best algorithms are CoDE, JADE, and b6e6rl60 in this order, followed by the medium-performing algorithms (b6e6rl480, EPSDE, and jDE), SADE appeared the worst performing.

A summarized comparison of the algorithms on 25 benchmark functions is shown in Table 4. The rank of the algorithm with respect to average function error was assigned to each algorithm for each function. All the error values less than 1×10^{-10} are again considered equal and were assigned by their average rank. The algorithms are presented in the ascending order of their average rank. We can see that best performing algorithms are CoDE and JADE while other algorithms differ only very little and there is almost no difference in the performance of b6e6rl modifications.

Similar summarized comparison on the subset of easier test functions (F1 to F14) is depicted in Table 5. The order of the algorithms differs a little from the comparison on the whole set of functions. CoDE and JADE are again best performing but b6e6rl480 modification is the third best algorithm followed by b6e6rl60 with almost equal average rank. The other algorithms have their average rank greater by more than one.

The modifications of b6e6rl algorithm were also compared in the second set of experiments in order to investigate the ability of the algorithm to find an acceptable solution of the problem. The experiments were carried out on the easier subset of test functions (F1–F14), 25 runs were performed for each function. The higher maximum number of *FES* was allowed but a run was stopped if the solution found in the run was near to the correct solution of the problem. The stopping condition was set as follows:

$$(f(x_{min}) - f(x^*)) < \varepsilon \quad OR \quad FES > 3 \times 10^6$$

Table 2. Results of b6e6rl60, b6e6rl480, JADE, and jDE, $D = 30$, $FES = 300000$

F	b6e6rl60		b6e6rl480		JADE		jDE	
	mean	std	mean	std	mean	std	mean	std
F1	**0.00E+00**	0.0E+00	**0.00E+00**	0.0E+00	**0.00E+00**	0.0E+00	**0.00E+00**	0.0E+00
F2	**1.18E-13**	6.3E-14	**2.98E-13**	1.5E-13	**1.07E-28**	1.0E-28	1.11E-06	2.0E-06
F3	9.09E+04	5.3E+04	3.79E+05	2.2E+05	**8.42E+03**	7.3E+03	1.98E+05	1.1E+05
F4	**1.11E-13**	7.2E-14	5.77E-07	2.0E-06	**1.73E-16**	5.4E-16	4.40E-02	1.3E-01
F5	5.44E+02	5.4E+02	1.14E+03	6.7E+02	**8.59E-08**	5.2E-07	5.11E+02	4.4E+02
F6	**3.41E-14**	2.8E-14	**3.41E-14**	2.8E-14	1.02E+01	3.0E+01	2.35E+01	2.5E+01
F7	**6.30E-03**	7.7E-03	7.09E-03	8.6E-03	8.07E-03	7.4E-03	1.18E-02	7.8E-03
F8	2.10E+01	5.7E-02	2.09E+01	7.1E-02	2.09E+01	1.7E-01	2.09E+01	4.9E-02
F9	**0.00E+00**	0.0E+00	**0.00E+00**	0.0E+00	**0.00E+00**	0.0E+00	**0.00E+00**	0.0E+00
F10	6.38E+01	1.0E+01	4.79E+01	9.0E+00	**2.41E+01**	4.6E+00	5.54E+01	8.5E+00
F11	2.66E+01	2.1E+00	2.75E+01	1.6E+00	2.53E+01	1.7E+00	2.79E+01	1.6E+00
F12	1.48E+04	6.3E+03	1.16E+04	4.6E+03	6.15E+03	4.8E+03	8.63E+03	8.3E+03
F13	1.42E+00	1.2E-01	**1.25E+00**	8.3E-02	1.49E+00	1.1E-01	1.66E+00	1.4E-01
F14	1.26E+01	2.3E-01	1.25E+01	3.0E-01	**1.23E+01**	3.1E-01	1.30E+01	2.0E-01
F15	3.64E+02	1.2E+02	3.88E+02	8.3E+01	3.51E+02	1.3E+02	3.77E+02	8.0E+01
F16	1.32E+02	1.0E+02	9.46E+01	7.0E+01	1.01E+02	1.2E+02	7.94E+01	3.0E+01
F17	1.61E+02	7.1E+01	1.15E+02	2.6E+01	1.47E+02	1.3E+02	1.37E+02	3.8E+01
F18	9.05E+02	1.2E+00	9.06E+02	1.5E+00	9.04E+02	1.0E+00	9.04E+02	1.1E+01
F19	9.06E+02	1.7E+00	9.06E+02	2.0E+00	9.04E+02	8.4E-01	9.04E+02	1.1E+00
F20	9.05E+02	1.0E+00	9.06E+02	2.8E+00	9.04E+02	8.5E-01	9.04E+02	1.1E+00
F21	**5.00E+02**	1.2E-13	**5.00E+02**	1.2E-13	**5.00E+02**	4.7E-13	**5.00E+02**	4.8E-13
F22	8.82E+02	1.9E+01	8.88E+02	2.9E+01	8.66E+02	1.9E+01	8.75E+02	1.9E+01
F23	**5.34E+02**	3.6E-04	**5.34E+02**	3.9E-04	5.50E+02	8.1E+01	**5.34E+02**	2.8E-04
F24	**2.00E+02**	6.0E-13	**2.00E+02**	6.0E-13	**2.00E+02**	2.9E-14	**2.00E+02**	2.9E-14
F25	**2.11E+02**	1.1E+00	2.12E+02	1.1E+00	**2.11E+02**	7.9E-01	**2.11E+02**	7.3E-01
#wins	10		8		11		6	

Table 3. Results of SADE, EPSDE, and CoDE, $D = 30$, $FES = 300000$

	SADE		EPSDE		CoDE	
F	mean	std	mean	std	mean	std
F1	**0.00E+00**	0.0E+00	**0.00E+00**	0.0E+00	**0.00E+00**	0.0E+00
F2	8.26E-06	1.7E-05	**4.23E-26**	4.1E-26	**1.69E-15**	4.0E-15
F3	4.27E+05	2.1E+05	8.74E+05	3.3E+06	1.05E+05	6.3E+04
F4	1.77E+02	2.7E+02	3.49E+02	2.2E+03	5.81E-03	1.4E-02
F5	3.25E+03	5.9E+02	1.40E+03	7.1E+02	3.31E+02	3.4E+02
F6	5.31E+01	3.3E+01	6.38E-01	1.5E+00	1.60E-01	7.9E-01
F7	1.57E-02	1.4E-02	1.77E-02	1.3E-02	7.46E-03	8.6E-03
F8	2.09E+01	5.0E-02	2.09E+01	5.8E-02	**2.01E+01**	1.4E-01
F9	2.39E-01	4.3E-01	3.98E-02	2.0E-01	**0.00E+00**	0.0E+00
F10	4.72E+01	1.0E+01	5.36E+01	3.0E+01	4.15E+01	1.2E+01
F11	1.65E+01	2.4E+00	3.56E+01	3.9E+00	**1.18E+01**	3.4E+00
F12	**3.02E+03**	2.3E+03	3.58E+04	7.1E+03	3.05E+03	3.8E+03
F13	3.94E+00	2.8E-01	1.94E+00	1.5E-01	1.57E+00	3.3E-01
F14	1.26E+01	2.8E-01	1.35E+01	2.1E-01	**1.23E+01**	4.8E-01
F15	3.76E+02	7.8E+01	**2.12E+02**	2.0E+01	3.88E+02	6.9E+01
F16	8.57E+01	6.9E+01	1.22E+02	9.2E+01	**7.37E+01**	5.1E+01
F17	7.83E+01	3.8E+01	1.69E+02	1.0E+02	**6.67E+01**	2.1E+01
F18	8.68E+02	6.2E+01	**8.20E+02**	3.4E+00	9.04E+02	1.0E+00
F19	8.74E+02	6.2E+01	**8.21E+02**	3.4E+00	9.04E+02	9.4E-01
F20	8.78E+02	6.0E+01	**8.22E+02**	4.2E+00	9.04E+02	9.0E-01
F21	5.52E+02	1.8E+02	8.33E+02	1.0E+02	**5.00E+02**	4.9E-13
F22	9.36E+02	1.8E+01	**5.07E+02**	7.3E+00	8.63E+02	2.4E+01
F23	**5.34E+02**	3.6E-03	8.58E+02	6.8E+01	**5.34E+02**	4.1E-04
F24	**2.00E+02**	6.2E-13	2.13E+02	1.5E+00	**2.00E+02**	2.9E-14
F25	2.14E+02	2.0E+00	2.13E+02	2.6E+00	**2.11E+02**	9.0E-01
#wins	4		7		12	

Table 4. Comparison of algorithms performance according to their average rank on 25 benchmark functions

Algorithm	Average rank
CoDE	2.74
JADE	3.12
jDE	4.12
b6e6rl60	4.24
b6e6rl480	4.36
SaDE	4.44
EPSDE	4.98

Table 5. Comparison of algorithms performance according to their average rank on functions F1 to F14

Algorithm	Average rank
CoDE	2.61
JADE	2.68
b6e6rl480	3.61
b6e6rl60	3.64
jDE	4.75
SaDE	5.04
EPSDE	5.68

Table 6. Required accuracy ε for the benchmark functions

Functions	ε
F1 – F5	1×10^{-6}
F6 – F14	1×10^{-2}

Table 7. Reliability and average *FES* in successful runs for b6e6rl modifications on the easier subset of test functions

Function	b6e6rl60		b6e6rl480	
	R	Av. FES	R	Av. FES
F1	100	48682	100	44174
F2	100	69984	100	87403
F4	100	129533	100	229982
F6	100	140374	100	163003
F7	84	57483	72	59480
F9	100	53794	100	45209
F12	12	862940	4	1835400

where x_{min} is the best solution found in a run, x^* is the global minimum of the function, ε is the required accuracy prescribed in [9], and *FES* is the current number of function evolutions. The values of ε used in the experiments are given in Table 6. The results of this comparison obtained on the F1–F14 subset of test functions are shown in Table 7, the reliability of the search is given in the column R, which is the percentage of the runs that found a solution satisfying the condition $(f(x_{min}) - f(x^*)) < \varepsilon$. Only functions with $R > 0$ are presented in the table, in the other test problems no acceptable solution was found in 3×10^6 function evaluations. The results show if the b6e6rl algorithm can find an acceptable solution, it is able to find it faster than in 3×10^5 *FES* in most test problems. However, in 7 out of 14 problems the algorithm is not able to find any good solution even in 3×10^6 *FES*.

It is seen from the table that the preference of successful strategies for a longer period in b6e6rl480 does not bring better efficiency of the search. The average *FES* of b6e6rl480 is higher in 5 out of 7 functions. As it was expected, the preference of successful strategies for a longer period decreased the reliability (in 2 out of 7 functions).

5   Conclusion

The performance of b6e6rl variant of competitive DE appeared to be comparable with the state-of-the-art adaptive versions of differential evolution algorithm when applied to the hard benchmark problems [9]. With respect the number of wins, it was outperformed by CoDE and JADE only. The average ranks of both b6e6rl modifications are in the middle of the algorithms in the experimental comparison. In the subset of easier test functions, the performance of b6e6rl modifications is even better in their average ranks, outperformed only by CoDE and JADE.

The preference of successful strategies for a longer period by setting the control parameter of competition to $\delta = 1/480$ does not bring better efficiency of the search compared to its former recommended value $\delta = 1/60$. The recommended values of the parameters controlling the competitive adaptation should be used when considering the application of b6e6rl algorithm.

Acknowledgement. This work was supported by the European Regional Development Fund in the IT4Innovations Centre of Excellence project (CZ.1.05/1.1.00/02.0070) and partly supported by University of Ostrava, the project SGS13/PřF/2012.

References

1. Brest, J., Greiner, S., Boškovič, B., Mernik, M., Žumer, V.: Self-adapting control parameters in differential evolution: A comparative study on numerical benchmark problems. IEEE Transactions on Evolutionary Computation 10, 646–657 (2006)
2. Das, S., Suganthan, P.N.: Differential evolution: A survey of the state-of-the-art. IEEE Transactions on Evolutionary Computation 15, 27–54 (2011)
3. Kaelo, P., Ali, M.M.: A numerical study of some modified differential evolution algorithms. European J. Operational Research 169, 1176–1184 (2006)
4. Mallipeddi, R., Suganthan, P.N., Pan, Q.K., Tasgetiren, M.F.: Differential evolution algorithm with ensemble of parameters and mutation strategies. Applied Soft Computing 11, 1679–1696 (2011)
5. Neri, F., Tirronen, V.: Recent advances in differential evolution: a survey and experimental analysis. Artificial Intelligence Review 33, 61–106 (2010)
6. Price, K.V., Storn, R., Lampinen, J.: Differential Evolution: A Practical Approach to Global Optimization. Springer (2005)
7. Qin, A.K., Huang, V.L., Suganthan, P.N.: Differential evolution algorithm with strategy adaptation for global numerical optimization. IEEE Transactions on Evolutionary Computation 13, 398–417 (2009)
8. Storn, R., Price, K.V.: Differential evolution - a simple and efficient heuristic for global optimization over continuous spaces. J. Global Optimization 11, 341–359 (1997)

9. Suganthan, P.N., Hansen, N., Liang, J.J., Deb, K., Chen, Y.P., Auger, A., Tiwari, S.: Problem definitions and evaluation criteria for the CEC 2005 special session on real-parameter optimization (2005), http://www.ntu.edu.sg/home/epnsugan/

10. Tvrdík, J.: Competitive differential evolution. In: Matoušek, R., Ošmera, P. (eds.) MENDEL 2006, 12th International Conference on Soft Computing, pp. 7–12. University of Technology, Brno (2006)

11. Tvrdík, J.: Adaptation in differential evolution: A numerical comparison. Applied Soft Computing 9, 1149–1155 (2009)

12. Tvrdík, J.: Self-adaptive variants of differential evolution with exponential crossover. Analele of West University Timisoara, Series Mathematics-Informatics 47, 151–168 (2009), http://www1.osu.cz/~tvrdik/down/global_optimization.html

13. Wang, Y., Cai, Z., Zhang, Q.: Differential evolution with composite trial vector generation strategies and control parameters. IEEE Transactions on Evolutionary Computation 15, 55–66 (2011)

14. Zaharie, D.: A comparative analysis of crossover variants in differential evolution. In: Markowska-Kaczmar, U., Kwasnicka, H. (eds.) Proceedings of IMCSIT 2007, pp. 171–181. PTI, Wisla (2007)

15. Zaharie, D.: Influence of crossover on the behavior of differential evolution algorithms. Applied Soft Computing 9, 1126–1138 (2009)

16. Zhang, J., Sanderson, A.C.: JADE: Adaptive differential evolution with optional external archive. IEEE Transactions on Evolutionary Computation 13, 945–958 (2009)

Differential Evolution and Perceptron Decision Trees for Fault Detection in Power Transformers

Freitas A.R.R., Pedrosa Silva R.C., and Guimarães, F.G.

[1] Programa de Pós-Graduação em Engenharia Elétrica, Universidade Federal de Minas Gerais,
Av. Antônio Carlos 6627, Belo Horizonte, MG, 31270-901 Brazil
alandefreitas@gmail.com
[2] Departamento de Engenharia Elétrica, Universidade Federal de Minas Gerais, Av. Antônio
Carlos 6627, Belo Horizonte, MG, 31270-901 Brazil
fredericoguimaraes@ufmg.br

Abstract. Classifying data is a key process for extracting relevant information out of a database. A relevant classification problem is classifying the condition of a transformer based on its chromatography data. It is a useful problem formulation as its solution makes it possible to repair the transformer with less expenditure given that a correct classification of the equipment status is available. In this paper, we propose a Differential Evolution algorithm that evolves Perceptron Decision Trees to classify transformers from their chromatography data. Our approach shows that it is possible to evolve classifiers to identify failure in power transformers with results comparable to the ones available in the literature.

1 Introduction

Classification problems can aid the process of decision making in many real world applications. Among those is the extraction of relevant characteristics about clients or preventing failure in equipments. As an example of a classification problem, the composition of gases present in a transformer can indicate its condition of operation (Section 2.1). By obtaining the classification rules we can then infer information on classes from the data related to the problem.

In this paper we describe an approach to generate classification rules with Differential Evolution (DE) algorithms. DE is a metaheuristic that iteratively searches high quality solutions (Section 2.2). Each candidate solution used by our DE algorithm is a Perceptron Decision Tree (PDT), which are decision trees that consider all the attributes of the data in each of its nodes (Section 2.3). This paper introduces the following contributions:

- An algorithm based on Differential Evolution for evolving Perceptron Decision Trees (Section 3)
- An approach for representation and manipulation of PDT in the context of DE solutions (Section 3.1-3.2)
- A strategy for replacing solutions (Section 3.3), in which the worst classifiers give place to new ones
- A method for controlling the legitimacy of the evaluation (Section 3.4) that makes evaluation more legitimate when it seems to be necessary

V. Snasel et al. (Eds.): SOCO Models in Industrial & Environmental Appl., AISC 188, pp. 143–152.
springerlink.com   © Springer-Verlag Berlin Heidelberg 2013

The results of our final algorithm for different databases are compared (Section 4). Those results are also compared to works with similar databases (Section 5). We conclude that even though a more extensive study is needed, the proposed algorithm is efficient for solving the problem with chromatography database (Section 6).

2   Background

2.1   Dissolved Gas Analysis in Power Transformers

Dissolved Gas Analysis (DGA) is a usual method for detecting faults in power transformers as it allows diagnosis without de-energizing the transformer [23]. For the experiments of this work we have used three databases of chromatography related to power transformers. In order to find potential failure in transformers, we can analyze the gases produced in the equipment oil as it is exposed to heat [15]. The problem consists in diagnosing the fault from the data describing the gases generated under that condition. Thus, the problem of analyzing dissolved gases consists in finding rules that can identify failures in the transformers. The possible classifications of a transformer are (i) normal, (ii) electrical failure, and (iii) thermal failure.

2.2   Differential Evolution

Differential Evolution [14,22] is an algorithm developed to improve the quality of a solution over many iterations. A population of candidate solutions is kept and new solutions are generated and tested from a simple combination of preexistent solutions. The best solutions are then kept. DE is an algorithm employed for real valued functions but it does not use the gradient of the function to be optimized. Many books have been published approaching theoretical aspects of DE, see for instance [9,11].

The basic iterative process of DE is described below:

- Generate many solutions **x** with random positions in the search space.
- Initialize the crossover probability CR and differential weight F
- Until a halting criterion is not met, the following steps are repeated:
  - For each solution **x**:
    - · Choose other three solutions **a**, **b** and **c** from the population that are different from **x**.
    - · Choose a random index R between 1 and the population size.
    - · In order to generate a new solution **y**, for each position i of the solution:
      - · Generate a real valued number r between 0 and 1 with uniform distribution
      - · If $r < CR$, $\mathbf{y}_i = \mathbf{a}_i + F(\mathbf{b}_i - \mathbf{c}_i)$, else $\mathbf{y}_i = \mathbf{x}_i$
    - · If **y** is better than **x**, **x** is replaced by **y**
- Return the best solution found

The values F, CR and the population size must be chosen by the person using the algorithm. Given a suitable formulation of a problem of data classification, DE can then evolve classification rules for the test database.

2.3 Perceptron Decision Trees

A common approach for classifiers is to employ binary classification trees. In those trees, the condition of one of the attributes is tested according to a threshold. The difference between the attribute value and the threshold defines the next tree node to be considered. This process repeats until a leaf node is reached. In those nodes, there is information on the class that should correctly classify the data.

In this work, we propose a different approach for defining each tree node. With the equation $\mathbf{wx} + \theta = z$, where $\mathbf{w}$ as weight scalars, $\mathbf{x}$ are the data attributes and θ is a constant, we can linearly divide the search space into areas where $z > 0$ and areas where $z < 0$. With this approach, we define a PDT.

Many authors have employed similar concepts of aggregating linear classifiers or perceptrons in a tree under different names [4,5,6,7], including PDT [3].

3 Methodology

In this section we describe how DE was used for evolving PDT. The problem presents many possible parameters and a brief study of those is presented.

The database employed in this work consists in chromatography data of transformers. The data has 5 attributes, representing the quantity of each of the following gases: (i) Hydrogen, (ii) Methane, (iii) Acetylene, (iv) Ethane, and (v) Ethylene. Besides the value of each of those attributes, there is also the classification of the transformer into 3 possible classes: Normal (Class 1), Electrical Failure (Class 2), and Thermal Failure (Class 3).

Three different unbalanced databases were used for the tests. A brief description of the databases is shown on Table 1. We can notice that most samples belong to class 1 and that is mostly due to the unbalancing of the database 3.

Table 1. Databases

Database	Number of samples	Class 1	Class 2	Class 3
DB 1	52	30,77%	42,31%	26,92%
DB 2	232	39,22%	26,29%	34,48%
DB 3	224	81,70%	5,80%	12,50%
Total	508	57,09%	18,90%	24,02%

The information was split into two sets. The first set, with 70% of the samples, contains the training set for the generation of the classifier while the second dataset, with 30% of the samples, contains the validation data, for testing the generalization capacity of the classifier.

3.1 Representation of a Classifier

As seen in Section 2.3, each PDT contains many linear classifiers, being each of those defined by a vector of weights $\mathbf{w}$ of size n and a constant θ, being n the number of attributes of the classification problem. The number of classifiers in a PDT is $2^{(h-1)} - 1$, being h a variable defined as the depth of the PDT.

Besides the classifiers, each PDT contains $2^{(h-1)}$ leaf nodes, which contain a possible classification for a sample. Those nodes are defined in the discrete space. A complete PDT can be defined by the matrices of size $n \times 2^{h-1} - 2$ for the weights, $1 \times 2^{h-1} - 1$ for the constants and $1 \times 2^{h-1}$ for the leaf nodes. This is the size of each solution employed by the DE.

Each of the $2^{h-1} - 1$ columns of the classifier matrices defines a classifier. After using the linear classifier at position i of n, the classifier at position $2i$ is used if the output of the classifier is < 0 or the classifier $2i + 1$ is used otherwise.

As the total error will be the standard of comparison between the algorithms, the objective function value of each of the solutions is defined as the number of misclassifications, including false positives and false negatives. This objective function, naturally, must be minimized by the DE.

Three initial values for h were tested. While the individual size is $O(2^{h-1} - 1)$, the evaluation cost is still $O(h)$ because only one possible path is searched through the PDT. The values h of 3, 5 and 7 were tested for evolving PDT. The PDT had representation size 3, 15 and 63, respectively. Where $h = 3$, 333 generations were executed, with $h = 5$, 200 generations, and with $h = 7$, 142 generations. That was meant to keep at least the evaluation cost similar throughout the generations. Ten executions using all the databases with $h = 5$ achieved an initial average training error of 17.65%, with $h = 3$ had 18.57% and with $h = 7$ had 20.01%.

3.2 Operators

Simple operators, as shown in Section 2.2 were employed for the generation of new solutions. Values of F and CR are defined as random values between 0.4 and 1 and between 0.9 and 1, respectively [22]. Those values are altered at each iteration of a generation.

As for the leaf nodes, a discrete approach must be used for the crossover of solutions [17]. The approach used is inspired in the idea that DE uses a vector of differences to alter the solution, however now this vector represents swap movements between two possible positions. In doing so, the movement described as $\mathbf{b}_i - \mathbf{c}_i$ is only performed if $\mathbf{a}_i = \mathbf{b}_i$. If $\mathbf{a}_i = \mathbf{c}_i$, the movement $\mathbf{c}_i - \mathbf{b}_i$ is applied. In a third possible case, where $\mathbf{a}_i \neq \mathbf{b}_i$ and $\mathbf{a}_i \neq \mathbf{c}_i$, no movement is performed. The value F is used to define if the operation should happen. The operation occurs if a randomly generated number is less than F.

3.3 Replacement of Individuals

As the replacement of individuals is made one by one, it may happen that some solutions are stagnant in bad points of the search space. In order to avoid this problem, a new individual survival operator was developed to keep always new candidates in the population.

At each generation, new random individuals are generated to replace the $x\%$ worst individuals. Moreover, the individuals with an error rate greater than $1/n_{classes}$ are also replaced.

Tests with 3 levels for x were performed. When the value of x was 0%, the response in 200 generations had an average error of 19.01%, when $x = 10\%$ the final classifier had 16.34% and when $x = 20\%$ the answer had error 18.11%. The individual replacement value was then adjusted to $x = 10\%$.

3.4 Legitimacy

In the initial generations, where all the individuals are still very random, not many comparisons are needed to perceive that some are better than the others. In most cases, with the employment of classifiers in less than 10 samples it is possible to define clearly how some PDT are better than the others. The same does not apply after many generations, when most solutions classify the samples with a low error rate. Having this in mind, and the high cost of evaluating solutions, which may take as much as 10 seconds per generation, we propose an approach to reduce the time spent to evaluate solutions in the first generations.

The key idea is that initial solutions do not need to be evaluated with the same legitimacy as the solutions in the last generations, where a more refined analysis is necessary to distinguish good solutions. The value l defines the legitimacy utilized in the evaluation of individuals. At a generation with legitimacy l, only l samples are tested in the evaluation of each solution.

The initial value of this parameter was defined as twice as the number of attributes of each samples. After each generation, the value l is updated accordingly to the equation $l = \lceil \min(N, l * \alpha) \rceil$ where N is the number of available samples for test and α is the rate of increase of the legitimacy for each generation.

The parameter α was defined as $1 + \delta/\sigma$, where δ is the speed of legitimization, defined as 10^{-2}, and σ is the standard deviation of the objective function value of the solutions. Thus, when the solutions are still very diverse, the parameter α is smaller.

Some individuals can occasionally have an objective function value smaller than it is due in cases where the legitimization is low. For this reason, whenever the objective function value is calculated, the legitimacy value used to obtain the objective function value is also stored. Thus, if the individual is not replaced by a new one after t generations and the legitimacy value has already increased, the individual is reevaluated with the new legitimacy value l. The value of t was defined as 10.

In Figure 1, we have the relation between execution time and those legitimacy values. In this plot, the dashed line represents the legitimacy l divided by N and the continuous line represents the best objective function value found so far in a given generation. The time spent by each generation is directly linked to the legitimacy of that generation. We can perceive the direct relation between the evaluation legitimacy and the time spent per generation by comparing the graphs.

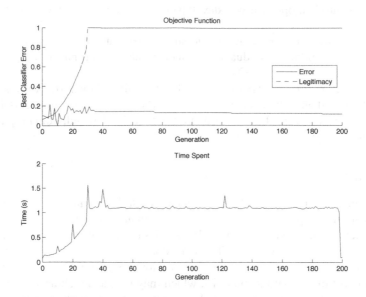

Fig. 1. Relation between evaluation legitimacy, objective function and time

The best objective function value known also varies much before the legitimacy value reaches its maximum value. That happens because some solutions can obtain good objective function values which are not so appropriate and this is corrected only after t generations. After full legitimacy has been reached, there is no other referential to evaluate the individuals because all the database is already being used. Therefore the best objective function value will only decrease after this.

4   Results

In computational tests, the average time for executing a DE with 200 generations for a joint database with all the samples was 17 minutes. Having the number of factors in mind, we performed 3 experiments in each of the 3 databases. That was defined as experiment 1, where the number of generations is 200.

In Table 2 we have the results of the tests with 200 generations. In the Table, we have the average time spent, the average and minimum error obtained for the training set, and the average and minimum error obtained for the validation set.

The algorithm is using the replacement operator and it does not present final convergence after 200 iterations. For this reason, in order to have a reference in relation to the capacities of the algorithm in a real situation of search for a good classifier, we performed a second experiment. In this new test, we performed 15 independent executions of the algorithm for each database with 2000 generations. Thus, we have an idea of the capabilities of the algorithm with more generations. This test was defined as experiment 2.

Table 2. Results of tests with 200 generations

Database	Time (s)	Training Error	Minimum Error	Validation Error	Minimum Error
DB1	59	1,85%	0%	20,83%	6,25%
DB2	224	26,13%	25,31%	38,57%	34,29%
DB3	218	11,46%	8,92%	17,91%	14,93%

Table 3 shows the results of this test. The figures show the minimum, average and maximum error of the 15 executions in the validation set. For comparison, we also present the error in the validation set of other well known classifiers, such as: J48 [18], BFTree [21], RandomTree [12], IB1 [1], MLPs [19] and Naive Bayes [25]. The results of these methods were obtained from the use of the Weka Framework[1] [12]. A Kruskal-Wallis significance test was performed on the results and no significant difference ($p < 0.05$) was found. Thus, it can be observed that the results achieved by the proposed approach were competitive, in addition, looking through the best results of the method (Min.) its potential in generating good classifiers could be verified.

Table 3. Results of the test with 2000 generations

Database	PDT Min.	PDT Avg.	PDT Max.	j48	BFTree	RandTree	NaiveBayes	IB1	MLP
DB1	0%	13.33%	31.25%	31.25%	18.75%	25%	25%	6.25%	18.75%
DB2	22.86%	34.29%	40%	22.39%	23.88%	23.88%	41.79%	26.86%	35.82%
DB3	11.94%	19.60%	28%	18.57%	18.57%	21.42%	24.28%	24.28%	21.42%

5   Discussion

Some works have been done to diagnose failure in transformers including Artificial Neural Networks [26], Neural Networks with expert systems [24], decision trees [8] and genetic algorithms with niches [16].

In other databases for the same problem, Pereira and Vasconcelos [16] obtain a right classification rate of 91%, comparable to many cases of this algorithm. In a profound analysis of the problem, he defines many important criteria for classification. In his work, an approach based in niches leads to different solutions that value correct classification in each of the classes.

A deeper study with benchmark databases is needed to have more general conclusions on the behavior of the proposed algorithm. Although other works have used other databases, if we assume similar complexity of the data to be classified, the results obtained in this work are satisfactory.

---

[1] Available in http://www.cs.waikato.ac.nz/ml/weka/

Castanheira [8] presents also good results for the problem with the use of neural networks and decision trees, having the last ones obtained better results. The modeling of the problem through PDT have the capacity to classify data in many situations and it has clear advantages in relation to classification power when compared to simple decision trees. The properties of the PDT with DE still need to be deeply studied such that better results can be obtained.

Despite the classification power of PDT, the computational time spent with the operators in the case of a large tree grows exponentially and therefore it must be carefully adjusted.

6 Conclusion and Future Work

The halting criterion of 200 generations was decided in regard to the computational time used for each experiment, with the intention that many tests should be possible. However, it is likely that a greater number of generations is more suitable for the algorithm to have time to converge almost completely. When the tests were performed with 2000 generations, the validation error of the algorithm kept falling for the databases, indicating that the algorithm was not yet presenting overfitting problems. Recognizing the number of generations needed for the validation error to begin to rise is important because from this point on, new classifiers that model only noise of the training set are being generated.

The algorithm certainly is suitable for solving the problem. Nevertheless, for a comparison with the algorithms in the literature, it would be important to have an analysis with other benchmark databases.

All the parameters of the algorithm were adjusted separately, considering that there are no interaction between the factors. Of course, this adjustment has led to a condition of execution of the algorithm which is better than the initial one. However, before implementing new features in the algorithm, it would be important to adjust the parameters with factorial experiments where it is possible to better understand the interaction between the parameters. As each iteration takes much time, this experiment would have to be very well planned.

In the evolution of the classifiers, only the training error was used. In spite of the fact that we can not use the validation set throughout the evolution, a strategy of objective function that maximizes the separation margin of the data can be used to broaden the capacity of generalization of the classifiers, alike the way Support Vector Machines work [2,10]. A boosting strategy can be applied to increase the margin by including a rise in the probability of classifying the most difficult samples [20,13].

Another important issue in the definition of the objective function is to define the costs involved in the process. A low error rate may not simply represent good solutions for practical applications because they do not represent the specific errors of each class and the most important: the cost involved in each sort of error.

In any of the problems mentions here, however, the approach based on DE and PDT has been shown to be useful for the solution of the problem.

Acknowledgement. This work has been supported by the Brazilian agencies CAPES, CNPq, and FAPEMIG; and the Marie Curie International Research Staff Exchange Scheme Fellowship within the 7th European Community Framework Programme.

References

1. Aha, D.W., Kibler, D., Albert, M.K.: Instance-based learning algorithms. Machine Learning 6(1), 37–66 (1991)
2. Angulo, C., Gonzalez-Abril, L.: Support vector machines. Pattern Recognition, 1 (2012)
3. Bennett, K., Cristianini, N., Shawe-Taylor, J., Wu, D.: Enlarging the margins in perceptron decision trees. Machine Learning 41(3), 295–313 (2000)
4. Bennett, K., Mangasarian, O.: Robust linear programming discrimination of two linearly inseparable sets. Optimization Methods and Software 1(1), 23–34 (1992)
5. Bennett, K., Mangasarian, O.: Multicategory discrimination via linear programming. Optimization Methods and Software 3(1-3), 27–39 (1994)
6. Breiman, L.: Classification and regression trees. Chapman & Hall/CRC (1984)
7. Brodley, C., Utgoff, P.: Multivariate decision trees. Machine Learning 19(1), 45–77 (1995)
8. Castanheira, L.: Aplicação de mineração de dados no auxílio à tomada de decisão em engenharia. Master's thesis, Universidade Federal de Minas Gerais (2008)
9. Chakraborty, U.: Advances in differential evolution, vol. 143. Springer (2008)
10. Cortes, C., Vapnik, V.: Support-vector networks. Machine Learning 20(3), 273–297 (1995)
11. Feoktistov, V.: Differential evolution: in search of solutions, vol. 5. Springer-Verlag New York Inc. (2006)
12. Frank, E., Hall, M., Holmes, G., Kirkby, R., Pfahringer, B., Witten, I.H., Trigg, L.: Weka. In: Maimon, O., Rokach, L. (eds.) Data Mining and Knowledge Discovery Handbook, pp. 1305–1314. Springer US (2005)
13. Freund, Y., Schapire, R., Abe, N.: A short introduction to boosting. Journal-Japanese Society For Artificial Intelligence 14(771-780), 1612 (1999)
14. Mallipeddi, R., Suganthan, P., Pan, Q., Tasgetiren, M.: Differential evolution algorithm with ensemble of parameters and mutation strategies. Applied Soft Computing 11(2), 1679–1696 (2011)
15. Morais, D., Rolim, J.: A hybrid tool for detection of incipient faults in transformers based on the dissolved gas analysis of insulating oil. IEEE Transactions on Power Delivery 21(2), 673–680 (2006)
16. Pereira, M.A., Vasconcelos, J.A.: A niched genetic algorithm for classification rules discovery in real databases. In: Anais do XVIII Congresso Brasileiro de Automática (2010)
17. Prado, R.S., Silva, R.C.P., Guimarães, F.G., Neto, O.M.: Using differential evolution for combinatorial optimization: A general approach. In: SMC, pp. 11–18 (2010)
18. Quinlan, J.R.: C4.5: Programs for Machine Learning (Morgan Kaufmann Series in Machine Learning). Morgan Kaufmann (1993)
19. Rumelhart, D.E., Hinton, G.E., Williams, R.J.: Learning internal representations by error propagation. In: Parallel Distributed Processing: Explorations in the Microstructure of Cognition, vol. 1, pp. 318–362. MIT Press, Cambridge (1986)
20. Schapire, R., Freund, Y.: Boosting: Foundations and algorithms (2012)
21. Shi, H.: Best-first decision tree learning. Master's thesis, University of Waikato, Hamilton, NZ (2007)

22. Storn, R., Price, K.: Differential evolution–a simple and efficient heuristic for global optimization over continuous spaces. Journal of Global Optimization 11(4), 341–359 (1997)
23. Sun, H., Huang, Y., Huang, C.: A review of dissolved gas analysis in power transformers. Energy Procedia 14, 1220–1225 (2012)
24. Wang, Z., Liu, Y., Griffin, P.: A combined ann and expert system tool for transformer fault diagnosis. IEEE Transactions on Power Delivery 13(4), 1224–1229 (1998)
25. Zhang, H.: Exploring conditions for the optimality of naïve bayes. International Journal of Pattern Recognition and Artificial Intelligence (IJPRAI) 19(2), 183–198 (2005)
26. Zhang, Y., Ding, X., Liu, Y., Griffin, P.: An artificial neural network approach to transformer fault diagnosis. IEEE Transactions on Power Delivery 11(4), 1836–1841 (1996)

Multiobjective Pareto Ordinal Classification for Predictive Microbiology

M. Cruz-Ramírez[1], J.C. Fernández[1], A. Valero[2],
P.A. Gutiérrez[1], and C. Hervás-Martínez[1]

[1] Dept. of Computer Science and Numerical Analysis, U. of Córdoba, Rabanales Campus,
Albert Einstein Building, 14071, Córdoba, Spain
{mcruz,jcfernandez,pagutierrez,chervas}@uco.es
[2] Dept. of Food Science and Technology, U. of Córdoba, Rabanales Campus,
Darwin Building, 14071, Córdoba, Spain
bt2vadia@uco.es

Abstract. This paper proposes the use of a Memetic Multiobjective Evolutionary Algorithm (MOEA) based on Pareto dominance to solve two ordinal classification problems in predictive microbiology. Ordinal classification problems are those ones where there is order between the classes because of the nature of the problem. Ordinal classification algorithms may take advantage of this situation to improve its classification. To guide the MOEA, two non-cooperative metrics have been used for ordinal classification: the Average of the Mean Absolute Error, and the Maximum Mean Absolute Error of all the classes. The MOEA uses an ordinal regression model with Artificial Neural Networks to classify the growth classes of microorganisms such as *Listeria monocytogenes* and *Staphylococcus aureus*.

1 Introduction

Knowing how different food product properties, environments and their history can influence the micro-flora developed when food is stored is an important first step towards forecasting its commercial shelf-life, alterations and safety. In order to be able to predict microbial behavior in each new situation and estimate its consequences with respect to the safety and quality of food, there has to be an exact definition of the food environment and how it will influence microbial growth and survival. The need to learn more about microbial behavior in limiting conditions that prevent growth could be met by using mathematical models. Binary logistic regression has frequently been applied to determine the probability of growth under a given set of conditions. However, microbial responses in limiting environmental conditions (i.e. low pH, temperature, water activity, etc.) are subject to several variable sources, often not experimentally controlled. This can result in biased growth/no growth estimations when the probability of growth approaches 0.5 [9].

One of the first published works related to the development of multiclassification models showed a high degree of accuracy when estimating growth/no growth boundaries of *S. aureus* [11]. However, although this paper provided a categorical classification into three classes, by adding new information about the probability of growth

V. Snasel et al. (Eds.): SOCO Models in Industrial & Environmental Appl., AISC 188, pp. 153–162.
springerlink.com

associated to stringent conditions, growth/no growth models can provide alternative and more accurate estimations.

For this reason, this paper addresses growth/no growth models from the perspective of ordinal classification into four classes. A classification problem occurs when an object needs to be assigned to a predefined group or class based on a number of observed attributes related to that object. Ordinal classification is the problem where the variable to be predicted is not of a numeric or nominal type but is instead ordinal, so that the categories have a logical order. In our case, the predictive microbiology problem to be solved has clearly ordinal behavior (see Section 3). Artificial Neural Networks (ANNs) [1] have been an important tool for classification since recent research activities identified them as a promising alternative to traditional classification methods such as logistic regression. On the other hand, Evolutionary Algorithms (EAs) [16] are global search heuristics and one of the main alternatives to local search algorithms for training ANNs. Obtaining ANN models using EAs is known as Evolutionary Artificial Neural Networks [16]. These methodologies maintain a population of ANNs that are subject to a series of transformations in the evolutionary process so as to obtain acceptable solutions to the problem.

Often a great number of objectives must be processed to obtain a viable solution to a problem, usually without any a priori knowledge of how the objectives interact with each other. This is known as a Multiobjective Optimization Problem, and the most popular methods are based on Pareto dominance. The training of ANNs by evolutionary Pareto-based algorithms is known as Multiobjective Evolutionary Artificial Neural Networks, and has been in use for the last few years in the resolving classification tasks [13]. Hybridization of intelligent techniques, coming from different computational intelligence areas, is a common solution, because of the growing awareness that such combinations frequently perform better than the individual techniques coming from computational intelligence [6].

This study deals with learning and the improvement in the generalization of classifiers designed using a MOEA with ANNs to determine growth limits in two important microorganisms in predictive microbiology, *Listeria monocytogenes* and *Staphylococcus aureus*.

The rest of the paper is organized as follows. Section 2 covers background materials, explaining the classification model used, the MOEA and the ordinal metrics used for guiding the MOEA. Section 3 presents the *L. monocytogenes* and *S. aureus* microorganisms. Section 4 shows the experimental design and results and finally Section 5 shows the conclusions obtained.

2 Background

2.1 Ordinal Model

The big problem in ordinal classification is that there is no notion of the precise distance between classes. The samples are labeled by a set of ranks with different categories given an order. Nominal classification algorithms can also be applied to prediction problems involving ordinal information but obviating the order of the classes. However, this process loses information that could improve the predictive ability of the classifier.

Although there are some other approaches for ordinal regression [14] (mainly based on reducing the problem to binary classification, or simplifying it to regression or cost-sensitive classification), the majority of proposals can be grouped under the term *threshold methods*. These methods are based on the idea that, in order to model ordinal ranking problems from a regression perspective, one can assume that some underlying real-valued outcomes exist, although they are unobservable. Consequently, two different things are estimated:

- A function $f(\mathbf{x})$ that predicts the real-valued outcomes and tries to discover the nature of the assumed underlying outcome.
- A threshold vector $\mathbf{b} \in \mathbb{R}^{J-1}$ to represent the intervals in the range of $f(\mathbf{x})$, where $b_1 \leq b_2 \leq \ldots \leq b_{Q-1}$ (possible different scales around different ranks).

We propose an adaption of the classical Proportional Odd Model (POM) model [18] for ANNs. Since we are using the POM model and ANNs, our proposal does not assure monotonicity. The POM model works based on two elements: the first one is a linear layer with only one node (see Fig. 1) whose inputs are the non-linear transformations of a first hidden layer. The task of this node is to stamp the values into a line, to give them an order, which facilitates ordinal classification. After this one node linear layer, an output layer is included with one bias for each class whose objective is to classify the patterns into their corresponding class. This classification structure corresponds to the POM model, which, like the majority of existing ordinal regression models, can be represented in the following general form:

$$
C(\mathbf{x}) = \begin{cases}
c_1, & \text{if } f(\mathbf{x}, \theta) \leq \beta_0^1 \\
c_2, & \text{if } \beta_0^1 < f(\mathbf{x}, \theta) \leq \beta_0^2 \\
\ldots \\
c_J, & \text{if } f(\mathbf{x}, \theta) > \beta_0^{J-1}
\end{cases},
\tag{1}
$$

where $\beta_0^1 < \beta_0^2 < \cdots < \beta_0^{J-1}$ (this will be the most important constraint in order to adapt the nominal classification model to ordinal classification), J is the number of classes, $\mathbf{x}$ is the input pattern to be classified, $f(\mathbf{x}, \theta)$ is a ranking function and θ is the vector of parameters of the model. Indeed, the analysis of (1) reveals the general idea previously presented: patterns, $\mathbf{x}$, are projected to a real line by using the ranking function, $f(\mathbf{x}, \theta)$, and the biases or thresholds, β_0^j, separating the ordered classes, where $\beta_0^0 = -\infty$ and $\beta_0^J = \infty$.

The standard POM model approximates $f(\mathbf{x}, \theta)$ by a simple linear combination of the input variables, while our model considers a non-linear basis transformation of the inputs. Let us formally define the model for each class as $f_l(\mathbf{x}, \theta, \beta_0^l) = f(\mathbf{x}, \theta) - \beta_0^l$; $1 \leq l \leq J$, where the projection function $f(\mathbf{x}, \theta)$ is estimated with the following S sigmoidal basis functions $f(\mathbf{x}, \theta) = \beta_0 + \sum_{j=1}^{S} \beta_j B_j(\mathbf{x}, \mathbf{w}_j)$, replacing $B_j(\mathbf{x}, \mathbf{w}_j)$ by sigmoidal basis functions:

$$
B_j(\mathbf{x}, \mathbf{w}_j) = \frac{1}{1 + exp\left(-\left(\sum_{i=1}^{k} w_{ji} x_i\right)\right)}.
$$

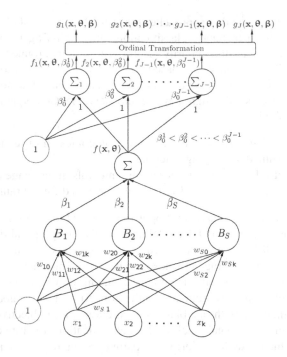

Fig. 1. Proposed sigmoidal model for ordinal regression

By using the POM model, this projection can be used to obtain cumulative probabilities, cumulative odds and cumulative logits of ordinal regression in the following way [18]:

$$P(Y \leq l) = P(Y = 1) + \cdots + P(Y = l),$$

$$odds(Y \leq l) = \frac{P(Y \leq l)}{1 - P(Y \leq l)},$$

$$logit(Y \leq l) = \ln\left(\frac{P(Y \leq l)}{1 - P(Y \leq l)}\right) = f(\mathbf{x}, \theta) - \beta_0^l \equiv f_1(\mathbf{x}, \theta, \beta_0^1),$$

$$P(Y \leq l) = \frac{1}{1 + \exp(f(\mathbf{x}, \theta) - \beta_0^l)} \equiv \frac{1}{1 + \exp(f_l(\mathbf{x}, \theta, \beta_0^l))}; \ 1 \leq l \leq J,$$

where $P(Y = j)$ is the probability a pattern $\mathbf{x}$ has of belonging to class j, $P(Y \leq l)$ is the probability a pattern $\mathbf{x}$ has of belonging to class 1 to l and the logit is modeled by using the ranking function, $f(\mathbf{x}, \theta)$, and the corresponding bias, β_0^l. We can come back to $P(Y = l)$ from $P(Y \leq l)$:

$$P(Y = l) = g_l(\mathbf{x}, \theta, \beta) = P(Y \leq l) - P(Y \leq l - 1), \ l = 1, \ldots, J,$$

and the final model can be expressed as:

$$g_l(\mathbf{x}, \Theta, \beta) = \frac{1}{1 + \exp(f_l(\mathbf{x}, \Theta, \beta_0^l))} - \frac{1}{1 + \exp(f_{l-1}(\mathbf{x}, \Theta, \beta_0^{l-1}))}, l = 1, \ldots, J.$$

2.2 Metrics and Methodology

There are many ordinal measures to determine the efficiency of a g classifier, but not all pairs formed by these metrics are valid to guide a MOEA. To guide the learning of our MOEA for designing ANN models to determine the growth limits of microorganisms in predictive microbiology, the two metrics used are the Average Mean Absolute Error (*AMAE*) and the Maximum Mean Absolute Error (*MMAE*). These metrics are explained in detail in [7]. In general, these two ordinal measures are non-cooperative [7]: when the value of one of them increases the value of the other decreases. Thus the use of a MOEA based on Pareto dominance is justified.

The MOEA used in this work is called MPENSGA2 (Memetic Pareto Evolutionary NSGA2). This algorithm is based on the original algorithm NSGA2, and is described in detail in [10]. The framework of the MPENSGA2 is shown in Fig. 2, and the main differences with respect to [10] are:

- The metrics for guiding the algorithm are *AMAE* and *MMAE*.
- The delete links mutator and the parametric mutator have been modified to take into account the constraints of the ordinal model POM, β_0^i bias.
- The local search algorithm, *iRprop*$^+$ is used in all generations, not only in three generations of the evolutionary process. This is because the datasets used in this work have a relatively small size, so that the computational cost does not increase dramatically. The function with respect to which the local search is made is the cross-entropy function, as well as *MMAE* and *AMAE*, which are not derivable. Since patterns, according to the ordinal model, are projected onto a straight line, an improvement in entropy is an improvement in *AMAE* and the *MMAE* metrics.

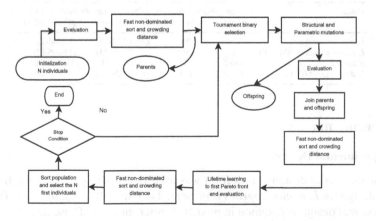

Fig. 2. Framework of the MPENSGA2 Algorithm

3   Description Problem

In this paper, four observed microbial responses are obtained based on the growth probability of a microorganism: $p = 1$ (growth); $0.5 \leq p < 1$ (high growth probability); $0 < p < 0.5$ (low growth probability); and $p = 0$ (no-growth). These four observed microbial responses were used in [8]. There is an intrinsic ranking of the different classes, so this paper tackles the problem of microbial growth from an ordinal point of view. This paper addresses the following two microbiological problems:

Listeria monocytogenes have been a serious problem for food industries due to their ubiquity in the natural environment and the specific growth conditions of the pathogen that lead to its high prevalence in different kinds of food products. One impetus for this research has been the problem of listeriosis, and different strategies have been proposed to limit the levels of contamination at the time of consumption to less than 100 CFU/g (European Commission [5]). *L. monocytogenes* data were gathered at 7°C in Nutrient Broth with different combinations of environmental factors pH (5.0–6.0, six levels at regular intervals), water activity (0.960–0.990, six levels) and acetic acid concentration (0–0.4% (w/w), three levels), as can be seen in [20]. Thus, 108 different conditions were tested with 20 replicates per condition.

Staphylococcus aureus has been recognized as an indicator of deficient food and processing hygiene and is a major cause of food gastroenteritis worldwide [19]. A fractional factorial design was followed in order to ascertain the growth limits of *S. aureus* [19] by carefully choosing a subset (fraction) of the experimental runs of a full factorial design in order to reduce experimental time and resources. The selection was based on restricting the levels of the environmental factors studied for the growth/no-growth domain of *S. aureus*. Since no growth was detected at 7.5°C or below, data were collected at 8°, 10°, 13°, 16°and 19°C, at pH levels from 4.5 to 7.5 (0.5 intervals) and at $19a_w$ levels (from 0.856 to 0.999 at regular intervals). In this study, there are 30 replicates per condition, more than other studies obtaining the growth/no-growth transition.

Table 1. Features of the datasets

Dataset	#Training patterns	#Generalisation patterns	#Patterns per class in training	#Patterns per class in generalisation
L. monocytogenes	72	36	(29-10-10-23)	(15-3-6-12)
L. monocytogenes (SMOTE)	92	36	(29-20-20-23)	(15-3-6-12)
S. aureus	146	141	(60-22-7-57)	(57-23-5-56)
S. aureus (SMOTE)	167	141	(60-22-28-57)	(57-23-5-56)

4   Experiments

4.1   Experimental Design

A fractional factorial design matrix form was used in this study (in [20] the fractional factorial design for *L. monocytogenes* is presented and in [19], for *S. aureus*). This type of experimental design is common in predictive microbiology. These designs consider values close to the frontiers to train, and central values to generalize.

Table 1 shows the characteristics of the original datasets, where their unbalanced nature can be appreciated. In order to deal with this unbalanced nature, a pre-processing method has been applied to each dataset. The method applied is the well-known Synthetic Minority Over-sampling Technique algorithm (SMOTE) [2]. This method is applied twice to the minority class in such a way that each application of SMOTE duplicates its number of patterns. The synthetic generated patterns are only used to train the model, not to test it, as they cannot be considered real data. These synthetic patterns were generated using information from the five nearest neighbors. This method has been configured and run using WEKA software [15]. The final distribution of the datasets after applying this pre-processing can be seen in Table 1. This study only presents the results obtained after applying the SMOTE procedure.

Once the Pareto front is built into the last generation of training, two selection strategies are used to choose the individuals. The first strategy selects the best model in $AMAE$, which is the upper individual from the Pareto front. This selection method is called MPENSGA2-A. The second strategy selects the best model in $MMAE$, which is the bottom individual from the Pareto front. This selection method is called MPENSGA2-M. Because the MPENSGA2 algorithm is stochastic, the algorithm was run 30 times and the mean and standard deviation was obtained from the 30 individuals for the upper and lowest extremes.

In all experiments, the population size for the MPENSGA2 algorithm is established at 100. The probability of choosing a type of mutator and applying it to an individual is equal to 1/5. For the $iRprop^+$ algorithm, the number of epochs is established at 20. The other configured parameters are $\eta^+ = 1.2$, $\eta^- = 0.5$, $\Delta_0 = 0.0125$, $\Delta_{min} = 0$ and $\Delta_{max} = 50$, based on previous works.

4.2   Comparison Methods

For comparison purposes, different nominal and ordinal classification methods from the literature have been included in the experimentation.

The nominal classification methods are: MLogistic, SLogistic, MLP, C4.5 and LibSVM. These methods have been configured and run using WEKA software [15].

The ordinal classification methods are: SVMRank [17], ASAOR(C4.5) [12], GPOR [3], SVOR-EX and SVOR-IM [4]. These methods have been configured and run using the code provided by the authors. The corresponding hyper-parameters for these methods were adjusted using a grid search with a 10-fold cross-validation.

4.3   Results

Table 2 shows the results obtained in generalization for the two pre-processed datasets. For the MPENSGA2 method, the results presented correspond to mean values and standard deviation (mean$_{SD}$) for the 30 extreme models of the Pareto fronts generated in 30 runs (one Pareto front for each run). In addition, the best models obtained in each of the Pareto front extremes are presented in order to make comparisons with other methods. The measures used for this comparison are: the Correctly Classified Rate, CCR; the Mean Absolute Error, MAE; the Average MAE, $AMAE$; the Maximum MAE, $MMAE$; and the Kendall's τ_b. The CCR is the classic nominal metric to evaluate a classifier,

Table 2. Statistical results in generalization for the pre-processing datasets

Dataset	Method	CCR	MAE	AMAE	MMAE	τ_b
	Mlogistic	0.750	0.250	0.279	0.333	0.844
	Slogistic	0.750	0.305	0.454	0.833	0.817
	MLP	0.722	0.277	0.233	0.333	0.845
	C4.5	0.805	0.222	0.195	0.333	0.862
	LibSVM	0.833	*0.166*	0.287	0.666	0.892
	SVMRank	0.805	0.194	0.175	0.333	0.884
L. monocytogenes	ASAOR(C4.5)	0.750	0.277	0.300	0.333	0.823
	GPOR	0.750	0.250	0.279	0.333	0.874
	SVOR-EX	0.750	0.250	0.191	0.333	0.871
	SVOR-IM	0.861	*0.166*	0.141	0.333	0.883
	Best MPENSGA2-A	**0.916**	**0.111**	**0.079**	*0.250*	*0.920*
	Best MPENSGA2-M	*0.888*	**0.111**	*0.100*	**0.166**	**0.922**
	MPENSGA2-A	$0.837_{0.037}$	$0.168_{0.037}$	$0.161_{0.048}$	$0.302_{0.070}$	$0.895_{0.021}$
	MPENSGA2-M	$0.839_{0.038}$	$0.165_{0.038}$	$0.159_{0.049}$	$0.297_{0.074}$	$0.896_{0.022}$
	Mlogistic	0.687	0.475	0.750	1.217	0.718
	Slogistic	0.673	0.553	0.766	1.304	0.656
	MLP	**0.773**	*0.340*	0.646	1.200	0.794
	C4.5	0.716	0.496	0.732	1.400	0.649
	LibSVM	0.758	0.347	0.476	0.800	0.781
	SVMRank	0.695	0.347	0.450	0.652	0.794
S. aureus	ASAOR(C4.5)	0.702	0.390	0.497	0.800	0.760
	GPOR	0.581	0.929	1.137	1.600	0.382
	SVOR-EX	0.680	0.361	0.459	0.652	0.791
	SVOR-IM	0.723	**0.319**	**0.419**	0.600	**0.806**
	Best MPENSGA2-A	*0.764*	**0.319**	*0.426*	*0.571*	*0.802*
	Best MPENSGA2-M	0.652	0.425	0.500	**0.414**	0.745
	MPENSGA2-A	$0.695_{0.053}$	$0.366_{0.054}$	$0.486_{0.048}$	$0.849_{0.138}$	$0.783_{0.029}$
	MPENSGA2-M	$0.554_{0.073}$	$0.518_{0.083}$	$0.519_{0.092}$	$0.641_{0.117}$	$0.722_{0.043}$

The best result is in **bold** face and the second best result in *italics*.

the *MAE*, *AMAE* and *MMAE* are ordinal metrics that depend on the distance between the ranking of two consecutive classes, and the τ_b measure is another ordinal metric independent on the values chosen for the ranks representing the classes.

The best models are generated by the MPENSGA2-A method, this method obtaining the best results in all metrics. Thus, we propose the MPENSGA2 algorithm to solve the two problems of predictive microbiology, specifically the upper best model of the Pareto front, which maximizes the value of *AMAE*.

EAs, and more specifically the MOEA, are computationally expensive especially when compared to local search algorithms, but the evolution of architectures enables ANNs to adapt their topologies to the different datasets without human intervention. This thus provides an approach to automatic ANN design as both ANN connection weights and structures can be evolved [21]. It is clear that the different non-evolutionary methods considered in this study demand a lower computational cost than the MPENSGA2 algorithm. However, the obtained models benefit clearly from the optimized structure learned by the EA, which allows achieve better results in generalization.

5 Conclusions

The proposed models for obtaining ordinal classification by using a generalized POM model with sigmoidal basis functions present competitive results with respect to other nominal and ordinal classifiers considering the five metrics used.

This modeling approach can provide a new insight for the predictive microbiology field, since it could directly determine, and with a high confidence level, if a pathogenic or spoilage microorganism could flourish. For stakeholders, the application of this kind of tools could be very useful in order to set microbiological criteria or to determine microbial shelf life under a given set of conditions. Implementation of risk management measures in food industries based on qualitative approaches (i.e. a combination of factors which limit microbial growth to below 0.01 or which fall into the "low growth probability" class) will suppose a breakthrough in guaranteeing microbial food safety.

Acknowledgement. This work was subsidized in part by the Spanish Inter-Ministerial Commission of Science and Technology under Project TIN2011-22794, the European Regional Development fund, and the "Junta de Andalucía" (Spain), under Project P08-TIC-3745. M. Cruz-Ramírez's research has been subsidized by the FPU Predoctoral Program (Spanish Ministry of Education and Science), grant reference AP2009-0487.

References

1. Bishop, C.: Neural Networks for Pattern Recognition. Oxford University Press (1995)
2. Chawla, N., Bowyer, K., Hall, L., Kegelmeyer, W.: SMOTE: Synthetic minority oversampling technique. Journal of Artificial Intelligence Research 16, 321–357 (2002)
3. Chu, W., Ghahramani, Z.: Gaussian processes for ordinal regression. Journal of Machine Learning Research 6, 1019–1041 (2005)
4. Chu, W., Keerthi, S.S.: Support Vector Ordinal Regression. Neural Computation 19(3), 792–815 (2007)
5. Commission, E.: Opinion of the scientific committee on veterinary measures relating to public health on listeria monocytogenes (1999),
   http://www.europa.eu.int/comm/food/fs/sc/scv/out25
6. Corchado, E., Abraham, A., Ponce, A.C., Ferreira de Carvalho, L.: Hybrid intelligent algorithms and applications. Information Sciences 180(14), 2633–2634 (2010)
7. Cruz-Ramírez, M., Hervás-Martínez, C., Sánchez-Monedero, J., Gutierrez, P.A.: A preliminary study of ordinal metrics to guide a multi-objective evolutionary algorithm. In: 11th International Conference on Intelligent Systems Design and Applications, ISDA 2011, Córdoba, Spain, pp. 1176–1181 (2011)
8. Cruz-Ramírez, M., Sánchez-Monedero, J., Fernández-Navarro, F., Fernández, J., Hervás-Martínez, C.: Memetic Pareto differential evolutionary artificial neural networks to determine growth multi-classes in predictive microbiology. Evol. Intelligence 3(3-4), 187–199 (2010)
9. Fernández, J., Hervás, C., Martínez-Estudillo, F., Gutierrez, P.: Memetic pareto evolutionary artificial neural networks to determine growth/no-growth in predictive microbiology. Applied Soft Computing 11, 534–550 (2011)
10. Fernández, J., Martínez, F., Hervás, C., Gutiérrez, P.: Sensitivity versus accuracy in multi-class problems using memetic pareto evolutionary neural networks. IEEE Transactions on Neural Networks 21(5), 750–770 (2010)

11. Fernandez-Navarro, F., Valero, A., Hervás-Martínez, C., Gutiérrez, P.A., Gimeno, R.G., Cosano, G.Z.: Development of a multi-classification neural network model to determine the microbial growth/no growth interface. I. J. of Food Microbiology 141(3), 203–212 (2010)
12. Frank, E., Hall, M.: A simple approach to ordinal classification. In: Proceedings of the 12th European Conference on Machine Learning, EMCL 2001, pp. 145–156 (2001)
13. Goh, C., Teoh, E., Tan, K.: Hybrid multiobjective evolutionary design for artificial neural networks. IEEE Transactions on Neural Networks 19(9), 1531–1548 (2008)
14. Gutiérrez, P.A., Pérez-Ortíz, M., Fernández-Navarro, F., Sánchez-Monedero, J., Hervás-Martínez, C.: An experimental study of different ordinal regression methods and measures, Salamanca, Spain, pp. 296–307 (2012)
15. Hall, M., Frank, E., Holmes, G., Pfahringer, B., Reutemann, P., Witten, I.H.: The weka data mining software: an update. Special Interest Group on Knowledge Discovery and Data Mining Explorer Newsletter 11(1), 10–18 (2009)
16. Kordík, P., Koutník, J., Drchal, J., Kovárik, O., Cepek, M., Snorek, M.: Meta-learning approach to neural network optimization. Neural Networks 23, 568–582 (2010)
17. Li, L., Lin, H.T.: Ordinal regression by extended binary classification. Advances in Neural Information Processing Systems 19, 865–872 (2007)
18. McCullagh, P.: Regression models for ordinal data. Journal of the Royal Statistical Society, Series B (Methodological) 42(2), 109–142 (1980)
19. Valero, A., Pérez-Rodríguez, F., Carrasco, E., Fuentes-Alventosa, J., García-Gimeno, R., Zurera, G.: Modelling the growth boundaries of staphylococcus aureus: Effect of temperature, ph and water activity. International Journal Food Microbiology 133, 186–194 (2009)
20. Vermeulen, A., Gysemans, K.P.M., Bernaerts, K., Geeraerd, A.H., Impe, J.F.V., Debevere, J., Devlieghere, F.: Influence of ph, water activity and acetic acid concentration on *Listeria monocytogenes* at 7°c: Data collection for the development of a growth/no growth model. International Journal of Food Microbiology 114(3), 332–341 (2007)
21. Yao, X.: Evolving artificial neural networks. Proceedings of the IEEE 87, 1423–1447 (1999)

Evaluation of Novel Soft Computing Methods for the Prediction of the Dental Milling Time-Error Parameter

Pavel Krömer[1,2], Tomáš Novosád[1], Václav Snášel[1,2], Vicente Vera[4],
Beatriz Hernando[4], Laura García-Hernández[7], Héctor Quintián[3],
Emilio Corchado[2,3], Raquel Redondo[5], Javier Sedano[6], and Alvaro E. García[4]

[1] Dept. of Computer Science, VŠB-Technical University of Ostrava, Czech Republic
[2] IT4Innovations, Ostrava, Czech Republic
{pavel.kromer,tomas.novosad,vaclav.snasel}@vsb.cz
[3] Departamento de Informática y Automática, Universidad de Salamanca, Spain
escorchado@usal.es
[4] Facultad de Odontología, UCM, Madrid, Spain
{vicentevera,aegarcia}@odon.ucm.es
[5] Department of Civil Engineering, University of Burgos, Burgos, Spain
rredondo@ubu.es
[6] Dept. of AI & Applied Electronics,
Castilla y León Technological Institute, Burgos, Spain
javier.sedano@itcl.es
[7] Area of Project Engineering, University of Cordoba, Spain
ir1gahel@uco.es

Abstract. This multidisciplinary study presents the application of two well known soft computing methods – flexible neural trees, and evolutionary fuzzy rules – for the prediction of the error parameter between real dental milling time and forecast given by the dental milling machine. In this study a real data set obtained by a dynamic machining center with five axes simultaneously is analyzed to empirically test the novel system in order to optimize the time error.

Keywords: soft computing, dental milling, prediction, evolutionary algorithms, flexible neural trees, fuzzy rules, industrial applications.

1 Introduction

Accurate scheduling and planning becomes increasingly important part of modern industrial processes. To optimize the manufacturing of products and schedule the utilization of devices, the product manufacturing time has to be known in advance. However, the predictions given by traditional methods and tools are often less accurate. Precise prediction of product manufacturing time is important for industrial production planning in order to meet, industrial, technological, and economical objectives [2,16]. One of the main goals of a production process is to deliver products on time and utilize the resources at maximum during

V. Snasel et al. (Eds.): SOCO Models in Industrial & Environmental Appl., AISC 188, pp. 163–172.
springerlink.com

production cycles. The production time estimate provided either by production models (i.e. by auxiliary software) or human experts are often less accurate than desirable [2]. Soft computing techniques can be used for flexible and detailed modelling of production processes [5]. The area of soft computing represents a set of various technologies involving non-linear dynamics, computational intelligence, ideas drawn from physics, physicology and several other computational frameworks. It investigates, simulates and analyzes very complex issues and phenomena in order to solve real-world problems: such as the failures detection in dental milling process, which requires a multidisciplinary approach [13].

In this study, a real data set obtained by a dynamic machining center with five axes simultaneously is analyzed by means of two soft computing techniques to empirically test the system in order to optimize the time error. The rest of this paper is organized as follows. Section 2 and section 3 present the background on the methods used to predict dental time-error. Section 4 introduces the experimental application and in section 5 conclusions are drawn .

2   Flexible Neural Tree

Flexible neural tree (FNT) [3] is a hierarchical neural network, which is automatically created in order to solve given problem. Its structure is usually determined using some adaptive mechanism and it is intended to adapt to the problem and data under investigation [11,10,4]. Due to this property of the FNTs, it is not necessary to setup some generic static network structure not related to the problem domain beforehand.

A general and enhanced FNT model can be used for problem solving. Based on the predefined instruction/operator sets, a FNT model can be created and evolved. In this approach, over-layer connections, different activation functions for different nodes and input variables selection are allowed. The hierarchical structure could be evolved by using genetic programming. The fine tuning of the parameters encoded in the structure could be accomplished by using parameter optimization algorithms. The FNT evolution used in this study combines both approaches. Starting with random structures and corresponding parameters, it first tries to improve the structure and then as soon as an improved structure is found, it fine tunes its parameters. It then goes back to improving the structure again and, provided it finds a better structure, it again fine tunes the rules'

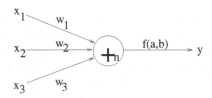

Fig. 1. A flexible neuron operator (instructor)

parameters. This loop continues until a satisfactory solution is found or a time limit is reached. A tree-structural based encoding method with specific instruction set is selected for representing a FNT model in this research. The reason for choosing the representation is that the tree can be created and evolved using the existing or modified tree-structure-based approaches. The function set F and terminal instruction set T that can be used to build a FNT model can be described as follows:

$$S = F \cup T = \{+_2, +_3, \ldots, +_N\} \cup \{x_1, x_2, \ldots, x_n\} \tag{1}$$

where $+_i$ $(i = 2, 3, \ldots, N)$ denote non-leaf nodes' instructions and taking i arguments. Input variables $x_1, x_2, \ldots, x_n$ are leaf nodes' instructions and taking no argument each. The output of a non-leaf node is calculated as a flexible neuron model. From this point of view, the instruction $+_i$ is also called a flexible neuron operator (instructor) with i inputs. A schematic view of the flexible neuron instructor is shown in fig. 1. In the *creation process* of neural tree, if a non-terminal instruction, i.e., $+_i$ is selected, i real values are randomly generated and used for representing the connection strength between the node $+_i$ and its children. In addition, two adjustable parameters a_i and b_i are randomly created as flexible activation function parameters. Activation function can vary according to given task. In this work we use following classical Gaussian activation function:

$$f(a_i, b_i, x) = e^{-(\frac{x - a_i}{b_i})^2} \tag{2}$$

The output of a flexible neuron $+_n$ can be calculated as follows. The total excitation of the $+_n$ is

$$net_n = \sum_{j=1}^{n} w_j \times x_j \tag{3}$$

where x_j $(j = 1, 2, \ldots, n)$ are the inputs to node $+_n$. The output of the node $+_n$ is then calculated by

$$out_n = f(a_n, b_n, net_n) = e^{-(\frac{net_n - a_n}{b_n})^2} \tag{4}$$

A typical evolved flexible neural tree model is shown in fig. 2. The overall output of a flexible neural tree can be computed from left to right by depth-first method, recursively.

The fitness function maps the FNT to a scalar, real-valued fitness values that reflect the FNT's performances on a given task. Firstly the fitness functions should be seen as error measures, i.e. mean square error (MSE) or root mean square error (RMSE). A secondary non-user-defined objective for which algorithm always optimizes FNTs is FNT size as measured by number of nodes. Among FNTs with equal fitness smaller ones are always preferred. MSE and RMSE are given by:

$$MSE(i) = \frac{1}{P} \sum_{j=1}^{P} (y_1^j - y_2^j)^2, \quad RMSE(i) = \sqrt{MSE(i)} \tag{5}$$

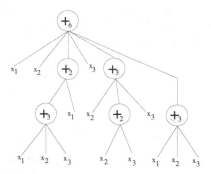

Fig. 2. A typical representation of neural tree with function instruction set $F = \{+_2, +_3, +_4, +_5, +_6\}$, and terminal instruction set $T = \{x_1, x_2, x_3\}$

where P is the total number of samples, y_1^j and y_2^j are the actual time-series and the FNT model output of j-th sample. $MSE(i)$ and $RMSE(i)$ denotes the fitness value of i-th individual.

Finding an optimal or near-optimal flexible neural tree can be accomplished by various evolutionary and bio-inspired algorithms [11,10,4]. The general learning procedure for constructing the FNT model can be described in high level as follows [3]:

1. Set the initial values of parameters used in the GA algorithms. Set the elitist program as NULL and its fitness value as a biggest positive real number of the computer at hand. Create a random initial population (flexible neural trees and their corresponding parameters)
2. Structure optimization by genetic algorithm, in which the fitness function is calculated by MSE or RMSE
3. If a better structure is found and no better structure is found for certain number of generations, then go to step (4), otherwise go to step (2)
4. Parameter optimization by genetic algorithms. In this stage, the tree structure or architecture of flexible neural tree model is fixed, and it is the best tree taken from the sorted population of trees. All of the parameters used in the best tree formulated a parameter vector to be optimized by local search
5. If the maximum number of local search is reached, or no better parameter vector is found for a significantly long time then go to step (6); otherwise go to step (4);
6. If satisfactory solution is found, then the algorithm is stopped; otherwise go to step (2).

Evolutionary methods [1] are in this study used for FNT structure optimization as well as for activation function parameters and tree nodes weights optimization. The selection, crossover and mutation operators used are the same as those of standard genetic programming [1]. A genetic algorithm starts with selection of two parents from current population. The product of crossover operator can be one or more offspring - two in this study. The mutation of offspring is performed at the last step of genetic algorithm. After these three steps we have new offspring which is placed into a newly created population. The process is repeated until desired new population is built. As soon as the new population is built, the new population is evaluated and sorted according to the fitness function.

Selection is in the FNT evolution implemented using the weighted roulette wheel algorithm and the tree structure crossover is implemented as an exchange of randomly selected subtrees of parent chromosomes. The crossover of node weights and activation function parameters is done in a similar way as in previous studies applying genetic algorithms to neural network training [6]. A variety of FNT mutation types were used:

1. Changing one terminal node: randomly select one terminal node in the neural tree and replace it with another terminal node.
2. Changing one function node: randomly select one function node and replace it with a newly generated subtree.
3. Growing: select a random function node in hidden layer of the neural tree and add newly generated subtree as a new child.
4. Pruning: randomly select a node in the neural tree and delete it in the case the parent node has more than two child nodes.

The mutation of tree weights and activation function parameters is the same as in the genetic algorithms for artificial neural networks [6].

3   Fuzzy Rules Evolved by Genetic Programming

Fuzzy rules (FR) [7,8,15] inspired by the area of fuzzy information retrieval (IR) [9] and evolved by genetic programming have been shown to achieve interesting results in the area of data mining and pattern analysis.

The fuzzy rules use similar data structures, basic concepts, and operations as the fuzzy information retrieval but they can be used for the analysis (i.e. classification, prediction) of general data. A fuzzy rule has the form of a weighted symbolic expression roughly corresponding to an extended Boolean query in the fuzzy IR analogy. The rule consists of weighted feature (attribute) names and weighted aggregation operators. The evaluation of such an expression assigns a real value from the range $[0, 1]$ to each data record. Such a valuation can be interpreted as an ordering or a fuzzy set over the data records. The fuzzy rule is a symbolic expression that can be parsed into a tree structure. The tree structure consists of nodes and leaves (i.e. terminal nodes). An example of fuzzy rule is give below:

feature1:0.5 and:0.4 (feature2[1]:0.3 or:0.1 ([1]:0.1 and:0.2 [2]:0.3))

In the fuzzy rule syntax can be seen three types of nodes: the feature node is defined by feature name and its weight (*feature1:0.5*) and represents a requirement on current value of a feature, past feature node is defined by feature name, index of previous record, and weight (*feature2[1]:0.3*) and it is requirement on previous value of a feature. Finally, the past output node is defined by the index of previous output and weight (*[1]:0.5*) and represents a requirement on previous value of the predicted output variable. Clearly, such a fuzzy rule can be used for the analysis of both, data sets consisting of independent records and time series.

The fuzzy rules are evaluated using the formulas and equations from the area of fuzzy IR and fuzzy sets (see e.g. [7,8,15]). The terminal node weights are interpreted as threshold for data feature values and operator nodes are mapped to fuzzy set operators. The fuzzy rule predicting certain value for a given data set is found using standard genetic programming that evolves a population of tree representations of the rules in a supervised manner. The whole procedure is very similar to the evolution of the FNT structure described in section 2 but it differs in the choice of the fitness function which is taken from the area of fuzzy IR. The correctness of search results in IR can be evaluated using the measures precision P and recall R. Precision corresponds to the probability of retrieved document to be relevant and recall can be seen as the probability of retrieving a relevant document. Precision and recall in the extended Boolean IR model can be defined using the Σ−count $\|A\|$ [19]:

$$\rho(X|Y) = \begin{cases} \frac{\|X \cap Y\|}{\|Y\|} & \|Y\| \neq 0 \\ 1 & \|Y\| = 0 \end{cases}, \quad P = \rho(REL|RET), \quad R = \rho(RET|REL) \quad (6)$$

where REL stands for the fuzzy set of all relevant documents, RET for the fuzzy set of all retrieved documents, and $\|A\|$ is the Σ−count, i.e. the sum of the values of characteristic function μ_A for all members of the fuzzy set $\|A\| = \sum_{x \in A} \mu_A(x)$ [19]. The F-score F is among the most used scalar combinations of P and R:

$$F = \frac{(1 + \beta^2)PR}{\beta^2 P + R} \quad (7)$$

For the evolution of fuzzy rules [7,8,15] we map the prediction given for training data set by the fuzzy rule to RET and the desired values to REL. F corresponds to the similarity of two fuzzy sets and a fuzzy rule with high F provides good approximation of the output value.

4   Dental Miling Time-Error Prediction in Industry

FNTs and FRs were used for the estimation of the time-error parameter in a real dental milling process. The data was gathered by means of a Machining Milling Center of HERMLE type-C 20 U (iTNC 530), with swivelling rotary (280 mm), with a control system using high precision drills and bits.

The models were trained using an initial data set of 98 samples obtained by the dental scanner in the manufacturing of dental pieces with different tool types (plane, toric, spherical and drill). The data set contained records consisting of 8 input variables (Tool, Radius, Revolutions, Feed rate X, Y and Z, Thickness, Initial Temperature) and 1 output variable (Time Error for manufacturing) as shown in table 1. Time error for manufacturing is the difference between the

Table 1. Description of variables in the data set

Variable (Units)	Range of values
Type of tool	Plane, toric, spherical and drill
Radius (mm.)	0.25 to 1.5
Revolutions per minute (RPM)	7,500 to 38,000
Feed rate X (mm. by minute)	0 to 3,000
Feed rate Y (mm. by minute)	0 to 3,000
Feed rate Z (mm. by minute)	50 to 2,000
Thickness (mm.)	10 to 18
Temperature ($^\circ$C)	24.1 to 31
Real time of work (s)	6 to 1,794
Time errors for manufacturing (s)	-28 to -255

time estimated by the machine itself and real production time. Negative values indicate that real time exceeds estimated time. The goal of this study was to evaluate the ability of evolutionary evolved FNTs and FRs to predict the dental milling time-error from the data. The parameters used for the evolution of the FNT and FR are shown in table 2. They were selected on the basis of initial experiments and past experience with the methods.

Because the number of records in the data set was small, a 10-fold cross-validation schema was selected. The final model is obtained using the full data set. Next, several different indexes were used to validate the models [18,17] such as the percentage representation of the estimated model, the loss (error) function ($\mathcal{V}$) and the generalization error value.

The percentage representation of the estimated model was calculated as the normalised mean error for the prediction (FIT1, FIT) using the validation data set and full data set respectively. The loss function $\mathcal{V}$ is the numeric value of the MSE that was computed using the training data set, the generalisation error value is the numeric value of the normalised sum of square errors (NSSE) that was computed using the test data set [12,14].

The results of both methods are shown in table 3. The presented values are averages after 10 independent runs for each of the 10 folds. Clearly, the FNT method was significantly better than FRs which in turn delivered results similar to those by the previously used soft computing methods [16]. Visual illustration of the time-error prediction by FNT and FR for first fold is shown in fig. 3 and fig. 4 respectively. Note that both methods are stochastic and the results may vary for independent runs.

FNT and FR have shown a good ability to learn the relations hidden in the data as shown in fig. 3a and fig. 4a and indicated by high FIT and low $\mathcal{V}$ in table 3 [16]. The good generalization ability of the methods is illustrated in fig. 3b and fig. 4b and supported by low NSSE and high FIT1 in fig. 3b. The results obtained by the FNT model are best-so-far for the dental milling time-error parameter prediction.

Table 2. FNT and FR evolution parameters

Method	Parameters
FNT	pop. size 100, crossover probability P_C 0.8, mutation probability P_M 0.2, limiting number of 10 generations, fitness function $RMSE$, Gaussian activation function with a, b, and weights from the range $[0, 1]$
FR	pop. size 100, crossover probability P_C 0.8, mutation probability P_M 0.2, limiting number of 1000 generations, no past feature nodes and no past output nodes allowed, fitness function F-Score with $\beta = 1$

Table 3. Dental milling time-error prediction indexes

Method	FIT1[%]	FIT[%]	$\mathcal{V}$	NSSE
FNT	95.89	92.02	0.0041	0.0150
FR	86.80	86.75	0.0079	0.0888

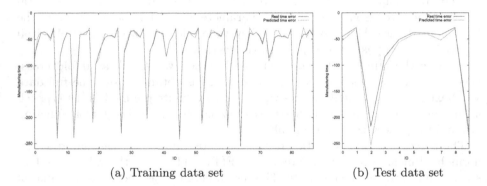

(a) Training data set   (b) Test data set

Fig. 3. Example of visual results of training and prediction by FNT (fold 1)

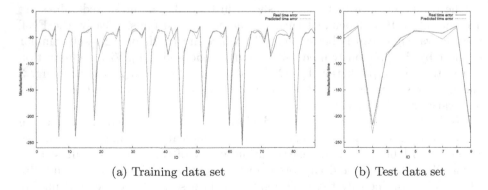

(a) Training data set   (b) Test data set

Fig. 4. Example of visual results of training and prediction by FRs (fold 1)

5   Conclusions

This study presents the comparison of some performance indexes of two well known soft computing methods for the prediction of the dental milling time-error parameter. Both soft computing models were trained on a real-world data set describing the production of a dental milling machine and their ability to adapt to the data was compared. To provide a good analysis of the performance of the methods, a 10-fold cross-validation was performed. The results of the cross-validation showed that the FNT managed to find models with significantly better average accuracy in terms of FIT, FIT1, V, and NSSE. The FNTs will be further studied as predictors of the dental milling time-error and other parameters such as accuracy.

Acknowledgements. This research is partially supported through a projects of the Spanish Ministry of Economy and Competitiveness [ref: TIN2010-21272-C02-01] (funded by the European Regional Development Fund). The authors would also like to thank to ESTUDIO PREVIO and TARAMI (both from Madrid, Spain) for their collaboration in this research. This work was also supported by the European Regional Development Fund in the IT4Innovations Centre of Excellence project (CZ.1.05/1.1.00/02.0070) and by the Bio-Inspired Methods: research, development and knowledge transfer project, reg. no. CZ.1.07/2.3.00/20.0073 funded by Operational Programme Education for Competitiveness, co-financed by ESF and state budget of the Czech Republic.

References

1. Affenzeller, M., Winkler, S., Wagner, S., Beham, A.: Genetic Algorithms and Genetic Programming: Modern Concepts and Practical Applications. Chapman & Hall/CRC (2009)
2. Chang, P., Liao, T.: Combining som and fuzzy rule base for flow time prediction in semiconductor manufacturing factory. Applied Soft Computing 6(2), 198–206 (2006)
3. Chen, Y., Abraham, A.: Flexible Neural Tree: Foundations and Applications. In: Chen, Y., Abraham, A. (eds.) Tree-Structure Based Hybrid Computational Intelligence. ISRL, vol. 2, pp. 39–96. Springer, Heidelberg (2010)
4. Chen, Y., Yang, B., Meng, Q.: Small-time scale network traffic prediction based on flexible neural tree. Appl. Soft Comput. 12(1), 274–279 (2012)
5. Custodio, L.M.M., Sentieiro, J.J.S., Bispo, C.F.G.: Production planning and scheduling using a fuzzy decision system. IEEE Transactions on Robotics and Automation 10(2), 160–168 (1994)
6. Ding, S., Li, H., Su, C., Yu, J., Jin, F.: Evolutionary artificial neural networks: a review. Artificial Intelligence Review, 1–10 (2011), doi:10.1007/s10462-011-9270-6
7. Krömer, P., Platoš, J., Snášel, V., Abraham, A.: Fuzzy classification by evolutionary algorithms. In: IEEE International Conference on Systems, Man, and Cybernetics, pp. 313–318. IEEE System, Man, and Cybernetics Society (2011)

8. Krömer, P., Platoš, J., Snášel, V., Abraham, A., Prokop, L., Mišák, S.: Genetically evolved fuzzy predictor for photovoltaic power output estimation. In: 2011 Third International Conference on Intelligent Networking and Collaborative Systems (INCoS), pp. 41–46. IEEE (2011)

9. Pasi, G.: Fuzzy sets in information retrieval: State of the art and research trends. In: Bustince, H., Herrera, F., Montero, J. (eds.) Fuzzy Sets and Their Extensions: Representation, Aggregation and Models. STUDFUZZ, vol. 220, pp. 517–535. Springer, Heidelberg (2008)

10. Peng, L., Yang, B., Zhang, L., Chen, Y.: A parallel evolving algorithm for flexible neural tree. Parallel Computing 37(10-11), 653–666 (2011)

11. Qi, F., Liu, X., Ma, Y.: Synthesis of neural tree models by improved breeder genetic programming. Neural Computing & Applications 21, 515–521 (2012), doi:10.1007/s00521-010-0451-z

12. Sedano, J., Corchado, E., Villar, J., Curiel, L., de la Cal, E.: Detection of heat flux failures in building using a soft computing diagnostic system. Neural Network World 20(7), 883–898 (2010)

13. Sedano, J., Curiel, L., Corchado, E., de la Cal, E., Villar, J.R.: A soft computing method for detecting lifetime building thermal insulation failures. Integr. Comput.-Aided Eng. 17(2), 103–115 (2010)

14. Sedano, J., Curiel, L., Corchado, E., de la Cal, E., Villar, J.R.: A soft computing method for detecting lifetime building thermal insulation failures. Integr. Comput.-Aided Eng. 17(2), 103–115 (2010)

15. Snášel, V., Krömer, P., Platoš, J., Abraham, A.: The Evolution of Fuzzy Classifier for Data Mining with Applications. In: Deb, K., Bhattacharya, A., Chakraborti, N., Chakroborty, P., Das, S., Dutta, J., Gupta, S.K., Jain, A., Aggarwal, V., Branke, J., Louis, S.J., Tan, K.C. (eds.) SEAL 2010. LNCS, vol. 6457, pp. 349–358. Springer, Heidelberg (2010)

16. Vera, V., Corchado, E., Redondo, R., Sedano, J., Garcia, A.: Applying soft computing techniques to optimize a dental milling process. Neurocomputing (submitted)

17. Vera, V., Garcia, A.E., Suarez, M.J., Hernando, B., Corchado, E., Sanchez, M.A., Gil, A.B., Redondo, R., Sedano, J.: A bio-inspired computational high-precision dental milling system. In: NaBIC, pp. 423–429. IEEE (2010)

18. Vera, V., Garcia, A.E., Suarez, M.J., Hernando, B., Redondo, R., Corchado, E., Sanchez, M.A., Gil, A.B., Sedano, J.: Optimizing a dental milling process by means of soft computing techniques. In: ISDA, pp. 1430–1435. IEEE (2010)

19. Zadeh, L.A.: Test-score semantics dor natural languages and meaning representation via Pruf. In: Empirical Semantics. Quantitative Semantics, vol. 1, pp. 281–349. Studienverlag Brockmeyer, Bochum (1981)

Self-organizing Migration Algorithm
on GPU with CUDA

Michal Pavlech

Tomas Bata University in Zlin, Faculty of Applied Informatics
nám. T.G.Masaryka 5555, 760 01 Zlín
Czech Republic
pavlech@fai.utb.cz

Abstract. A modification of Self-organizing migration algorithm for general-purpose computing on graphics processing units is proposed in this paper. The algorithm is implemented in C++ with its core parts in c-CUDA. Its implementation details and performance are evaluated and compared to previous, pure C++ version of algorithm. 6 commonly used artificial test functions are used to test the performance. The test results clearly show significant speed gains without a compromise in convergence quality.

Keywords: SOMA, CUDA, GPGPU, evolutionary algorithm.

1 Introduction

Modern graphics cards (GPUs) are capable of much more than processing and displaying of visual data. Their multiprocessor architecture is suitable for parallel algorithms which can benefit from the SIMD architecture. GPUs have generally lower clock frequency than modern processors but rely on parallel execution of instructions over large blocks of data.

Evolutionary algorithms (EAs) are in their very essence parallel processes and as such are suitable for implementation on such multiprocessor devices with little or no modifications to their functionality. First attempts on utilizing the processing power of GPUs for EAs were done before the release of general purpose computing APIs. Researchers had to modify algorithms to fit them into specialized GPU processing units – the shaders and data structures had to be translated into textures. An example of this approach is the work of Wong , Wong and Fok who used GPU to increase the performance of genetic algorithm and reported speedup of up to 4.42 times [1, 2].

With the emergence of general purpose computing frameworks for GPUs, like Compute Unified Device Architecture (CUDA), Open Computing Language (OpenCL) and DirectCompute, new possibilities have risen and general purpose computing on graphics processing units (GPGPUs) became available to a broader audience.

A large number of evolutionary algorithms has been ported to GPUs, with CUDA being arguably the most common API used. Some of these algorithms include: genetic algorithm [3, 4], genetic programming [5], differential evolution [6], ant colony optimization [7] and particle swarm optimization [8, 9].

V. Snasel et al. (Eds.): SOCO Models in Industrial & Environmental Appl., AISC 188, pp. 173–182.
springerlink.com © Springer-Verlag Berlin Heidelberg 2013

Self-organizing migration algorithm (SOMA) was chosen for a subset of its characteristics which are not common in other EAs and make it suitable for parallel architecture with limited communication between processing units.

The paper is divided as follows: first part describes SOMA, modifications and steps necessary for its implementation using CUDA. Next part describes methodology used in testing the algorithms performance followed by results of the tests and conclusion.

2 Methods

2.1 SOMA

SOMA was first introduced by Zelinka [10]. It is modeled after behavior of intelligent individuals working cooperatively to achieve common goal, for example pack of animals working together to find food source. This behavior is mimicked by individuals moving towards another individual, known as the leader, which generally has the best fitness value. Efficiency and convergence ability of SOMA was proved in numerous applications [11, 12].

SOMA uses slightly different nomenclature in comparison to other evolutionary algorithms, one round of the algorithm is called migration. There is a set of three control parameters which control the algorithm's behavior and significantly affect its performance:

- *PathLength* ∈ [1; 5]: Specifies how far from the leader will the active individual stop its movement.
- *Step* ∈ [0.11; *PathLength*]: Defines the size of discreet steps in solution space.
- *PRT* ∈ [0; 1]: Perturbation controls creation of perturbation vectors which influence the movement of active individual.

The movement of individuals through error space is altered by a random perturbation. In order to perturb movement of individuals, boolean vector **PRTVector** is generated according to equation:

$$\mathbf{PRTVector}_j \quad = \quad 1 \text{ if } \text{rand}_j(0,1) < PRT$$
$$= \quad 0 \text{ otherwise} \tag{1}$$

where $j = 0,1,...,$ *Dimension*-1. The **PRTVector** is generated before movement of active individual, and is generated for each individual separately. Value 0 in **PRTVEctor** means, that corresponding dimension of individual is locked – individual cannot change its value during this migration. If all elements of **PRTVector** are set to 1, individual moves straight towards the leader.

Creation of new individuals for next population is implemented using vector operations. Active individual moves towards the leader according to equation:

$$\mathbf{x}_{i,j,t}^{ML+1} = \mathbf{x}_{i,j,start}^{ML} + \left(\mathbf{x}_{L,j}^{ML} - \mathbf{x}_{i,j,start}^{ML}\right).t.\mathbf{PRTVector}_j \tag{2}$$

where *ML* is the number of current migration round, $x^{ML}_{i,j,start}$ is position of active individual at beginning of current migration, $x^{ML}_{L,j}$ is the position of the leader, $t \in [0; pathLength]$, $t = 0$, *Step*, $2*Step$,...

This equation is applied to all individuals except the leader, which does not move. For each step in solution space (denoted by *t*) newly created individual is evaluated. At the end of migration (individual made all steps lower than *PathLength*) individual is set to a position which had the best fitness value during the current migration. Therefore the fitness value of an individual cannot deteriorate.

2.2 cuSOMA

The aim of this work was to create the fastest possible implementation of SOMA using the CUDA toolkit without compromising its functionality and convergence ability. This algorithm was named cuSOMA. cuSOMA is implemented as a C++ class with its core components in c-CUDA.

CUDA C extends C by allowing the programmer to define C functions, called kernels, that, when called, are executed N times in parallel by N different CUDA threads. The CUDA threads execute on a physically separate device that operates as a coprocessor to the host running the C program, the host and the device maintain their own separate memory spaces in DRAM, referred to as host memory and device memory, respectively [13].

Early implementations of EAs for GPUs reported problems with generation of random numbers which was time consuming on GPU hardware [6]. More recent nVidia CUDA SDK features a library called CURAND which deals with efficient pseudorandom generators on GPUs [14]. CURAND random generators need to preserve their states using a data structure *curandState* in order to avoid generating the same number sequences for each thread and for each kernel call. Separate states for each thread are stored in an array and have to be copied from host to device prior to kernel calls. Initialization of states is done only once during the class initialization and uses standard C++ random generator. After a seed is generated a kernel is launched, which initializes random generators for each thread in parallel.

Population of candidate solutions for cuSOMA is stored in one dimensional array on the device. To minimize the impact of memory transfers on algorithm performance the population and states are copied from device only after a user definable amount of migrations finished. The migration of individuals requires 2 additional arrays to store intermediary positions and the best position discovered during the migration. Both arrays for temporary values have the same structure as the original population. The fitness values of each individual are stored inside a population after the phenotype. The scheme of population is depicted in Fig 1.

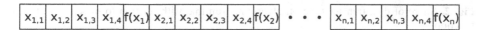

Fig. 1. Scheme of population stored on device

One possible bottle neck of this solution is the transfer of data between device and host. To quantify this problem a test was conducted which identified the amount of time needed for data transfers and time for actual computation on device. The data transfer time should be proportional to a size of copied data and there are two parameters which influence this amount: population size and dimension of a solution. Data transfer times include: allocation of arrays on device, copying of population and states to and from device and freeing of allocated memory. Two tests were conducted to explore the influence of these two parameters with all the time measurements conducted with CUDA events. First test was run with dimension locked at 50 and population size growing steadily from 100 to 2,500 with a step of 100 and with population size from 5,000 to 25,000 with step of 5,000. All attempts were run for 100 migration rounds. Second test was run with population size and migration rounds locked to 1,000 and 100 respectively and with dimension of the test functions growing from 25 to 250 with step of 25 and additionally from 500 to 2,500 with step of 500. All tests were repeated for 10 times and averaged in order to avoid random glitches. The share of data transfer times on total run time can be seen in Fig. 2 and Fig. 3 where it is displayed as a percentage of total runtime with x axis displayed in logarithmic scale for better readability.

It can be seen that data transfer times are smaller than 1% of the overall computation time for all test cases. Increase in dimension resulted in longer computation times and thus the influence of data transfers decreases. Increase in population size lead to opposite results up to 2,500 individuals where data transfer share rose steadily up to 0.36% for De Jong function. For larger population sizes, on the contrary, this share declined. Results show that computational time rose faster than data transfer time for a number of individuals which was significantly larger than number of thread processors on GPU and thus the share of data transfer time decreased.

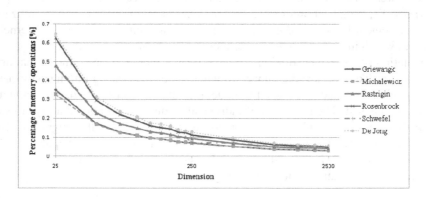

Fig. 2. Share of data transfer time on algorithms runtime with relation to dimension of cost function

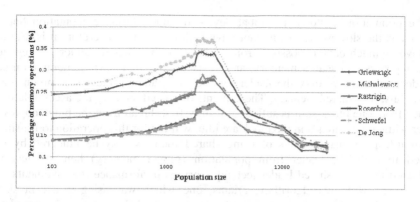

Fig. 3. Share of data transfer time on algorithms runtime with relation to population size

For larger populations it is not possible to execute all the operations using only one CUDA block because the thread per block limit is set to 1024 (512 for devices with compute capacity lower than 2). Large populations require use of several separate blocks. Number of blocks is decided using equation:

$$blocks = \frac{(population\_size + threads\_per\_block - 1)}{threads\_per\_block} \tag{3}$$

During the computation each thread calculates the index of an individual which it should migrate according to equation:

$$index = (block\_id \times threads\_per\_block) + thread\_id \tag{4}$$

where *block_id* is the index of current block in a grid and *thread_id* is thread index in current block.

The influence of threads per block on performance of cuSOMA was investigated in three tests which had population size set to 1,000, 3,000 and 5,000, in order to simulate spawning of more blocks, dimension set to 50 and number of migrations to 100.

Table 1 shows the best performing number of threads per block for 3 different population sizes. The number of threads has considerable impact on cuSOMA performance. The test showed that 16 threads per block was the setting with the most consistent performance across population sizes and cost functions and therefore was used in further tests.

Table 1. Number of threads with fastest execution with relation to population size and test function

Population	Griewangk	Michalewicz	Rastrigin	Rosenbrock	Schwefel	De Jong
1,000	16	16	16	16	16	16
3,000	32	32	32	16	32	16
5,000	16	256	16	16	16	16

The population is stored in global device memory during computation but this memory is the slowest type of memory available on device. Therefore it is desirable to move as much data as possible from global to either shared memory or registers. Both, registers and shared memory, are too small to contain whole population so it was decided to move only the leader for each migration, which is common for all individuals, into shared memory. The contents of shared memory are accessible by all threads from one block so it is sufficient if only first thread from the block copies the leader while other threads in the block are idle, waiting for this operation to finish.

Possible performance gains of using shared memory may be hindered by time needed to copy the leader from population (global memory) hence a test was conducted to find if shared leader technique brings performance improvements. The test was run with two algorithm variants, one which was using leader in shared memory and one which was accessing leader from global memory. The population size was set to 3,000, dimension to 50 and number of migrations to 100, number of threads per block was changed in powers of 2. Fig. 4 shows the difference in time between the two algorithm's versions. All values above zero mean that version with shared leader was faster for a given number of threads per block.

The leader in shared memory turned out to be a questionable improvement. There are certainly performance benefits for all test functions, but they are in range of milliseconds. For 16 threads the highest improvement was 3.25ms (Griewangk function) and the highest performance loss was 7.19ms (Michalewicz function). Overall results of shared leader test can be summed as follows: algorithm with shared leader was faster in 33 cases but slower in 15. Although it adds only minor speed improvements the shared leader was left intact as a part of cuSOMA.

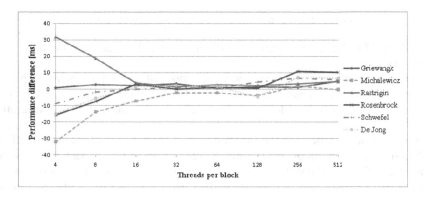

Fig. 4. Performance gains from leader in shared memory with relation to threads per block

In order to determine which individual will be the leader for next migration round it is necessary to search the whole population for individual with the best fitness. Because the population during migrations stays in the global memory of GPU it is possible to perform parallel search for the best individual. cuSOMA uses the parallel reduction to find the index of the leader and store it in the global memory where the threads performing migration can access it.

2.3 Performance Tests

The most important reason for implementing SOMA on GPU is the promise of possible speedup when compared to running the algorithm on CPU. To test these possible gains two types of tests were conducted with 6 different artificial test functions. Both tests were conducted on the same hardware: nVidia Tesla C2075, 448 thread processors at 1150 MHz and Intel Xeon E5607, at 2.26 GHz.

All the test cases were run for 100 migration rounds, repeated 10 times and their results averaged. The measured parameter was the speedup of cuSOMA when compared to CPU implementation.

First test was aimed at how well the cuSOMA scales to increase in population size.

Second test used constant population size and the dimension of the test functions was increasing. The purpose of this test was to decide if cuSOMA is suitable for functions which require higher computational power for evaluation.

In addition to speedup tests a simple test to determine if the convergence ability of SOMA was not compromised was conducted with all 6 test functions set to 25 dimensions, population size to 1,000 and number of migrations to 100.

Detailed setup of all tests is in Table 2.

Table 2. details of performance tests

Test	Population size	dimension
1	100-25,000	50
2	1,000	25-2500
3	1,000	25

3 Results

Fig. 5 shows that cuSOMA scales extremely well to the size of population in values from 100 to 2500. Each increase of population size enlarged the performance gap between CPU and GPU implementation of SOMA. Further increase in population size showed that the initial steady growth of speedup is much less significant for higher volumes of individuals. However no test function showed substantial decrease in speedup and cuSOMA stayed superior to CPU implementation.

The highest recorded speedup was 126.9 for Michalewicz function and 25,000 individuals. cuSOMA seems to be more suitable for more computationally demanding functions as can be seen from comparison of results for De Jong and Michalewicz functions. This effect is probably caused by higher clock speed of CPU in comparison to thread processor clock speed of GPU, with this difference becoming more obvious for less demanding functions.

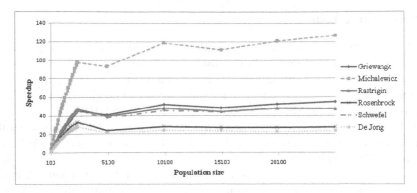

Fig. 5. Speedup measured with relation to population size

Fig. 6 shows that cuSOMA does not scale very well to increase in cost function dimension, which can be also viewed as an increase in computational complexity of cost function evaluation. The GPU implementation is considerably faster than its CPU counterpart with speedups ranging from 13 (De Jong function) to 56 (Michalewicz function). Results from higher dimensions show that the value of speedup remained almost constant witch exception of Michalewicz function which constantly showed improvements with each dimension increase.

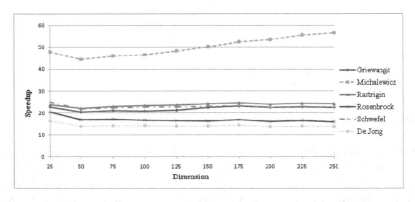

Fig. 6. Speedup measured with relation to test function's dimension

As shown in Table 3, cuSOMA was able to find known global minima [15] for 5 test functions with Rosenbrock's function being an exception. Further test runs with more migration rounds were able to find global optimum even for this function. Value of global extreme of Michalewicz's function for 25 dimensions is not known.

Table 3. The best found values in 10 test runs

	Griewangk	Michalewicz	Rastrigin	Rosenbrock	Schwefel	De Jong
Best value	0	-24.633	0	0.0114101	-10474.6	1.66679E-13
Best known value	0	Not known	0	0	-10474.6	0

4 Conclusion

An implementation of self-organizing migration algorithm for GPUs using nVidia CUDA was presented. This new version was named cuSOMA and provides significant improvements in computation time. The test with artificial test functions showed high speedups in comparison to CPU implementation of SOMA with highest recorded speedup of 126.9. cuSOMA scales very well to large population sizes and therefore it should be especially suitable for functions where finding global minimum requires a high number of individuals in order to satisfactory search the solution space. On the other hand, increase in computational complexity of test functions did not produce further speedup gains in comparison to CPU version but nevertheless it was still considerably faster.

The influence of block size on algorithm performance was investigated and it can be seen, that it can have significant impact on runtime but it is also dependent on the nature of optimized function. Also the share of data transfer times on overall runtime was investigated and tests showed that for large populations this time grows slower that actual computational time.

Further research will focus on fine tuning cuSOMA and finding other optimization algorithms which could be even better suited for general computing on graphical processing units.

Acknowledgments. This paper was created as a part of internal grant agency project number IGA/FAI/2012/033 at Tomas Bata University in Zlin, Faculty of Applied Informatics and of the European Regional Development Fund under project CEBIA-Tech No. CZ.1.05/2.1.00/03.0089.

References

1. Wong, M.L., Wong, T.T., Fok, K.L.: Parallel evolutionary algorithms on graphics processing unit. In: Proc. IEEE Congress Evolutionary Computation, vol. 3, pp. 2286–2293 (2005)
2. Fok, K.L., Wong, T.T., Wong, M.L.: Evolutionary Computing on Consumer Graphics Hardware. IEEE_M_IS 22, 69–78 (2007)
3. Pospichal, P., Jaros, J., Schwarz, J.: Parallel Genetic Algorithm on the CUDA Architecture. In: Di Chio, C., Cagnoni, S., Cotta, C., Ebner, M., Ekárt, A., Esparcia-Alcazar, A.I., Goh, C.-K., Merelo, J.J., Neri, F., Preuß, M., Togelius, J., Yannakakis, G.N. (eds.) EvoApplicatons 2010, Part I. LNCS, vol. 6024, pp. 442–451. Springer, Heidelberg (2010)
4. Zhang, S., He, Z.: Implementation of Parallel Genetic Algorithm Based on CUDA. In: Cai, Z., Li, Z., Kang, Z., Liu, Y. (eds.) ISICA 2009. LNCS, vol. 5821, pp. 24–30. Springer, Heidelberg (2009)
5. Langdon, W.B.: Large Scale Bioinformatics Data Mining with Parallel Genetic Programming on Graphics Processing Units. In: de Vega, F.F., Cantú-Paz, E. (eds.) Parallel and Distributed Computational Intelligence. SCI, vol. 269, pp. 113–141. Springer, Heidelberg (2010)
6. de Veronese, L.P., Krohling, R.A.: Differential evolution algorithm on the GPU with C-CUDA. In: Proc. IEEE Congress Evolutionary Computation (CEC), pp. 1–7 (2010)

7. Fu, J., Lei, L., Zhou, G.: A parallel Ant Colony Optimization algorithm with GPU-acceleration based on All-In-Roulette selection. In: 2010 Third International Workshop on Advanced Computational Intelligence (IWACI), pp. 260–264 (2010)

8. Zhou, Y., Tan, Y.: GPU-based parallel particle swarm optimization. In: Proc. IEEE Congress Evolutionary Computation CEC 2009, pp. 1493–1500 (2009)

9. de Veronese, L.P., Krohling, R.A.: Swarm's flight: Accelerating the particles using C-CUDA. In: Proc. IEEE Congress Evolutionary Computation CEC 2009, pp. 3264–3270 (2009)

10. Zelinka, I.: SOMA—self organizing migrating algorithm. In: Onwubolu, G.C., Babu, B.V. (eds.) New Optimization Techniques in Engineering. Springer, Berlin (2004)

11. Senkerik, R., Zelinka, I., Oplatkova, Z.: Comparison of Differential Evolution and SOMA in the Task of Chaos Control Optimization - Extended study. In: Complex Target cf 2009 IEEE Congress on Evolutionary Computation, vols. 1-5, pp. 2825–2832. IEEE (2009)

12. Tupy, J., Zelinka, I., Tjoa, A., Wagner, R.: Evolutionary algorithms in aircraft trim optimization. In: Dexa 2008: 19th International Conference on Database and Expert Systems Applications, Proceedings, pp. 524–530. IEEE Computer Soc. (2008)

13. NVIDIA CUDA C Programming Guide. NVIDIA Developer Zone,
http://developer.download.nvidia.com/compute/DevZone/docs/
html/C/doc/CUDA_C_Programming_Guide.pdf (accessed March 13, 2012)

14. CUDA Toolkit 4.1 CURAND Guide. NVIDIA Developer Zone,
http://developer.download.nvidia.com/compute/DevZone/docs/
html/CUDALibraries/doc/CURAND_Library.pdf (accessed March 13, 2012)

15. Molga, M., Smutnicki, C.: Test functions for optimization needs (2005),
http://www.zsd.ict.pwr.wroc.pl/files/docs/functions.pdf
(accessed March 13, 2012)

Urban Traffic Flow Forecasting
Using Neural-Statistic Hybrid Modeling

M. Annunziato[1], F. Moretti[2], and S. Pizzuti[1,2]

[1] Energy New technologies and sustainable Economic development Agency (ENEA),
'Casaccia' R.C. via anguillarese 301,
00123 Rome, Italy
{mauro.annunziato,stefano.pizzuti}@enea.it
[2] University Roma Tre,
Dept. of Computer Science and Automation,
via della vasca navale 79, 00146 Rome, Italy
{moretti,pizzuti}@dia.uniroma3.it

Abstract. In this paper we show a hybrid modeling approach which combines Artificial Neural Networks and a simple statistical approach in order to provide a one hour forecast of urban traffic flow rates. Experimentation has been carried out on three different classes of real streets and results show that the proposed approach clearly outperforms the best of the methods it combines.

1 Introduction

Transportation is a wide human-oriented field with diverse and challenging problems waiting to be solved. Characteristics and performances of transport systems, services, costs, infrastructures, vehicles and control systems are usually defined on the basis of quantitative evaluation of their main effects. Most of the transport decisions take place under imprecision, uncertainty and partial truth. Some objectives and constraints are often difficult to be measured by crisp values. Traditional analytical techniques were found to be not-effective when dealing with problems in which the dependencies between variables were too complex or ill-defined.

Moreover, hard computing models cannot deal effectively with the transport decision-makers' ambiguities and uncertainties.

In order to come up with solutions to some of these problems, over the last decade there has been much interest in soft computing applications of traffic and transport systems, leading to some successful implementations[3].

The use of Soft Computing methodologies (SC) is widely used in several application fields [1][10][24][25]. In modeling and analyzing traffic and transport systems SC are of particular interest to researchers and practitioners due to their ability to handle quantitative and qualitative measures, and to efficiently solve complex problems which involve imprecision, uncertainty and partial truth. SC can be used to bridge modeling gaps of normative and descriptive decision models in traffic and transport research.

V. Snasel et al. (Eds.): SOCO Models in Industrial & Environmental Appl., AISC 188, pp. 183–190.
springerlink.com

Transport problems can be classified into four main areas : traffic control and management, transport planning and management, logistics, design and construction of transport facilities.

The first category includes traffic flow forecasting which is the topic tackled in this work. This issue has been faced by the soft computing community since the nineties [8, 11, 15, 22, 27, 28, 29] up today [7, 9, 16, 21] with Artificial Neural Networks (ANN) [2,13]. As example, among the most recent work [16] focuses on traffic flow forecasting approach based on Particle Swarm Optimization (PSO) with Wavelet Network Model(WNM). [21] reviews neural networks applications in urban traffic management systems and presents a method of traffic flow prediction based on neural networks. [7] proposes the use of a self-adaptive fuzzy neural network for traffic prediction suggesting an architecture which tracks probability distribution drifts due to weather conditions, season, or other factors. Among the other techniques SVR[14], Adaptive Hinging Hyperplanes[20] and Multivariate State Space [26] are worth mentioning.

All the mentioned applications have one feature in common : they use one single global model in order to perform the prediction. Therefore, the main novelty of the proposed work is to combine different heterogeneous models in order to get a meta-model capable of providing predictions more accurate than the best of the constituent models. In particular, we compose a neural networks ensemble with a simple statistical model and compare the results over the one hour forecast.

2 Methods

2.1 Naïve

In order to perform a meaningful comparison for the forecasting, a naïve model should be introduced in order to quantify the improvement given by more intelligent and complex forecasting techniques. For seasonal data a naïve model might be defined as:

$$x_t = x_{t-s} \tag{1}$$

with S the appropriate seasonality period. This model gives a prediction at time t presenting the value observed exactly a period of S steps before. For this work we put the value of S = 1 which corresponds to the previous hour. It means that to predict the flow rate of the following hour it is used the current flow measure.

2.2 Statistical

One the simplest and most widely used models when dealing with regular time series (as urban traffic flows) is to build an average weekly distribution of the traffic flow sampled hourly. Thus, from the data we compute for each day the average flow rate hour by hour in such a way that we get an average distribution made of 24X7=168 points.

2.3 Neural Networks Ensembling

The other method, as suggested also by literature, we applied is ANN and ANN ensembling. The term 'ensemble' describes a group of learning machines that work together on the same task, in the case of ANN they are trained on some data, run together and their outputs are combined as a single one. The goal is obtain better predictive performance than could be obtained from any of the constituent models. In the last years several ensembling methods have been carried out [6, 18, 19]. The first one, also known as Basic Ensemble Method (BEM), is the simplest way to combine M neural networks as an arithmetic mean of their outputs. This method can improve the global performance [5, 23] although it does not takes into account that some models can be more accurate than others. This method has the advantage to be very easy to apply. A direct BEM extension is the Generalised Ensemble Method (GEM) [5, 23] in which the outputs of the single models are combined in a weighted average where the weights have to be properly set, sometimes after an expensive tuning process. Other methods are Bootstrap AGGregatING (BAGGING) [17] and Adaboost [4, 12].

2.4 Hybrid Model

Hybrid models are an extension of the ensembling approach in the sense that the final goal is to combine different models in such a way that the accuracy of the composition is higher than the best of the single models. The difference is that the combination is performed among highly heterogeneous models, that is models generated by different methods with different properties and thus the composition among them is a complex rule taking into account the peculiarities of the models and/or of the problem itself.

Therefore, in this work we propose a novel hybrid model which combines an ANN ensemble with the statistical model.

The composition rule is the following :

"IF the statistical model has a high error (meaning that for some reason we are out of a normal situation) THEN use the neural model ELSE use the statistical one"

This criterion is based on the absolute error of the statistical model, thus the composition rule turns into

$$
\begin{aligned}
|x^t - y^t_s| > \varepsilon &\quad \Rightarrow \quad y^{t+1} = y^{t+1}_n \\
|x^t - y^t_s| \leq \varepsilon &\quad \Rightarrow \quad y^{t+1} = y^{t+1}_s
\end{aligned}
\tag{2}
$$

Where y^{t+1} is the outcome (one hour prediction) after the composition rule, y^{t+1}_n is the prediction of the neural ensemble, y^t_s is the current outcome of the statistical model and y^{t+1}_s is its prediction.

This basically means that if we are in normal statistical conditions (where the statistical model makes a small error) then use as prediction model the statistical one (which is very accurate in this condition), else (when out of normal statistical situations) take the neural ensembling estimation.

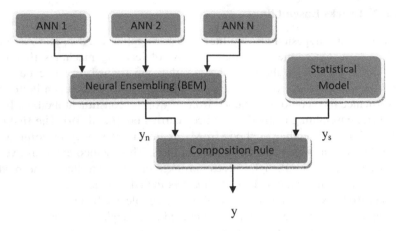

Fig. 1. Proposed hybrid modeling approach

3   Experimentation

In this paragraph we test and compare the methods presented in the previous section. The test case has concerned the short term traffic flow rate of three different streets (tab.1) located in the town of Terni (about 90km north of Rome). The data set is made of 3 months (13 weeks) of measurement corresponding to 2184 hourly samples. The data set has been partitioned into training/testing and validation made respectively of 10 and 3 weeks each.

Table 1. Street features

	Maximum traffic flow rate
Street 1	600
Street 2	800
Street 3	950

The ANN are feed-forward MLP with 10 hidden neurons and one output (the one hour flow forecast) with sigmoid as activation function for all the neurons. The number of inputs N has been chosen with a preliminary analysis by calculating the validation prediction error after ensembling for different values of N (tab.2). By this analysis it turned out the optimal number of input neurons (namely the length of the history window) to be eight.

Training has been performed through the Back-Propagation algorithm with adaptive learning rate and momentum stopping after 100000000 iterations and a 'save best' strategy to avoid overfitting. The reported result are averaged over 10 different runs (with standard deviation in brackets) and the ensemble is therefore made by the same 10 models.

The reported errors are measured as

$$e = |x-y|/(M-m) \tag{3}$$

Where x is the real value to be predicted, y is the output model, M is the real maximum value and m is the minimum.

Table 2. History length selection

N (hours)	Street 1	Street 2	Street 3
3	5.72%	6.88%	5.81%
5	3.9%	5.07%	3.99%
8	3.29%	3.43%	3.02%
10	3.54%	4.12%	3.74%

Afterwards, it has been tuned (tab.3) the parameter ε of the hybrid model (2).

Table 3. Hybrid model parameter ε tuning

	$\varepsilon=10$	$\varepsilon=20$	$\varepsilon=30$	$\varepsilon=40$	$\varepsilon=50$	$\varepsilon=60$
Street 1	2.98%	2.83%	2.81%	2.8%	2.88%	2.99%
Street 2	2.85%	2.69%	2.65%	2.66%	2.68%	2.75%
Street 3	3.25%	3.13%	3.08%	3.04%	3.03%	3.04%

At last, the following table shows the comparison of the models considered in this work in terms of prediction accuracy over the validation set and figure 2 shows a graphical comparison.

Table 4. Model comparison

	Naive	Statistic	ANN	ANN Ensembling	Hybrid
Street 1	8.92%	5.90%	3.74% ±0.10%	3.29%	
					2.8%
Street 2	9.99%	7.14%	4.00% ±0.10%	3.43%	
					2.65%
Street 3	7.66%	5.56%	3.48% ±0.09%	3.02%	3.02%
Average	8.86%	6.20%	3.74% ±0.10%	3.25%	2.82%

From this analysis it is clear that in general the proposed hybrid approach outperforms the best of the 'classical' models (which turns out to be ANN ensembling) providing a remarkable improvement in prediction accuracy. Such level of precision is very important when dealing with applications like traffic and lighting control where the higher the model accuracy is the more effective the control system is.

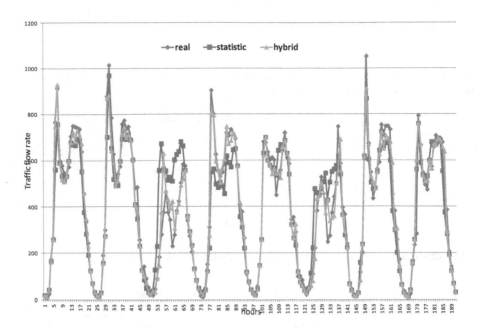

Fig. 2. Model comparison

From this graph it is clear that the hybrid model performs much better than the statistical model because, when out of normal conditions, it switches to the neural ensembling method which takes into account the real traffic dynamics.

4 Conclusion

In this paper we showed a novel hybrid modeling approach which combines Artificial Neural Networks and a simple statistical approach in order to provide a one hour forecast of urban traffic flow rates. Experimentation has been carried out on three different classes of real streets and results showed that the proposed approach clearly outperforms the best of the methods it puts together achieving a prediction error lower than 3%. The reason for that is that the neural ensembling model is capable to provide more reliable estimations when out of standard conditions because it considers the real traffic dynamics.

The accuracy of the proposed hybrid modeling approach is such that it can be applied for intelligent monitoring, diagnostic systems and optimal control.

Future work will focus on further modeling improvements using more sophisticated ensembling methods as well as different composition methods for the hybrid model based on fuzzy sets rather than fixed thresholds. Moreover, the proposed method will be compared to other approaches already used in the field as Wavelets, SVR, Adaptive Hinging Hyperplanes and Multivariate State Space.

As application, we are going to use these models in public lighting control in order to reduce energy consumption.

References

1. Abraham, A.: Hybrid Soft Computing and Applications. International Journal of Computational Intelligence and Applications 8(1), 5–7 (2009)
2. Arbib, M.A.: The Handbook of Brain Theory and Neural Networks. The MIT Press, Cambridge (1995)
3. Avineri, E.: Soft Computing Applications in Traffic and Transport Systems: A Review. Advances in Soft Computing 1, 17–25 (2005)
4. Avnimelech, R., Intrator, N.: Boosting regression estimators. Neural Computation 11, 491–513 (1999)
5. Bishop, C.M.: Neural Networks for Pattern Recognition, pp. 364–369. Oxford University Press (1995)
6. Breiman, L.: Combining Predictors. In: Sharkey, A.J.C. (ed.) Combining Artificial Neural Nets – Ensemble and Modular Multi-net Systems, pp. 31–50. Springer, Berlin (1999)
7. Bucur, L., Florea, A., Petrescu, B.S.: An adaptive fuzzy neural network for traffic prediction. In: 18th Mediterranean Conference on Control & Automation (MED), Marrakech, pp. 1092–1096 (2010)
8. Canca, D., Larrañeta, J., Lozano, S., Onieva, L.: Traffic intensity forecast in urban networks using a multilayer perceptron. In: Joint International Meeting EURO XV - INFORMS XXXIV, Barcelona, Spain (1997)
9. Çetiner, B.G., Sari, M., Borat, O.: A Neural Network Based Traffic-Flow Prediction Model. Mathematical and Computational Applications 15(2), 269–278 (2010)
10. Corchado, E., Arroyo, A., Tricio, V.: Soft computing models to identify typical meteorological days. Logic Journal of the IGPL 19(2), 373–383 (2011)
11. Dougherty, M.S., Cobbett, M.: Short term inter-urban traffic forecasts using neural networks. In: Proc. Second DRIVE-II Workshop on Short-Term Traffic Forecasting, Delft, The Netherlands, pp. 65–79 (1994)
12. Drucker, H.: Improving regressors using boosting techniques. In: Fisher, D.H. (ed.) ICML, pp. 107–115. Morgan Kaufmann (1997)
13. Haykin, S.: Neural Networks, a comprehensive foundation, 2nd edn. Prentice Hall, New Jersey (1999)
14. Hong, W.: Traffic flow forecasting by seasonal SVR with chaotic simulated annealing algorithm. Neurocomputing 74(12-13), 2096–2107 (2011)
15. Ishak, S., Kotha, P., Alecsandru, C.: Optimization of Dynamic Neural Network Performance for Short-Term Traffic Prediction. Transportation Research Record 1836, 45–56 (2003)
16. Jawanjal, S., Bajaj, P.: A Design Approach to Traffic Flow Forecasting with Soft Computing Tools. In: 2010 3rd International Conference on Emerging Trends in Engineering and Technology (ICETET), Goa, pp. 81–84 (2010)
17. Kohavi, R., Bauer, E.: An empirical comparison of voting classification algorithms: Bagging, boosting and variants. Machine Learning 36, 105–142 (1999)
18. Krogh, A., Vedelsby, J.: Neural network ensembles, cross validation and active learning. In: Tesauro, G., Touretzky, D.S., Leen, T.K. (eds.) Advances in Neural Information Processing Systems, vol. 7, pp. 231–238. MIT Press (1995)
19. Liu, Y., Yao, X.: Ensemble learning via negative correlation. Neural Networks 12(10), 1399–1404 (1999)
20. Lu, Y., Hu, J., Xu, J., Wang, S.: Urban Traffic Flow Forecasting Based on Adaptive Hinging Hyperplanes. In: Deng, H., Wang, L., Wang, F.L., Lei, J. (eds.) AICI 2009. LNCS, vol. 5855, pp. 658–667. Springer, Heidelberg (2009), doi:10.1007/978-3-642-05253-8_72

21. PamułA, T.: Road Traffic Parameters Prediction in Urban Traffic Management Systems using Neural Networks. Transport Problems 6(3), 123–128 (2011)
22. Park, D., Rilett, L.R., Han, G.: Spectral Basis Neural Networks for Real-Time Travel Time Forecasting. Journal of Transportation Engineering 125(6), 515–523 (1999)
23. Perrone, M.P., Cooper, L.N.: When networks disagree: ensemble methods for hybrid neural networks. In: Mammone, R.J. (ed.) Neural Networks for Speech and Image Processing. Chapman-Hall (1993)
24. Sedano, J., Curiel, L., Corchado, E., de la Cal, E., Villar, J.R.: A soft computing method for detecting lifetime building thermal insulation failures. Integrated Computer-Aided Engineering 17(2), 103–115 (2010)
25. Zhao, S.-Z., Iruthayarajan, M.W., Baskar, S., Suganthan, P.N.: Multi-objective robust PID controller tuning using two lbests multi-objective particle swarm optimization. Inf. Sci. 181(16), 3323–3335 (2011)
26. Stathopoulos, A., Karlaftis, M.G.: A multivariate state space approach for urban traffic flow modeling and prediction. Transportation Research Part C 11, 121–135 (2003)
27. Taylor, C., Meldrum, D.: Freeway traffic data prediction using neural networks. In: Proc. 6th VNIS Conference, Seattle, WA, pp. 225–230 (1995)
28. van Lint, J.W.C., Hoogendoorn, S.P., van Zuylen, H.J.: Freeway Travel time Prediction with State-Space Neural Networks. Transportation Research Record 1811, 30–39 (2003)
29. Zheng, W., Der-Horng, L., Shi, Q.: Short-Term Freeway Traffic Flow Prediction: Bayesian Combined Neural Network Approach. Journal of Transportation Engineering, 114–121 (2006)

Gravitational Search Algorithm Design
of Posicast PID Control Systems

P.B. de Moura Oliveira[1], E.J. Solteiro Pires[2], and Paulo Novais[3]

[1] INESC TEC - INESC Technology and Science (formerly INESC Porto)
Department of Engineering, School of Sciences and Technology,
5001–801 Vila Real, Portugal
{oliveira,epires}@utad.pt
[2] Departamento de Informática, Universidade do Minho 4710-057 Braga, Portugal
pjon@di.uminho.pt

Abstract. The gravitational search algorithm is proposed to design PID control structures. The controller design is performed considering the objectives of set-point tracking and disturbance rejection, minimizing the integral of the absolute error criterion. A two-degrees-of-freedom control configuration with a feedforward prefilter inserted outside the PID feedback loop is used to improve system performance for both design criteria. The prefilter used is a Posicast three-step shaper designed simultaneously with a PID controller. Simulation results are presented which show the merit of the proposed technique.

1 Introduction

Controllers based on proportional, integrative and derivative (PID) modes are applied within the majority of industrial control loops. Despite the development of more complex control methodologies, there are several reasons for the success and resilience of PID control, such as simplicity, performance and reliability in a wide range of system dynamics [1]. Thus, the development of new PID control based schemes and design methodologies are relevant research issues. Two classical control system design goals are input reference tracking and disturbance rejection. Optimal PID controller settings for set-point tracking can result in poor disturbance rejection and vice-versa, i.e., optimal disturbance rejection PID settings can result in poor set-point tracking. The design of PID controllers both for set-point tracking and disturbance rejection can be improved using two-degrees-of-freedom (2DOF) configurations [2]. A well known 2DOF configuration uses a feedforward prefilter applied to the input reference signal and a PID controller within the feedback loop. The ideal design of such 2DOF controllers requires simultaneous optimization of system response both for set-point tracking and disturbance rejection.

The GSA algorithm was proposed by Rashedi et al. [3] which reported the advantages of using this algorithm in optimizing a set of benchmark unimodal and multimodal functions. In [3] a comparison was presented between GSA, particle swarm optimization (PSO) and a real genetic algorithm (RGA), showing that GSA performs better that PSO and RGA in the tested function set. Since its proposal GSA has been reported successfully in solving several problems [4,5,6]. In this paper the

V. Snasel et al. (Eds.): SOCO Models in Industrial & Environmental Appl., AISC 188, pp. 191–199.
springerlink.com

gravitational search algorithm (GSA) is proposed to design 2DOF control configuration in which the prefilter is a three-step Posicast input shaper and the feedback loop is a PID controller.

2   Gravitational Search Algorithm

The GSA was proposed originally by [3], and it is inspired in the natural interaction forces between masses. Accordingly to Newton's law of gravity, the gravitational force, F, between two particles in the universe can be represented by:

$$F = G\frac{M_1 M_2}{R^2} \tag{1}$$

where: M_1 and M_2 are the two particles masses, G is the gravitational constant and R is the distance between the two particles. Newton's well known second law relates force with acceleration, a, and mass, M, as:

$$F = Ma \Leftrightarrow a = \frac{F}{M} \tag{2}$$

Considering a swarm of particles (or population), X, of size s, in which every element represents a potential solution for a given search and optimization problem, moving in a n-dimensional space, with vector x representing the particle position. The force between particles i and j, for dimension d and iteration t is represented by [3]:

$$F_{ij}^d = G(t)\frac{M_i(t)M_j(t)}{R_{ij}(t)+\varepsilon}(x_j^d(t) - x_i^d(t)) \tag{3}$$

where: ε is a small constant, and the gravitational constant can be defined in every iteration by:

$$G(t) = G(t_0)\frac{t_0^\beta}{t} \quad \beta < 1 \tag{4}$$

with: $G(t_0)$ representing the initial gravitational constant, and R representing the Euclidian distance between the two particles. The use of R instead of R^2 in (1) was proposed by [3] based on experimental tests. The total force that acts in each particle i for a certain dimension, d, is evaluated by:

$$F_i^d(t) = \sum_{j \in K_{best}, j \neq i}^s \varphi_{1j} F_{ij}(t) \tag{5}$$

in which, φ_{1j} represents an uniform randomly generated number in the interval [0,1] and K_{best} is the set of best particles, with size set to k_0 at the beginning of the search procedure and decreased linearly over time. The acceleration of mass i, called law of motion [3], is represented by:

$$a_i = \frac{F_i^d(t)}{M_i(t)} \tag{6}$$

with M_i representing the inertia mass for particle i, evaluated with:

$$M_i = \frac{m_i(t)}{\sum\limits_{j=1}^{s} m_j(t)} \tag{7}$$

and

$$m_i(t) = \frac{fit_i(t) - worst(t)}{best(t) - worst(t)} \tag{8}$$

where: *fit*, *best*, *worst* represent respectively the: current, best and worst fitness values for particle i in iteration t. The velocity and position of each particle are up-dated accordingly to the following equations:

$$v_i^d(t+1) = \varphi_{2i} \, v_i^d(t) + a_i^d(t) \tag{9}$$

$$x_i^d(t+1) = x_i^d(t) + v_i^d(t+1) \tag{10}$$

with φ_{2j} representing an uniform randomly generated number in the interval [0,1].

3 PID Control Design: Problem Statement

A general PID control structure for single-input single-output systems can be illustrated using the classical block diagram presented in Figure 1. Two of the more relevant control design objectives are set-point tracking and disturbance rejection.

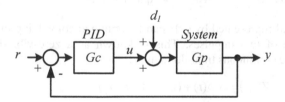

Fig. 1. PID control configuration

For some types of system dynamics optimum set-point tracking can be achieved by using an open-loop control feedforward configuration. The modification of the reference input in order to improve system tracking can be implemented by using command shaping techniques [8,13]. The input shaping concept was originally proposed [7] to control underdamped systems. However the same technique can be used for other non-oscillatory system dynamics, as reported in [9].

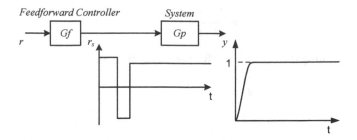

Fig. 2. Feedforward input command shaping

The pre-filter or shaper modifies the input step reference input, r, into another signal, r_s, appropriate to cancel some of the system dynamics, in order to achieve a dead-beat response, as illustrated in Figure 2. Cancelling the underdamped complex poles with a feedforward controller was originally proposed by Smith [7], and termed Posicast control. Considering a unit step reference input, the half-cycle Posicast control signal is represented by:

$$r_s(t) = A_1 r(t) + A_2 r(t - t_1) \tag{11}$$

with:

$$A_1 + A_2 = 1$$
$$t_1 = \frac{T_d}{2} \tag{12}$$

and: r representing an unit reference input step, r_s the shaped signal, A_1 and A_2 the first and second step amplitudes and T_d the undamped time period. Equation (11) represented in the Laplace complex domain results in the following shaper transfer function:

$$G_f(s) = A_1 + A_2 e^{-t_1 s} = A_1 + (1 - A_1) e^{-t_1 s} \tag{13}$$

The control signal represented by (11) can be obtained convolving the unit step input with a sequence of two impulses. This is known as a zero-vibration shaper [10], represented in the continuous time-domain as:

$$ZV(t) = A_1 \delta(t) + (1 - A_1) \delta(t - t_1) \tag{14}$$

where $\delta(t)$ is the Dirac delta function. The amplitude of the first step or impulse is a function of the overshoot, Mp:

$$A_1 = \frac{1}{1 + M_p} = \frac{1}{1 + e^{-\frac{\zeta \pi \omega_n}{\omega_d}}} \tag{15}$$

with: ζ representing the damping factor, ω_n and ω_d representing the undamped and damped natural frequencies, respectively. This study considers a Posicast shaper with three steps, represented by:

$$G_{fts}(s) = A_1 + A_2 e^{-t_1 s} + A_3 e^{-t_2 s} \quad 0 < t_1 < t_2 \tag{16}$$

$$A_1 + A_2 + A_3 = 1 \quad A_1 \geq 1, A_2 < 0 \tag{17}$$

As systems can be subjected to disturbances and model uncertainty, the ideal feed-forward control configuration presented in Figure 2 is usually combined with a feedback control loop. This combination can be accomplished using several control configurations, incorporating the feedforward controller (called hereby shaper) inside or outside the feedback loop. This problem has been addressed by [8], showing that using the shaper inside the loop is not advantageous for rejecting input disturbances. Thus, the control configuration used in this study is a two-degrees-of-freedom (2DOF) configuration presented in Figure 3, with the Posicast input command shaper (PICS) outside the loop.

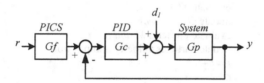

Fig. 3. Two-degrees-control configuration with input command shaping

The 2DOF controller design can be accomplished using several methodologies. The PID controller can be designed first for achieving good disturbance rejection, and then the PICS can be designed to enhance set-point tracking. The former is a sequential design procedure. However, this can result in low performance. Indeed, both feedforward input shaper and feedback controller should be designed simultaneously. This type of methodology, also called concurrent design [11], is particularly useful when both design objectives are conflicting.

4 GSA Design of PID Control Structures

4.1 PID Design for Set-Point Tracking

The PID controller used in this study is governed by the following equation:

$$G_c(s) = \frac{s^2 K_d + s K_p + K_i}{s} \left(\frac{1}{1 + s T_f} \right) \tag{18}$$

where: K_p, K_i and K_d represent the proportional, integrative and derivative gains, respectively, and T_f the filter time constant. The GSA is proposed to design the control structures described in the previous section. The first case is the PID control design considering the control configuration presented in Figure 1, for set-point tracking minimizing the integral of absolute error criterion:

$$IAE = \int_0^T |e(t)| dt, \quad with \quad e(t) = r(t) - y(t) \tag{19}$$

The algorithm used is presented in Figure 4, based on the original GSA [3], with some minor adaptations. In this case, if the swarm initialization is performed using a totally random procedure, some controller PID gains will make the system unstable. In these unstable cases the IAE value is disproportional high compared with stable cases, which makes the GSA to perform badly. Thus, to avoid unstable particles incorporating the first population, the swarm is initialized randomly using a candidate interviewing procedure. A randomly generated particle is allowed to be part of the initial swarm if it fulfills a predefined minimum IAE threshold.

$t = 0$
initialize swarm $X(t)$
while(!(termination criterion))
    evaluate $X(t)$
    update G, *best*, *worst*
    evaluate particles M *and* a
    update particles velocity and position
    $t = t + 1$
end

Fig. 4. Gravitational Search Algorithm for PID controller design

The gravitational constant is updated using:

$$G(t) = G_0 \left(1 - \frac{t}{n\_it} \right) \tag{20}$$

where n_it represents the total number of iterations. This linear decreasing equation was found better for this application, among other possibilities tested by experimentation. The evaluation of each particle mass and acceleration are evaluated with equations (7) and (6), respectively. Particle velocity and position are updated with (9) and (10), respectively. The simulation experiment considers the control of a fourth order system with time delay represented by the model:

$$G_p(s) = \frac{1}{(1+s)^4} e^{-s} \tag{21}$$

Each swarm particle encodes the PID gains parameters $\{K_p, K_i, K_d\}$ and the filter constant was set to 0.1. The search interval was equal both for initialization and search defined by: $0.1 \leq K_p, K_i\ K_d \leq 5$. The initialization threshold was set to an IAE of 1000. The total number of iterations ($n\_it=150$) was the search termination criterion used. This number was deliberately set low as the aim here is not to achieve the optimal PID settings but good settings in a short evolutionary time period. The value used for the initial gravitational constant was $Go=0.5$. The best PID gains achieved were $\{K_p=1.06,\ K_i=0.34,\ K_d=1.63\}$, results in an IAE=361. Figure 5-a) presents the unit step system response and respective control signal.

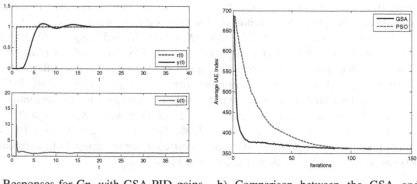

a) Responses for Gp, with GSA PID gains for set-point tracking.

b) Comparison between the GSA and PSO.

Fig. 5. Simulation results for set-point tracking

Figure 5-b) presents a comparison between GSA algorithm and a standard Particle Swarm Optimization (PSO) algorithm, showing the evolution of the mean value of the ITAE index from a set of 20 runs. The PSO parameters in terms of swarm size and number of objective function evaluations were the same as used in the GSA. The cognitive and social constants were set to 2 and the inertia weight was linearly decreased between 0.9 and 0.4 over the 150 iterations. For this parameters set Figure 5-b) clearly shows that the GSA convergence rate is faster in an early stage of the run. However, the PSO could be set to a faster convergence rate by reducing the higher limit for the inertia weight. The achieved value for the average fitness value is the same, which indicates that both algorithms converged for the same value in all trial runs. An interesting feature shown in Figure 5-b) is that the average fitness trend for GSA is more irregular than the PSO. This may prove relevant in escaping search traps such as local optima.

4.2 PID Design for Set-Point Tracking and Disturbance Rejection

If the PID gains derived for set-point tracking are applied to disturbance rejection the performance achieved is not good, as illustrated in Figure 6.a). To improved disturbance rejection, the 2DOF configuration presented in Figure 3 is used, with and three-step input shaper represented by (16) and a PID controller. The design is performed considering the simultaneous optimization of both pre-filter and PID controller. The optimization procedure considers an input step applied to the reference input first, and an input step applied to the input disturbance input, d_l, when the system as settled its tracking (in this case t=35s). The cost function used is the ITAE and each swarm particle encodes both the prefilter parameters and PID gains $\{A_1, A_2, t_2, \mu, K_p, K_i, K_d\}$ subjected to the amplitude constraint (17). The search intervals for the PID gains are the same as before and for the three-step Posicast shaper: $1 \leq A_1 \leq 2$, $-2 \leq A_2 \leq 0$, $0.5 \leq t_1 \leq 6$ and $1.1 \leq \mu \leq 4$. Parameter A_3 is evaluated using the amplitude constraint (17) and $t_1 = t_2/\mu$. Figure 6-a) presents the simulation results comparing the PID configuration with the 2DOF with Posicast input shaping. No limits were imposed to the actuator signal and the parameters achieved for the prefilter were $\{A_1 = 2.0, A_2 = -1.59, A_3 = 0.59, t_1 = 0.61, t_2 = 2.43\}$, and PID gains $\{K_p = 1.55, K_i = 0.55, K_d = 2.45\}$, IAE=538. The results with the 2DOF clearly improved the single's PID, with an IAE=687, accounting with the unit step disturbance. Figure 6-b) presents the simulation results comparing the 2DOF with actuator saturation limits $-15 \leq u(t) \leq +15$, and the PID controller was implemented using a anti-windup scheme based on the conditioning technique [12,13]. The parameters achieved for the prefilter were $\{A_1 = 2, A_2 = -0.1, A_3 = -0.9, t_1 = 0.45, t_2 = 0.5\}$, and PID gains $\{K_p = 1.96, K_i = 0.68, K_d = 2.68\}$, with IAE=450. The plots presented in Figure 6-b) show an significant improvement both compared to the PID as well as the PID without the anti-windup scheme.

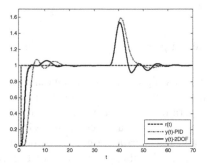
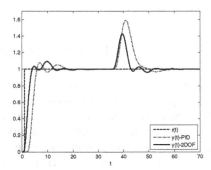

a) Responses for Gp, for single PID and 2DOF Posicast PID.

b) Responses for Gp, for single PID and 2DOF Posicast PID with saturation and anti-windup.

Fig. 6. Simulation results for disturbance rejection

5 Conclusions

The GSA was proposed to optimize PID control structures using the integral of absolute error criterion. Two control configurations were addressed: i) classical feedback loop with PID controller for set-point tracking ii) 2DOF configuration using a feedforward Posicast input command shaper, placed outside the feedback PID loop. Both three-step Posicast parameters and PID gains were designed simultaneously, both for the objectives of set-point tracking and disturbance rejection. The same relevance was given to both objectives. The results presented show that GSA has a faster convergence rate than PSO algorithm for PID design and it can conveniently design both the input shaper and PID controller in the 2DOF configuration, with and without considering controller variable saturation levels. Further research will explore the proposed technique for other process dynamics.

References

1. Åström, K.J., Hägglund, T.: The Future of PID Control. Control Engineering Practice 9(11), 1163–1175 (2001)
2. Araki, M., Taguchi, H.: Two-Degree-of-Freedom PID Controllers. International Journal of Control Automation, and Systems 1(4), 401–411 (2003)
3. Rashedi, E., Nezamabadi-pour, H., Saryazdi, S.: GSA: A Gravitacional Search Algorithm. Information Sciences 179, 2232–2248 (2009)
4. Precup, R.E., David, R.C., Petriu, E.M., Preitl, S., Răda, M.C.: Gravitational Search Algorithms in Fuzzy Control Systems Tuning. Preprints of the 18th IFAC World Congress, pp. 13624–13629 (2011)
5. Khajehzadeh, M., Raihan Taha, M., El-Shafie, A., Eslami, M.: Search for critical failure surface in slope stability analysis by gravitational search algorithm. International Journal of the Physical Sciences 6(21), 5012–5021 (2011); Academic Journals
6. Duman, S., Sonmez, Y., Guvenc, U., Yorukeren, N.: Application of Gravitational Search Algorithm for Optimal Reactive Power Dispatch Problem. In: IEEE Symposium Innovations in Intelligent Systems and Applications (INISTA), pp. 519–523 (2011)
7. Smith, O.J.M.: Posicast Control of Damped Oscillatory Systems. Proc. IRE 45(9), 1249–1255 (1957)
8. Huey, J.R., Sorensen, K.L., Singhose, W.E.: Useful applications of closed-loop signal shaping controllers. Control Engineering Practice 16, 836–846 (2008)
9. Tuttle, T.D.: Creatig Time-Optimal Commands for Linear Systems, PhD Thesis, MIT (1997)
10. Singer, N.C., Seering, W.P.: Preshaping command inputs to reduce system vibration. Journal of Dynamic Systems Measurement and Control 112(3), 76–82 (1990)
11. Chang, P.H., Park, J.: A concurrent design of input shaping technique and a robust control for high-speed/high-precision control of a chip mounter. Control Engineering Practice 9(2001), 1279–1285 (2001)
12. Hanus, K.M., Henrotte, J.L.: Conditioning technique, a general anti-windup and bumpless transfer method. Automatica 23(6), 729–739 (1987)
13. Moura Oliveira, P.B., Vrančić, D.: Underdamped Second-Order Systems Overshoot Control. Accepted for publication in the IFAC Conference on Advances in PID Control, PID 2012, Brecia (2012)

A Web Platform and a Decision Model for Computer-Interpretable Guidelines

Tiago Oliveira, Paulo Novais, and José Neves

University of Minho, Braga, Portugal
{toliveira,pjon,jneves}@uminho.pt

Abstract. Situations of medical error and defensive medicine are common in healthcare environments and have repercussions in the quality of care under offer. The occurrence of adverse events and the increase of healthcare expenses are some of the consequences of medical malpractice. Indeed, these situations may be prevented by encouraging the compliance with Clinical Guidelines (CGs). However, the current format of CGs proved to be disadvantageous for real-time application, i.e., they may not provide recommendations to healthcare professionals when required, and on time. The introduction of Computer-Interpretable Guidelines (CIGs) may provide a solution to this problem, however they are not widely implemented and there are some issues that need to be contemplated. Indeed, in this paper it is presented the CompGuide project for guideline representation and sharing, combined with the handling of incomplete information in that context.

1 Introduction

Clinical Guidelines (CGs) [1] are documents based on scientific evidence and consensus among experts that provide recommendations to deal with specific clinical cases. Their main objective is to structure the tasks of a clinical process according to the health condition of a patient.

Official development programs of CGs started between the late 70s and the early 80s. Since then, CGs have progressed significantly. The medical community has continuously addressed the weaknesses of CGs and their development. Initially they were solely based on informal consensus among clinical experts, working on a regional base, and the guideline development group was exclusively composed by physicians. Now guidelines are based on rigorous scientific evidence evaluated by multidisciplinary groups of professionals, including a wide range of expertise from different scientific fields, namely from management of human resources to exam pactice. In fact, the need for a standardized evidence grading system led to the creation, in the year 2000, of the Grading of Recommendations Assessment, Development and Evaluation (GRADE) [2] project, an initiative for evidence grading. The guideline development programs evolved to national development programs and have spread across the world with the help of the Guidelines International Network (G-I-N) [3], created in 2002 and with a membership of 85 organizations from 43 countries.

V. Snasel et al. (Eds.): SOCO Models in Industrial & Environmental Appl., AISC 188, pp. 201–210.
springerlink.com

Development of CGs is a central subject in the medical community, since increased compliance with standards may provide a solution to mitigate the effects of medical errors and defensive medicine. Medical errors are mistakes committed by healthcare professionals that result in harm to the patient. It is a universal problem, namely in the United States (US), as it is shown in the bar graph of Fig. 1, but also in Europe. Studies show that the rate of adverse outcomes in London hospitals is 10.8% and the fatality rate resulting from medical error is 8% [4]. These numbers may not seem too adverse, but from the perspective of patients that put their lives on the hands of healthcare professionals, they are quite expressive and object of concern.

On the other hand, defensive medicine occurs when a healthcare professional avoids treating certain patients or orders treatments and exams to avoid criticism and eventual lawsuits. Studies about defensive medecine in breast cancer detection estimate that nearly 50% of the tested women will receive a false positive [6]. This puts them in emotional distress and may create a state of pseudo-disease. In some cases, the trust in their physician may be seriously undermined. The effectiveness of CGs in addressing these issues can be seen in a case study for ischemic heart disease, in which an increase of 10% in guideline compliance was associated with a decrease in 10% in mortality [7].

CGs have come a long way since they were idealized. However there are still some issues to be addressed in order to be able to effectively mitigate medical errors and defensive medicine. In the following sections we will point the shortcomings of paper-based CGs as well as related work in the field. In the last section we will present the CompGuide project, which is aimed at the representation of clinical knowledge contained in CGs, increasing their availability and the handling of incomplete information in the clinical process [1]. Soft computing techniques in the form of clinical decision support systems have been used by researchers in this field to manage the issue of incomplete information in the clinical process [9].

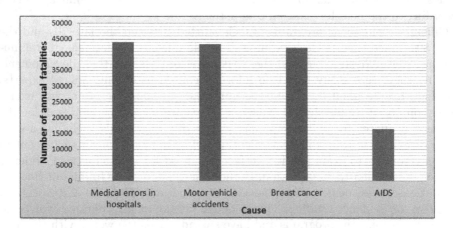

Fig. 1. Bar graph showing the annual fatalities in the US caused by medical errors in hospitals, motor vehicle accidents, breast cancer and AIDS [5]

2   Shortcomings of Paper-Based Clinical Guidelines

There are some issues with the text format of CGs, pointing out that long textual documents are difficult to consult. The information takes too long to be obtained and the texts are susceptible to ambiguous interpretations, given the lack of accuracy of the medical concepts that are used and their unstructured nature [8]. To a healthcare professional that has to enforce and perfect his/her clinical practice, it is nearly impossible to collect, assess and interpret the recommendations of these CGs at the moment of the delivery of care. The maintenance of this type of documents is problematic, since a modification usually implies restructuring the whole document. This is the reason why most guidelines are only revised, thus making it difficult to keep up with the rapid development of scientific knowledge.

Healthcare professionals are also concerned that compliance with CGs may lead to an inflexible clinical practice, too focused on rules to follow or respect. They fear that this will restrain their ability to adjust medical procedures to the context under which they are inserted, by reducing their decision-making capability. Justified variability in clinical practice is necessary when there is the need to accommodate differences in healthcare systems, in the characteristics of the populations (e.g., social, demographic, cultural, health condition) or when the patient and the healthcare professionals have preferences among clinical procedures that are logically acceptable [8].

From the analysis of the current limitations of CGs, it is evident that they should develop an interaction with the user, which in this case is the healthcare professional. Guidelines should provide real time recommendations as the clinical process unfolds, taking into account the state of the patient, and the preferences of the healthcare professionals in control. In order to facilitate their update they should present a modular structure, so that only a portion of the guideline may be adjusted without revising the whole document. These perceptions of interactivity and modularity are essential for the creation of living guidelines, the next stage in the evolution of CGs.

3   Computer-Interpretable Guidelines

CIGs are representations of CGs in a digital format. A CIG system is essentially a Clinical Decision Support System that integrates some basic features, namely a guideline depiction model and an execution engine [10]. Trying to stay up-to-date, in this review we will mention the current trends in the development of CIGs and afterwards the insight mechanims that enhance this field.

3.1   Current Trends

Currently there are few CIG systems available. However we will address them by their depiction models and mention the execution engines available for each one. The depiction models present in this review are Arden Syntax [11], Guideline Interchange Format (GLIF) [12], PROforma [13] and SAGE [14].

Arden Syntax was developed in 1989 and is now a standard of Health Level 7 (HL7) [11]. The current version of Arden Syntax is Arden Syntax 2.8. This approach focuses on sharing simple and independent guidelines as modules. Each CG is modeled as a Medical Logic Module (MLM), which comprises relevant knowledge.

GLIF represents an effort of Intermed Collaboratory in the development of a sharable CG representation model [12]. The GLIF depiction model dates from 1998 and its current version is GLIF3. It consists of a set of five classes, each one representing a step in the clinical process. This approach is task-based and follows the Task Network Model (TNM), so every moment of the clinical process is labeled as a Decision Step, Patient State Step, Branch Step, Synchronization Step or Action Step. There is not a formal method of representation of temporal constraints between steps in GLIF, however this is assured by a subset of Asbru temporal language, which is the strong argument of this approach. At Columbia University, GLIF is being integrated with the Clinical Event Monitor and the Computerized Physician Order Entry (CPOE) system to provide clinical decision support. The GLIF3 Guideline Execution Engine (GLEE) is a tool for executing guidelines in this format.

In 1998, the Advanced Computation Laboratory of Cancer Research of the United Kingdom initiated the development and assessment of the PROforma depiction model [13]. The objective of this model was the construction of guidelines as flowcharts where the nodes are instances of pre-defined classes of tasks. The classes are Plans, Actions, Decisions and Questions. Each class has a set of attributes that reflects its information needs. Among the execution engines for PROforma, Arezzo and HeCaSe2 are to be highlighted [13].

The SAGE (Standards-Based Sharable Active Guideline Environment) project is a collaboration of six research groups (IDX Systems, University of Nebraska Medical Center, Intermountain Health Care, Apelon Inc., Stanford Medical Informatics and the Mayo Clinic) [14]. SAGE includes a guideline depiction model and a guideline execution engine. Its objective is to establish an infrastructure to enable sharing guidelines in heterogeneous clinical information systems. SAGE is involved with organizations of healthcare standards (mainly HL7) to bridge the gap between guideline logic and real life implementations, and it is considered the evolutionary successor to EON and GLIF. The SAGE depiction model for Clinical Guidelines consists of Guideline Recommendation Sets, which are composed as a graph of Context Nodes. These Context Nodes can be Action Nodes, Decision Nodes and Routing Nodes. The patient state is retrieved directly from the electronic health record of the healthcare entity. SAGE makes use of terminologies and ontologies such as SNOMED-CT and LOINC. However SAGE is a relatively recent approach and shows some deficiencies concerning the integration of standards.

Recently, new approaches for guideline modeling, aimed to improve the aspects of the previously mentioned ones, are emerging, of which the Guideline Acquisition, Representation and Execution (GLARE) [15] is to be noticed.

3.2   Development Perspectives

From the study of the different CIG systems, it is possible to extract some common features that should be in mind in the development of a CIG system, namely:

- a guideline repository with different versions of guidelines;
- a guideline editor that enables the acquisition of new guidelines;
- a guideline representation language with a set of primitives of the tasks of the clinical process;
- access to the Electronic Medical Record (EMR) and to a Clinical Management System (CMS); and
- use of terminology and information standards.

It is common sense to state that a paper format cannot be compared to a computerized guideline, since the first cannot be processed electronically. But, our perspective is how an electronic format can be more advantageous and provide a new set of tools to facilitate the work of healthcare professionals. From this point of view, besides addressing the drawbacks of the paper format, CIGs may have a positive impact in the development process of guidelines. The computerized format and an underlying development framework enable the implementation of features for collecting evidence and to grade them, as well as for group decision making. The application of formal methods, based on Mathematical Logic [19,20] may be used to structure the development process and thus to prevent the elaboration of weak guidelines. Although these are not the main goals of the present work, they are interesting possibilities brought by electronic guidelines.

We have also identified two aspects where CIG systems are lacking. A good feature that could be implemented in these systems is a web-based version of the guideline editor, thus enabling healthcare professionals to freely build their guidelines online. This would be advantageous since it would allow the development of collaborative features of guideline development among different clinical experts scattered across a wide geographical area. It would solve one of the major problems of guideline distribution, which is the choice of the most suitable mean to deliver these recommendations to care workers.

Another aspect is the handling of incomplete information that occurs in the clinical process and the impact it has in decision making. Cases of uncertainty, inexactitude and incoherence in the clinical process may stop the flow of information from the observation phase to the decision phase, thus preventing a healthcare professional from devising a suitable treatment plan for a patient.

4   The CompGuide Project

The CompGuide Project is an initiative with the following objectives:

- the development of a web platform for acquisition and execution of CGs in a digital format;

- the development of a new guideline depiction model that captures all the information needs of the clinical process; and
- the development of a clinical decision model that combines guideline recommendations with incomplete information that may arise from the clinical process.

In the following sections we will address the different components of this project, like features, and discuss how they may be implemented , with some detail.

4.1 Architecture of the Web Application

Web applications are the ideal support for delivering and gathering information. They are platform independent and are available at any time at any local access. This ensures that our purpose covers a wide range of devices, enabling people who want to collaborate to share their work and vision. The application uses the JavaServer Pages (JSP) [16] technology to add dynamic content to html pages. The architecture of the application is displayed in Fig. 2. It follows the Model-View-Controller (MVC) design pattern, where there is a separation between request handling, business logic and interface. Under this model, a servlet handles all the requests, manages the logic and instantiates the Java beans. The Java beans contain the Guideline Constructor and the Guideline Inference Engine, which have access to the Data Sources, i.e., a database containing data about the patient, a repository of guidelines in a MySQL database, the Unified Medical Language System (UMLS) and Terminology Services (UTS) [17]. The UMLS integrates and distributes key terminology and has three knowledge sources: the Metathesaurus that maps medical terms synonymous of the same medical concept (e.g., SNOMED CT, LOINC), the Semantic Network that establishes associative connections between terms (e.g., cause and effect) and the SPECIALIST lexicon, for syntactical, morphological and orthographic analysis of the terms. The connection to the UTS is possible through a Java API provided by UMLS. Finally, the JSP obtains the response from the beans and formats the response accordingly.

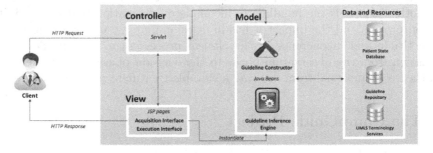

Fig. 2. The architecture of the CompGuide web application

4.2   Guideline Representation

Since there is not a standard model for guideline representation, we intend to develop a depiction model capable of performing that task,that is integrated with a standard terminology of clinical terms, the UMLS, and is in accordance with standard models of clinical information, such as the HL7. This will establish the necessary infrastructure to deal with interoperability issues with applications that are already used in clinical settings. The approach to guideline modeling of CompGuide presents an abstract view of decision making processes and task management during a clinical action or process [18]. The model is depicted in Fig. 3. A CG is viewed as a set of tasks, to which is given the designation of *plan*.

A *plan* contains instances of primitive classes that reflect the assignments of a CG. An *action* is an undertaking that represents a clinical procedure to be performed by the healthcare professional. To feed inputs to the system we use the *question task*. When a decision point is reached in the guideline workflow, it is used the decision task, which contains rules that associate conclusions to the parameters and values of the state of a patient. *Action, question* and *decision* are the atomic tasks of the model. It was considered that any type of atomic task gravitates around a clinical term, either it designates a parameter of the patient state, a clinical procedure or a clinical exam. The scheduling constraints are defined by attributes such as previous and next, that contain the id(entification) of the tasks that come before and after the present duty.

The other types of tasks defined in a plan are aimed at controlling special cases of the clinical workflow. The *aggregation module* groups tasks that are part of a cycle or iteration, creating the conditions for the user to define their periodicity, duration and objective. It is also used to represent tasks that belong to alternative pathways of the clinical workflow, like the ones that follow a decision task, in which the system chooses the next undertaking of the clinical process according to the conclusion reached at the decision step. The *aggregation module* can also group simultaneous tasks.

Another relevant aspect of the model is the *terminology* subclass of *plan*. *Terminology* comprehends the terms used in all the tasks of the *plan* along with their Concept Unique Identifier (CUI), which is a code used in the UMLS Metathesaurus to identity a concept and associate the different terms that may be used as a synonymous.

4.3   Clinical Decision Model

Before applying a clinical decision model that includes incomplete information, it is necessary to represent it in an appropriate way. Extensions to the Language of Logic Programming (ELP) [19,20] is one of the few techniques that enable this representation, using Mathematical Logic. ELP uses negation-by-failure and classic negation to represent explicit negative information. From this point of view the absence of information is also taken into account in the decision model. ELP enables the representation of cases of incomplete information

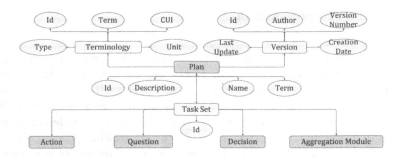

Fig. 3. The CompGuide guideline model and its different types of tasks

about the state of a patient. For instance, in cases of inexactitude where there are different possibilities for the value of a clinical parameter, these possibilities are represented as abducibles or exceptions. In cases of uncertainty, if the value of the clinical parameter is unknown, this is represented as a null value.

The example of Fig. 4 is a simplified fragment of the ATP III guideline for Detection and Treatment of High Blood Cholesterol in Adults (developed by the US National Heart, Lung and Blood Institute) that is responsible for the detection of metabolic syndrome (ms). The information about the patient John is a typical case of incomplete information.

Decision making in these situations requires the use of an information quantification method. The Quality-of-Information (QoI) [19,20] is a methodology associated with ELP. It is defined in terms of truth values taken in the interval [0,1] that are attributed to the clinical parameters of the patient according to their number of abducibles and null values. Given this, it is possible to calculate the QoI of each condition in a decision and calculate scores for each conclusion with the relative weights of its conditions. The decision model of CompGuide is based upon ELP and QoI [18]. The first stage of the decision model is the *Formulation of Hypotheses* where it is carried out a survey of the available options in terms of a decision. The following stage is *Voting*, which includes the *Evaluation of Conditions* and the *Evaluation of the QoI*. The scores of the options are

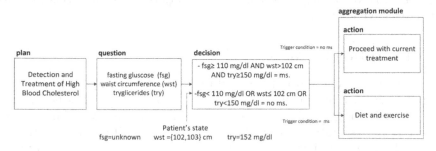

Fig. 4. Fragment of the guideline for Detection and Treatment of High Blood Cholesterol in Adults

calculated and, in the *Clinical Rule Selection*, the option with the best score is selected. In the *Clinical Task Selection*, the next task in the clinical workflow is selected through a matching with its trigger condition.

5   Conclusions and Future Work

The proposed system gathers the main features of the available CIG systems and goes beyond by trying to develop a web collaborative platform.The work requires further development of the web application and continuous improvement of the guideline model and the decision model. As observed the guideline model should be expressive enough to allow the definition of the different types of tasks that compose a guideline, as well as scheduling constraints on those tasks, without increasing the complexity of the model, in order to allow an intuitive acquisition of guidelines in the web application. The QoI approach really enhances our methodology for problem solving, since it offers a way to deal with incomplete information concerning the cases of missing information and conflicting/redundant and contradictory information. Indeed, uncertainty is dealt with in terms of the causality between symptoms and diseases, measured by the different scenarios that model and drive the universe of discourse.

What this approach offers is an intuitive environment for building and executing guidelines, through an expressive model, accessible to any healthcare professional and patient, as well as a decision model capable of processing incomplete and uncertain information. These features will offer a solution to current problems of guideline development, which none of the existent approaches managed to address.

Acknowledgements. This work is funded by National Funds through the FCT - Fundação para a Ciência e a Tecnologia (Portuguese Foundation for Science and Technology) within project PEst-OE/EEI/UI0752/2011". The work of Tiago Oliveira is also supported by a doctoral grant by CCTC - Computer Science and Technology Center (UMINHO/BI/004/2012).

References

1. Rosenbrand, K., Croonenborg, J., Wittenberg, J.: Guideline Development. In: Teije, A., Miksch, S., Lucas, P. (eds.) Computer-based Medical Guidelines and Protocols: A Primer and Current Trends, pp. 3–22 (2008)
2. Kavanagh, B.: The GRADE System for Rating Clinical Guidelines. PLoS Medicine 6, 5 (2009)
3. Ollenschläger, G.: International Guideline Network G-I-N (Guidelines International Network): Background and goals. In: Medizinische Klinik Munich Germany 1983, vol. 98(7), pp. 411–412 (2003)
4. Kalra, J.: Medical errors: an introduction to concepts. Clinical Biochemistry 37(12), 1043–1051 (2004)
5. Brennan, T.: The Institute of Medicine report on medical errors-could it do harm? The New England Journal of Medicine 96(6), 1123–1125 (2000)

6. Chawla, A., Gunderman, R.: Defensive medicine: prevalence, implications, and recommendations. Academic Radiology 15(7), 948–949 (2008)
7. Asher, E., Parag, Y., Zeller, L., Yerushalmi, R., Reuveni, H.: Unconscious defensive medicine: The case of erythrocyte sedimentation rate. European Journal of Internal Medicine 18(1), 35–38 (2007)
8. Woolf, S., Grol, R., Hutchinson, A., Eccles, M., Grimshaw, J.: Potential benefits, limitations, and harms of clinical guidelines. British Medical Journal 318(7182), 527–530 (1999)
9. Abraham, A.: Hybrid Soft Computing and Applications. International Journal of Computational Intelligence and Applications 8(1), 5–7 (2009)
10. Isern, D., Moreno, A.: Computer-based execution of clinical guidelines: a review. International Journal of Medical Informatics 77(12), 787–808 (2008)
11. Hripcsak, G., Ludemann, P., Pryor, T., Wigertz, O., Clayton, P.: Rationale for the Arden Syntax. Computers and Biomedical Research an International Journal 27(4), 291–324 (1994)
12. Ohno-Machado, L., et al.: The guideline interchange format. Journal of the American Medical Informatics Association 5(4), 357 (1998)
13. Vier, E., Fox, J., Johns, N., Lyons, C., Rahmanzadeh, A., Wilson, P.: PROforma: systems. Computer Methods and Programs in Biomedicine 2607(97) (1997)
14. Tu, S., et al.: The SAGE Guideline Model: achievements and overview. Journal of the American Medical Informatics Association 14(5), 589–598 (2007)
15. Bottrighi, A., Terenziani, P., Montani, S., Torchio, M., Molino, G.: Clinical guidelines contextualization in GLARE. In: AMIA Annual Symposium Proceedings AMIA Symposium AMIA Symposium, vol. 2006, p. 860 (2006)
16. Geary, D., Horstmann, C.: Java Server Pages documentation. Prentice-Hall (2007)
17. US National Library of Medicine, Unified Medical anguage System (UMLS) (2011)
18. Oliveira, T., Neves, J., Costa, A., Novais, P., Neves, J.: An Interpretable Guideline Model to Handle Incomplete Information. In: Omatu, S., Santana, J., González, S., Molina, J., Bernardos, A., Corchado, E. (eds.) Distributed Computing and Artificial Intelligence - 9th International Conference (DCAI 2012), vol. 151, pp. 437–444 (2012)
19. Novais, P., Salazar, M., Ribeiro, J., Analide, C., Neves, J.: Decision Making and Quality-of-Information. In: Corchado, E., Novais, P., Analide, C., Sedano, J. (eds.) SOCO 2010. AISC, vol. 73, pp. 187–195. Springer, Heidelberg (2010)
20. Neves, J., Ribeiro, J., Pereira, P., Alves, V., Machado, J., Abelha, A., Novais, P., Analide, C., Santos, M., Fernández-Delgado, M.: Evolutionary intelligence in asphalt pavement modeling and quality-of-information. Progress in Artificial Intelligence 1(1), 119–135 (2012)

Smart Time Series Prediction

Eva Volna, Michal Janosek, Vaclav Kocian, and Martin Kotyrba

University of Ostrava, 30. dubna 22, 70103 Ostrava, Czech Republic

Abstract. This article deals with a smart time series prediction based on characteristic patterns recognition. Our goal is to find and recognize important patterns which repeatedly appear in the market history for the purpose of prediction of subsequent trader's action. The pattern recognition approach is based on neural networks. We focus on reliability of recognition made by developed algorithms with optimized patterns which also causes the reduction of the calculation costs.

1 Pattern Recognition in Time Series

Market systems are chaotic systems from their nature. One of the essential characteristics of a chaotic system is its extreme sensitivity to initial conditions. A tiny change in values at the beginning of the time series produces drastic changes in behavior later on [9]. For example [10], stocks are being sold or bought based on their prices. The price depends on how much has been bought or sold. The feedback loop has both positive and negative effects. The law of supply and demand implies a negative feedback loop, because the higher the price, the lower the demand, which, in fact, causes a lower price in the future. However, a parallel speculation mechanism implies a positive feedback loop, because an increasing price makes an assumption that the price will increase in the future and thus motivate the traders to buy more stocks. As we do not know delay the between these two effects we are not able to predict anything well. These nonlinear effects are common in the markets [10]. Nevertheless, it is fair to say that the markets are not purely chaotic. Although a chaotic system is a collection of orderly, simple behaviors, modelling the market has turned out to be more difficult than anticipated. The problem is that chaotic systems can be unusually flexible and rapidly switch between their many different behaviors. One way to isolate these individual behaviors of the market might lie in a presumption that when the market is perturbed in just the right way (e.g. a large drop in price), it would exhibit one of its many regular behaviors for a short time. In the meantime, constructing leading indicators is difficult, that is why a forecast horizon for any market time series is limited.

Recent studies show that market patterns might implicate useful information for stock price forecasting. Currently, there are mainly two kinds of market pattern recognition algorithms: an algorithm based on rule-matching [1] and an algorithm based

V. Snasel et al. (Eds.): SOCO Models in Industrial & Environmental Appl., AISC 188, pp. 211–220.
springerlink.com   © Springer-Verlag Berlin Heidelberg 2013

on template-matching [6]. Nonetheless, both of these two categories have to design a specific rule or template for each pattern. However, both types of algorithms require participation of domain experts, and lack the ability to learn. For the last few decades, neural networks have shown to be a good candidate for solving problems with the market analysis. A typical illustration is the study conducted in [4], where a recognition algorithm for triangle patterns based upon a recurrent neural network was introduced. Market patterns can be classified into two categories: continuation patterns and reversal patterns. Continuation patterns indicate that the market price is going to keep its current movement trend; while reversal patterns indicate that the market price will move to the opposite trend. More than sixty important technical patterns are detailed in [2]. Patterns can be seen as some sort of maps which helps us to orientate in certain situations and navigate us to profitable trades.

We focus on a smart time series prediction based on characteristic patterns recognition. The article proposes the market pattern recognition approach based on neural networks. All developed algorithms are implemented in the Java language and experiments' data is placed on web (http://www1.osu.cz/~r09728).

2   Time Series Prediction Based on Pattern Recognition Algorithm

The whole course of our experiment can be divided into several tasks:

- Determination of training patterns - using knowledge of finance
- Patterns' binarization to obtain training sets
- Adaptation of neural networks
- Data analysis - pattern recognition
- Results evaluation of data analysis - time series prediction

2.1   Training Sets Preparation

It is necessary to remark that determination of training patterns is one of the key tasks that needed our attention. Improperly chosen patterns can lead to confusion of neural networks. A neural network "adapted" on incorrect patterns can give meaningless responses.

The search for the patterns is a complicated process which is usually performed manually by the user. In order to test the efficiency of the pattern recognition system we applied data from X-Trade Brokers [8] that is a set of data that reflect the situation on the market. We used time series which shows the development of market values of EURUSD, which reflect the exchange rate between EUR and USD. The testing time scale was four months from April to June 2011 on a 5-minute chart. That means that every 5 minutes a new record is created in the table. We used for experiments 17697 records totally that are indicated in table 1.

Table 1. The experiment's data sample

Date	Time	Open	High	Low	Close	V		C1	C2	C3
2011.04.01	00:00	1.4172	1.4172	1.4166	1.4166	10	I			
2011.04.01	00:10	1.4167	1.4169	1.4167	1.4168	12	I			
-	-	-	-	-	-	-	-	-	-	-
2011.04.28	18:30	1.4808	1.4812	1.4802	1.4805	28	I		27	28
2011.04.28	18:35	1.4803	1.4804	1.4799	1.4801	24	I		27	28
2011.04.28	18:40	1.4802	1.4804	1.4798	1.4800	32	I			28
2011.04.28	18:45	1.4801	1.4802	1.4791	1.4792	28	I			28
2011.04.28	18:50	1.4791	1.4794	1.4789	1.4792	30	I			28
-	-	-	-	-	-	-	-	-	-	-
2011.04.30	23:50	1.4500	1.4500	1.4493	1.4495	13	I			
2011.04.30	23:55	1.4494	1.4495	1.4491	1.4491	18	I			

Columns description:

Date - corresponding date
Time - start time of the 5 minutes interval
Open - open value of the time interval
High - maximum value of the time interval
Low - minimum value of the time interval
Close - close value of the time interval
V - trade volume in the time interval

These attributes are provided by the X-Trade Brokers Company. Other columns starting with the vertical line and labeled C1, C2, C3 are user defined columns and serve to mark the known patterns by the expert. These known patterns are designated as our training set for neural networks. Numbers in these columns represent an order how patterns were marked by the expert.

We choose the flip pattern for our experiments, see Figure 1. This pattern occurs in markets with both decreasing and increasing tendency. We can split this pattern to two subpatterns; a pattern anticipating price decrease (column C1) and a pattern anticipating price increase (column C2). Column C3 determines the pattern which doesn't belong either to C1 or C2.

In Figure 1 is shown Flip pattern: a sequence of three or more constrain actions that alters the subset of information being visually analyzed. This pattern is based on the support and the resistance pattern again. Sometimes the pattern is called as Role Reversal pattern or simply RR. Now, we are going to speculate on role reversal. Support will change to resistance and vice versa. First a market will create a new support (1). Then it will breach that support (2) and in the end bounce from the S/R level (3). In fact the support (1) has been changed to resistance (3). In Figure 1 we speculate on the price decrease. The speculation on the price increase looks very similar, it is mirrored only.

Fig. 1. Flip pattern

During experiments, we used 47 patterns found by expert, representing classes with denotations "sell", "buy", and "wait". We have marked classes of patterns with corresponding symbols "sell" - C1, "buy" - C2, and "wait" - C3. The training set contained 19 patterns "sell", 20 patterns "buy", and 8 patterns "wait". The third "wait" class was established on the basis of initial experiments, which showed that a neural network that is adapted only for patterns „buy" and "sell" tends to include any input to the first or the second class. It was undesirable, therefore we created patterns, which our network should interpreted as "wait".

Pattern Binarization

Pattern binarization lies in a conversion of training patterns (vector values of quantities) into one-dimensional array of binary or bipolar values. Then such a field is an input into a neural network. Binarization is directly related to neural network adaptation and follow-up analysis of results. Successful data analysis requires to be training and test patterns binarized in a same way. At first we drew all vector components into bitmap of height h and width w as a black and white graph in our experiments. Then the bitmap is spread over rows into one-dimensional array of size $w \times h$. In the course of binarization, it is necessary to successively solve the following partial tasks:

- Choice of data components (columns in table) that will be used in the experiment.
- Template binarization schemes (neural network works with inputs of constant length).
- Normalization range of values (height normalization).
- Normalization number of values (width normalization).
- Determination of binarization method.
- Conversion of n-dimensional data into one-dimensional array.

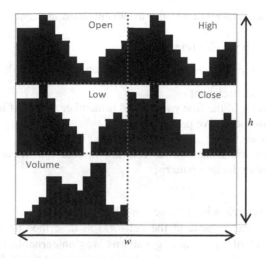

Fig. 2. Bitmaps input quantities: "Open", "High", "Low", "Close, and "Volume"

Each value in columns "Open", "High", "Low", "Close", and "Volume" in table 1 was binarized separately and created bitmaps were finally integrated into only one bitmap. It is possible to notice that the first four quantities "Open", "High", "Low", and "Close" from table 1 contain almost identical information - in contrast to "Volume", the graph of which is different. All 5 quantities are inserted in the 2x3 grid, where (one) sixth of bitmap remains unused. (One) more of any variable can be placed in the free position in order to strengthen its influence on adaptation results, see Fig. 2.

2.2 Modification of Hebb Rule

Hebb network belongs to the very simple artificial neural networks by reason that it contains minimum parameters and is adapted in one cycle. The Hebb rule [7] determines the change in the weight connection from u_i to u_j by $\Delta w_{ij} = r * y_i * y_j$, where r is the learning rate and y_i, y_j represent the activations of u_i and u_j respectively. Thus, if both u_i and u_j are activated the weight of the connection from u_l to u_j should be increased. Examples can be given of input/output associations which can be learned by a two-layer Hebb rule pattern associator. In fact, it can be proved that if the set of input patterns used in training are mutually orthogonal, the association can be learned by a two-layer pattern associator using Hebbian learning. However, if the set of input patterns are not mutually orthogonal, interference may occur and the network may not be able to learn associations. This limitation of Hebbian learning can be overcome by using the delta rule.

Based on the experimental study we suggested new principles of Hebb adaptation that are the following [5]:

1. Before adaptation, algorithm walks through the training set and identify as irrelevant all the items, which value in all patterns is the same.

2. Weights of connections related to the irrelevant items are ignored during the adaptation.
3. Thanks to that, such weights remain '0'.

Algorithm that marks irrelevant items can be written as follows [5]:

1. Mark all items as irrelevant.
2. Load input vector of the first pattern and remember values of its items.
3. Repeat with all successive patterns:
   a) Load input vector.
   b) Mark every irrelevant item as relevant in case that its actual value differs from that in the first pattern.
4. End.

Hebb network configuration is presented in table 2. We had to create a reduced set of training patterns for the purpose of the network, as described above. In our experimental study, the set of used training patterns was unlearnable for Hebb network. Therefore it was necessary to create a reduced training set for it, because 9 patterns from used 47 patterns have been identified as problematic (blocking adaptation). One pattern "buy" (C2) and all (!) patterns "wait", in which too many pixels were overlap from different classes. While backpropagation rule should be also able to express relations between inputs, Hebb rule not. As a matter of fact, this did not work and different approach had to be developed. Therefore, we designed *Hebb algorithm - active mode modification*, which runs as follows:

Hebb algorithm tends to find almost all introduced patterns familiar. A new modification has been designed, tested and used for the active mode of Hebb rule. In a common operation, the output value of each output neuron is derived from its net value in a very simple way (eq. 1):

$$y_i = -1, \text{ if } net_i < 0, \text{ and } y_i = 1, \text{ if } net_i \geq 0$$
$$net_i = \sum_j w_{ij} y_j \tag{1}$$

The problem there is that y_i takes the same output no matter how big its net_i value is (apart from sign rules). In other words, we do not have any information about the certainty of the network result. Our network output function was replaced with simple equivalence: $y_i = net_i$.

Then we can use parameter which means minimum $|y_i|$ value required to accept the neural network result. In case, that $|y_i| < confidence$ for some output, the neural network result is ignored. Using the *confidence* parameter, we can easily and accurately regulate count (and a minimum of quality) of found patterns. As a matter of course it is not easy to find an optimal value for the *confidence* parameter. One of the reasons is the fact that the optimal value vary with patterns' size, patterns amount and analysis results requirements.

Table 2. Hebb network configuration

Network topology	*Input layer:* x neurons, where x=number of bits in the input bitmap
	Output layer: 2 neurons
	Interconnection: fully connected
	confidence parameter: 3174
Type of I/O values:	bipolar

3 Results Evaluation

3.1 Data Analysis - Pattern Recognition

The data analysis is the main task of the experiment. All previous jobs can be summarized as "preparation". Data files were presented to the learned networks during the analysis. The aim of the analysis was to obtain a list of occurrences of learned patterns in data. Columns containing occurrences of the patterns were added to original data during the analysis. The # character serves as a separator of those columns (table 3).

Data analysis was conducted as a simulation of real operation. Entries were submitted to the networks successively, so as they accumulate in time. Every time, familiar patterns were found, their class number and length (number of records) were recorded behind the # character in the current last record.

Searching always proceeded within the designated permissible minimum and maximum length (number of records) of pattern. Minimal length *pMin*=20rec and maximal length *pMax*=70rec were used during our experiment. These values were received from minimal and maximal lengths of the training patterns. Searching began at the length of *pMin*, which gradually grew larger until it reached the length of *pMax*. If more than one pattern was found, then only one with the greatest degree of learned pattern fitness was chosen, see Figure 3.

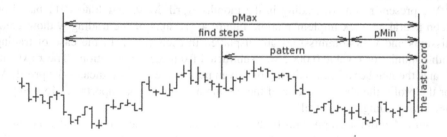

Fig. 3. The-searching strategy of pattern finding

3.2 Time Series Prediction

Results obtained using Hebb algorithm were used for evaluation of the analysis. Table 3 shows small fragment of the analysis table results. Confidence parameter affects the number of found patterns. We experimentally set up its value 3174.

- the first 7 columns hold original OHLC data
- I column serves as a separator for training patterns definition
- C1 and C2 columns hold definitions of the learning patterns
- # column serves as a separator for found patterns declaration
- P column, where found patterns are written

As we can see, neural network tends to find clusters of patterns. The size of these clusters (the cluster from table 3 has size equal to 6) depends on the parameter *confidence*. It would be useful to filter out one (the best) pattern from such a cluster, but the filtering method is not developed yet.

Table 3. The analysis output sample. Value of *confidence* was 3174

Date	Time	Open	High	Low	Close	V		C1	C2	-	P
2011.04.01	17:35	1.4183	1.4188	1.4179	1.4183	66	I			#	
2011.04.01	17:40	1.4184	1.4203	1.4183	1.4199	60	I			#	
2011.04.01	17:45	1.4198	1.4210	1.4196	1.4204	91	I		2	#	
2011.04.01	17:50	1.4203	1.4214	1.4199	1.4212	104	I		2	#	
-	-	-	-	-	-	-	-	-	-	-	-
2011.04.01	19:45	1.4216	1.4217	1.4214	1.4217	8	I		2	#	2-22
2011.04.01	19:50	1.4218	1.4218	1.4216	1.4216	3	I		2	#	2-21
2011.04.01	19:55	1.4215	1.4219	1.4215	1.4218	6	I		2	#	2-26
2011.04.01	20:00	1.4219	1.4220	1.4217	1.4218	8	I			#	2-27
2011.04.01	20:05	1.4217	1.4220	1.4217	1.4220	8	I			#	2-28
2011.04.01	20:10	1.4221	1.4224	1.4220	1.4223	17	I			#	2-29

To interpret the data from the analysis it is necessary to create a basic trading system and trade all the recommendations provided by the neural network. Particular examples are demonstrated in the table 3.

We present results of trading in the months April, May, and June 2011 based on Hebb neural network implementation, Figure 4. Results of the trading without commissions and with commissions are displayed in each graph. In the case of trading with commissions, value 0.0002 was subtracted from each transaction. The x axis indicates the number of trades during the given period, the y axis indicates the profit. As for the profit, the absolute value of the trading account is not important. More important is the trading equity itself.

Figure 4shows a graph with trading results based on the recommendations using outputs from our Hebb network. As we can see in the graph without commission, the result of trading settled just below 0.05. In a contrast to it, the result with commission dropped to below -0.15 for the given period. This is in particular due to the large number of trades.

Hebb network has surprised us because it was able to stay in positive account numbers in business without commissions. The disadvantage of this approach was stagnation after reaching the value 0.05 and also excessive overtrading. Due to high number of transactions it is not quite good from real trading because for each transaction we have to pay a commission fee. We would like to propose some method of control of the number of trades for adaptation algorithms in our future work.

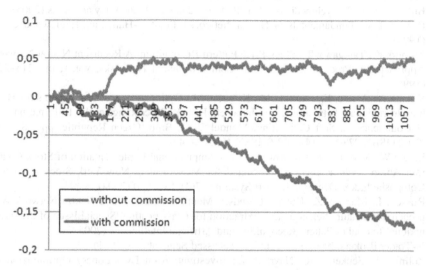

Fig. 4. Hebb algorithm - managed trading

4   Conclusion and Future Work

In this article, a short introduction into the field of pattern recognition in time series with prediction of subsequent trader's action has been given. We focus on reliability of recognition made by the described algorithms with optimized patterns based on artificial neural networks. Our experimental study confirmed that for the given class of tasks can be acceptable simple classifiers (we tested the simplest type of Hebb learning). The advantage of simple neural networks is its very easy implementation and quick adaptation. Easy implementation allows implementing them at low-performance computers (PLC) and their fast adaptation facilitates the process of testing and finding the appropriate type of network for the given application. According to our results of experimental studies, it can be stated that pattern recognition in time series using chosen Hebb network was successful. Nowadays, we would like to develop mechanism for filtering the best items of clusters of patterns, which emerge during the data analysis.

Acknowledgments. The research described here has been financially supported by University of Ostrava grant SGS2/PRF/2012. Any opinions, findings and conclusions or recommendations expressed in this material are those of the authors and do not necessarily reflect the views of the sponsors.

References

[1] Anand, S., Chin, W.N., Khoo, S.C.: Chart Patterns on Price History. In: Proc. of ACM SIGPLAN Int. Conf. on Functional Programming, Florence, Italy, pp. 134–145 (2001)

[2] Bulkowski, N.: Encyclopedia of Chart Patterns, 2nd edn. John Wiley and Sons (2005)

[3] Fausett, L.V.: Fundamentals of Neural Networks. Prentice-Hall, Inc., Englewood Cliffs (1994)

[4] Kamijo, K., Tanigawa, T.: Stock Price Pattern Recognition: A Recurrent Neural Network Approach. In: Proc. of the Int. Joint Conf. on Neural Networks, vol. 1, pp. 215–221 (1990)

[5] Kocian, V., Volna, E., Janosek, M., Kotyrba, M.: Optimizatinon of training sets for Hebbian-learningbased classifiers. In: Matoušek, R. (ed.) Proceedings of the 17th International Conference on Soft Computing, Mendel 2011, Brno, Czech Republic, pp. 185–190 (2011) ISBN 978-80-214-4302-0, ISSN 1803-3814

[6] Leigh, W., Modani, N., Hightower, R.: A Computational Implementation of Stock Charting: Abrupt Volume Increase As Signal for Movement in New York Stock Exchange Composite Index. Decision Support Systems 37(4), 515–530 (2004)

[7] Russell, I., Markov, Z., Holder, L. (eds.): Machine Learning and Neural Network Approaches to Feature Selection and Extraction for Classification, Special Issue of the International Journal of Pattern Recognition and Artificial Intelligence (2005)

[8] X-Trader Brokers, http://xtb.cz (accessed September 3, 2011)

[9] Zelinka, I., Senkerik, R., Navratil, E.: Investigation on Evolutionary Optimitazion of Chaos Control. Chaos, Solitons & Fractals (2007), doi:10.1016/j.chaos.2007.07.045

[10] Gershenson, C.: Design and Control of Self-organizing Systems. Dissertation, Vrije Universiteit Brussel (2007)

Soft Computing Testing in Real Industrial Platforms for Process Intelligent Control

Larzabal E., Cubillos J.A., Larrea M., Irigoyen E., and Valera J.J.

University of the Basque Country (UPV/EHU), Intelligent Control Research Group
(GICI - gici.drupalgardens.com)
{ekaitz.larzabal,eloy.irigoyen}@ehu.es

Abstract. By testing advanced control techniques based on Soft Computing into industrial platforms is possible to analyse the feasibility and reliability of these implementations for being subsequently used in real industrial processes. In many cases, this fact is not taken into account for several reasons concerning with the complexity of performing hardware implementations. Hence, simulation testing becomes the last step before showing an implemented solution. The main objective of this work is to give a step beyond for achieving a more realistic test of the Intelligent Control techniques. For this reason, a first approximation of a Genetic Algorithm controller (NSGA-II) is implemented, tested, studied and compared in the stages of the controller design, and simultaneously in different industrial platforms. Most relevant results obtained in software simulation and in Hardware In the Loop (HIL) implementation are finally shown and analysed.

1 Introduction

Since many years, Intelligent Systems area has been presented as a new chance to solve tricky problems, being particularly relevant in coping with the intricacy and the complexity of the real world industrial process control [1]. As Rudas and Fodor present in their work Intelligent Systems [1], new fields emerged in this area developing computational solutions for the new approaches based on intelligence, such as Computational Intelligence [2], Soft Computing [3], and combining techniques from both fields, Hybrid Systems [4].

In both fields, Computational Intelligence (CI) and Soft Computing (SC), the GAs have recently appeared in the developed solutions in industrial control applications [5]. These solutions might be classified into two categories: one group for analysing and off-line design; other group for adapting and on-line controller tuning.There exist a short number of proposals where GAs directly calculate an on-line and real time control action, due to: (1) non-reliable computation, (2) high computational cost, and (3) problems in obtaining convergence and stability, such as discussed works in Fleming et al. [6] and Valera et al [7].

Currently, there exist lots of studies for solving industrial control problems based on ANN, as the work presented by Bose in Motor Drives [8], and based on FL, as the one presented by Precup in a survey on industrial applications of fuzzy control [9]. The principal gap of these works is to develop a real implementation in real industrial processes. Usually, all the performed studies are tested in an experimental stage with industrial

V. Snasel et al. (Eds.): SOCO Models in Industrial & Environmental Appl., AISC 188, pp. 221–230.
springerlink.com   © Springer-Verlag Berlin Heidelberg 2013

process models which can include complex dynamics, but they are carried out in a simulation framework. Although there are multiple proposals based on SC techniques in the literature, few real developments can be found [8]. All of them are developed with laboratory equipment, without using real industrial platforms, such as industrial computers, embedded computers or programmable logic controllers. Some works are laboratory implementations with FPGA/DSP, or with a Host-PC configuration, testing communication and control issues, but not proving the robustness of a laboratory equipment in real industrial processes [8].

This work takes a step forward using GAs for intelligent control, providing a framework for the rapid prototyping and testing using industrial and usually used HW platforms in the Industry. In subsequent sections, promising results with different real platforms are presented. These results were obtained in several tests for controlling processes with complex dynamics and in solving optimization problems based on GAs with high computational cost requirements. Real tests have been carried out with two different industrial platforms; a robust industrial PC controller with Peripheral Component Interconnect (PCI) bus, and a PAC (Programmable Automation Controller). Beyond using GAs for solving optimization problems focused on tasks as planning, scheduling, tracking and calibration, this work introduces first results in intelligent controlling of processes with complex dynamics. Specifically, this development computes future values for control actions by an optimization process using GAs. This future values are computed for a predictive control scheme where, depending on the sample time, the computational cost could be very demanding.

There exist two fundamental problems on applying Model Predictive Control (MPC) strategies: (1) the accuracy of the model to approach the process or plant to be controlled, and (2) the control optimization problem to solve (specially for non-linear model predictive control, NMPC) during each controller sample time. Therefore, GAs as a part of Evolutionary Algorithms can offer a relevant possibility in high complexity optimization problem solving when process models with high non linearity are chosen and the selected cost functions are non-quadratic and non-convex [7].

Moreover, if the used prediction horizon is long, new difficulties appear in NMPC problem solving which have to be considered in real applications. On the one hand, the computational cost in solving the control problem with such horizons is high. On the other hand, this control strategy produces convergence problems in the optimization process, so in a general control formulation for working out a solution the computational cost will grow up as well. Furthermore, if it is necessary to control process systems with faster dynamics, a new problem appears when trying to perform an implementation in a Real Time (RT) control scheme, because of time requirements to be reached in each short sample time. This work tries to show how a control algorithm based on GAs can be executed in RT, on several industrial platforms and in shorter sample times. To this end, a Hardware In the Loop (HIL) testing configuration was prepared in the laboratory where controllers were implemented into two different industrial platforms.

In this work we present in section 2 the multiobjective GA approach used in our NMPC strategy. Subsequently, in order to implement the GA in real time with Matlab/Simulink and xPC-Target tools, several modifications and adjustments have been carried out and presented in section 3. Section 4 shows in a simulation context, two different controlled systems which will be later used in real tests. The two real platforms tested in this work are presented in section 5. Results with both real platforms in RT tests are showed and explained in section 6. Finally, section 7 contains the last conclusions and future works.

2   The Genetic Algorithm NSGA-II

The optimization of multiple objectives problems, where any improvement in one of the objectives makes other objectives worse, has been a extensively explored research area. The optimal solutions obtained in such problems are denominated non-inferior solutions and all of them belong to the set of Pareto [10]. There exist several methods to search for the non-inferior (set of Pareto) solutions in the multi-objective optimization context. Among them the ones based on evolutionary algorithms stand out. Some of these contributions can be found in [11][6] and in [7]. Examples of efficient evolutionary algorithms such as Nondominated Sorting Genetic Algorithm (NSGA) and Micro-Genetic Algorithm (1-GA) are presented in [12] and in [13] respectively. The main drawbacks of these techniques are their high computational cost and the need of a decision maker to select one solution among the Pareto set. Other drawbacks related to the control context are the difficulty of demonstrating the stability, the convergence to a near global optimal and the robustness of the final solution

The NSGA-II used in this work is the one proposed by Deb et al. in [14]. The NSGA-II is the evolution of the NSGA originally proposed by Srinivas and Deb in [12]. This second version of the algorithm arose to answer the main criticisms (high computational complexity, lack of elitism and the need for specifying the sharing parameter) the NSGA received [14].

The possibility to tackle the multi-objective problems in the context of NMPC makes very interesting the NSGA-II algorithm. The introduction of the elitist mechanisms in the NSGA-II algorithm that improves the convergence time of the Pareto solutions is especially useful for control, where there exist specific problems as the time requirements (directly linked to the computational complexity of the controller algorithm). In [15] and [7], the authors takes the multi-objective NMPC scheme approach by using NSGA-II [12] to search the set of Pareto non-inferior solutions at each sampling time of the controller. For these applications, the NSGA-II works as the optimization solver of the NMPC problem at each sampling time.

The NSGA-II flowchart explained briefly starts with the initialization of the population (size N), evaluates the objective functions and ranks the population. The next step consists of a loop that ends when the stopping criteria is met. Inside the loop the following steps are taken in the following order; selection, crossover, mutation, evaluation of the objective functions, combination of population, ranking and finishes with the selection of N individuals. If the stopping criteria is met the final population is presented. See [14] for a detailed description of the NSGA-II algorithm.

NSGA-II was proposed as a Multiobjective GA although for this article it has been used with SISO plants with only one objective. It should be addressed that the scope of this work has been the preparation of a framework for the rapid prototyping and testing of intelligent control techniques in industrial control platforms. Future work will lead to the implementation of multiobjective control strategies using GAs in RT as other use cases.

3   RT NSGA-II Programming for Predictive Control

The NSGA-II original code provided by Deb et al. in [14] was written in C. This code has been rewritten in a new s-function code for implementing in the Matlab/Simulink development environment. Some code reduction was made in order to minimize the coding size: e. g. lines related to the binary coding was removed because only real number coding was required. Also the controller time performance has been improved by making some modifications to adequate the NSGA-II in order to enhance its execution in short control sampling times. Finally, it is important to note that a new stop criteria has been added to include in the control strategy not only the GA iterations, but also the execution time.

In order to perform the predictive control with the NSGA-II algorithm, the objective function should be evaluated for the entire prediction horizon. The function evaluation is related to the error produced in the controlled variable of the control loop. To calculate this error the algorithm needs a model of the system to be controlled that can be obtained by mathematical approximation, neural network identification, etc. In this work, the first RT implementations have been performed by using mathematical models presented below. Furthermore, some tests have been made with neural network models performing a satisfactory control as well. Once the GA ends the searching (with a time based or a limited generation number stopping criteria) the last solution is used as predictive control action at each controller sampling time.

4   Simulation Results

In this section some simulation results are presented. For these experiments the two following nonlinear systems have been chosen from different benchmark systems:

System 1 (This is a modification of the system presented in [16]):

$$y_{k+1} = 0.8 \left[u_k^3 + \frac{y_k y_{k-1}}{1 + y_k^2} \right] \tag{1}$$

This system has been created for increasing the non-linear dynamics near the origin of coordinates of the original system.

System 2 (This system is presented in [17]):

$$y_{k+1} = \frac{1.5 y_k y_{y-1}}{1 + y_k^2 + y_{k-1}^2} + 0.7 \sin [0.5 (y_k + y_{k-1})] \cos [0.5 (y_k + y_{k-1})] + 1.2 u_k \tag{2}$$

For the simulation stage, the controller has been implemented in Matlab/Simulink. A batch of 50 simulations have been performed for each system. Figure 2 and Figure 3 show the evolution of the mean value in each batch of simulations. The system 1 performs with a 25% overshoot (M_p) while the system 2 performs with a 45% M_p. It is obvious this is not the best possible performance, but the implemented controller is quite simple and does not include any kind of constraint.

The system 1 output signal shows the following statistical values; $M_{p\_Max} = 0.3674$; $M_{p\_Min} = 0.3648$; $M_{p\_Mean} = 0.3661$; $M_{p\_Variance} = 2.537e - 7$. It should be mentioned that the control action max. value never exceeds 0.7223.

The system 2 output signal shows the following statistical values; $M_{p\_Max} = 1.4984$; $M_{p\_Min} = 1.4941$; $M_{p\_Mean} = 1.4960$; $M_{p\_Variance} = 6.8509e - 7$. It should be mentioned that the control action max. value never exceeds 0.8344.

The simulations show that both systems can be controlled with the proposed NSGA-II predictive controller. The simulation results are very satisfactory despite the implemented controller is quite simple and does not include any kind of restriction.

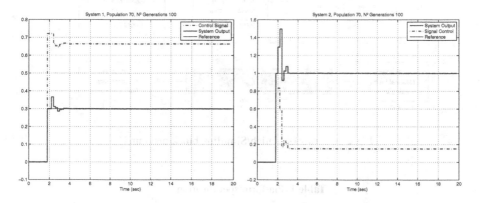

Fig. 1. Sample time 0.2 seg, Initial population 70, n generations 100. a) System 1 and b) System 2

Although the results may not be as satisfactory as they should, the scope of this work should not be lost. The HIL implementation of this first controller is challenging enough to continue with this first controller before trying to develop a very accurate one. All in all, more research has to be done in order to improve the controller and the results that it provides.

5   Rapid Control Prototyping and HIL Testing

Nowadays, the software development platforms provide accurate simulation results, but these simulations are not always enough to understand the real behaviour of the system. The next step to take should be the testing of the development in a real system. But the risk of a malfunction during the real testing is still there, and sometimes that risk is simply unacceptable. Hardware In the Loop arises to fill the gap between the software simulations and the real system implementation. The simulation gives a step towards the real implementation using the HIL test with the externalization of the signals. The HIL can be used to simulate a single component or even the whole system replacing some of the parts by mathematical models [18]. The benefits of the HIL have been tested in many different industrial fields as automotive [19], electronics [20], wind energy systems [21], etc. Therefore, the Intelligent Control Research Group (GICI) of the UPV/EHU has adopted HIL methodology to test the different controllers under more real situations.

In the case of study of this article, the HIL test implies the externalization of the control signal u(k) and the system output signal y(k), as shown in Figure 2. The controller is going to be hosted in one platform and the system to be controlled will be placed in another platform. The configuration of the platforms and the data acquisition systems are listed in the table 1.

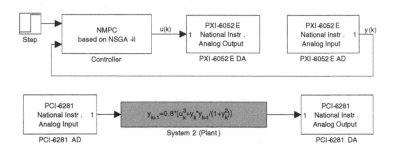

Fig. 2. Mathworks Matlab/Simulink block diagram

Table 1. Platform specification

Industrial PC / Kontron			
Host	Kontron	xPC-Target 4.0	NI pxi-6052 (16bits)
Target	PC	xPC-Target 4.0	NI pci-6281(18bits)
PAC / Beckhoff			
Host	Beckhoff	Xubuntu 8.04 / RTAI 3.8	K-Bus / KL3404 (AI) / KL4034 (AO) (12bits).
Target	PC	xPC-Target 4.0	NI pci-6221(16bits)

Both platforms have in common the target configuration for the system to be controlled. This system is composed by a PC with a National Instruments PCI Data Acquisition card described above. All the configurations and the SW/HW installation steps are explained in the *howtos* hosted in the GICI webpage.

6   Results (HIL Real Time Simulations)

In this section some results are shown. The HIL experiments have also been performed with the two systems (System 1 and System 2) in 50 simulations, both running in the two platforms presented in the previous section: Industrial PC and PAC.

It should be remarked the difference in the bit resolution of the two platforms. The industrial PC analog I/O card has a 16bit resolution, whereas the PAC K-Bus analog I/O cards have a 12bit resolution. Also the NI PCI-6281 DAQ card has a 18bit resolution, while the NI PCI-6221 DAQ card has 16bit resolution. This difference is one of the reasons, but not the unique, of the noisier signal of the PAC platform.

Note that the system to be controlled is already running before the controller starts in all the results shown above.

6.1   Industrial PC (PCI Bus Based On)

The Industrial PC is a very powerfull and robust platform, with a similar core as the PCs but with an industrial bus (PCI). The results of this platform with the two systems under testing (1) and (2) can be seen in Figure 3(a) and Figure 3(b).

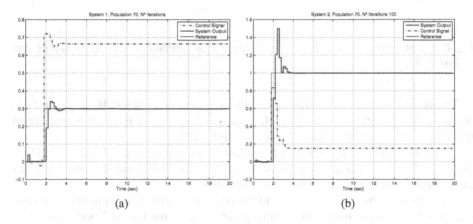

(a)   (b)

Fig. 3. Sample time 0.2 seg, Initial population 70, n generations 100. a) System 1 and b) System 2

The system 1 output signal shows the following statistical values; $M_{p\_Max} = 0.3769$; $M_{p\_Min} = 0.3622$; $M_{p\_Mean} = 0.3649$; $M_{p\_Variance} = 1.3517e - 5$. It should be mentioned that the control action max. value never exceeds 0.7229.

The system 2 output signal shows the following statistical values; $M_{p\_Max} = 1.7932$; $M_{p\_Min} = 1.4679$; $M_{p\_Mean} = 1.5673$; $M_{p\_Variance} = 0.0163$. It should be mentioned that the control action max. value never exceeds 0.8478.

The results of the Industrial PC are very similar to the ones presented in the simulation section.

6.2   PAC / Beckhoff: K−Bus

The PAC can be described as a less powerfull platform in comparison with the industrial PC, but the differences between both platforms are becoming less and less significant.

The PAC has the capability of being programmed with PLCs programming language, and it can be seen as a bridge between the PLCs and the industrial PCs. The results of this platform with the two systems can be seen in Figure 4(a) and Figure 4(b).

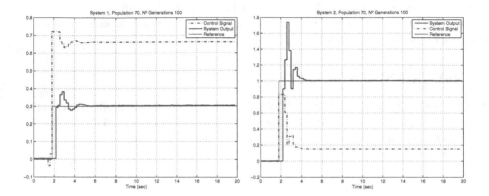

Fig. 4. Sample time 0.2 seg, Initial population 70, n generations 100. a) System 1 and b) System 2

The system 1 output signal shows the following statistical values; $M_{p\_Max} = 0.4006$; $M_{p\_Min} = 0.3713$; $M_{p\_Mean} = 0.3829$; $M_{p\_Variance} = 2.7136e - 5$. It should be mentioned that the control action max. value never exceeds 0.7223.

The system 2 output signal shows the following statistical values; $M_{p\_Max} = 1.8117$; $M_{p\_Min} = 1.5871$; $M_{p\_Mean} = 1.7777$; $M_{p\_Variance} = 0.0020$. It should be mentioned that the control action max. value never exceeds 0.8404.

The results of the PAC are similar to the ones presented in the simulation section. Compared with the Industrial PC results, it can be seen that the PAC performs with a noisier signal. All in all, the results of the two platforms and the simulations are very similar.

The two platforms are capable of performing the 100 generations of the GA in the required time each sample time. In case of not performing the 100 generations a stop criteria has been implemented based in the required time to guarantee the RT and deterministic performance of the controller.

7   Conclusions

In this work the first step to implement Soft Computing techniques in industrial platforms have been taken in a Hardware In the Loop structure. A basic NMPC controller has been prototiped and tested in real industrial hardware platforms with promising results. These results demonstrate that present industrial platforms have enough computational capability to run advanced control strategies using intelligent and Soft Computing computation techniques. The rapid prototyping and testing framework presented in this article is very useful to make fast improvements in the algorithms in order to satisfy the hard real time and computational cost requirements.

The application of the NSGA-II algorithm in a NMPC strategy has been also presented as an example of prototyping and testing. Different steps for developing the control algorithm, from off-line simulations to rapid control prototyping and HiL testing, have been done, laying the basis for the implementation of future intelligent control strategies over real industrial controllers. All these steps have shown satisfactory results as the ones presented in this paper.

Future work will lead us to the implementation of multivariable and multiobjective intelligent-expert controls for highly non-linear and complex controlled plants. The combination of Neural Networks, Genetic Algorithms and Fuzzy Logic in advanced strategies is being a promise solution to optimize highly complex control problems. With the framework presented in this article the hybridization of those techniques is being investigated, implemented and tested easier.

Another research line will guide us to the implementation of a neural network identification system. Firstly, the offline identification should be implemented in RT (it has already been implemented in simulations) and secondly the identification should be done online. This improvement will help the controller to be less dependant of a mathematical model of the system.

Acknowledgement. This work comes under the framework of the project titled Implementation / Integration of Intelligent Control Techniques on Real Time Control Systems ATICTA-2 with reference S-PE10UN70 granted by the Basque Regional Government (GV / EJ). The work has also been funded by the Education, University and Research Department of the Basque Government (GV/EJ) BFI-2010-118 research fellowship.

References

1. Rudas, I.J., Fodor, J.: Intelligent systems. Proceedings of Int. J. of Computers, Communications & Control 3, 132–138 (2008)
2. Duch, W.: What is computational intelligence and where is it going. Challenges for Computational Intelligence 63, 1–13 (2007)
3. Bonissone, P.P.: Soft computing: the convergence of emerging reasoning technologies. Soft Computing 1, 6–18 (1997)
4. Sahin, S., Tolun, M., Hassanpour, R.: Hybrid expert systems: A survey of current approaches and applications. Expert Systems with Applications 39(4), 4609–4617 (2012)
5. Saridakis, K., Dentsoras, A.: Soft computing in engineering design – a review. Advanced Engineering Informatics 22(2), 202–221 (2008)
6. Fleming, P., Purshouse, R.: Evolutionary algorithms in control systems engineering: A survey. Control Engineering Practice 10(11), 1223–1241 (2002)
7. Valera, J., Irigoyen, E., Gómez, V., Artaza, F., Larrea, M.: Intelligent multi-objective nonlinear model predictive control (imo-nmpc): Towards the 'on-line' optimization of highly complex control problems. Expert Systems with Applications 39(7), 6527–6540 (2012)
8. Bose, B.K.: Neural network applications in power electronics and motor drives- an introduction and perspective. IEEE Trans. on Industrial Electronics 54(1), 14–33 (2007)
9. Precup, R.E., Hellendoorn, H.: A survey on industrial applications of fuzzy control. Computers in Industry 62(3), 213–226 (2011)
10. Pareto, V.: Cours d'Économie Politique, vol. I & II. Université de Lausanne (1897)

11. Deb, K.: Multi-objective optimization using evolutionary algorithms. John Wiley & Sons, Ltd. (2001)
12. Srinivas, N., Deb, K.: Multiobjective optimization using nondominated sorting in genetic algorithms. Evolutionary Computation 2, 221–248 (1994)
13. Toscano Pulido, G., Coello Coello, C.A.: The Micro Genetic Algorithm 2: Towards Online Adaptation in Evolutionary Multiobjective Optimization. In: Fonseca, C.M., Fleming, P.J., Zitzler, E., Deb, K., Thiele, L. (eds.) EMO 2003. LNCS, vol. 2632, pp. 252–266. Springer, Heidelberg (2003)
14. Deb, K., Pratap, A., Agarwal, S., Meyarivan, T.: A fast and elitist multiobjective genetic algorithm: Nsga-ii. IEEE Trans. on Evolutionary Computation 6(2), 182–197 (2002)
15. Laabidi, K., Bouani, F., Ksouri, M.: Multi-criteria optimization in nonlinear predictive control. Mathematics and Computers in Simulation 76(5-6), 363–374 (2008)
16. Narendra, K.S., Parthasarathy, K.: Identification and control of dynamical systems using neural networks. IEEE Trans Neural Networks 1(1), 4–27 (1990)
17. Harris, C. (ed.): Advances in Intelligent Control. CRC Press (1994)
18. Hanselmann, H.: Hardware-in-the-loop simulation as a standard approach for development, customization, and production test of ecu's. Technical Report 931953, SAE Int. (1993)
19. Kendall, I., Jones, R.: An investigation into the use of hardware-in-the-loop simulation testing for automotive electronic control systems. Control Engineering Practice 7(11) (1999)
20. Lu, B., Wu, X., Figueroa, H., Monti, A.: A low-cost real-time hardware-in-the-loop testing approach of power electronics controls. IEEE Trans. on Industrial Electronics 54(2) (2007)
21. Li, H., Steurer, M., Shi, K., Woodruff, S., Zhang, D.: Development of a unified design, test, and research platform for wind energy systems based on hardware-in-the-loop real-time simulation. IEEE Trans. on Industrial Electronics 53(4), 1144–1151 (2006)

Human Resource Allocation in Process Simulations Based on Competency Vectors

Štěpán Kuchař and Jan Martinovič

VŠB Technical University in Ostrava,
IT4Innovations,
17. listopadu 15/2172, 708 33 Ostrava - Poruba,
Czech Republic
{stepan.kuchar,jan.martinovic}@vsb.cz

Abstract. This paper specifies a method to describe human resources' skills and competencies for the use in automatic simulations of operative business processes to enhance the precision of such simulations. A way to convert this description into the vector space model is also provided and used to find and evaluate adequate resources for performing process activities based on their competency requirements. Two different competency vector representations are introduced and experimentally compared on the software process of a local software development company using the F-measure.

Keywords: Human Resource Allocation, Business Process Simulation, Competency Model, Vector Space Model.

1 Introduction

A large number of business processes are based on activities performed by human resources (HR) and skilled and properly trained employees are one of the main sources of the company's competitive advantage [6,12]. But they have to be managed well so that their potential can be fully utilized. This is normally done by managers and team leaders but in automatic simulations, managers are not available to plan these allocations because these simulations are fully performed by the machine. This article introduces a solution to the problem of evaluating available employees to the current activity in the ongoing process based on the employee's competencies. This evaluation will enhance the simulation by the means of finding the appropriately skilled worker that can be allocated to the activity based on the actual availability of evaluated workers. It can be also used to support the decisions for training or hiring new employees based on the requirements of the process activities and predicted workload of the process.

Existing business process simulation models are not very concerned with accurate human resources modeling and description [2,20]. Many simulation models suppose that all employees in one role have the same skills and that it does not matter which one of these workers will be allocated to the activity [17,21]. But clearly each worker in the process is unique with its own set of skills and experiences; each one has specific working habits and performance [4]. In our paper we try to use the competency-based

V. Snasel et al. (Eds.): SOCO Models in Industrial & Environmental Appl., AISC 188, pp. 231–240.
springerlink.com   © Springer-Verlag Berlin Heidelberg 2013

approach to differentiate individual human resources in the process and how to use it to correctly allocate resources to the process activities.

Our evaluation approach is inspired by the recommender systems used for human resources selection in e-recruitment. The goal of these systems is to find the right applicant for given job offer and are used in personal agencies and HR departments to support the recruitment decisions of HR consultants. Research concerning these systems is however focused mainly on mining required data about applicants and job offers from their textual representation using ontologies [7,10,18], learning models [3] or n-grams [14] and only a few are concerned with the search and evaluation methods [15,16]. We localize this problem to the simulation of internal processes so we can rely on structured information from internal company's HR management systems and focus on the evaluation method. In addition our approach is concerned with matching resources to atomic process activities, not the whole job offer which consists of many activities and so it can be used for more precise capability and training assumptions and for estimating performance indicators.

2   Competency-Based Description of Human Resources

The first problem that has to be solved is the description of the employees' skills in the process. This is the human resources management area of expertise where the competency models [8,9,23] and skills frameworks (e.g. Skills Framework for the Information Age [22]) are used. Competency models define various competencies which are important for the company and its processes. Competencies are defined as sets of knowledge, abilities, skills and behavior that contribute to successful job performance and the achievement of organizational results [23]. Skills frameworks have the same purpose, but they describe skills particular for one domain rather than general competencies. But in fact skills are just a special type of competencies.

Competency models and skills frameworks also describe how to measure and evaluate individual competencies. In most cases competencies are measured by a number of advancing stages where higher levels of competency include everything from their lower levels. The first competency model had five stages [8] and later models used the same system, but they did not keep the number of stages. There is no standard for how many stages should a competency model have and every model defines its own set of stages.

Competencies of a specific human resource can therefore be described by the competency level acquired by the resource. This also means that this resource has mastered this given level and all lower levels of the competency. This way it is not important how many levels does the competency model have because the computing model can assume, that the highest acquired level of the best resource is also the highest level of the competency model.

Let's have a small example of one HR consultant working in a personal agency. His competencies in a 10-level model could look like this:

- Personnel knowledge in IT - 5. level,
- personnel knowledge in Management - 3. level,
- internal IS user skill - 6. level,

- psychology - 4. level,
- customer knowledge of VSB-TUO - 4. level,
- customer knowledge of MyCompany - 0. level.

Domain specific skills (personnel knowledge, internal IS user skill), general competencies (communication, psychology) and knowledge of the environment (customer knowledge) are contained in this example. It is clear that competencies in the model have to be based on the company's requirements and professional domain.

3   Competency-Based Description of Process Activity Requirements

All activities in the process also have competency-based requirements that describe what competencies should the worker performing the activity know. Each activity will therefore be defined by the set of competency levels for each required resource type entering the activity specifying that only workers with given or higher level will do the activity as planned. Resources with lower competencies are able to finish the activity, but it will take additional time to learn how to perform the activity and their work is prone to contain more errors. A simple example of requirements for the activity of leading an interview with an applicant in the personal agency follows:

- Personnel knowledge - 5. level,
- internal IS user skill - 0. level,
- psychology - 5. level,
- customer knowledge - 4. level.

If we compare this example with the worker example from previous chapter, one can notice the generalization of some requirements (personnel knowledge and customer knowledge). When assessing the employee's competencies, it is better to define the competency levels in specific parts of the domain so that the resources are assessed as precisely as possible. On the other hand the activity requirements should only define a level for the whole competency category and relevant part of the domain will be specified by actual process case. In other words, if the personal agency tackles with a case where they have to find a programmer for the company VSB-TUO, then the requirements in this case will be refined as personnel knowledge in IT and customer knowledge of VSB-TUO. This specification has to be supported by a parameterization of the process case in the process simulation tool. This can be easily done with discrete event simulations by employing the Colored Petri Nets approach [1,13]. This approach allows assigning various parameters to every token in the Petri net and these parameters could be seeded with case specific information either in the initial marking or even during the process run.

4   Direct Vector Representation

To evaluate human resources based on the process activity requirements we utilize the vector space model that is very often used in document searches [5]. One of this models advantages is the option of ranking vectors according to their decreasing similarities to the query vector.

To use this model for the competencies a way to describe the resource competencies as vectors has to be found. This can be solved directly by creating a vector of the competency levels for given resource. If the set of competencies determined for all resources is $c_1, c_2, ..., c_m$ and resource r_i has mastered these competencies at levels $l_{i,1}, l_{i,2}, ..., l_{i,m}$, then the vector for this resource will be:

$$r_i = (l_{i,1}, l_{i,2}, \ldots, l_{i,m}) \tag{1}$$

Competency vectors in this format can be directly used as the document vectors in the vector space model [5]. To be able to create such vectors it is crucial that all resources have levels assigned to all competencies $c_1, ..., c_m$. If this is not the case, then the competencies have to be unified, missing competencies of the resources can then be filled with level 0 or some competencies can be merged into one with their level defined by their weighted average.

Activity requirements for every resource type entering the activity will serve as a query for the model and can be represented in the same manner. A vector representation of the activity a with required competencies $rc_{a,1}, rc_{a,2}, ..., rc_{a,p}$ and required levels $rl_{a,1}, rl_{a,2}, ..., rl_{a,p}$ is:

$$a = (rl_{a,1}, rl_{a,2}, \ldots, rl_{a,p}) \tag{2}$$

Activity requirements vector in this format can be directly used as the query vector in the vector space model [5]. Following conditions concerning the comparability of resource and activity competencies have to be met in order to use this vector representation:

1. Resource and activity competencies are the same and are in the same order:

$$(\forall a \in A) \, [m = p \wedge (\forall j \in \{1, 2, \ldots, m\}) \, [c_j = rc_{a,j}]], \tag{3}$$

   where A is a set of all observed activities in the process, m is the number of resource competencies and p is the number of activity competencies.

2. Resource and activity competencies share the same scale for their levels:

$$(\forall a \in A) \, (\forall i \in \{1, 2, \ldots, n\}) \, (\forall j \in \{1, 2, \ldots, m\})$$
$$[l_{i,j} \in \langle minlevel(j), maxlevel(j) \rangle \Rightarrow rl_{a,j} \in \langle minlevel(j), maxlevel(j) \rangle], \tag{4}$$

   where A is a set of all observed activities in the process, m is the number of resource competencies, $minlevel(j)$ is the lowest possible level of competency c_j and $maxlevel(j)$ is the highest possible level of c_j.

3. All levels of corresponding competencies c_j and $rc_{a,j}$ have to share the same meaning for all activities in the process and all $j = 1, ..., m$.

5   Fragmented Vector Representation

The biggest problem of the direct representation is that it does not take the specifics of competency and requirements description into account. Resource competencies are specific by the fact, that the resource mastered all lower levels of the competency, but requirements define that all higher levels are required. This difference can be encoded into the representation by fragmenting the competencies to a number of elements equal to the highest possible level of the competency. The level of one competency can therefore be divided into more vector elements containing 1 if the resource knows this level or 0 if he does not. The bordering element will be able to hold a rational number to enable more accurate evaluation.

Every competency $c_1, ..., c_m$ will be split into fragmented competencies:

$$fc_{j,1}, fc_{j,2}, ..., fc_{j,maxlevel(j)},$$ (5)

where $maxlevel(j)$ is the highest possible level of c_j. Vector representation for the fragmented competencies $fc_{j,k}$ and resource r_i is:

$$fl_{i,j} = \left(fl_{i,j,1}, fl_{i,j,2}, ..., fl_{i,j,maxlevel(j)} \right),$$ (6)

where the values $fl_{i,j,k}$ will be determined from level $l_{i,j}$ as:

$$fl_{i,j,k} = \begin{cases} 1 & ,k \leq l_{i,j} \\ l_{i,j} + 1 - k & ,l_{i,j} < k < l_{i,j} + 1 \\ 0 & ,k \geq l_{i,j} + 1 \end{cases}$$ (7)

The resource r_i will be described by the vector of vectors:

$$r_i = (fl_{i,1}, fl_{i,2}, ..., fl_{i,m})$$ (8)

This representation needs to be adapted to the vector space model that works with vectors containing atomic elements [5]. This can be easily done by replacing the $fl_{i,j}$ vectors directly by their elements $fl_{i,j,k}$ creating the vector:

$$r_i = \left(fl_{i,1,1}, ..., fl_{i,1,maxlevel(1)}, ..., fl_{i,m,1}, ..., fl_{i,m,maxlevel(m)} \right)$$ (9)

The same conditions for all resource competencies from the direct representation have to be met.

Vector representation of the activities will follow the same pattern but with its own specifics. In contrary to the resource competencies the activity requirements describe the lowest required level of the competency for given activity, but it could be useful to limit the highest required level. This will reduce the actual value of over skilled resources that can be left to do more difficult tasks. The fragmented competency levels will then be set to 0 below the lower limit, to a non-positive number above the upper limit and to a number from the interval $\langle 0, 1 \rangle$ in between the lower and upper limits. This value describes the importance of actual competency for performing actual activity, the most important competencies having this value set to 1.

If all the conditions for comparability defined in the direct representation are met, the vector representing the activity a is:

$$a = \left(frl_{a,1,1}, ..., frl_{a,1,maxlevel(1)}, ..., frl_{a,m,1}, ..., frl_{a,m,maxlevel(m)} \right),$$ (10)

where $frl_{a,j,k}$ is defined as:

$$frl_{a,j,k} = \begin{cases} 0 & ,k < lrl_{a,j} \\ rcw_{a,j} & ,lrl_{a,j} \leq k \leq hrl_{a,j} \\ hlp_{a,j} & ,k > hrl_{a,j} \end{cases} \tag{11}$$

where $lrl_{a,j}$ is the lower limit of the required competency rc_j for activity a, $hrl_{a,j}$ is the upper limit, $rcw_{a,j}$ is the importance of required competency rc_j for activity a and $hlp_{a,j} \leq 0$ is a penalty for exceeding the upper limit for required competency rc_j and activity a where $hlp_{a,j} = 0$ corresponds to no penalty and higher negative numbers mean greater the penalty. All competencies rc_j that are not required for activity a have $rcw_{a,j}$ set to 0 and $hlp_{a,j}$ also set to 0.

A simple example with only two competencies and their maximal level set to 5 can help to better understand this representation. Resource r_1 with mastered competency levels at $(1,2)$ is represented as $r_1 = (1,0,0,0,0,1,1,0,0,0)$ and resource r_2 with competencies at $(5,4)$ is represented as $r_2 = (1,1,1,1,1,1,1,1,1,0)$. If an activity a has competency requirements set to $lrl_{a,1} = 2$, $hrl_{a,1} = 5$, $lrl_{a,2} = 4$, $hrl_{a,2} = 5$ then its representation is $a = (0,1,1,1,1,0,0,0,1,1)$ (both competencies are very important for this activity, hence $rcw_{a,j}$ is set to 1 for both).

To evaluate resources against the activity requirements a referential resource has to be added to the model. This referential resource r_r will have exactly the same competencies as the activity requirements (i.e. $l_{r,j} = lrl_{a,j}$). Vector representation of the referential resource for this example will therefore be $r_r = (1,1,0,0,0,1,1,1,1,0)$. By comparing resources with the referential resource we can see how good the resource is at performing the activity and how much can it influence the activity's performance. Resources with higher similarities than the referential resources similarity are suitable to perform the activity; resources with lower similarities will have troubles with the activity.

6 Experiment

To compare the two representations introduced in this paper we performed an experiment on a software process of a local middle-sized software development company. Their 153 employees were evaluated in 22 competencies important in the software process on a 10-level scale. 52 basic activities were identified in the process and their requirements were specified for the same 22 competencies to ensure their compatibility. All competencies were set to be equally significant ($rcw_{a,j} = 1$) to ensure comparability with the direct vector representation that has not been able to model varying competency importance. The values for $hlp_{a,j}$ were set to 0 for all activities in one configuration and to -30 in another configuration to penalize the resources with very high competencies.

The comparison in this experiment is based on evaluating the precision and recall [19] and combining them with the widely used F-measure [19] for all activities in the process using both vector representations introduced in this paper. To use the precision-recall method, relevant resources for each activity had to be identified and this was done by expert estimation by the process manager. Next task was to find a suitable measure to evaluate the similarity in the model and the cosine and Euclid measures [5] were employed for this comparison.

The final task was to decide how many of the top evaluated resources for each activity should be considered in the computation of precision and recall. It was obvious from the set of relevant workers for each activity that this cannot be a single number, because the number of relevant workers varied significantly. It also was not feasible to derive this number directly from the number of relevant workers to avoid unnecessary relation with the expert estimation that is not normally present during the application of the model. Referential resource can be used to solve this problem, because it represents the lowest competency levels that are needed to perform the activity (even slightly worse resources can perform the activity, but they were omitted from the results in this experiment). Thus only resources with higher similarity level than the referential resource for each activity were declared as found. This procedure could not be used for the direct vector representation, because the referential resource, having the exactly same vector as the activity vector, was always the resource with the highest similarity. Therefore the number of top evaluated resources for the fragmented representation was also used for the direct representation.

Table 1 shows the results of the evaluation based on the role specific model. Each activity is linked to one role in the process and only workers from this role are considered in the search. Precision and recall averages over all activities are stated along with two F-measure averages. The first one has β set to 1 which means that both precision and recall are equally significant. But precision is more important in simulation allocations, because it is usually more important to allocate the correctly capable worker than finding all capable workers, and so the second F-measure average is for $\beta = 0.5$ which determines that precision is 2 times more important than recall.

Table 1. Role Specific Results

Representation	Precision Avrg.	Recall Avrg.	F-measure Avrg. $\beta = 1$	F-measure Avrg. $\beta = 0.5$
Cosine measure				
Direct	75.28	84.05	74.42	73.33
Fragmented $hl p_{a,j} = 0$	72.94	79.29	71.97	71.68
Fragmented $hl p_{a,j} = -30$	84.54	79.29	78.68	81.12
Euclid measure				
Direct	74.32	83.09	73.53	72.44
Fragmented $hl p_{a,j} = 0$	76.53	64.86	64.36	68.67
Fragmented $hl p_{a,j} = -30$	86.58	64.86	69.52	76.07

These results show that direct representation is slightly better than the basic fragmented one but the latter is more flexible in its use and extension. Only by increasing the penalty for high competency levels the results significantly improved and further upgrades to the representations are possible (weighing individual competencies, weighing individual competency levels etc.). Cosine measure is also more useful than the Euclid measure even though the latter has slightly better precision but at the expense of significantly lower recall. The proposed model can also be used for all resources regardless of their roles. This can be very useful to identify resources that are linked to wrong roles

Table 2. Role Independent Results

Representation	Precision Avrg.	Recall Avrg.	F-measure Avrg. $\beta = 1$	F-measure Avrg. $\beta = 0.5$
Cosine measure				
Direct	31.04	83.88	41.12	33.95
Fragmented $hlp_{a,j} = 0$	31.66	70.68	40.95	34.58
Fragmented $hlp_{a,j} = -30$	40.43	70.68	48.25	42.93
Euclid measure				
Direct	24.20	68.19	32.49	26.59
Fragmented $hlp_{a,j} = 0$	26.56	59.31	30.91	27.45
Fragmented $hlp_{a,j} = -30$	30.45	59.31	34.01	30.98

and can lead to better restructuring of the resources. Results for the role independent configurations are in Table 2.

These results have similar properties as the previous ones but the final F-measures are significantly lower. Some further enhancements could improve the results but the results show that this model in its current form is not applicable for precise allocations of resources with widely different competencies.

7 Conclusion and Future Work

This paper proposed a method for evaluating the suitability of employees to perform various activities in business processes based on their competencies. This evaluation can be used in automatic process simulations to allocate the best fitting worker to current activity in the simulation so that the simulation can take into account all specifics of this worker. Such simulations can also be used to identify possible areas for training or hiring new employees with specific competency sets and how will these personnel enhancements influence the process indicators.

One of the useful specifics of the fragmented representation is weighing of individual activity requirement levels. Introducing lower or even negative weights to high activity requirement levels could solve a problem of allocating over skilled employees and also enhance the precision of the results. This weighing could also be used for estimating the performance indicators because the growth of resources capabilities is slower at higher levels [11].

Availability of workers in the process should also be taken into consideration and unavailable resources should be omitted from actual allocations. On the other hand there is a possibility that a suitable unavailable resource will be available soon and the method should be able to consider if it is better to wait for this suitable resource or to allocate one that is less skilled but immediately available. This problem can be solved with the help of the previous improvement by estimating the performance indicators of the employee and comparing the waiting and processing time of currently unavailable worker with the processing time of the available worker.

Acknowledgement. This work was supported by the Bio-Inspired Methods: research, development and knowledge transfer project, reg. no. CZ.1.07/2.3.00/20.0073 funded by Operational Programme Education for Competitiveness, co-financed by ESF and state budget of the Czech Republic and by the European Regional Development Fund in the IT4Innovations Centre of Excellence project (CZ.1.05/1.1.00/02.0070).

References

1. van der Aalst, W.M.P.: The application of Petri nets to workflow management. The Journal of Circuits, Systems and Computers 8(1), 21–66 (1998)
2. van der Aalst, W.M.P., Nakatumba, J., Rozinat, A., Russell, N.: Business process simulation: How to get it right. BPM Center Report BPM-08-07, BPMcenter.org (2008)
3. Aiolli, F., De Filippo, M., Sperduti, A.: Application of the preference learning model to a human resources selection task. In: IEEE Symposium on Computational Intelligence and Data Mining, CIDM 2009, pp. 203–210 (April 2009)
4. Andre, M., Baldoquin, M.G., Acuna, S.T.: Formal model for assigning human resources to teams in software projects. Inf. Softw. Technol. 53(3), 259–275 (2011)
5. Berry, M.W.: Survey of text mining: clustering, classification, and retrieval, vol. 1. Springer-Verlag New York Inc. (2004)
6. Chang, W.A., Huang, T.C.: Relationship between strategic human resource management and firm performance: A contingency perspective. International Journal of Manpower 26(5), 434–449 (2005)
7. Crow, D., DeSanto, J.: A hybrid approach to concept extraction and recognition-based matching in the domain of human resources. In: 16th IEEE International Conference on Tools with Artificial Intelligence, ICTAI 2004, pp. 535–541 (November 2004)
8. Dreyfus, S.E., Dreyfus, H.L.: A five-stage model of the mental activities involved in directed skill acquisition. Technical report, DTIC Document (1980)
9. Ennis, M.R.: Competency models: a review of the literature and the role of the employment and training administration (ETA). US Department of Labor (2008)
10. Gómez-Pérez, A., Ramírez, J., Villazón-Terrazas, B.: An Ontology for Modelling Human Resources Management Based on Standards. In: Apolloni, B., Howlett, R.J., Jain, L. (eds.) KES 2007, Part I. LNCS (LNAI), vol. 4692, pp. 534–541. Springer, Heidelberg (2007)
11. Hanne, T., Neu, H.: Simulating human resources in software development processes. Berichte des Fraunhofer ITWM 64(64), 83–87 (2003)
12. Hatch, N.W., Dyer, J.H.: Human capital and learning as a source of sustainable competitive advantage. Strategic Management Journal 25(12), 1155–1178 (2004)
13. Jensen, K.: Coloured Petri Nets: Basic Concepts, Analysis Methods, and Practical Use. Springer (1996)
14. Kessler, R., Béchet, N., Roche, M., El-Bèze, M., Torres-Moreno, J.-M.: Automatic Profiling System for Ranking Candidates Answers in Human Resources. In: Meersman, R., Tari, Z., Herrero, P. (eds.) OTM-WS 2008. LNCS, vol. 5333, pp. 625–634. Springer, Heidelberg (2008)
15. Kessler, R., Béchet, N., Torres-Moreno, J.-M., Roche, M., El-Bèze, M.: Job Offer Management: How Improve the Ranking of Candidates. In: Rauch, J., Raś, Z.W., Berka, P., Elomaa, T. (eds.) ISMIS 2009. LNCS, vol. 5722, pp. 431–441. Springer, Heidelberg (2009)
16. Malinowski, J., Keim, T., Wendt, O., Weitzel, T.: Matching people and jobs: A bilateral recommendation approach. In: Hawaii International Conference on System Sciences, vol. 6, p. 137c. IEEE Computer Society, Los Alamitos (2006)
17. Pesic, M., van der Aalst, W.M.P.: Modelling work distribution mechanisms using colored petri nets. Int. J. Softw. Tools Technol. Transf. 9(3), 327–352 (2007)

18. Radevski, V., Trichet, F.: Ontology-Based Systems Dedicated to Human Resources Management: An Application in e-Recruitment. In: Meersman, R., Tari, Z., Herrero, P. (eds.) OTM 2006 Workshops. LNCS, vol. 4278, pp. 1068–1077. Springer, Heidelberg (2006)
19. van Rijsbergen, C.J.: Information Retrieval. Butterworth (1979)
20. Rozinat, A., Wynn, M.T., van der Aalst, W.M.P., ter Hofstede, A.H.M., Fidge, C.J.: Workflow simulation for operational decision support. Data Knowl. Eng. 68(9), 834–850 (2009)
21. Russell, N., van der Aalst, W.M.P., ter Hofstede, A.H.M., Edmond, D.: Workflow Resource Patterns: Identification, Representation and Tool Support. In: Pastor, Ó., Falcão e Cunha, J. (eds.) CAiSE 2005. LNCS, vol. 3520, pp. 216–232. Springer, Heidelberg (2005)
22. SFIA Foundation. Framework reference SFIA version 4G (2010)
23. Sinnott, G., Madison, G., Pataki, G.: Competencies: Report of the competencies workgroup, workforce and succession planning work groups (September 2002)

Soft Computing for the Analysis
of People Movement Classification

Javier Sedano[1], Silvia González[1], Bruno Baruque[2],
Álvaro Herrero[2], and Emilio Corchado[3]

[1] Instituto Tecnológico de Castilla y León. C/ López Bravo 70, Pol. Ind. Villalonquejar,
09001 Burgos, Spain
javier.sedano@itcl.es

[2] Civil Engineering Department, University of Burgos. C/ Francisco de Vitoria s/n,
09006 Burgos, Spain
{bbaruque,ahcosio}@ubu.es

[3] Departamento de Informática y Automática, University of Salamanca, Plaza de la Merced
s/n, 37008 Salamanca, Spain
escorchado@usal.es

Abstract. This article presents a study of the best data acquisition conditions
regarding movements of extremities in people. By using an accelerometer, there
exist different ways of collecting and storing the data captured while people
moving. To know which one of these options is the best one, in terms of classification, an empirical study is presented in this paper. As a soft computing
technique for validation, Self-Organizing maps have been chosen due to their
visualization capability. Empirical verification and comparison of the proposed
classification methods are performed in a real domain, where three similar
movements in the real-life are analyzed.

1 Introduction

Over recent years, the use of Artificial Intelligence (AI) and miniaturized sensors for
real world problem solutions has undergone a significant growth.

Some devices integrate sensors to measure movement in the extremities. These are
able to record acceleration patterns by means of wearable accelerometer devices, and
enable daily physical activity measurement and analysis by applying AI techniques.

Up to now, some solutions have been developed using sensors to measure variables
and movement behaviour of the human being, such as the analysis of a cow daily activity data [10], the delivery of patients' information to health care personnel using
mobile phones [7], stress detection system [13], etc...

The use of accelerometers in order to objectively measure body movement is displayed in several studies together with some AI techniques, such as Support Vector
Machine (SVM) [3]. Other AI techniques applied for human activity recognition are
Hierarchical Hidden Markov Models (HHMM) [16], Fuzzy Basic Functions (FBF)
[5], Decision Tree (DT) [8, 3], Neural Network Classifiers such as Resilient Backpropagation (RPROP) [12, 18], K-Nearest Neighbour (K-NN) [4], Dynamic Time Warping (DTW) [11], Kohonen self-organizing maps (SOM) [14], or Naive Bayes (NB)

V. Snasel et al. (Eds.): SOCO Models in Industrial & Environmental Appl., AISC 188, pp. 241–248.
springerlink.com   © Springer-Verlag Berlin Heidelberg 2013

[3]. Most of these techniques use statistical measures like mean, standard deviation, variance, interquartile range, energy, correlation between axes and entropy, for the different features of the data.

This study focuses on determining the way of collecting accelerometer data to improve movement analysis performance. The main point in present study is to find the way of collecting accelerometer data to know the best sampled interval in order to obtain the best divided classes of movements. Thereby, a system could register the data set with the best possible period, optimizing its consumption while improving the performance. On the other hand, this on-going research aims at knowing the number of variables that the model should incorporate to get as low as possible classification error. The final objective is to obtain the furthest classes from the movement data. In order to do so, data are gathered with different time periods and number of variables. Then, in order to obtain some guidelines for the data gathering design, the most relevant group in the data are identified with a SOM.

The remaining sections of this paper are structured as follows: section 2 presents the proposed soft computing approach and the neural projection techniques applied in this work. Some experimental results are presented and described in section 3; the conclusions of this study are discussed in section 4, as well as future work.

2   Soft-Computing Techniques

This study proposes to apply soft computing models to people's movement detection. As a first stage, the study aims to investigate which measurement conditions are the best to distinguish correctly the activity in which the user is engaged. To perform this identification automatically, the Self-Organizing Map, has been used. This model was selected because, although: it is originally designed to cluster data samples; but it can also serve to obtain a proper data classification as well. Differentiating from some other models, it can handle unknown or not clearly classifiable data samples and it can also give a visual (and quantifiable) hint of where the new data can be represented with respect of previously presented samples.

2.1   Self-Organizing Maps

Topology preserving mapping comprises a family of techniques with a common target: to produce a low-dimensional representation of the training samples that preserves the topological properties of the input space. From among the various techniques, the best known is the Self-Organizing Map (SOM) algorithm [9].

SOM aims to provide a low-dimensional representation of multi-dimensional data sets while preserving the topological properties of the input space. The SOM algorithm is based on competitive unsupervised learning; an adaptive process in which the neurons in a neural network gradually become sensitive to different input categories, which are sets of samples in a specific domain of the input space. The update of neighbourhood neurons in SOM is expressed as:

$$w_k(t+1) = w_k(t) + \alpha(t) \cdot \eta(v, k, t) \cdot (x(t) - w_k(t)) \tag{1}$$

Where, x denotes the network input, w_k the characteristics vector of each neuron; α, is the learning rate of the algorithm; and $\eta(v, k, t)$ is the neighbourhood function, in which v represents the position of the winning neuron (Best Matching Unit or BMU) in the lattice, and k the positions of the neurons in its neighbour-hood.

2.2 Quality Measures

In order to compare the results obtained by the different maps, three of the most widespread measures are used in this study:

- **Classification Error** [15]. Topology preserving models can be easily adapted for classification of new samples using a semi-supervised procedure. A high value in the classification accuracy rate implies that the units of the map are reacting in a more consistent way to the classes of the samples that are presented. As a consequence, the map should represent the data distribution more precisely.
- **Topographic Error** [2]. It consists on finding the first two best matching units (BMU) for each entry of the dataset and testing whether the second is in the direct neighbourhood of the first or not.
- **Goodness of Map** [1]. This measure combines two different error measures: the square quantization error and the grid distortion. It takes account of both the distance between the input and the BMU and the distance between the first BMU and the second BMU in the shortest path between both along the grid map units, calculated solely with units that are direct neighbours in the map.

3   Data Gathering and Analysis

The data collection device consists of a Microchip PIC18F2580 microcontroller and an Analog Device ADXL335 [6] 3D accelerometer [17] that integrates MEMS and SMT technologies. The physical magnitude sampling is performed by the accelerometer and the movement module (M) is calculated on the three axes: X, Y, Z.

Fig. 1. DXL335 Accelerometer

One female volunteer, aged 25, was enrolled in the study. The accelerometer is placed on the volunteer's right wrists. In this study, a real life dataset was created by performing several activities with similar behaviour in order to provide classes of movements that are very similar. The performed activities have been: walking, walking while holding a bag in the right hand and walking while keeping the right hand into the trousers pocket.

Acceleration vector -R(t)- was obtained at the beginning of the cycle and subsequently each vector value was collected every 62.5 ms. The values to create the test dataset (M) were calculated in each period of time, with the Euclidean distance [9] between the actual position -R(t)- and the previous position -R(t-1)-. Many samples of data from each activity have been created according to different period of time. 2, 4, 8 and 16 data samples from the same activity are grouped by means of simple addition in the data set. For further details check Table 1.

Table 1. Values of the groups

N° of samples (for each of the 3 classes)	N° of measurements grouped per data sample
1000	2/16
1000	4/16
1000	8/16
605	1

Finally, three different options for building each of the datasets from these data are considered:

- Option 1 is formed from 4 data sets, each one with two instants of time: previous and current.
- Option 2 is formed from 4 data sets, each one with three instants of time: two instants before current one, previous and current.
- Option 3 is formed from 4 data sets, each one with four instants of time: three instants before current one, two instants before current one, previous and current. Further details on the analyzed datasets can be found in Table 2.

Table 2. Analyzed datasets

Option	N° of columns	Columns (time instants)			
1	2			t-1	t
2	3		t-2	t-1	t
3	4	t-3	t-2	t-1	t

So in this case, there is an array of 12 different combinations available to assemble the datasets for the tests: four different ways of summarizing the measures to construct each sample (summarizing 2, 4, 8 or 16) and three combinations for the number of dimensions considered in each sample (t-3 to t, t-2 to t and t-1 to t).

3.1 Experimental Results

Experiments have been performed with all datasets previously explained. All data were gathered when the volunteer performed the same three activities, but the data were summarized in different ways (See previous page). The purpose is therefore, to find which of those combinations for data gathering could be considered the best one to separate activities, in order to be used in further analyses. Experiments have been performed using the SOM as a classifier, but two other measures have been performed on the resulting map. A 10-fold cross-validation schema has been used in order to obtain the most significant results as possible.

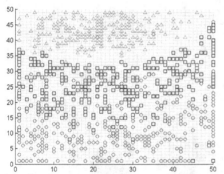

Fig. 2a. 16 measurements and 4 dimensions in each sample (t-3 to t)

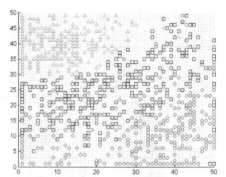

Fig. 2b. 16 measurements and 2 dimensions in each sample (t-1 to t)

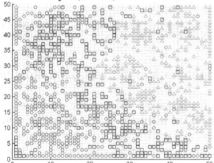

Fig. 2c. 2 measurements and 4 dimensions in each sample (t-3 to t)

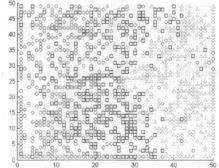

Fig. 2d. 2 measurements and 2 dimensions in each sample (t-1 to t)

Fig. 2. Different maps obtained for the most distinctive dataset combinations

Figure 2 shows the maps obtained for each dataset. Each one of the three movements (classes) is depicted in a different way: green triangles, red circles and blue squares. As can be seen, maps obtained when summarizing samples in 16 measures a sample, yield maps separating in a clearer way the three different classes of the dataset. On the contrary, when using only 2 measurements to obtain a sample for the dataset, the classes in the maps appear more mixed. This is especially true for classes "walking" (red circle) and "walking holding a bag" (blue square). Also, the results can be considered better when 4 dimensions are used in each sample (Figure 2a) than when using only two (Figure 2b). In this case the third class ("walking with a hand in the pocket") can be easily separated from the other two, as there is a clear gap of blank space separating them. Also, although there is a certain overlapping between the other two classes in all cases, the organization seems to be more stratified when using 4 dimensions in each sample (Figure 2a).

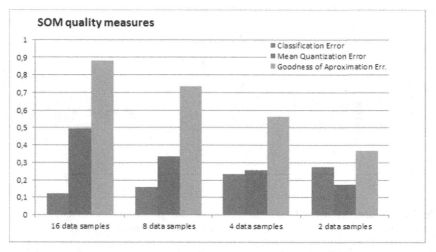

Fig. 3. Comparison of the SOM quality obtained when varying the number of measures summarized in each sample

Quantifiable results have been also obtained for each of the maps trained over the different datasets.

Firstly, a comparison of the three measures when modifying the frequency in which the data samples are considered   is included in Figure 3 All measures represented are error measures; so the lower the value, the best the model is performing. The Classification error is a percentage measure, while the other two are dimensionless measures. As can be seen, the classification performance clearly degrades when reducing the number of measurements summarized in each data sample (from 16 to 2). The other two measures seems to improve with this reduction. This is due to the fact that the dataset becomes sparser when using more frequently acquired measurements, therefore, decreasing the quantization error of the maps. In the case of this study, this is not the desired effect: a situation where samples are clearly separated (i.e. distinguishable) is the preferred one. As Goodness of Approximation Error accounts also for data quantization error, it behaves in a very similar way to the MQE.

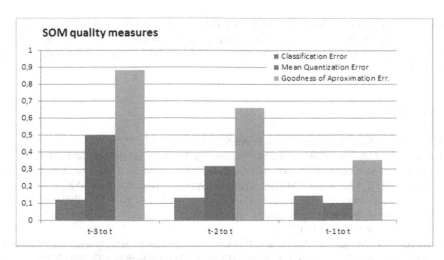

Fig. 4. Comparison of the SOM quality obtained when varying the number of time instants considered for each sample

Finally, the comparison of the results when considering the measures registered on 4, 3 or 2 previous time instants is included in Figure 4. Although it is no as significantly as in the previous comparison, classification error increases when decreasing the number of dimensions used to characterize a sample (from 11% with 4 dimensions to 14% with only 2). The other two measures decrease for similar reasons as explained before.

4   Conclusions and Future Research Lines

This work presents a first approximation to the problem of the automatic identification of the activities performed by a human user by means of a movement sensor. Several datasets have been registered in order to check which of the measuring conditions are the most appropriate to perform a simple identification of activities. Results point to the fact that the best settings are the ones that group more measures in each training sample (16 measures by sample in the experiments) as this tends to generate less sparse data in the data space, making classification easier. Also, including multiple dimensions to characterize each sample (4 dimensions in the experiments) in order to favour the correct classification; since the majority of automated learning algorithms rely on multi-dimensional analysis to perform their training.

Future work will be aimed at improving the classification results currently presented. Other automatic classification methods such as Artificial Neural Networks or Classification Trees could be used to further improve the results obtained by the SOM by constructing a hierarchical structure.

Acknowledgments. This research has been partially supported through the Spanish Ministry of Science and Innovation (MICINN) under project CIT-020000-2009-12 (funded by the European Regional Development Fund); project of the Spanish Ministry of Science and Innovation TIN2010-21272-C02-01 (funded by the European

Regional Development Fund) and MICINN PID 560300-2009-11. The authors would also like to thank the vehicle interior manufacturer, Grupo Antolin Ingenieria S.A., within the framework of the MAGNO2008 - 1028.- CENIT Project also funded by the MICINN.

References

[1] Comparing Self-Organizing Maps, ICANN 1996, vol. 1112 (1996)

[2] Bação, F., Lobo, V., Painho, M.: Self-organizing Maps as Substitutes for K-Means Clustering. In: Sunderam, V.S., van Albada, G.D., Sloot, P.M.A., Dongarra, J. (eds.) ICCS 2005, Part III. LNCS, vol. 3516, pp. 476–483. Springer, Heidelberg (2005)

[3] Banos, O., Damas, M., Pomares, H., Prieto, A., Rojas, I.: Daily living activity recognition based on statistical feature quality group selection. Expert Systems with Applications 39(9), 8013–8021 (2012)

[4] Bicocchi, N., Mamei, M., Zambonelli, F.: Detecting activities from body-worn accelerometers via instance-based algorithms. Pervasive and Mobile Computing 6(4), 482–495 (2010)

[5] Chen, Y.-P., Yang, J.-Y., Liou, S.-N., Lee, G.-Y., Wang, J.-S.: Online classifier construction algorithm for human activity detection using a tri-axial accelerometer. Applied Mathematics and Computation 205(2), 849–860 (2008)

[6] Analog Devices. Adxl335, Accelerometer (2012)

[7] Fuentes, D., Gonzalez-Abril, L., Angulo, C., Ortega, J.A.: Online motion recognition using an accelerometer in a mobile device. Expert Systems with Applications 39(3), 2461–2465 (2012)

[8] Hong, Y.-J., Kim, I.-J., Ahn, S.C., Kim, H.-G.: Mobile health monitoring system based on activity recognition using accelerometer. Simulation Modelling Practice and Theory 18(4), 446–455 (2010)

[9] Kohonen, T.: Self-Organizing Maps. Springer Series in Information Sciences. Springer (2001)

[10] Martiskainen, P., Järvinen, M., Skön, J.-P., Tiirikainen, J., Kolehmainen, M., Mononen, J.: Cow behaviour pattern recognition using a three-dimensional accelerometer and support vector machines. Applied Animal Behaviour Science 119(1-2), 32–38 (2009)

[11] Muscillo, R., Schmid, M., Conforto, S., D'Alessio, T.: Early recognition of upper limb motor tasks through accelerometers: real-time implementation of a dtw-based algorithm. Computers in Biology and Medicine 41(3), 164–172 (2011)

[12] Riedmiller, M., Braun, H.: A direct adaptive method for faster backpropagation learning: the rprop algorithm. In: IEEE International Conference on Neural Networks, vol. 1(3), pp. 586–591 (1993)

[13] Sedano, J., Chira, C., González, J., Villar, J.R.: Intelligent system to measuring stress: Stresstic. DYNA 87(3), 336–344 (2012)

[14] Van Laerhoven, K., Cakmakci, O.: What shall we teach our pants? (2000)

[15] Vesanto, J.: Data mining techniques based on the self-organizing map. Master's thesis, Helsinki University of Technology (May 1997)

[16] Wang, J., Chen, R., Sun, X., She, M.F.H., Wu, Y.: Recognizing human daily activities from accelerometer signal. Procedia Engineering 15(0), 1780–1786 (2011)

[17] Xie, H., Fedder, G.K., Sulouff, R.E.: 2.05 - accelerometers. In: Gianchandani, Y., Tabata, O., Zappe, H. (eds.) Comprehensive Microsystems, pp. 135–180. Elsevier, Oxford (2008)

[18] Yang, J.-Y., Wang, J.-S., Chen, Y.-P.: Using acceleration measurements for activity recognition: An effective learning algorithm for constructing neural classifiers. Pattern Recognition Letters 29(16), 2213–2220 (2008)

Intelligent Model to Obtain Current Extinction Angle for a Single Phase Half Wave Controlled Rectifier with Resistive and Inductive Load

José Luis Calvo-Rolle[1], Héctor Quintián[2], Emilio Corchado[2], and Ramón Ferreiro-García[1]

[1] University of Coruña, Department of Industrial Engineering
Avda. 19 de febrero, s/n, 15405, Ferrol, A Coruña, Spain
{jlcalvo,ferreiro}@udc.es
[2] University of Salamanca, Departmento de Informática y Automática
Plaza de la Merced s/n, 37008, Salamanca, Spain
{escorchado,hector.quintian}@usal.es

Abstract. The present work show the model of regression based on intelligent methods. It has been created to obtain current extinction angle for a half wave controlled rectifier. The system is a typically non-linear case of study that requires a hard work to solve it manually. First, all the work points are calculated for the operation range. Then with the dataset, to achieve the final solution, several methods of regression have been tested from traditional to intelligent types. The model is verified empirically with electronic circuit software simulation and analytical methods. The model allows obtaining good results in all the operating range.

Keywords: Half wave controlled rectifier, regression, non-linear model, MLP, SVM, polynomial models, LWP, K-NN.

1 Introduction

Electronic rectifiers are one of the most common circuits in electronic topics [1]. Its objective in general terms is to convert an alternate signal into continuous signal [2]. This type of circuits can be used in applications like DC power supplies [3], peak signal detectors [4], and so on.

Despite the simplest configuration of a rectifier circuit can be easy to understand and solve, for more complicate configurations they can be very difficult. Essentially, the difficulty depends of: the number of phases of the source, the characteristic of the load and if it is controlled or not controlled type [2].

The present research is focused on the single phase half wave controlled rectifier for resistive and inductive load. As would be seen in the case of study section, this variety of rectifier is not easy to solve in analytically form. The main reason is due to non-linear nature of this topology [3].

V. Snasel et al. (Eds.): SOCO Models in Industrial & Environmental Appl., AISC 188, pp. 249–256.
springerlink.com © Springer-Verlag Berlin Heidelberg 2013

Usually to take a solution, to solve non-linear systems are used traditional methods with the help of computation tools. In this sense as methods to obtain a solution can be cited: Trust-region dogled [5], Trust-region-reflective [6], Levenberg-Marquardt [7] [8] [9]. It is possible with these methods achieve good results in general terms with relatively few iterations.

Taking into account the fact that the operating range of the single phase half wave controlled rectifier for resistive and inductive load is known, it is possible to obtain the results of the angle extinction for the range. This fact is possible due the value do not depends the other parameters like voltage peak, frequency and so on. With the dataset is feasible apply regression and obtain automatically the extinction angle.

The traditional approaches for regression are based on Multiple Regression Analysis (MRA) methods [10]. MRA-based methods are very popular among others because have application in a lot of fields. It is well knowledge these methods have limitations [10] [11]. The limitations result from troubles associated with MRA-based methods, such as the inability of MRA to adequately deal with interactions between variables, nonlinearity, and multicollinearity [12] [10] [13]. More recently Softcomputing (SC)-based methods have been proposed as an option for many contributors as [11] [14] [15] [16] [17] [18].

Take into account all explained, a novel approach is proposed. The aim of the study is to develop an intelligent model that allows obtaining angle extinction of the single phase half wave controlled rectifier for resistive and inductive load. Many Softcomputing techniques have been tested in order to obtain the best fitness of the created model.

The work is structured as follow. It starts with a brief description of the single phase half wave controlled rectifier for resistive and inductive load. Afterwards intelligent regression methods tested in this study are described. Next section shows the model developed and the results for the different methods. Finally conclusions and future works are exposed.

2 Case of Study

2.1 Single Phase Half Wave Controlled Rectifier for Resistive and Inductive Load

The single phase half wave controlled rectifier circuit with resistive and inductive load is showed in figure 1.

The load (resistance $R1$ and inductance $L1$) is fed for the voltage source $V1$ via thyristor $X1$. The principal characteristic of the circuit is that the current is not in phase with the voltage. This fact is due the inductive component of the load. The firing pulse is applied to the gate of thyristor $X1$ (trigger pin) at instant defined as α. It remains in the ON state until the load current tries to go to a negative value. This instant is defined as angle extinction β. The current $i(\omega t)$ is defined by equation 1, where Vm is the peak voltage, ω is the frequency, t is the time, τ is L/R and θ is $tan\text{-}1(\omega L/R)$.

$$i(\omega t) = \frac{V_m}{Z}\left[\sin(\omega t - \theta) + \sin(\theta).e^{\frac{-\omega t}{\omega \tau}} \right] \tag{1}$$

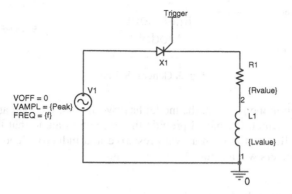

Fig. 1. Single phase half wave controlled rectifier for resistive and inductive load circuit

Figure 2 show the voltage (dots in blue) and current in the load (continuous line in red). As can be shown, the voltage exists in the load when current exist. Angle extinction is specified at figure.

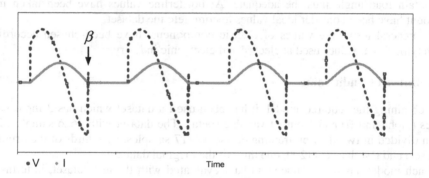

Fig. 2. Load Voltage and current

To obtain angle extinction value β it is necessary to solve the equation 2 [2].

$$i(\beta) = \frac{V_m}{Z}\left[\sin(\beta - \theta) + \sin(\theta).e^{\frac{-\beta}{\omega\tau}}\right] = 0 \qquad (2)$$

There is not any analytical solution for the equation 2, and it is necessary numeric methods to solve the equation like the mentioned above in the introduction.

2.2 Novel Approach

The general schema of the proposed topology where intelligent model is used to obtain the value of angle extinction is illustrated in figure 3.

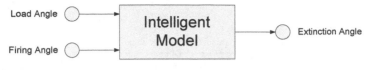

Fig. 3. General Schema

As can be seen in the figure 2, the model has two inputs: the load angle and the firing angle. As an output, the model provide the extinction angle that the single phase half wave controlled rectifier circuit with resistive and inductive load has for the two mentioned parameters when it run.

3 Model Approach

3.1 Obtaining Dataset

For the dataset creation several things have been taken into account. Firstly it is necessary to include all the range, i.e. load angle should go from 0 degrees to 90 degrees. Know that this fact the combination of the values of the resistance and the inductance to obtain load angle must be adequate. At borderline values have been taken into account have been consider ideal values to complete the dataset.

In general terms the values of electric components have been chosen according with typical real values used at electric and electronic industry.

3.2 Dataset Conditioning

For obtaining the required model, it has been used a dataset which has 2 input variables (alpha and fi) and 1 output variable (beta). The dataset with 3626 samples, has been divided in two dataset, training dataset (2417 samples, two thirds of the original dataset) and test dataset (1209, one third of the original dataset).

Each model has been trained and later evaluated with the test dataset, in terms of the MSE (Mean Square Error). In all techniques has been trained using cross validation with 10 folds.

3.3 Used Techniques

Following each of the used techniques and its parameters are described.

Artifical Neural Networks (ANN). Multilayer Perceptron (MLP): A multilayer perceptron is a feed forward artificial neural network [19] [27]. It is one of the most typical ANNs due to its robustness and relatively simple structure. However the ANN architecture must be well selected to obtain good results. Tests were performed using 1 hidden layer, from 10 to 30 neurons in the hidden layer and log sigmoid and tangent sigmoid activation functions.

Finally the best results were obtained using 20 neurons and log sigmoidal transfer function for the hidden layer.

Support Vector Regression (SVR). Support Vector Regression (SVR) is a modification of the algorithm of the Support Vector Machines (SVM) [20] for classification. In SVR the basic idea is to map the data into a high-dimensional feature space F via a nonlinear mapping and to do linear regression in this space.

LS-SVM. Least Square Support Vector Machine (LS-SVM). Least Square formulation of SVM, are called LS-SVM, in the approximation the solution is obtained by solving a system of linear equations, and it is comparable to SVM in terms of generalization performance [21]. The application of LS-SVM to regression is known as LS-SVR (Least Square Support Vector Regression). In LS-SVR, the ε-insensitive loss function is replaced by a classical squared loss function, which constructs the Lagrangian by solving the linear Karush-Kuhn-Tucker (KKT) system:

$$\begin{bmatrix} 0 & I_n^T \\ I_n & K + \gamma_{-1}I \end{bmatrix} \begin{bmatrix} b_0 \\ b \end{bmatrix} = \begin{bmatrix} 0 \\ y \end{bmatrix} \tag{3}$$

Where I_n is a [nx1] vector of ones, T means transpose of a matrix or vector, γ a weight vector, b regression vector and b_0 is the model offset. In LS-SVR, only 2 parameters (γ, σ) are needed. Where σ is used the width of the kernel [22].

LS-SVM [23] Matlab toolbox has been used. In this toolbox, the tuning of the parameters $(\gamma, \sigma$ equation 3) is conducted in two steps. First, a state-of-the-art global optimization technique, Coupled Simulated Annealing (CSA) [24], determines suitable parameters according to some criterion. Second, these parameters are then given to a second optimization procedure (simplex or gridsearch) to perform a fine-tuning step.

The optimal parameters obtained by the previous method are: $\gamma = 2.1515\,e^7, \sigma^2 = 1.99477$

Polynomial Models. Generally, a polynomial regression model may also be defined as a linear summation of basis functions:

$$F(x) = \sum_{i=1}^{k} a_i f_i(x) \tag{4}$$

where k is the number of basis functions (equal to the number of model's parameters); $f_i(x)$ ($I = 1,2,..., k$) is a predefined polynomial basis function. The number of the basis functions in polynomial model of degree p is:

$$m = \prod_{i=1}^{p}(1 + d/i) \tag{5}$$

The estimation of model's parameters is made based on the training data typically using the Ordinary Least-Squares (OLS) method, minimizing:

$$a = \arg\min_a \sum_{i=1}^{n} \left(y_{(i)} - F(x_{(i)})\right)^2 \tag{6}$$

where $x_{(i)}$ is the vector of input variables' values of the i_{th} data point and $y_{(i)}$ is the output value of that point. In this polynomials method as well as all the other regression modeling methods, the systems of linear equations in OLS are solved using Gaussian elimination and back substitution.

The best approximation has been obtained using Degree = 5.

Locally Weighted Polynomials (LWP). Locally Weighted Polynomial approximation [25] is designed to address situations in which the models of global behaviour do not perform well or cannot be effectively applied without undue effort. The LWP approximation is carried out by point wise fitting of low-degree polynomials to localized subsets of the data.

The coefficients of the polynomial are calculated, minimizing:

$$a = \arg\min_a \sum_{i=1}^{n} w(x_{query}, x_{(i)}) \big(F(x_{(i)} - y_{(i)})\big)^2 \tag{7}$$

where w is a weight function; x_{query} is the query point nearest neighbours of which will get the largest weights. The weight function w depends on the Euclidean distance (in scaled space) between the point of interest x_{query} and the points of observations x. The kind of weighting for Gaussian weight function implemented is:

$$w(x_{query}, x_{(i)}) = \exp\left(-\alpha\mu_{(i)}^2\right) \tag{8}$$

where α is a coefficient (bandwidth is $n\alpha$, where n is the number of total points) and the $\mu_{(i)}$ is a scaled distance from the query point to the i_{th} point in the training data set:

$$\mu_i = \frac{\|x_{query} - x_{(i)}\|}{\|x_{query}, x_{farthest}\|} \tag{9}$$

where $\|\cdot\|$ is the Euclidean norm; $x_{farthest}$ is the farthest training point from the point x_{query}. The locality of the approximation is controlled by varying the value of the coefficient α. If α is equal to zero then local approximation transforms into global approximation.

The best approximation has been obtained using degree 6 and bandwidth 58.75.

K-Nearest Neighbours (K-NN). K-NN technique [26] is a nonparametric method, which generally, can be written as:

$$F(x) = \frac{\sum_{i=1}^{k} w_i y_i'}{\sum_{i=1}^{k} w_i} \tag{10}$$

where y_i' is the output value of i_{th} point from the k points nearest to the query point x and w_i is the weight of that point. Here closeness implies the Euclidean distance. In this case the distance weighting schemes used is:

$$w_i = 1 - \|x - x_i'\| \; (i = 1,2,\dots,k) \tag{11}$$

where $x_{(i)}'$ is the input value of i_{th} point from the nearest ones.

The best approximation has been obtained using number of nearest neighbours = 6.

4   Results

Once the models were trained, they were tested getting the following results in terms of MSE:

With showed results, it is possible to conclude that the best models are ANN and SMV-LS. They are more than 10 times better than the other tested methods. ANN and SVM-LS obtain similar results, but the computational cost required for training process of SVM-LS is so much higher than ANN. The main reason is due to the required optimization process of SVM-LS for getting the best parameters (γ,σ) necessary for the training process, but once it has been trained, the computational cost is the same as ANN.

Table 1. Results for each of the applied methods

Method	Parameters	MSE (Test dataset)
ANN	1 hidden layer, 20 neurons, log sigmoidal activation function	0.000031147
SVM	$\gamma = 2.1515\ e^7, \sigma^2 = 1.99477$	0.000031634
Polynomial	Degree = 5	0.033615344
LWP	Degree 6 and bandwidth 58.75	0.00022042078
KNN	Number of nearest neighbours = 6	0.49652762

5 Conclusions

With the novel approach to obtain the extinction angle, as a principal conclusion, put in highlight that very good results have been achieved in general terms. With the intelligent model created based on Soft computing techniques is not necessary to solve the non-linear equation well analytically or by simulation.

As can be seen in results section the best approximation is obtained with ANN, where the MSE has a value minor than 0.000032. Other techniques allow good results like SVM or LWP. With any of them, the angle extinction would be calculated, among others because the real components have tolerances around 5%.

References

1. Malvino, A.P., Bates, D.J.: Electronic principles. Recording for Blind & Dyslexic, Princeton (2008)
2. Hart, D.W.: Power Electronics. McGraw-Hill, New York (2011)
3. Rashid, M.H.: Power electronics handbook: devices, circuits, and applications. Butterworth-Heinemann, Burlington (2011)
4. Mohan, N.: Power electronics: a first course, Hoboken, N.J. (2012)
5. Coleman, T.F., Li, Y.: An Interior, Trust Region Approach for Nonlinear Minimization Subject to Bounds. SIAM Journal on Optimization 6, 418–445 (1996)
6. Coleman, T.F., Li, Y.: On the Convergence of Reflective Newton Methods for Large-Scale Nonlinear Minimization Subject to Bounds. Mathematical Programming 67(2), 189–224 (1994)
7. Levenberg, K.: A Method for the Solution of Certain Problems in Least-Squares. Quarterly Applied Mathematics 2, 164–168 (1944)
8. Marquardt, D.: An Algorithm for Least-squares Estimation of Nonlinear Parameters. SIAM Journal Applied Mathematics 11, 431–441 (1963)
9. Moré, J.J.: The Levenberg-Marquardt Algorithm: Implementation and Theory. In: Watson, G.A. (ed.) Numerical Analysis. Lecture Notes in Mathematics, vol. 630, pp. 105–116. Springer (1977)
10. Mark, J., Goldberg, M.: Multiple Regression Analysis and Mass Assessment: A Review of the Issues. Appraisal Journal 56(1), 89–109 (1988)
11. Do, A.Q., Grudnitski, G.: A Neural Network Approach to Residential Property Appraisal. The Real Estate Appraiser 58(3), 38–45 (1992)
12. Larsen, J.E., Peterson, M.O.: Correcting for Errors in Statistical Appraisal Equations. The Real Estate Appraiser and Analyst 54(3), 45–49 (1988)

13. Limsombunchai, V., Gan, C., Lee, M.: House Price Prediction: Hedonic Price Model Vs. Artificial Neural Network. American Journal of Applied Sciences 1(3), 193–201 (2004)
14. Worzala, E., Lenk, M., Silva, A.: An Exploration of Neural Networks and Its Application to Real Estate Valuation. Journal of Real Estate Research 10, 185–202 (1995)
15. Guan, J., Levitan, A.S.: Artificial Neural Network Based Assessment of Residential Real Estate Property Prices: A Case Study. Accounting Forum 20(3/4), 311–326 (1997)
16. Taffese, W.Z.: Case-Based Reasoning and Neural Networks for Real Estate Valuation. In: Proceedings of 25th International Multi-Conference: Artificial Intelligence and Applications, Innsbruck, Austria, pp. 84–89 (2007)
17. Guan, J., Zurada, J., Levitan, A.S.: An Adaptive Neuro-Fuzzy Inference System Based Approach to Real Estate Property Assessment. Journal of Real Estate Research 30(4), 395–420 (2008)
18. Peterson, S., Flanagan, A.B.: Neural Network Hedonic Pricing Models in Mass Real Estate Appraisal. Journal of Real Estate Research 31(2), 147–164 (2009)
19. Bishop, C.M.: Pattern recognition and machine learning. Springer, New York (2006)
20. Cristianini, N., Shawe-Taylor, J.: An Introduction to Support Vector Machines. Cambridge University Press, Cambridge (2000)
21. Ye, J., Xiong, T.: Svm versus least squares svm. In: The 11th International Conference on Artificial Intelligence and Statistics (AISTATS), pp. 640–647 (2007)
22. Yankun, L., Xueguang, S., Wensheng, C.: A consensus least support vector regression (LS-SVR) for analysis of near-infrared spectra of plant samples. Talanta 72, 217–222 (2007)
23. De Brabanter, K., Karsmakers, P., Ojeda, F., Alzate, C., De Brabanter, J., Pelckmans, K., De Moor, B., Vandewalle, J., Suykens, J.A.K.: LS-SVMlab Toolbox User's Guide version 1.7 (2010), http://www.esat.kuleuven.be/sista/lssvmlab/
24. Xavier de Souza, S., Suykens, J.A.K., Vandewalle, J., Bolle, D.: Coupled Simulated Annealing. IEEE Transactions on Systems, Man and Cybernetics - Part B 40(2), 320–335 (2010)
25. Cleveland, W.S., Devlin, S.J.: Locally weighted regression: An approach to regression analysis by local fitting. Journal of the American Statistical Association 83, 596–610 (1988)
26. Duda, R.O., Hart, P.E., Strork, D.G.: Pattern Classification, 2nd edn. Wiley, Chichester (2001)
27. David, J., Henao, V.: Neuroscheme: A modeling language for artificial neural networks. Dyna-Colombia 147, 75–82 (2005)

A Statistical Classifier for Assessing the Level of Stress from the Analysis of Interaction Patterns in a Touch Screen

Davide Carneiro[1], Paulo Novais[1], Marco Gomes[1],
Paulo Moura Oliveira[2], and José Neves[1]

[1] Department of Informatics, University of Minho, Portugal
{dcarneiro,pjon,jneves}@di.uminho.pt, pg18373@alunos.uminho.pt
[2] University of Tras-os-Montes e Alto Douro, Portugal
oliveira@utad.pt

Abstract. This paper describes an approach for assessing the level of stress of users of mobile devices with tactile screens by analysing their touch patterns. Two features are extracted from touches: duration and intensity. These features allow to analyse the intensity curve of each touch. We use decision trees (J48) and support vector machines (SMO) to train a stress detection classifier using additional data collected in previous experiments. This data includes the amount of movement, acceleration on the device, cognitive performance, among others. In previous work we have shown the co-relation between these parameters and stress. Both algorithms show around 80% of correctly classified instances. The decision tree can be used to classify, in real time, the touches of the users, serving as an input to the assessment of the stress level.

1 Introduction

There are many scenarios in which the use of stress-aware applications could be of interest to improve the performance and quality of work of organizations. In general there is an interest in the scientific community for applications that can acquire and use meaningful information from the user's context. In [7], the authors provide a review of several context-aware applications published in conferences and journals between 2000 and 2007. Moreover, the authors also suggest a new classification framework of context-aware systems and explore each of its features. In this scope and given the nature of stress, soft-computing approaches can be very useful [11].

Stress is evidently part of this context information and can be quite important, depending on the scope of the application. In [8], a system to support tacit communication between fire-fighters with multiple levels of redundancy in both communication and user alerts is presented. This system supports decision and planning based on the level of stress of the fire-fighters, in real time, allowing a better management and security of the personal in the field. Applications for domestic environments also exist. In [9], a Conflict Manager to resolve conflicts for

V. Snasel et al. (Eds.): SOCO Models in Industrial & Environmental Appl., AISC 188, pp. 257–266.
springerlink.com

context-aware applications in smart home environments is presented. Conflicts arise when multiple users access an application or when various applications share limited resources to provide services. In order to resolve conflicts between users the Conflict Manager looks at parameters such as their levels of stress.

In this paper, we exploit the fact that tactile devices are nowadays relatively common and available. Moreover, many professions require or welcome their use, such as medical personnel, the military, fire-fighters, among many others. We propose a statistical classifier that is able to assess the level of stress of the users by analysing their touch patterns. The two features considered are the variation of the intensity and the duration of the touch.

2   Background

The word *stress* has many connotations and definitions based on various perspectives of the human condition. Many experts endorse the original definition of *stress* concept to the one proposed by Hans Selye [6]. He defined stress as a non-specific response of the body to any demand placed upon it. Selye defined external demands as *stressors* (the load or stimulus that triggers a response) and the internal body changes that they produce as the *stress response*.

However, specialists have expanded the previous concept of *stress*. Now, it is seen as the inability to cope with a perceived threat to one's mental, physical, or emotional well-being, which results in a series of physiological responses of adaptation. Researchers started to focus on the cognitive and behavioural causes of stress, and stress became viewed as a mind-body, psychosomatic, or psycho-physiologic phenomenon. A free interpretation of this phenomenon could refer stress as a physico-physiologic arousal response occurring in the body as result of stimuli by virtue of the cognitive interpretation of the individual.

Given the complexity of stress and its effects, a multi-modal approach is applied to obtain a more complete schematic description, that accounts for its known or inferred properties. These modalities include quantifiable measurements on the

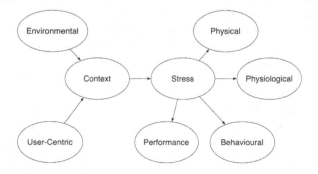

Fig. 1. A high-level information model about stress

user's physical appearance, physiology, behaviours and performance. Figure 1 depicts the multi-modal approach used.

The context node is divided in two types according to the source of contextual information, namely, the user-centric and the environmental context. User-centric information is composed of two categories: the background and the dynamic behaviour. The background is composed by several attributes that can be extracted from the users profile. These attributes are the age, gender, working area, social status, personality traits among others. The dynamic behaviour reflects the contextual attributes related to the users activity. The environmental information fuses the physical environment characteristics, social environment information and computational environment measurements. Physical environment includes attributes such as the time, temperature, location, noise level, and luminance. High levels of noise, extreme temperatures and low levels of luminance are well known potential stressors. The social environment includes issues such as the population density around the user. The computation environmental context can be characterized by the measurement of the electromagnetic field and the number of surrounding electronic devices.

Among the features that can reveal stress, those that can characterize the behavioural node are the user's interactions with the computer, the mouse/touch screen pressure from clicks/touches, his/her agitation level (through the sensory data from the accelerometer placed in mobile devices or by analyzing movement), as well as input frequency and speed. Also, the performance node is depicted in terms of accuracy and response, where the accuracy feature is related to the precision of the touch and the response feature corresponds to the analysis (qualitative and temporally) of the users responses to the platform demands. The physiological variables provide observable features about the users stress state [10]. These features include the Galvanic Skin Response (GSR), that assesses the electrical proprieties of the skin in response to different kinds of stimuli, the General Somatic Activity (GSA) that assesses the minute movement of human body and others such as respiratory rate or pupilographic activity. Physical appearance includes the visual features that characterize the users eyelid movement such as pupil movement (e.g. eye gaze, papillary response), facial expression or head movement.

3   A Non-invasive Environment for Detecting Stress

The experiment detailed in section 5 was undertaken in the Intelligent Systems Lab, in the University of Minho. In this lab, we built a closed environment with the main objective of monitoring several interaction parameters that could be related to stress: the stress lab. This stress lab allows for a user to interact with specific applications while being fully monitored in a non invasive way. It is composed of several devices that can acquire information from the user's context (Figure 2). Table 1 presents a brief description of the main functionalities of these devices.

Fig. 2. Devices used to study the effects of stress in the interaction parameters

We are interested in devices that can provide us with some information about the user but without interfering with the interaction or with the activities being performed. This environment allows us to capture information about the following parameters: Touch Accuracy, Touch Intensity, Touch Duration, Amount of Movement, Acceleration and Cognitive Performance. In previous work we have performed significance tests on these parameters to find differences at the level of the user due to stress (see for example [1,2]). In this paper, we use this test environment and, supported by the results previously achieved, we train a classifier that can distinguish between "stressed" and "calm" touch patterns.

Table 1. Brief description of the devices that compose the stress detection environment

Device	Brief description	Main features
HP Touchs-mart	All-in-one PC	touchscreen, web cam, large screen
Samsung Galaxy Tab	Tablet PC	touchscreen, web cam, accelerometer, relatively large screen, mobile, Android OS
HTC PDAs	Smartphones	touchscreen, camera, accelerometer, mobile, Android OS
Sony FCB-EX780BP	25x Super HAD PAL Color Block Camera with External Sync	25x Optical Zoom, Image stabilizer, Day/Night Mode, Privacy Zone Masking

4   Feature Extraction

To assess the level of stress of each touch, out system relies on the event listeners provided by the Android framework. An event listener is an interface in the `View` class that contains a single callback method that will be called by the Android framework when the View to which the listener has been registered is triggered by user interaction with the item in the UI. For the purpose of this paper, our system uses the `onTouch()` callback method, which is called when the user performs an action qualified as a touch event, including a press, a release, or any movement gesture on the screen (within the bounds of the item). Thus, in each touch of the user on an item of the UI several touch events are fired: one when the finger of the user first touches the screen (identified by the action event ACTION_DOWN), several while the user is touching (depending on the duration of the touch) and one when the finger releases the screen (identified by the action event ACTION_UP).

Each of these events has information about the intensity of the touch (via the getPressure() method) quantifying the pressure exerted on the screen and about the position of the event. Moreover, when each event is fired our system registers it with the current time. This allows to visualize the evolution of a touch in terms of its intensity over time. Moreover, from this information we can also extract the duration and intensity features.

Duration Features. The duration of a touch is defined as the difference between the time-stamps of the action events ACTION_UP and ACTION_DOWN. One of the hypothesis being tested is that the stress of a user will have an influence on the duration of the touch, hence our interest. The duration of the touch can however be influenced by factors other than the stress. Namely, the type of item of the UI being touched. In that sense, we do not use for this purpose events fired by items such as sliders or by scrolling pages. For the purpose of this experiment, we are just interested in the standard touches used to interact with buttons, inputting text and similar actions.

Intensity Features. The intensity of a touch event depicts the force exerted by finger of the user when touching the device. Given that each touch event includes a pressure and that each touch fires several touch events (as described in section 4), it is possible to analyse the variation of the intensity throughout all the touch, from the moment the finger touches the screen to the moment it releases it.

Concerning this temporal evolution of the intensity of the touch, we are interested in the initial and final value of intensity of each touch as well as its maximum and mean values. First we will thus test the hypothesis that stress can influence these parameters and that there are significant differences in at least some of them between a stressed and a calm user.

5   Experiment Design and Methods

The main goal of this experiment is to investigate if it is possible to build a classifier for touch patterns that can be used in real-life applications to provide some information about the level of stress of the user. Our approach is to use two standard and well known machine learning tools: a decision tree constructor and a support vector machine. As the decision tree constructor we use the J48 algorithm - the java implementation of the C4.5 [3]. As support vector machine, we use the SMO function, which implements John Platt's sequential minimal optimization algorithm for training a support vector classifier [4]. These experiments were performed using the Weka workbench (Weka 3.6.3) [5].

The results of the two classifiers will be compared by looking at some performance measures such as the percentage of correctly classified instances, the Kappa statistic (which is a chance-corrected measure of agreement between the classifications and the true classes) and the ROC area.

We have a particular interest in using decision trees since a model of a decision tree can then be used to classify, in real time and in a real life application, the level of stress of a user, by following the explicit rules defined by the model.

5.1   Dataset

The dataset used during this experiment was collected in the Intelligent Systems Lab of the Department of Informatics, at the University of Minho. To collect the data, a group of 18 users was asked to play a game that included performing mental calculations and could also include memorizing some intermediary results for posterior use. During the game, the users could be subject to stress in the form of unexpected repetitive and annoying sounds, vibration of the handheld device or a time limit.

With this setting it was possible, in previous experiments, to build the dataset depicted in table 2, that allowed us to determine how each user is affected by stress through significance tests, and develop personalized stress models. Moreover, a generic model was also developed that can be used, although with expected smaller accuracy, when no personalized data about a user is available.

Table 2. Description of the dataset used as a basis for building the classifier

data	brief description	size
Acceleration	Data concerning the acceleration felt on the handheld device while playing the game	27291
Maximum intensity of touch	Data about the maximum intensity of each touch in a touchscreen	1825
Mean intensity of touch	This dataset contains data about the mean intensity of each touch event in a touchscreen	1825
Amount of movement	A dataset containing information about the amount of movement during tests	25416
Touches on target	This dataset contains information about the accuracy of the touches	1825
Stressed touches	A dataset containing information that allows to classify each touch as stressed or not stressed	1825
Score	A dataset describing the performance of the user playing the game, during the tests	321
Touch duration	A dataset containing the duration of each touch event	1825

5.2   Experiment Design

All the parameters in table 2 are correlated with stress, with some users show-ing more significant results than others. In previous work we have studied this relation. We will now focus on how we use this information to build a classifier.

As previously described, each touch in the screen results in several touch events that are fired during the time of the touch. This number varies according to the duration of the touch. In that sense, this data for each touch, as it is, cannot be used to build a classifier (each touch would have a potentially different list of values of intensity, one for each touch event). Figure 3 (a) highlights this by depicting different types of touches.

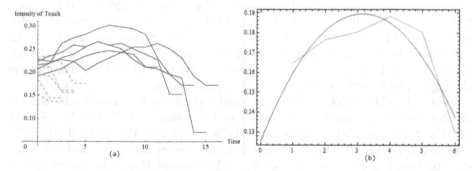

Fig. 3. (a) 10 different touch patterns from users: touches can be composed of a different number of touch events. The orange lines depict touches classified as "calm" whereas the blue lines belong to touches classified as "stressed". (b) Fitting a polynomial curve (blue curve) to a given touch (orange line).

To tackle this problem we explored the fact that the intensity from all the touches follows a similar shape: a convex curve that grows to a maximum point and then decreases. Thus, the approach was to fit a second polynomial degree curve to each touch pattern. To perform this fitting in real-time the proposed system uses J/Link, the Mathematica's Java interface that allows for controlling Mathematica Kernels from Java programs. Specifically, we use the Fit[*data, funs, vars*] function which finds a least-squares fit to a list of data as a linear combination of the functions *funs* of variables *vars*. To implement this we are using Mathematica® v8.0. An example of this approach is depicted in Figure 3 (b). Given that the second degree polynomial curves are of the type $y = ax^2 + bx + c$ we can compare the parameters of the curve of each touch pattern: similar values of a,b and c indicate similar curves, thus similar touch patterns. Hence, the input for the classifier are three numeric attributes a,b and c (the independent variables) and a nominal attribute that describes the state of the user at the time of the touch as "stressed" or "not stressed" (the dependent variable). The classifiers were trained using this data, comprising a total of 349 instances.

6   Results and Conclusions

Since selecting the optimal parameters for an algorithm may be a rather time-consuming process, to implement this experiments we used a meta-classifier provided by weka that allows to optimize a given base-classifier. Specifically, we used the `weka.classifiers.meta.CVParameterSelection`. After finding the best possible configuration of parameters, the meta-classifier then trains an instance of the base classifier with these parameters and uses it for subsequent predictions. The meta-classifier was used with lower bound 0.01, upper bound 0.5 and 10 optimization steps.

When using the J48 classification tree as the base classifier for the meta-classifier, the model is able to correctly classify 271 out of the 349 instances, which amounts to 77.6504%. The Kappa statistic for this model is 0.5434 and the value of the ROC area is 0.796. The constructed tree has a size of 15 nodes and a total of 8 leaves (Figure 4 (a)). In this tree, attributes x0,x1 and x2 correspond to the values of a, b and c of the polynomial curve, respectively. Given this, it is possible to use the rules of this tree to build a classifier for distinguishing between stressed and calm touch curves.

When the SMO function is used to build a classifier, the results achieved are similar. In fact, the correctly classified instances amount to 79.9427% (279 out of 349), the value of the Kappa statistics is 0.5809 and the value of the ROC area is 0.781. These results also show that a classifier can be trained with this data to distinguish between stressed and calm touches. Given that the results of both classifiers are similar, we decided on using the J48 tree since it can easily be used by our system to classify touches in real time.

To evaluate the performance of the tree, we used it to classify, in real-time, the touches of 16 users during one of the experiments performed. In short, each user had to perform mental calculations under different levels of stress ranging from 1 (with no stressors) to 5 (with maximum level of stress induced). While the touches were being classified in real time, the remaining of the parameters described in section 3 were also under monitoring. This allowed us to ensure that there were significant differences on other parameters due to stress as well. In the worst case only one parameter showed significant differences and, in the best case, 5 different parameters showed significant differences for the same user. Concerning all the data, in average each user shows significant differences in 3 out of 6 parameters, which allowed us to conclude, in previous work, that stress does have an effect on these behavioural parameters.

Thus, what we did in this experiment was to analyse the behaviour of the classifier for each user under each level of stress and determine if the results of the classifier were in line with the results of the remaining parameters. Concerning the data collected from the 16 users, 13 show an increase in the touches classified as stressed when comparing the data from level 1 with the data from level 5. The minimum value of increase detected was of 6%, the maximum value of increase was of 60% and the mean increase of touches classified as stressed, for all users, was of 32.3077%. The three users for which the classifier reported a decreasing percentage of stressed touches for increased levels of stress have shown relatively low values of decrease (-2.5%, -5% and -1%). This means that the results of the classifier are consistent with the ones previously achieved in 81.25% of the cases. Figure 4 (b) depicts the mean increase of the touches classified as "stressed" in each of the five levels of stress of the experiment.

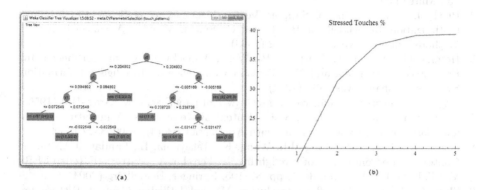

Fig. 4. (a) J48 prunned tree generated by the algorithm. This tree can be used to classify touches in real time as stressed ("yes" leaves) or not stressed ("no" leaves). (b) Mean increase in the percentage of touches classified as "stressed" in each of the five levels of stress concerning all the users.

Moreover, we have to state that the classifier was built as a generic model, i.e., we used data from all the users. We believe that better results would still be achieved if we were to develop personalized classifiers. This, however, was not the objective of the paper. Given this, we can conclude that it is possible to build a classifier for assessing the touches of users in a touch screen, based on their shape, and classify them as "stressed" or "not stressed". This, by itself is not enough to describe the level of stress of a user but can certainly be a significant input that can be used with that purpose, in conjunction with other inputs, as shown in this paper.

Acknowledgments. This work is funded by National Funds through the FCT - Fundação para a Ciência e a Tecnologia (Portuguese Foundation for Science and Technology) within project PEst-OE/EEI/UI0752/2011. The work of Davide Carneiro is also supported by a doctoral grant by FCT (SFRH/BD/64890/2009).

References

1. Carneiro, D., Montotya, J.C.C., Novais, P., Fernández-Caballero, A., Neves, J., Bonal, M.T.L.: Stress Monitoring in Conflict Resolution Situations. In: Novais, P., Hallenborg, K., Tapia, D.I., Rodríguez, J.M.C. (eds.) Ambient Intelligence - Software and Applications. AISC, vol. 153, pp. 137–144. Springer, Heidelberg (2012)

2. Carneiro, D., Novais, P., Costa, R., Neves, J.: Enhancing the Role of Multi-agent Systems in the Development of Intelligent Environments. In: Demazeau, Y., Dignum, F., Corchado, J.M., Bajo, J., Corchuelo, R., Corchado, E., Fernández-Riverola, F., Julián, V.J., Pawlewski, P., Campbell, A. (eds.) Trends in PAAMS. AISC, vol. 71, pp. 123–130. Springer, Heidelberg (2010)
3. Quinlan, R.: C4.5: Programs for Machine Learning. Morgan Kaufmann Publishers, San Mateo (1993)
4. Platt, J.: Fast Training of Support Vector Machines using Sequential Minimal Optimization. In: Schoelkopf, B., Burges, C., Smola, A. (eds.) Advances in Kernel Methods - Support Vector Learning (1998)
5. Holmes, G., Donkin, A., Witten, I.H.: Weka: A machine learning workbench. In: Proc. Second Australia and New Zealand Conference on Intelligent Information Systems, Brisbane, Australia (1994)
6. Selye, H.: The stress of life, vol. 5. McGraw-Hill paperbacks. McGraw-Hill (1956)
7. Hong, J.-Y., Suh, E.-H., Kim, S.-J.: Context-aware systems: A literature review and classification. Expert Systems with Applications 36(4), 8509–8522 (2009)
8. Jiang, X., Chen, N.Y., Hong, J.I., Wang, K., Takayama, L., Landay, J.A.: Siren: Context-aware Computing for Firefighting. In: Ferscha, A., Mattern, F. (eds.) PERVASIVE 2004. LNCS, vol. 3001, pp. 87–105. Springer, Heidelberg (2004)
9. Shin, C., Woo, W.: Conflict Resolution Method Utilizing Context History for Context-Aware Applications. Cognitive Science Research (577), 105–110 (2005)
10. Picard, R.W.: Affective computing. MITPress, Cambridge (1997)
11. Abraham, A.: Hybrid Soft Computing and Applications. International Journal of Computational Intelligence and Applications 8(1), 5–7 (2009)

Model Driven Classifier Evaluation
in Rule-Based System

Ladislav Clementis

Institute of Applied Informatics, Faculty of Informatics and Information Technologies, Slovak
University of Technology, Ilkovičova 3, 842 16 Bratislava, Slovakia
clementis@fiit.stuba.sk

Abstract. Rule-based evolutionary systems like learning classifier system are
widely used in industry and automation. In some types of problems we have additional information about problem solution. This information can be used in the
process of problem solutions. A rule-based system can be augmented by additional information concerning the given problem to enrich the process of system
adaptation to solve the problem. In many pattern matching tasks we know the patterns and we are looking for a pattern identification in environment. We provide
representative of a rule-based learning classifier system augmented with information about a property of solution. The augmented system solves an example of
pattern matching problem of simple Battleship game. This modified learning classifier system provides better convergence results by using the probability model
of the Battleship game problem space.

1 Introduction

Rule-based systems are using strength of symbolic rules [7] to solve various complex
tasks. One of the main practical uses of rule-based systems is solving classification
problems. Often used rule-based systems are those where rules are *condition-action*
type. These rules contain conditions in the form of *if-then*.

Rule-based systems can be divided into two categories in terms of adaptation to
solve problems. The first category represents rule-based systems where the set of rules
is given explicitly. The second category represents rule-based systems where the set of
rules is developed somehow. The development of rules is not an easy task, especially if
dealing with complex problems. In addition to solving complex problems is the explicit
identification of these rules practically impossible.

A learning mechanism of rule-based system is executing adaptation to solve given
task properly. Learning adaptation process can be executed *online* while system practically works or *offline* before system is set to real environment. Offline learning process
is usually used if rule-based system is about to solve problem which occurs rarely.
Offline learning rule-based systems are widely used in industry and automation, for
example when detecting an error depending on multiple detector outputs.

Research on *Learning classifier systems* is mostly concerned with their application
in datamining. This research is provided by *Learning Classifier Systems Group*, University of the West of England, Bristol since 1999. In this paper we provide an effective
example of *LCS* architecture modofication.

V. Snasel et al. (Eds.): SOCO Models in Industrial & Environmental Appl., AISC 188, pp. 267–276.
springerlink.com   © Springer-Verlag Berlin Heidelberg 2013

2   Related Work on Soft Computing

Many related work is concerning hard problem solutions. These problems are very complex and in information system applications fast decision-making is essential. *Soft computing* [1] is a representative of science field concerning aspects of complex task solutions and applications since early 1990s [21]. Soft computing is currently modern field of interest because many tasks can be solved faster by using inexact approaches. Various practical application examples [5,16] and theoretical studies can be found in current field of scientific research works.

LCS is using heuristic information evaluation and non-deterministic processes to solve classification tasks. Valid classification results can be used to solve many complex tasks by providing heuristic information about current state of concerned environment. Soft computing aspects of *LCS* have been well discussed in recent years [3].

3   Simple Learning Classifier System

Learning classifier system (LCS) as described by Holland [9,10] and later by Butz [4], Bull [2,3] and others is rule-based [6,11] evolutionary learning system. *LCS* consists of set of rules (referred to as *rule set*) usually described as population of classifiers, learning mechanism which is usually a *genetic algorithm (GA)* and rule evaluation method, usually the *Reinforcement learning* [8,19] *(RL)* mechanism, typically the *Q-learning algorithm* [20]. We distinguish between *Michigan-style LCS* [13,14] where we are searching for single solution and *Pittsburgh-style LCS* [17] where we are searching for multiple solutions [4]. We focus on *Michigan-style accuracy-based LCS* usually referred as *XCS* [4].

3.1   Rule Set in LCS

Rule set in *LCS* consists of *classifiers* [4]. These classifiers represents the knowledge base of *LCS*. Classifier described as rule consists of *matching condition, action* executed and *reward prediction* value. Matching condition of classifier covers problem subspace. If matching condition is corresponding to actual state of environment action can be executed. Classifiers compete to be applied so their action which changes a current state of environment can be executed.

3.2   Genetic Algorithm in LCS

From the view of *GA*, rule set is the population of classifiers [4]. This population is evolved by typical *GA* mechanisms as *selection, reproduction, mutation, recombination* and *deletion* of classifiers. After new classifiers are created by evolutionary process, other classifiers can be deleted from population and newly created classifiers take their place by creating next population generation.

GA uses selection of classifiers like widely used techniques of *roulette selection* or *tournament selection*. In typical *GA*, selection is based on *fitness* function values. In

LCS, reward prediction value of classifier represents its fitness function value which is gained by *RL* mechanism as response from an environment.

After a subset of classifiers is selected it can be reproduced and evolution process can be executed. Classifier matching condition with action as *gene* can be divided into *building blocks* (*BBs*) [4]. Recombination of classifiers is usually executed by *crossover* of building blocks between multiple classifiers. To some of resulting classifiers, modification called *mutation* can be applied usually by slightly changing their matching conditions.

Many other techniques are used to evolve classifiers, for example Classifier Fusion Methods [15]. In this paper we provide model based heuristic to evaluate classifiers.

3.3 Q-learning as Reinforcement Learning Mechanism in LCS

RL mechanism in *LCS* is used to evaluate classifiers for proper classifier selection in *GA* and also in selection of classifier which is applied to change an environment by executing an action of classifier. The higher reward prediction value of classifier is calculated if action of classifier changes the state of environment to a state which is closer to problem solution.

Calculation of reward prediction value can be executed by *Q-learning algorithm* [20]. *Q-learning algorithm* uses *Q-values* as reward prediction values to evaluate a classifier by state *s* and action *a*. *Q-values* are updated by equation 1 [4].

$$Q(s,a) \leftarrow Q(s,a) + \beta(r + \gamma \max_{a'} Q(s',a') - Q(s,a)) \tag{1}$$

In *Q-learning algorithm* we use parameters like *learning rate* usually of value $\beta \in \langle 0.8, 1 \rangle$ to control classifier reward prediction values calculation properly.

3.4 How Learning Classifier System Works

Single iteration of *LCS* starts with an information based on actual state of an environment. Whole population of classifiers is filtered by selecting only those classifiers whose conditions are matching the actual state of an environment. This subset of classifiers produces the *match set*.

After the *match set* is created, reward-based action selection is made. There are several techniques we can use for this purpose like selecting an action of classifier with the highest reward prediction value. Other option is selecting an action of classifiers with the highest average reward prediction value. Usually the random action selection of classifier is executed with some small non-zero probability $\varepsilon \in (0, 0.1\rangle$.

As we have successfully determined action to be executed it can be executed in an environment. Now we filter *match set* by selecting classifiers corresponding to the action executed. Resulting set of classifiers produces an *action set*. Reward prediction values of classifiers in the action set are updated usually by *RL* mechanism (*Q-learning algorithm*) as response from an environment and updated reward values are propagated to original population of classifiers.

Finally, *GA* executes evolutionary process iteration and therefore population of classifiers is modified by creating next generation of classifiers. Simplified iteration between *LCS* components is shown in figure 1.

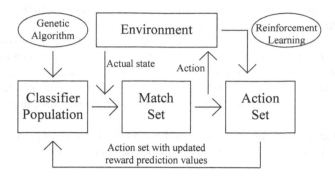

Fig. 1. Diagram showing interaction between *LCS* components in single iteration. Classifier population is filtered into *match set* corresponding to the actual environment state. *Match set* is filtered by selected action resulting into *action set*. Classifiers reward prediction values are updated by *RL* mechanism as an response from an environment. At the end of iteration *GA* is applied to handle classifier population.

Multiple iterations are executed one by one and population of classifiers is developed resulting into high quality rule set. Learning mechanism based on repeated iteration is used for *LCS* adaptation to solve the given problem.

In section 6 we provide the modified *LCS* architecture, augmented with the additional information about problem solution. This additional information is a model of simple pattern matching problem, described in section 4. In section 6 we provide the numeric comparison of these two approaches.

4   Model Driven Decision Process

Model driven architectures are widely used in many approaches for faster convergence to search for problem solution like *neural networks* [12,18] and many others.

We provide an example of simple pattern matching problem to describe the model driven decision process. Our problem space consists of two-dimensional discrete array of size $n \times m$, in our example 7×7. This problem space includes two patterns of size 1×3 which cannot overlap each other or have a common edge. In section 5 we provide a simple implementation example of this environment.

Generally each of $7 \times 7 = 49$ positions could be revealed or unrevealed. Initially all positions are unrevealed. We can reveal each position to find out if it hides part of pattern or not. Our goal is to completely match both patterns and successfully uncover them with minimum reveal steps taken.

This task can be described as the special case of the *Battleship game* from the view of single player. Our *problem space* we can describe as an enemy *battlefield* and two patterns as two enemy battleships placed in the enemy battlefield. Our goal is to completely sink both enemy battleships with minimum *hit* attempts performed. Unsuccessful shoot attempt we mark as *miss*.

We provide an example of a current state of environment after first hit. First successful hit attempt is completed and therefore first position of first pattern is revealed. Our goal is to increase probability of next hit attempt to be successful. This state of environment is shown in figure 2.

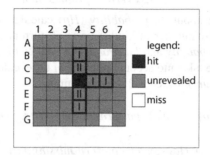

Fig. 2. View of the two dimensional problem space with one successful hit to black marked position *D4*. Unrevealed positions are marked grey and miss positions are marked white.

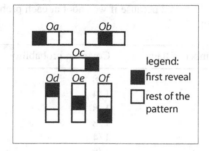

Fig. 3. Visual display of all possible pattern position options *Oa, Ob, Oc, Od, Oe* and *Of* according to first hit position. Options *Oa, Ob, Oc* are vertically oriented and options *Od, Oe* and *Of* are horizontally oriented.

If first part of pattern is hit, generally there are six possible options of pattern position according to first hit part of pattern. These options *Oa, Ob, Oc, Od, Oe* and *Of* are shown in figure 3. In current state of environment shown in figure 2 only possible pattern position options according to first hit position *D4* are *Oa, Od, Oe* and *Of*. Each unrevealed position which can be covering remaining hidden position of pattern is marked bold.

Figure 2 also shows number of flags. Single flag for single position marks that this position is covered by single pattern position option according to first hit position. Therefore the number of flags for each position means number of pattern position options covering this position in the environment.

From number of flags we can calculate *covering probability* (*Cp*) for each position *p*. *Cp* of position *p* is calculated by number of flags given to this position and total number of possible pattern position options of this state *s* given by equation 2.

$$Cp(p) = \frac{p\_flags\_count}{s\_coveringoptions\_count} \tag{2}$$

Cp gives us information about *hit probability* (*Hp*) calculated by equation 3 or 4 where *Cp* is multiplied by number of hidden parts of current pattern. Table 1 shows us non-zero *covering probability* and *hit probability* values of positions in our example shown in figure 2. We take into account number of unrevealed parts of current ship (*pattern_missingparts_count*), which is 2 in our example.

$$Hp(p) = Cp(p) \cdot pattern\_missingparts\_count \tag{3}$$

$$Hp(p) = \frac{p\_flags\_count \cdot pattern\_missingparts\_count}{s\_coveringoptions\_count} \tag{4}$$

Table 1. Positions with non-zero *covering probability* and *hit probability* values shown in table. Note that sum of *hit probabilities* is 2 because if we shoot at each probable position we hit both remaining parts of the pattern.

Hit attempt position	Number of flags	Covering probability	Hit probability
B4	1	1/8	1/4
C4	2	1/4	1/2
D5	1	1/8	1/4
D6	1	1/8	1/4
E4	2	1/4	1/2
F4	1	1/8	1/4
Σ	8	1	2

As shown in table 1, positions *C4* and *E4* have the highest probability to contain one of remaining parts of the pattern. Therefore choosing one of these positions to shoot is reasonable.

Next example shows uncovered part of second pattern at position *F1* after the first pattern successfully matched. Current state of the environment is shown in figure 4. Covering probability and hit probability values are shown in table 2.

According to probability values shown in table 2, position *E1* have the highest probability to successfully result into hit. Therefore it should be picked as the best candidate to reveal. Even if position *E1* is not including another part of pattern it is reasonable to choose this position at first.

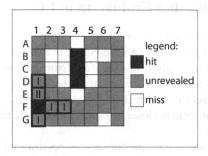

Fig. 4. View of the two dimensional problem space of second example of environment state with the first pattern successfully matched

Table 2. Position *covering probability* and *hit probability* values of the second state of an environment example

Hit attempt position	Number of flags	Covering probability	Hit probability
D1	1	1/6	1/3
E1	2	1/3	2/3
F2	1	1/6	1/3
F3	1	1/6	1/3
G1	1	1/6	1/3
Σ	6	1	2

5   Simple LCS Environment and Classifier Implementation

Current states of environment given as examples in section 4 are represented by an array of 7×7 symbols from three-character alphabet $\alpha = \{0, \#, 1\}$. Each position of current environment is represented by a single character. Complete state information is given as $s = [0, \#, 1]^{7 \times 7}$. Unrevealed position in an environment is represented by # character, miss position is represented by 0 character and successful hit position is represented by character 1.

As mentioned in section 3, classifier in *LCS* consists of matching condition, action and reward prediction value. In our case, the matching condition attribute we implement also like state information as an array of 7×7 characters of alphabet $\alpha = \{0, \#, 1\}$. Miss position is represented by character 0 attribute, hit position is represented by character 1 attribute and character # means that the corresponding position attribute value we do not take into account. Action attribute we represent as position identification or pointer which identifies exact position to be revealed if action executed. Reward prediction value we represent as a real-like number of real type variable.

6   Model Driven Classifier Evaluation in LCS

In classical *LCS* architecture, selection of action is based on reward prediction values of classifiers in match set. After action is selected, match set is filtered resulting into action set. Reward prediction values of classifiers from action set are updated by *Q-learning RL* mechanism. *Q-learning* update of classifiers reward prediction values is based on prediction how action of classifier changes a current state of an environment. If changed state of environment is closer to problem solution, higher reward prediction value is given to classifier.

In our modified *LCS*, we use *hit probability* of position which action of classifier points to as to be revealed if action executed. Hit probability of each position is calculated as shown in section 4. Classifiers whose actions are pointing at position with the highest hit probability are selected from match set resulting into action set.

We provide simulation results to compare strategy convergence of classical *LCS* approach where *Q-learning algorithm* is used as the *RL* mechanism for classifier evaluation, compared with our approach where classifier evaluation is based on hit probability model. Average simulation results are shown in table 3.

Table 3. Average simulation results showing classifier population quality during adaptation process. We have done *20* simulation runs. System was well adapted after *2000 LCS* iterations. Table shows average number of hit attempts of *LCS* with *Q-learning* algorithm and model driven heuristic classifier evaluation by matching two patterns in our simple example.

Iterations taken	Q-learning I.pattern	Q-learning II.pattern	Model-driven I.pattern	Model-driven II.pattern
0	28.6	41.3	26.7	42.5
200	28.1	40.5	27.9	40.9
400	26.3	40.9	26.4	36.2
600	27.9	36.3	24.1	35.7
800	26.6	37.1	22.2	32.4
1000	24.0	35.9	19.0	29.1
1200	23.2	31.4	17.4	23.0
1400	21.9	27.7	14.7	24.8
1600	14.4	22.3	15.6	20.6
1800	13.6	22.2	12.8	21.3
2000	13.7	21.7	13.8	22.8

Convergence speed is measured by number of iterations. Results are showing quality of rule-set measured by average hit attempts of system to match two patterns. According to problem definition, goal is to completely match two patterns of size 1×3. The best possible result of matching run is made by 6 successful hit attempts and no miss. The worst possible result of matching run is to uncover all positions with last remaining part of pattern uncovered as the last attempt. Therefore, the worst case of matching run is 49 hit attempts. Modified *LCS* enriched by probability model information have faster convergence speed than original *LCS* by solving our simple pattern matching example.

7 Conclusion

LCS by solving pattern matching task can use an additional information about problem solution to enrich the process of finding solution. *LCS* augmented by success probability information which is used as reward prediction value of *Q-learning algorithm* ensures system adaptation to problem solution search task.

This modified *LCS* architecture ensures more general emergence of reasonable decision strategy of rule set in *LCS*. This approach can be applied if model of problem can be described well. Therefore successful action result probabilities can be easily calculated and included into classifier selection process. If given problem has complex problem space and additional information about problem solution is available, it can be implemented into *LCS* to help to solve a pattern matching problem.

Acknowledgement. This contribution was supported by the VEGA (Slovak Scientific Grant Agency) of the Ministry of Education of the Slovak Republic (ME SR) and of the Slovak Academy of Sciences (SAS) under the contract No. VEGA 1/0553/12.

References

1. Abraham, A.: Hybrid soft computing and applications. International Journal of Computational Intelligence and Applications 8(1), 5–7 (2009)
2. Bull, L.: Learning classifier systems: A brief introduction. In: Applications of Learning Classifier Systems, p. 14. Springer (2004)
3. Bull, L., Kovacs, T.: Foundations of learning classifier systems. STUDFUZZ. Springer (2005)
4. Butz, M.V.: Rule-Based Evolutionary Online Learning Systems: A Principled Approach to LCS Analysis and Design. STUDFUZZ, vol. 109. Springer (2006)
5. Corchado, A., Arroyo, A., Tricio, V.: Soft computing models to identify typical meteorological days. Logic Journal of the IGPL 19(2), 373–383 (2011)
6. Drugowitsch, J.: Design and Analysis of Learning Classifier Systems: A Probabilistic Approach. SCI. Springer (2008)
7. Halavati, R., Shouraki, S.B., Lotfi, S., Esfandiar, P.: Symbiotic evolution of rule based classifier systems. International Journal on Artificial Intelligence Tools 18(1), 1–16 (2009)
8. Harmon, M., Harmon, S.: Reinforcement learning: A tutorial (1996),
   http://www.nbu.bg/cogs/events/2000/Readings/Petrov/
   rltutorial.pdf
9. Holland, J.H.: Adaptation in Natural and Artificial Systems. The University of Michigan Press, Ann Arbor (1975)
10. Holland, J.H.: Adaptation in Natural and Artificial Systems: An Introductory Analysis with Applications to Biology, Control and Artificial Intelligence. MIT Press, Cambridge (1992)
11. Kovacs, T., Llorà, X., Takadama, K., Lanzi, P.L., Stolzmann, W., Wilson, S.W. (eds.): IWLCS 2003-2005. LNCS (LNAI), vol. 4399. Springer, Heidelberg (2007)
12. Kriesel, D.: A Brief Introduction to Neural Networks, Zeta version (2007),
   http://www.dkriesel.com
13. Lanzi, P.L.: Learning classifier systems: then and now. Evolutionary Intelligence 1(1), 63–82 (2008)
14. Lanzi, P.L., Stolzmann, W., Wilson, S.W. (eds.): Learning Classifier Systems: From Foundations to Applications. LNCS (LNAI), vol. 1813. Springer, Heidelberg (2000)

15. Ruta, D., Gabrys, B.: An Overview of Classifier Fusion Methods (2000)
16. Sedano, J., Curiel, L., Corchado, E., de la Cal, E., Villar, J.R.: A soft computing method for detecting lifetime building thermal insulation failures. Integrated Computer-Aided Engineering 17(2), 103–115 (2010)
17. Sigaud, O., Wilson, S.W.: Learning classifier systems: A survey. Soft Computing 11(11), 1065–1078 (2007)
18. Smith, M.: Neural Networks for Statistical Modeling. Thomson Learning (1993)
19. Sutton, R.S., Barto, A.G.: Reinforcement learning: an introduction. In: Adaptive Computation and Machine Learning. MIT Press (1998)
20. Watkins, C.J.C.H., Dayan, P.: Q-learning. Machine Learning 8(3-4), 279–292 (1992), http://jmvidal.cse.sc.edu/library/watkins92a.pdf
21. Zadeh, L.A.: Fuzzy logic, neural networks, and soft computing. Communication of the ACM 37(3), 77–84 (1994)

Local Model of the Air Quality on the Basis of Rough Sets Theory

Filip Mezera and Jiří Křupka

Institute of System Engineering and Informatics, Faculty of Economics and Administration,
University of Pardubice, Studentská 84, 532 10 Pardubice, Czech Republic
st5360@student.upce.cz, jiri.krupka@upce.cz

Abstract. This article deals with the air quality modelling in two selected localities in the Czech Republic (CR). Data for the modelling were gained from the public sources. Primary source was the data server Czech Hydro Meteorological Institute (CHMI). Rough set theory (RST), Decision Trees (DTs) and Neutral Networks (NNs) were used for the analysis and the results comparison. At the end of the article there is the possible usage of the outputs of the models described. Outputs can help with the health protection of the inhabitants through the regulations set by the public administration authorities.

1 Introduction

Risks connected to the polluted air are the ones of the main environmental dangers [26,31], which are solved not only by regions, countries, but also by international organizations. This article deals with the synthesis and analysis of the air quality model in the selected localities of Czech Republic (CR). The model is aimed at dust particles (PM10) and weather character in two selected localities CR – Pardubice and Ostrava's neighbourhood. PM10 were selected, because they create an important part of the air quality and then they also carry the dangers of respiration diseases. Above all, small children can suffer from asthma or chronic inflammation of the upper respiratory tract [19]. Dust particles also carry carcinogenic substances which make higher the cancer risk significantly [2,3,28,31].

Using of methods of artificial inteligence in the weather conditions has been described in many articles, for example in [4]. Problematic of the air quality modelling in the CR regions was solved in the years 2007-2011 within the National Programme of Research of Ministry of Environment CR "The environment and natural resources protection". It was a project no. SP/4i2/60/07 titled "Indicators for Valuation and Modelling of Interactions among Environment, Economics and Social Relations". Suggested models used with data [21,22] from mobile and stationary meteorology stations in Pardubice's neighbourhood. Theories of Neutral Networks (NNs), fuzzy sets (FSs), Rough Sets Theory (RST) and Decision Trees (DTs) were used for creating the classificatory models [21,22].

Particular situation in every place is dependent on local conditions [6]. There are three main sources which participate at the overall level of the air pollution (pollution). First of them is the pollution from the big stationary sources (such as heating plants, power stations, ironworks etc.). Overall level of the pollution from these

V. Snasel et al. (Eds.): SOCO Models in Industrial & Environmental Appl., AISC 188, pp. 277–286.

sources is considered to be quite stable. Second source is the local heating. This source is distinctly dependent on the type of fuel which is used in the selected locality and then also on the weather. In the time of the decreasing temperatures the consumption of fuel and also the amount of substances in the air increases. The last part is the vehicular traffic, which significantly fluctuate between the main rush hours [2]. The public administration authorities can react directly with the effective remedies to decrease the level of pollution. E.g. in the years 2006, 2008, 2009 and 2010 were announced "regulation states" for the Moravian-Silesian region connected to the smog in the locality. This problem was, however, struggled in the big part of CR on 15th Nov. 2011, including Pardubice's region. Smog situation is defined in the regulation §8 art. 1 of the Act [32] as „state of the exceptional air pollution, where the level of air pollution by the polluting substance exceed the particular limit set by the implemented regulation" explained, with reference to the article 1, that „the exceptional limit is thought to be the level of the air pollution, while exceeded, where is a risk of health harm or harm of the ecosystem in very short exposition time" (regulation §8 art. 2 of the Act). The limit is $50\mu gm^{-3}$.

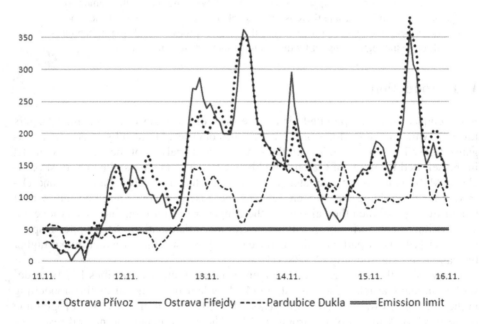

Fig. 1. Five-day concentration of PM10 in Ostrava (Přívoz and Fifejdy) and Pardubice (Dukla)

At Fig. 1 is possible to see that the limits in Pardubice are slightly exceeded (13th Nov. 2011). This did not lead to the signal of warning.

The warning is signalled unless the concentration of the suspended particles PM10 exceeds the limit 100 μgm^{-3} in average within the last 24 hours. In Ostrava (measuring stations Přívoz and Fifejdy) at that time reached values of 200 μgm^{-3} and more. Situation in Pardubice was quite worse during the afternoon on 12th Nov. 2011, when it momentarily exceeded the value of 140 μgm^{-3}. Values leading towards the warning signal were reached on 13th Nov. (3:00pm), when the average in the last 24 hours was

101.52 μgm^{-3}. Measured values are accessible with 3 hours delay on the CHMI portal [7]. But information about regulation state was publicly accessible with more than 18 hours delay. Pardubice's town hall published the smog announcement [18] the following day, i.e. 14th Nov. 2011. The next source of information for the inhabitants is regional, especially public broadcasting. That informed about the situation on 14th Nov. 2011 (10:36am) [5]. This situation does not correspond to the inhabitants needs. Mainly for seniors and young children exposed to the values higher than 100 μgm^{-3} (double as high as allowed norms) is harmfull. At the present delay of system of warning can reach up to two days, from the first contravention of level of 100 μgm^{-3}.

One of the options in an increase of the present state is the proceeding the air quality model, which can be used for prediction of the future state. It is possible to use the measured values of PM10, as well as the weather character. This has a huge impact to the air quality pollution [6]. If the model shows the high rate of accuracy, it could be possible to use it for informing the inhabitants.

2   Problem Formulation

Moravian-Silesian Region (Ostrava is its part) is the unique part in concentration of the large stationary sources of pollution (sources included to databases REZZO 1 and 2). This part of the pollution does not have any momentary differences. They form the local level of the background, thanks to which the states of the pollution are better identified. Due to the often occurrence of smog situations, Ostrava's neighbourhood is suitable for creation of the general model valid for the other CR regions. Air quality is monitored from the long-lasting point of view and in extreme situations can be regulated. Their portion of pollution is between 30 to 50 % in Ostrava [8]. Local sources (households, small polluters) have significant portion at the situation in the local neighbourhood and their regulation is very difficult. There is an assumption of correlation between the amount of giving off combustion products and the weather (especially the temperature). The portion at the pollution is 30-50 % in Ostrava, in the case of Pardubice it is estimated to 50 %. Other sources, such as traffic, can be regulated as well as momentarily and long-lastingly (travel by public transport for free, restrictions in entrance to problematic locations etc.). Their portion is between 15-40 % [2,3,8].

The weather plays the important role within the short-term air quality [8]. At the large sources we assume the independence in amount of harmful substances given off towards the weather [8]. At the small sources and traffic the correlation with the temperature characteristics can be estimated [2]. Moreover, the inverse character of the weather and wind speed will influence negatively the air quality. They cause that the harmful substances accumulate at the exposed places [10].

Considering the low number of measuring stations in the city with lower normal level of pollution [6], as they are Pardubice or Hradec Králové, the model must be counted with specific level of generalization. In every locality it is then possible to monitor certain differences of maximal pollution values in the time [13]. Considering levels of pollution, which are used for warning and regulation, the differences need not to be reflected in the model.

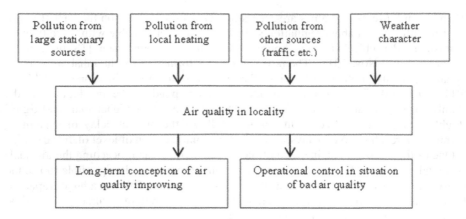

Fig. 2. Model of air quality in the selected location

2.1 Data Description and Pre-processing

The values measured at certain areas function as indicators of air quality. Data about the weather are gained from meteorological stations at the airports Ostrava – Mošnov and Pardubice. Moreover, the inputs from the stationary stations measuring the pollution were used. For Ostrava there were stations: Bartovice, Českobratrská, Fifejdy, Mariánské Hory, Poruba, Přívoz, and Zábřeh. In Pardubice there is one station measuring PM10 and that is at Dukla. Every factors and their impact to specific air pollution situation can be examined by modelling. In Pardubice the examination was done and outputs are in [12]. The examination was pursued with accurate data corresponding to the local conditions. Situation with excessive amount of PM10 is not typical in Pardubice. It is assume, that the model specified for Moravian-Silesian Region would be calibrated gradually to the local conditions. The model uses 17 input variables, which are described in the following Table 1.

For the dust particles classification the attributes l_1, l_2, m_1, m_2, ..., m_{11}, m_{12} are used. In the years 2006 to 2011 there was overall 451 observations selected. Training and testing sets contained 366 observations altogether from the year 2008. Validation set was created by the data from the years 2006, 2009, 2010 and 2011, when the air pollution limits were contravened. Data from the years 2007 and 2008 were not accessible.

When model was created, the results mentioned in [12] were evaluated. The variables, which are relevant for the PM10, were selected for next treatment. The variables l_1, l_2, c_1, c_2, m_9 and m_{10} were discarded, because they seemed to be a little relevant on the level of used generalization. From the parameters m_1 to m_6 were then derived 3 new variables d_1, d_2 and d_3, which record the occurrence of the inverse character of the weather. Then the scale of air quality mentioned by CHMI was also used [6], Table 2. The derived variable d_4 that describes the PM10 status in the time period t (i.e. minus 24 hours) was assigned by categories.

Table 1. Selected input variables of the model of the dust particles amount PM10

Name of variable	Variable	Type of variable
Station	l_1	set
Type of station	l_2	set
Day in the year	c_1	discrete
Day of the week	c_2	discrete
Average/Maximal/Minimal day temperature city	$m_1/m_2/m_3$	continuous
Average/Maximal/Minimal day temperature Lysá Hora	$m_4/m_5/m_6$	continuous
Average/ Maximal wind speed	m_7/m_8	continuous
Wind direction in the morning/in the afternoon	m_9/m_{10}	discrete
Air humidity/pressure	m_{11}/m_{12}	continuous
Difference between the average/maximal/minimal day temperature in the city and at Lysá Hora	$d_1/d_2/d_3$	continuous
PM10 in the time (last 24 hours average)	d_4	continuous

Table 2. Scale of air quality

Index	Air quality	PM10 (average value per 1h in μgm^{-3})
1	Very good	0 – 15
2	Good	>15 – 30
3	Satisfactory	>30 – 50
4	Convenient	>50 – 70
5	Bad	>70 – 150
6	Very bad	>150

3 Suggestion and Model Analysis

Design of the model at the basis of RST is shown in Fig. 3. RST [1,15,24,25] is based on searching common features from the data. It works with uncertainty within the upper and lower approximation and boundary region. Design of the model follows the work [12,13], which confirmed the ability of RST usage during the air quality modelling.

Categorized variables $\{k_1, k_2, k_3, k_4, k_5, k_6\}$ (Table 3) were derived from the former variables $\{m_7, m_8, d_1, m_{11}, m_{12}, d_4\}$ (Table 1) for the needs of RST usage thanks to the equidistant scaling, except k_6. Their values and amounts of categories were set experimentally or come out of the description of the phenomenon due to [6]. Estimated parameter v_y an average value of PM10 in the following 24 hours had only two categories. Results of RST model were compared with DTs (algorithms C5 a CRT) [14] and NNs. For algorithms DTs and NNs were used former continuous variables described in Table 1.

There were 23 rules generated for determining the output parameter of the model v_y by software Rough Sets Exploration System (RSES) [29]. The rules were generated using the algorithm Learning from Examples Module 2 (LEM2) [9]. Obtained rules show the importance of each attribute. The most important parameters are inversion k_3, wind speed k_1 and average amount of PM10 in the last 24 hours k_6. All ten rules

Data collection and data pre-processing:
- data set from CHMI portal
- data from meteorological stations
- data description and cleaning
- data deriving
- data categorization
- data partition (training, testing, validation)

RST model of the air quality
- training
- testing
- validation

Results evaluation and methods comparison
- Decision trees
- NNs

Fig. 3. Suggestion of the model via RST and its analysis

consist of these parameters which determine the negative value of v_y, i.e. $v_y = 1$. The most common negative rule is:

$$\text{IF } k_1=0 \text{ AND } k_3=1 \text{ AND } k_6=2 \text{ THEN } v_y = 1. \tag{1}$$

It means that, in case of the inverse character of the weather and slow wind speed, the air pollution stays concentrated at one place and makes the air quality worse in the selected locality.

Table 3. Categorized variables for RST

Attribute / Name of atribute	Values of attributes (0; 1; 2; 3; 4)
k_1 / Day average wind speed [ms^{-1}]	0 is <9; 1 is 9–13; 2 is 13–17; 3 is 17–21 and 4 is >21
k_2 / Maximal wind speed [ms^{-1}]	0 is < 9; 1 is 9–14; 2 is 14–19; 3 is 19–24; 4 is >24
k_3 / Inverse weather character [°C/100 above the sea level]	0 is <2; 1 is 2–5; 2 is >5
k_4 / Humidity [%]	0 is <66; 1 is 66–76; 2 is 76– 86; 3 is >86
k_5 / Pressure [hpsc m^{-1}]	0 is <1005; 1 is 1005–1012,5; 2 is 1012.5–1020; 3 is >1020
k_6 / Average amount of PM10 in the last 24 hours [μgm^{-3}]	0 is <22; 1 is 22–37; 2 is 37– 70; 3 is >70
v_y / Average amount of PM10 in the following 24 hours [μgm^{-3}]	0 is <70; 1 is >70

Accuracy of prediction reached 96.4 % (Table 4).

Table 4. Confusion matrix

		Predicted value	
		0	1
Detected value	0	65	2
	1	1	15

The result 96.4 % was compared with the results of DTs [13] and NNs. From the suggested DTs there were two, which have the accuracy higher than 90 %. One of them used the algorithm C5-boost and the second the algorithm CRT-boost. NNs with the good results were Multi Layer NN (MLP) and Radial Basis Function (RBF) NN. MPL NN has 6 neurons in the hidden layer and RBF NN 10 neurons. Comparison of the model results is in Table 5.

The verification of the results was done at the validation set (Table 5). Its specificity was much higher ratio of the days with negative air quality. This results in lower accuracy of the examined methods. Prediction ability of the RST and algorithm C5-boost and NNs methods is still quite high. Considering the complexity of the calculation and the following interpretation of the results, RST seems to be more robust and suitable for making a prediction model of the PM10 in the air.

There were not found the other considerable differences at the weather impact to the smog situation in Ostrava (accuracy 90.8 %) and Pardubice (accuracy 89.9 %). There is still the most important parameter the value of PM10, the inverse character of the weather and the average wind speed.

Table 5. Comparison of the results of the models at the basis of RST, DTs and NNs

Method	Accuracy (test set) [%]	Accuracy (validation set) [%]
RST	96.4	90.7
C5-boost	96.5	89.4
CRT-boost	93.8	84.7
MLP	96.2	91.3
RBF	95.2	90.8

If there were used the calculated rules at the basis of RST towards the predicted value v_y "Average amount of PM10 in the following 24 hours" during the warning against the harmful impact of the smog situation, it could be possible to shorten the length of the delay from the present average 36 hours to 3 hours. In that time are verifing data onto the CHMI server.

4 Conclusion

In this article there was introduced the problematic of the air quality and the solution of the warning system of the inhabitants against the negative impacts, which are connected to the stay at the exposed places. Nowadays, there is a average of 36 hours delay in the informing the public about the air pollution. This delay is too long, mainly with respect to the special groups (children, older people and people with respiration diseases). Creation the model, which identifies the bad quality of the air with the outlook of the following 24 hours, is the possibility how to prevent the selected groups of inhabitants from the health problems. Models are oriented to the amount of dust particles PM10 and the occurrence of smog in Ostrava and Pardubice's neighbourhood.

Firstly, it was necessary to find out, process and describe the data from the meteorological and air pollution stations. At the same time we have characterized both examined regions. RST work with categorized variables, therefore there were derived six variables. Data were then divided into three categories - training, testing and validating. Division of training and testing data was made by using the software IBM SPSS Modeler 14.2. For the rules calculation the software RSES 2.2.2 was used. The result rules were used for the prediction of "Average amount of PM10 in the following 24 hours".

Program RSES proved to be a suitable program for rules creation. The quality of the prediction at the tested set reached 96.4 %. Then there was a comparison of DTs and NNs made. The number of generated rules pointed out how robust RST is. During the validation by RST there were very good results reached (RST 90.7%, DTs 89.4% and MLP 91.3%).

Problems with air pollution belong to the important parts of the sustainable development of the regions from the long-lasting point of view. At the present trend of the larger cities development the solution and optimization of the air pollution will be the priority. The model of testing in other large cities of CR seems to be a good idea. With reference to the used methods the other high tech equipment is able to be used, e.g. rough-fuzzy access, case-based reasoning etc.

Acknowledgments. This work was supported by the project No. CZ.1.07/2.2.00/28.0327 Innovation and support of doctoral study program (INDOP), financed from EU and Czech Republic funds.

References

[1] Aviso, K.B., Tan, R.R., Culaba, A.B.: Application of Rough Sets for Environmental Decision Support in Industry. Clean Technologies and Environmental Policy 10, 53–66 (2007)

[2] Bellander, T., et al.: Using geographic information systems to assess individual historical exposure to air pollution from traffic and house heating in Stockholm. Environmental Health Perspectives 109(6), 363–369 (2001)

[3] Brauer, M., et al.: Air pollution from traffic and the development of respiratory infections and asthmatic and allergic symptoms in children. American Journal of Respiratory and Critical Care Medicine 166, 1092–1098 (2002)

[4] Corchado, E., Arroyo, A., Tricio, V.: Soft computing models to identify typical meteorological days. Logic Journal of the IGPL 19(2), 373–383 (2011)

[5] ČRo Pardubice (Český rozhlas): Pardubicko trápí smog (2011), http://www.rozhlas.cz/pardubice/zpravodajstvi/_zprava/975676 (accesed November 17, 2011)

[6] Český hydrometeorologický Ústav (2012), http://www.chmi.cz (accesed February 1, 2012)

[7] Český hydrometeorologický ústav: Data AIM v grafech (2012), http://pr-asv.chmi.cz/IskoAimDataView/faces/aimdatavw/viewChart.jsf (accesed February 1, 2012)

[8] Černikovský, L., Volný, R.: Znečištění ovzduší a jeho zdroje v Ostravě. In: Konference o kvalitě ovzduší v Ostravě (April 2, 2012), http://www.ostrava.cz/cs/o-meste/zivotni-prostredi/6.-konference-o-kvalite-ovzdusi-v-ostrave-2012 (accesed April 15, 2012)

[9] Grzymala-Busse, J.W., Wang, A.Z.: Modified algorithms LEM1 and LEM2 for Rule Induction from Data with Missing Attribute Values. In: 5th Int. Workshop on Rough Sets and Soft Computing (RSSC 1997) at the Third Joint Conference on Information Sciences (JCIS 1997), pp. 69–72. Research Triangle Park, NC (1997)

[10] Horák, et al.: Bilance emisí znečišťujících látek z malých zdrojů znečišťování se zaměřením na spalování tuhých paliv. Chemické Listy 105, 851–855 (2011)

[11] IBM SPSS Modeler 14.2 User's Guide (2012), ftp://ftp.software.ibm.com/software/analytics/spss/documentation/modeler/14.2/en/Users-Guide.pdf (accesed January 17, 2012)

[12] Jirava, P., Křupka, J., Kašparová, M.: System Modelling based on Rough and Rough-Fuzzy Approach. WSEAS Transactions on Information Science and Applications 10(5), 1438–1447 (2008)

[13] Kasparova, M., Krupka, J., Jirava, P.: Approaches to Air Quality Assessment in Locality of the Pardubice Region. In: 5th Int. Conf. Environmental Accounting Sustainable Development Indicators (EMAN 2009), Prague, Czech Repulbic, pp. 1–12 (2009)

[14] Kasparova, M., Krupka, J.: Air Quality Modelling by Decision Trees in the Czech Republic Locality. In: 8th WSEAS Int. Conf. on Applied Informatics and Communications (AIC 2008), pp. 196–201. WSEAS Press, Greece (2008)

[15] Komorowski, J., Pawlak, Z., Polkowski, L., Skowron, A.: Rough sets: A tutorial. In: Pal, S.K., Skowron, A. (eds.) Rough-Fuzzy Hybridization: A New Trend in Decision-Making, pp. 3–98. Springer, Singapore (1998)

[16] Kudo, Y., Murai, T.: A method of Generating Decision Rules in Object Oriented Rough Set Models. In: Rough Sets and Current Trends in Computing (RSCTC 2006), Kobe, Japan (2006)

[17] Maimon, O., Rokach, L.: Decomposition metodology for knowledge discovery and data mining. World Scientific Publishing, London (2005)

[18] Město Pardubice – Město je ohroženo smogem, byl vyhlášen signál upozornění (2011), http://www.pardubice.eu/urad/radnice/media/tiskove-zpravy/tz2011/tisk-111114.html (accesed November 19, 2011)

[19] Nařízení vlády 350/2002 Sb., kterým se stanovují imisní limity a podmínky a způsob sledování, posuzování, hodnocení a řízení kvality ovzduší, v platném znění (2002)

[20] Neri, M., et al.: Children's exposure to environmental pollutants and biomarkers of genetic damage: II. Results of a comprehensive literature search and meta-analysis. Mutation Research/Reviews in Mutation Research 612(1), 14–39 (2006)

[21] Olej, V., Obršálová, I., Křupka, J. (eds.): Modelling of selected areas of sustainable development by artificial intelligence and soft computing: regional level. Grada Publishing, The Czech Republic (2009)

[22] Olej, V., Obrsalova, I., Krupka, J. (eds.) Environmental Modeling for Sustainable Regional Development: System Approaches and Advanced Methods. IGI Global (2011)

[23] Pal, S.K., Skowron, A. (eds.): Rough-Fuzzy Hybridization: A New Trend in Decision Making. Springer, Singapore (1999)

[24] Pawlak, Z.: Rough Sets – Theoretical Aspects of Reasoning about Data. Kluwer, Boston (1991)

[25] Pawlak, Z.: Rough set approach to knowledge-based decision support. European Journal of Operational Research 99, 48–57 (1997)

[26] Portney, R.P., Stavins, R.N.: Public Policies for Enviromental Protection, Washington (2000)

[27] Rokach, L., Maimon, O.: Data mining with decision trees: Theory and applications. World Scientific Publishing, London (2008)

[28] Topinka, J., Binková, B., Mračková, G.: Influence of GSTM1 and NAT2 genotypes on placental DNA adducts in an environmentally exposed population. Environmental and Molecular Mutagenesis 30, 184–195 (1997), doi:10.1002/(SICI)1098-2280(1997)30:2< 184::AID-EM11>3.0.CO;2-9

[29] Skowron, A., Bazan, J., Szczuka, M.S., Wroblewski, J.: Rough Set Exploration System (version 2.2.2) (2009), http://logic.mimuw.edu.pl/~rses/ (accesed May 15, 2010)

[30] Stanczyk, U.: On Construction of Optimised Rough Set-based Classifier. Int. Journal of Mathematical Models and Methods in Applied Sciences 2, 533–542 (2008)

[31] WHO. IPCS: Environmental health criteria 210 - Principles for the assessment of risks to human health from exposure to chemicals, Geneva (1999)

[32] Zákon č. 86/2002 Sb., o ochraně ovzduší a o změně některých dalších zákonů (zákon o ochraně ovzduší), v platném znění

Modeling Forecast Uncertainty Using Fuzzy Clustering

Ashkan Zarnani[1] and Petr Musilek[1,2]

[1] Department of Electrical and Computer Engineering, University of Alberta, Edmonton, Alberta, Canada
{azarnani,petr.musilek}@ualberta.ca
[2] Department of Computer Science, VSB – Technical University of Ostrava, 17. Listopadu 15, Ostrava, Czech Republic

Abstract. Numerical Weather Prediction (NWP) systems are state-of-the-art atmospheric models that can provide forecasts of various weather attributes. These forecasts are used in many applications as critical inputs for planning and decision making. However, NWP systems cannot supply any information about the uncertainty of the forecasts as their immediate outputs. In this paper, we investigate the application of Fuzzy C-means clustering as a powerful soft computing technique to discover classes of weather situations that follow similar forecast uncertainty patterns. These patterns are then utilized by distribution fitting methods to obtain Prediction Intervals (PIs) that can express the expected accuracy of the NWP system outputs. Three years of weather forecast records were used in a set of experiments to empirically evaluate the applicability of the proposed approach and the accuracy of the computed PIs. Results confirm that the PIs generated by the proposed post-processing procedure have a higher skill compared to baseline methods.

1 Introduction

Although the deterministic interactions of physical simulations in Numerical weather prediction (NWP) models yield the expected values of different weather attributes in the mid-range future, such forecasts are uncertain due to the inaccuracy of initial conditions, low spatial resolution, and various simplifying assumptions [12][13]. Yet, such uncertainty information is not available in the immediate outputs of the system.

In many applications, it is desirable that forecasts be accompanied by the corresponding uncertainties. Information about forecast uncertainty can have important role in the planning and decision making processes that utilize the forecasts [2][8]. For instance, the expected accuracy of NWP temperature and wind speed forecasts can have crucial impact on the optimized operational planning and management of power grids using Dynamic Thermal Rating (DTR) systems which is the motivation of this study [6] [10]. The uncertainty of a forecast is typically presented using prediction intervals (PIs) that are accompanied by a percentage expressing the level of confidence, or expected nominal coverage rate (e.g., $T = [2°C, 14°C]$ conf = 95%) [2] [5].

A major common method to assess the uncertainty of weather forecasts is ensemble modeling. However, running multiple ensemble members to analyze the forecast accuracy can be very costly thus infeasible in many cases.

V. Snasel et al. (Eds.): SOCO Models in Industrial & Environmental Appl., AISC 188, pp. 287–296.
springerlink.com
© Springer-Verlag Berlin Heidelberg 2013

As an alternative, statistical post-processing methods can be applied on a result of an individual forecast. It is a well-known fact that the extent of forecast uncertainty varies with its context: the weather situation [13]. For example, low pressure systems are known to be less predictable than the more stable high pressure systems.

Soft computing techniques are increasingly applied in problems with large amount of data and uncertainty [3][14]. Lange *et al.* [9] used clustering over a historical performance data set of wind speed predictions and demonstrated a relationship between the forecast uncertainty and different meteorological situations. However, this analysis was not practically employed as a method of obtaining PIs for wind speed forecasts.

A practical application of weather classification to obtain PIs was proposed by Pinson et al. [17][18]. The authors used two predicted values of wind speed and wind power to categorize the historical forecast situations into four manually defined classes, each with different error distribution. The distribution of a new forecast case was then expected to follow the distribution of these classes based on an expert-based fuzzy membership definition. However, this method suffers from a major shortcoming of the manual grouping of predictions.

In this contribution, we use unsupervised learning over the historical performance of the NWP model to learn the patterns of forecast accuracy. To discover groups of forecast records that follow a similar prediction error distribution, Fuzzy C-means clustering algorithms is applied on a data set of past prediction accuracy records. Such objective-driven discovery of forecast situations is expected to find better groups compared to the manual definition of weather situations [15]. In addition, fuzzy association of forecast records with the discovered weather situations appears to be a more natural choice.

The process of evaluating of PIs forecasts, and probabilistic forecasts in general, is more complex compared to point forecasts. To empirically test the proposed approach, we apply the developed PI models to a large, real-world data set. We also develop a comprehensive PI evaluation framework. It not only covers all major measures from the PI evaluation literature, but also brings new insights to the PI verification process, leading to more accurate judgments.

The rest of this paper is organized as follows. Section 2 reviews the basic concepts and definitions of prediction intervals and forecast uncertainty modeling. Section 3 presents the proposed fuzzy clustering approach to discover forecast uncertainty patterns. The verification measures are described and analyzed in section 4. Experimental setup and results are provided in section 5. The final section 6 outlines main conclusions and indicates possible directions for future work.

2   Forecast Uncertainty and Prediction Intervals

The relation between the forecast $\hat{y}_t$ and its observation y_t can be described as:

$$y_t = \hat{y}_t + e_t, \tag{1}$$

i.e., each observation can be decomposed to the predicted value $\hat{y}_t$ for time t, and an error term e_t for the specific forecast instance.

Based on a probabilistic forecast, the cumulative distribution function (cdf) F_{y_t} is explicitly available. The prediction interval I_t^α is defined as $(1 - \alpha)$-confidence interval into which observation y_t is expected to fall with probability $1 - \alpha$. Therefore, it can be described as a range satisfying [5][15][17]:

$$P(y_t \in I_t^\alpha) = P(y_t \in [L_t^\alpha, U_t^\alpha]) = 1 - \alpha, \tag{2}$$

where L_t^α and U_t^α are, respectively, the lower and upper bound of prediction interval I_t^α defined by the corresponding distribution quantiles as:

$$L_t^\alpha = q_{y_t}^{\alpha_l=(\alpha/2)} = F_{y_t}^{-1}(\alpha/2), \quad U_t^\alpha = q_{y_t}^{\alpha_u=(1-\alpha/2)} = F_{y_t}^{-1}(1 - \alpha/2). \tag{3}$$

For instance, with $\alpha = 0.05$, the interval has a 95% confidence level bounded by quantiles $L_t^{0.05} = q_{y_t}^{0.025}$ and $U_t^{0.05} = q_{y_t}^{0.975}$, as $\alpha_l = 0.025$ and $\alpha_u = 0.975$.

Systematic characterization of forecast error can lead to modeling of forecast uncertainty for the target variable. This can be achieved by considering e_t in (1) as an instance of the random variable e, and associating F_t^e (or its estimate $\hat{F}_t^e$) as its cumulative distribution function, The corresponding estimated quantiles for the predictive distributions would hence be $\hat{L}_t^\alpha$ and $\hat{U}_t^\alpha$ [15][22][21]:

$$\hat{L}_t^\alpha = \hat{y}_t + \hat{q}_{e,t}^{(\alpha/2)}, \quad \hat{q}_{e,t}^{(\alpha/2)} = \hat{F}_t^{e-1}(\alpha/2), \tag{4}$$

$$\hat{U}_t^\alpha = \hat{y}_t + \hat{q}_{e,t}^{(1-\alpha/2)}, \quad \hat{q}_{e,t}^{(1-\alpha/2)} = \hat{F}_t^{e-1}(1 - \alpha/2), \tag{5}$$

where $\hat{q}_{e,t}^{(\alpha)}$ is the estimated α quantile of "error" based on the estimated forecast error distribution $\hat{f}_t^e$. The distribution of y_t, and hence the desired quantiles, are not explicitly known. Therefore, to find the $\hat{I}_t^\alpha$ prediction interval of y_t, the quantiles of e (i.e., the error associated with the forecast) are estimated and added to the predicted value $\hat{y}_t$ to obtain the lower and upper bounds for the original variable [17]. Thus, by finding quantiles over the forecast error distribution, one can find the quantiles over the forecast value that is expected to enclose the target observation.

3 Fuzzy C-means for Prediction Interval Modeling

A fine grouping of forecast situations can lead to clusters of predictions with a similar error behavior [13]. Such groupings can be found by clustering all available cases using the relative influential prediction variables as the features. Subsequently, each cluster can be independently analyzed by the method described in the previous section. This way, rather than considering all past errors together as a single set, the characteristics of error distribution within each cluster determines the prediction interval of that particular cluster.

In this study, we apply two different clustering algorithms to find optimal groupings of the NWP past forecasts: K-means [20] and Fuzzy C-means (FCM) clustering. K-means is a simple yet powerful clustering algorithm that has been used in many applications [19]. Consider a dataset $D = \{x_1, x_2, ..., x_N\}$, where each data point $x_j = \{x_j^1, x_j^2, ..., x_j^d\}$ represents d influential features (such as predicted temperature, wind speed and wind direction, precipitation, location, elevation, etc.), and N is the total number of available forecast cases. The algorithm finds the set of k cluster centers $C = \{c_1, c_2, \ldots, c_k\}$, and assigns a subset of points $D_i \in D$ to each cluster i. Each case j also has a forecast error e_j^y associated with the predictand y. Hence, each cluster has its own set of forecast errors for target y in the set E_i^y such that:

$$E_i^y = \{e_j^y | x_j \in D_i, j = 1..n^i\}, i = 1..k, \tag{6}$$

where n^i is the number of sample points in cluster i.

In the second stage of the process, a probability distribution $(\hat{F}_{i,t}^e)$ is fitted over each set of errors $E_i^y, i = 1..k$ to represent the forecast error characteristics of each cluster. We consider three fitting schemes: Gaussian distribution, Kernel Density Smoothing (using a Gaussian kernel) and Empirical distribution.

For instance, based on the Gaussian fitting method, each cluster i of forecast errors has its own estimated probability distribution described by its mean $\hat{\mu}_e^i$ and standard deviation $\hat{\sigma}_e^i$. When a new forecast x_{new} is made, the cluster to which it belongs can be identified by the nearest cluster center and boundaries of the corresponding prediction interval can be estimated using $\hat{F}_{i,t}^e$ instead of $\hat{F}_t^e$ in equations (4) and (5) which would provide $\hat{L}_i^\alpha$ and $\hat{U}_i^\alpha$ for each cluster independently.

Using K-means, each forecast case is assigned into a single cluster only. In a more natural approach, the forecast cases could be associated with various situations up to different degrees. This can be achieved using Fuzzy C-means algorithm that finds cluster patters based on fuzzy membership assumption of points over clusters [20]. The objective function of the clustering process is [1]:

$$J = argmin_C \sum_{i=1}^{N} \sum_{j=1}^{k} u_{ij}^m \| x_i - c_j \|^2, \tag{7}$$

where u_{ij} ($\sum_{l=1}^{k} u_{il} = 1$) represents the degree of membership of the point x_i in cluster j, and $m > 1$ is the fuzzification. The objective function can be minimized using gradient descent in an iterative process where the membership matrix and cluster centers are updated as follows:

$$u_{ij} = 1 / \sum_{l=1}^{k} \left(\frac{\|x_i - c_j\|}{\|x_i - c_l\|} \right)^{2/m-1} \tag{8}$$

As a result, a forecast case can simultaneously belong to more than one forecast situations. Many situations, such as transitions between different types of weather, can be better captured using this approach. Similarly to K-means, these fuzzy patterns of historical forecasts can be used to model forecast errors by fitting appropriate distributions. However, E_i^y is now a fuzzy set defined by membership of each error sample to the ith cluster (i.e. $u_{li}, l = 1..N$). Subsequently, the fitting methods must consider these membership values as the vector of sample weights in the process. Thus, error samples that have higher levels of association with a cluster have more impact on the corresponding error distribution.

In addition, any new forecast case x_{new} is now associated with all k clusters, but with different degrees of membership, $u_{new,j}, j = 1..k$. Hence, we need to devise a method to combine the error characteristics of different clusters, based on the new samples membership values. For this purpose, we apply a weighted opinion pool to consolidate the forecast error characteristics among the clusters. Because $\sum_{j=1}^{k} u_{new,j} = 1$, the weighted sum of the computed quantiles in each cluster based on the new forecast's levels of membership provides an intuitive method to obtain the final upper and lower quantiles:

$$\hat{L}_{new}^{\alpha} = \sum_{j=1}^{k} u_{new,j} \cdot \hat{L}_{j}^{\alpha}, \tag{9}$$

where $\hat{L}_{j}^{\alpha}$ represents the lower quantile of the prediction interval in the jth fuzzy cluster. The same method is used to compute the upper quantile.

4 Prediction Interval Verification

It is expected that, in a test setting, prediction interval forecasts will have empirical coverage of the observations as close as possible to their confidence level. This primary property of a PI forecaster M, called "reliability," is denoted Rel_M^{α}. [17]:

$$\bar{\xi}_{M}^{I\alpha} = \frac{1}{T}\sum_{i=1}^{T} \xi_{i}^{I\alpha}, \quad \text{where } \xi_{i}^{I} = \begin{cases} 1 & if \ \hat{L}_{\hat{y}_t}^{\alpha} \leq y_t \leq \hat{U}_{\hat{y}_t}^{\alpha} \\ 0 & empirical, \end{cases} \tag{10}$$

where T is the number of PIs in the evaluation data set, and ξ_{i}^{I} is an indicator of hit. ξ_{i}^{I} evaluates to one when the observation falls within the PI boundaries, otherwise it is set to zero, expressing a miss. Hence, Rel_M^{α} simply accounts for the difference between average hit of the forecasts (coverage rate) and the required nominal coverage defined for the PI.

A forecaster providing PIs with less vagueness, corresponding to the width of the PI, is clearly preferred. This leads to the second major measure of PI forecast quality called "sharpness" [11][17]:

$$Shp_M^{\alpha} = \overline{Width}_M^{\alpha} = \frac{1}{T}\sum_{i=1}^{T} Width_i^{\alpha} \tag{11}$$

where $Width_i^{\alpha} = \hat{U}_{\hat{y}_i}^{\alpha} - \hat{L}_{\hat{y}_i}^{\alpha}$ is the width of the ith prediction interval. Another important quality aspect of a PI computation method is its ability to provide intervals of variable width, depending on the forecast situation. A method with high "resolution" (Res_T^{α}) is capable of distinguishing forecasts with different amounts of uncertainty, and assign wider (high uncertainty) or narrower (low uncertainty) intervals accordingly. The standard deviation of PI widths is a natural choice to measure the method's resolution [15]:

$$Res_T^{\alpha} = \left[\frac{1}{T-1}\sum_{j=1}^{T}\left(\hat{U}_j^{\alpha} - \hat{L}_j^{\alpha} + Shp_M^{\alpha}\right)^2\right]^{\frac{1}{2}} \tag{12}$$

Having access to a single scalar summary measure of forecast quality is always attractive and useful for objective comparison of various methods. The most common prediction interval skill score is the Winkler's score [7], widely used as a conclusive objective evaluation measure for PI forecasting methods [11], [15], [18]. A comprehensive study performed by Gneiting and Raftery [4] prove that this score is "strictly proper" and would hence give the maximum score to a forecast that is actually the true belief of the forecaster and cannot be "hedged".

Using the notations defined above and the overall miss rate $(1 - \bar{\xi}_M^{I\alpha})$, the total score gained by a PI forecasting method M over the T cases in the test set can be expressed as:

$$SScore_M = T\left(-\frac{\alpha}{2}\overline{Width}_M^\alpha - (1 - \bar{\xi}_M^{I\alpha})\bar{\delta}_M^\alpha\right) = -T\left(\frac{\alpha}{2}\overline{Width}_M^\alpha + \bar{\Delta}_M^\alpha\right), \quad (13)$$

where $\bar{\delta}_M^\alpha$ is the average distance of an observation from the PI boundaries among the missed cases, and $\bar{\Delta}_M^\alpha$ is the average of this distance among all test cases owing to the fact that Δ_i is equal to zero for hit cases and δ_i for misses.

Due to availability of limited number of test cases in each cluster, the $SScore_i$ measurements incur some uncertainty as well. The width component of this score is constant in each cluster. However, the $\bar{\Delta}_M^{\alpha,j}$ measure's uncertainty (where $j=1..K$) decreases when evaluated by more test cases or when its sample values are closer to each other cluster j. To analyze the uncertainty of $\widehat{SScore}_M$, the one-sided confidence interval of the $\bar{\Delta}_M^{\alpha,j}$ measure with a specific confidence level is used to compute the skill score. After using this upper limit for all clusters, a lower limit on the $SScore_M$ with the desired confidence level can be determined:

$$P\left(\bar{\Delta}_M^{\alpha,j} < \bar{\Delta}_M^{\alpha,j^\beta}\right) = \beta \Rightarrow P\left(SScore_M > SScore_M^\beta\right) = \beta \quad (14)$$

where β is the desired confidence level over the measure as a percentage. Because $\bar{\Delta}_M^{\alpha,j}$ is a mean statistic, the Central Limit Theorem [21] can be used and hence its sampling distribution is essentially Gaussian. This leads to the following relation to obtain the one-sided confidence interval:

$$\bar{\Delta}_M^{\alpha,j^\beta} = \bar{\Delta}_M^{\alpha,j} + t(\beta, |T_j| - 1)\frac{s_{\Delta_M^{\alpha,j}}}{\sqrt{|T_j|}} \quad (15)$$

where $\bar{\Delta}_M^{\alpha,j}$ is the measured sample mean over the available sample test set, and $s_{\Delta_M^{\alpha,j}}$ is the sample standard deviation of individual $\Delta_i^{\alpha,j}$ values in cluster j. Hence, we can find the lower limit of the true $\bar{\Delta}_M^{\alpha,j}$ measure.

5   Experimental Evaluation

A hindcast data set of hourly predictions has been coupled with the respective observations of weather stations from the National Center for Atmospheric Research (NCAR) data repository. The WRF v3 simulations were run in three nested grids with resolutions of 10.8 km, 3.6 km and 1.2 km. The data set covers three years (2007, 2008 and 2009) of forecasts for two stations in BC. This data set contains about 51,000 records of historical performance of forecasts. There are total of 35 features available in this data set. The observations are used to derive the forecast error for temperature forecasts, and the described PI computation methods are applied to obtain prediction intervals for the forecasted temperature.

To investigate the role of influential variables and to select the optimal feature set in PI forecasts, 14 different subsets of the 25 available features were defined. These feature sets are combinations of BF1 (10 basic weather attributes), BF2 (a more complex feature set including attributes at different geopotential levels) and PG (derived features that represent the temporal gradient of surface pressure). The feature sets with letters PCx were obtained using Principal Component Analysis to decrease the dimensionality of the data to x. The results are based on three-fold cross-validation in which two years of data are used to train the PI model and the third year is set aside just to evaluate the trained model and calculate the quality measures of the resulting interval forecasts.

To compare the various proposed methods with baseline methods, some basic approaches are considered. The first baseline method is the *climatological* approach that considers all past error samples together (i.e. $K=1$) and computes the PI based on the fitted distribution. The second baseline method applies a manual grouping of the forecast situations based on the forecast *month*. In the evaluated approaches, the number of clusters was set in the range of 2 to 100, and the fuzzification parameter (m) in FCM was set to 1.2. Table 1 lists the PI quality details of the best performing setups for each algorithm. The results show that clustering methods considerably improve the skill of the PIs compared to the baseline methods and that FCM has a better performance compared to K-means.

Table 1. Top 4 setups from C-means and K-means along with baseline methods and detailed measures for temperature PIs based on Sscore$^{0.95}$ in 3-fold (yearly) cross validations

Algorithm	K	Features	Fit	Sharp-ness	Cov-erage	Cover-age$^{0.95}$	Reso-lution	RMS E	SScore	SScore$^{0.}_{95}$	Rank
FCM	45	BF2	Kernel	10.62	94.89	92.77	1.59	2.77	0.3220	0.3432	1
FCM	30	BF2PG	Kernel	10.91	94.93	93.26	1.65	2.86	0.3285	0.3452	2
FCM	50	BF2PG	Kernel	10.67	94.78	92.49	1.79	2.81	0.3231	0.3459	3
FCM	80	BF2PG	Kernel	10.25	94.58	91.53	1.74	2.71	0.3150	0.3460	4
K-means	50	BF2	Kernel	10.78	94.96	92.74	1.87	2.80	0.3254	0.3485	13
K-means	45	BF2	Kernel	10.86	94.89	92.78	1.87	2.83	0.3273	0.3492	15
K-means	40	BF2	Kernel	10.89	94.82	92.85	1.84	2.83	0.3303	0.3499	16
K-means	50	BF2PG	Kernel	10.94	94.87	92.60	2.20	2.87	0.3281	0.3506	18
Base-Month	12	Month	Kernel	12.21	95.12	94.10	1.91	3.12	0.3601	0.3704	943
Base-Clim.	1	-	Normal	12.17	94.78	94.49	0.00	3.11	0.3740	0.3774	1492

In Figure 1.a, a sample forecast error distribution is shown and the corresponding fitted kernel density distribution is also plotted. In the first stage of experiments the K-means algorithm was run with the different feature sets and fitting methods. Figures 1.b and 2.b show the box plots of the SScore$^{0.95}$ measure for these alternative choices. As can be seen, the Kernel fitting method and the BF2 feature sets can obtain PIs with higher skill. It must be noted here that when the measured *SScore* (and not its confidence bound) is used for evaluations, very large number of clusters (e.g. 200) would always achieve the best results. However, this is due to the fact that with such large values of K there would be very few test cases available to have a reliable measurement of the $\bar{\Delta}_M^{\alpha,j}$ statistic in individual clusters.

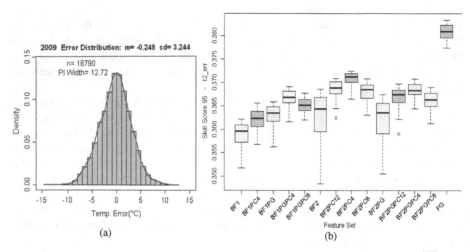

(a)

(b)

Fig. 1. (a) Forecast error distribution in 2009 and kernel fitted distribution (b) SScore$^{0.95}$ of the fourteen different feature sets using K-means in 3-fold (yearly) cross validations

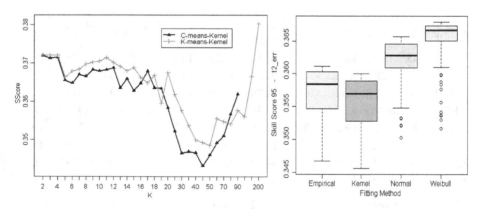

Fig. 2. (a) The trend of SScore$^{0.95}$ with increasing number of clusters (b) skill score of the four different fitting methods

Figure 2.a shows the trend of PI forecast skills for the best setups of K-means and FCM as the number of clusters is increasing. The curves also show the better performance of the Fuzzy C-means algorithm around $K=45$.

6 Conclusions

A new method is presented that can model forecast uncertainty from the historical performance of the NWP system and provide prediction intervals for new point forecasts. This is achieved using fuzzy clustering and density fitting methods over the prediction error records. The performance of this method was investigated through an experimental study employing an accurate evaluation framework. The availability of forecast uncertainty in the obtained PIs and their demonstrated higher skill compared to baseline methods suggests the effectiveness of this method. Due to the temporal nature of the weather attribute forecasts and their associated errors, application of time series analysis techniques in the PI forecasting methods can potentially improve the skill of the predictions in future work.

References

[1] Bezdek, J.C., Ehrlich, R., Full, W.: FCM: The fuzzy c-means clustering algorithm. Computers and Geosciences 10(2-3), 191–203 (1984)
[2] Chatfield, C.: Calculating Interval Forecasts. Journal of Business and Economics Statistics 11, 121–135 (1993)
[3] Corchado, E., Arroyo, A., Tricio, V.: Soft computing models to identify typical meteorological days. Logic Journal of the IGPL 19(2), 373–383 (2010)
[4] Gneiting, T., Raftery, A.E.: Strictly proper scoring rules, prediction and estimation. Journal of the American Statistical Association 102, 359–378 (2007)
[5] Hahn, G.J., Meeker, W.Q.: Statistical Intervals: A Guide for Practitioners. John Wiley, New York (1991)
[6] Hosek, J., Musilek, P., Lozowski, E., Pytlak, P.: Effect of time resolution of meteorological inputs on dynamic thermal rating calculations. IET GTD 5(9), 941–947 (2011)
[7] Jolliffe, I.T., Stephenson, D.B. (eds.): Forecast Verification: A Practitioner's Guide in Atmospheric Science. Wiley, Chichester (2003)
[8] Lange, M.: On the uncertainty of wind power predictions-Analysis of the fore-cast accuracy and statistical distribution of errors. Journal of Solar Energy Engineering 127, 177–184 (2005)
[9] Lange, M., Heinemann, D.: Relating the Uncertainty of Short-term Wind Speed Predictions to Meteorological Situations with Methods from Synoptic Climatology. In: Proceedings of the European Wind Energy Conference, EWEC, Madrid (2003)
[10] Khaki, M., Musilek, P., Heckenbergerova, J., Koval, D.: Electric Power System Cost/Loss Optimization Using Dynamic Thermal Rating and Linear Programming. In: EPEC 2010 (2010)
[11] Nielsen, H., Madsen, H., Nielsen, T.S.: Using quantile regression to extend an existing wind power forecasting system with probabilistic forecasts. Wind Energy 9, 95–108 (2006)
[12] Orrell, D., Smith, L., Barkmeijer, J., Palmer, T.: Model Error in Weather Forecasting. Nonlinear Proc. Geophys. 8, 357–371 (2001)
[13] Palmer, T.N.: Predicting Uncertainty in Forecasts of Weather and Climate. Reports on Progress in Physics 63, 71–116 (2000)
[14] Pedrycz, W.: Knowledge-Based Clustering: From Data to Information Granules. Wiley, Hoboken (2005)

[15] Pinson, P.: Estimation of the uncertainty in wind power forecasting. PhD Dissertation, Ecole des Mines de Paris (2006)

[16] Pinson, P., Juban, J., Kariniotakis, G.: On the quality and value of probabilistic forecasts of wind generation. In: Proceedings of the PMAPS Conference, IEEE Conference on Probabilistic Methods Applied to Power Systems, Stockholm, Sweden (2006)

[17] Pinson, P., Kariniotakis, G.: Conditional prediction intervals of wind power generation. IEEE Transactions on Power Systems 25(4), 1845–1856 (2010)

[18] Pinson, P., Nielsen, H.A., Møller, J.K., Madsen, H., Kariniotakis, G.N.: Nonparametric probabilistic forecasts of wind power: required properties and evaluation. Wind Energy 10, 497–516 (2007)

[19] Vejmelka, M., Musilek, P., Palus, M., Pelikan, E.: K-means clustering for problems with periodic attributes. International Journal of Pattern Recognition and Artificial Intelligence 23(4), 721–743 (2009)

[20] Xu, R., Wunsch II, D.: Survey of clustering algorithms. IEEE Transactions on Neural Networks 16(3), 645–678 (2005)

[21] Wilks, D.S.: Statistical Methods in the Atmospheric Sciences, 2nd edn. Academic Press, New York (2006)

[22] Wonnacott, T.H., Wonnacott, R.J.: Introductory Statistics. Wiley, New York (1990)

Proposing a New Method for Non-relative Imbalanced Dataset

Hamid Parvin, Sara Ansari, and Sajad Parvin

Nourabad Mamasani Branch, Islamic Azad University Nourabad Mamasani, Iran
{hamidparvin,s.ansari}@mamasaniiau.ac.ir

Abstract. A well-known domain in that it is highly likely for each exemplary dataset to be imbalanced is patient detection. In such systems there are many clients while a few of them are patient and the all others are healthy. So it is very common and likely to face an imbalanced dataset in such a system that is to detect a patient from various clients. In a breast cancer detection that is a special case of the mentioned systems, it is tried to discriminate the patient clients from healthy clients. It should be noted that the imbalanced shape of a dataset can be either *relative* or *non-relative*. The imbalanced shape of a dataset is *relative* where the mean number of samples is high in the minority class, but it is very less rather than the number of samples in the majority class. The imbalanced shape of a dataset is *non-relative* where the mean number of samples is low in the minority class. This paper presents an algorithm which is well-suited for and applicable to the field of *non-relative* imbalanced datasets. It is efficient in terms of both of the speed and the efficacy of learning. The experimental results show that the performance of the proposed algorithm outperforms some of the best methods in the literature.

Keywords: Imbalanced Learning, Decision Tree, Artificial Neural Networks, Breast Cancer Detection.

1 Introduction

In fact, each dataset that has an imbalanced distribution among the number of the data points in each of its classes can be considered as an imbalanced dataset. However in artificial intelligence communities, a dataset will be generally considered to be an imbalanced one if only if it has a very high-rated and sharp imbalanced distribution. We call this type of the mentioned imbalanced datasets, the imbalance between classes (e.g. consider the distribution of 10000:100 in a dataset with two classes where one class completely overshadows the other). Of course the imbalance concept is not dependent on the number of classes; it means that it is not only defined for or applicable to the datasets with two classes. It is highly likely that one faces an imbalance dataset having more than two classes. Thus in an imbalanced dataset it is required to use a classifier with a high accuracy in such a way that the minority class

V. Snasel et al. (Eds.): SOCO Models in Industrial & Environmental Appl., AISC 188, pp. 297–306.

detection is not affected by the majority class detection. It is obvious that the individual evaluation criteria such as overall accuracy or error rate do not provide sufficient information about the quality of learning in an imbalanced dataset.

Imbalanced shape of a dataset is called *intrinsic* where the nature of dataset source involves in being imbalanced. It should be noted that the imbalanced shape of a dataset can be either *relative* or *non-relative*. The imbalanced shape of a dataset is *relative* where the mean number of samples is high in the minority class, but it is very less rather than the number of samples in the majority class. The imbalanced shape of a dataset is *non-relative* where the mean number of samples is low in the minority class. This paper presents an algorithm which is well-suited for and applicable to the field of *non-relative* imbalanced datasets. It is efficient in terms of both of the speed and the efficacy of learning.

2 Backgrounds

A class of solutions to imbalanced datasets tries to apply some changes in dataset to be balanced and then uses a standard learning algorithm. Other class of solutions generally focuses on modifying the standard learning algorithms to be suited and adapted to learn in an imbalanced dataset [5]. In the first approach, there are two common ways: over-sampling and under-sampling. Random over-sampling method takes a set of samples from the minority class and then they are added to dataset. In fact, the number of samples in the minority class is enlarged in such a way that the number of data points in each class, either the minority class or the majority class, gets balanced. Alternatively there is another way to balance an imbalanced dataset named under-sampling method. Unlike the over-sampling method, the under-sampling method reduces a set of samples from the majority class in such a way that the number of data points in each class, either the minority class or the majority class, gets balanced. The over-fitting is the problem that challenges the over-sampling method. The concept losing is the main problem of the under-sampling method. An alternative to overcome the challenges is to turn to informed under-sampling methods. Two of the most well-known methods based on informed under-sampling are *EasyEnsemble* [2] and *BalanceCascade* [3]. Another example of the informed under-sampling methods is based on k-nearest neighbor [4].

In *EasyEnsemble* method it is tried to first produce many classifiers based on different runnings of the under-sampling method, and then to use them as an ensemble of classifiers. It is worthy to note that each mentioned classifier is produced by an AdaBoost mechanism. *EasyEnsemble* is an unsupervised strategy since it uses an independent random sampling with replacement in applying the under-sampling method. *BalanceCascade* method is very similar to *EasyEnsemble* method. *BalanceCascade* explores the sampling in a supervised manner. In *BalanceCascade* method it is tried to iteratively produce a classifier so as to improve the false positive rate of previously produced classifiers.

According to the research findings in the field of imbalanced learning, the criteria employed for assessing the quality of learning of a classifier in an imbalanced dataset are completely different from the common criteria used for evaluating the quality of learning of a classifier in a common dataset. So it is necessary to discuss the evaluation criteria suitable in the field of imbalanced learning. This section explains the approach how to assess the effectiveness of a model in learning of an imbalanced dataset. The common conventional measures to assess a classifier quality in learning of a dataset are the accuracy measure and the error rate measure. These criteria are used for a simple description of a learner (classifier) performance on a dataset but they are not suitable for imbalanced datasets.

Fig. 1 depicts the confusion matrix. In the confusion matrix the *True Positives* are the data points in dataset that have been assigned by classifier to the minority class (the patient class) and they really belong to the minority class. The *False Positives* are the data points in dataset that have been assigned by classifier to the minority class while they really belong to the majority class (the healthy class). The *False Negatives* are the data points in dataset that have been assigned by classifier to the majority class while they really belong to the minority class. The *True Negatives* are the data points in dataset that have been assigned by classifier to the majority class and they really belong to the majority class.

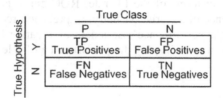

Fig. 1. Confusion Matrix

The performance criteria defined on the imbalanced datasets should be based on the mentioned confusion matrix to be unbiased to the majority class. Studying the confusion matrix makes it clear that the first column shows the number of positive samples (the number of samples in the minority class) and second column shows the number of negative samples (the number of samples in the majority class). It is also clear that the first row shows the number of the samples that classifier recognizes them as the minority class and the second row shows the number of the samples that classifier recognizes them as the majority class. Columns show the distribution of class samples. Indeed each metric using them simultaneously can't be free of sensitivity to class imbalancement. For example accuracy uses both columns and so it is sensitive to imbalancement, i.e. by changing the distribution of the number of data points of the classes of dataset the metric changes while the performance does not change. Some measures which are adjusted for evaluating the learning quality of a classifier at an imbalanced dataset are: (imbalanced) accuracy, precision, recall, F-measure and G-mean [1]. The accuracy of a classifier at an imbalanced dataset is obtained by equation 1.

$$Accuracy = \frac{TP + TN}{TP + TN + FP + FN} \qquad (1)$$

where *TP*, *TN*, *FP* and *FN* stand respectively for the number of *True Positives*, the number of *True Negatives*, the number of *False Positives* and the number of *False Negatives*. The precision is obtained by equation 2.

$$precision = \frac{TP}{TP + FP} \qquad (2)$$

where *TP* and *FP* are the same as equation 1. The recall is obtained by equation 3.

$$recall = \frac{TP}{TP + FN} \qquad (3)$$

where *TP* and *FN* are the same as equation 1. The F-measure is obtained by equation 4.

$$F - measure = 2 * \frac{precision * recall}{precision + recall} \qquad (4)$$

Evaluation based on receiver operating characteristic (ROC) curves, uses two criteria of the two single columns, TP rate and FP rate, of the Fig. 1 and draws a graph depicting the TP rate in terms of the FP rate. ROC curve is a powerful method to evaluate the performance of a learner visually. In precision-recall chart, one could get more information on the performance assessment of a learner [1]. These charts can be considered as the best way to display the performance of a learner in an imbalanced application.

3   Proposed Method

The structure of the proposed algorithm is similar to *EasyEnsemble*. The proposed algorithm initially takes a number of sub-samplings from the majority class with the size of the minority class. Considering each of the sub-sampled data from the majority class in addition to the data of the minority class as a temporal dataset, a decision tree or a multilayer perceptron is trained over the temporal dataset. Finally, all classifiers jointly work as an ensemble. Pseudo code of the proposed ModifiedBagging algorithm is presented in Fig. 2. Like Bagging, Boosting is another meta-algorithm in data mining that is more capable of learning hard problems. The main idea behind Boosting like Bagging is to learn a problem by a set of weak learners and then to create a single strong learner. A weak learner is defined to be a classifier which is only slightly correlated with the true classification or labels; it can label examples better than random guessing. In contrast, a strong learner is a classifier that is arbitrarily well-correlated with the true classification [12]. To complete our conclusion, the ModifiedBoosting is proposed based on the main Boosting algorithm proposed by Schapire [12]. Pseudo code of the proposed ModifiedBoosting is presented in Fig. 3. The proposed ModifiedBoosting is just like the proposed ModifiedBagging, except the majority class is subsampled based on a policy that the error-prone examples have more chance in subsequent samplings.

The ModifiedBagging algorithm pseudo code.
1. Input: A set of minority class examples $\mathcal{S}_{\emptyset|}$, a set of majority class examples $\mathcal{S}_{\emptyset-\emptyset}$, $\clubsuit\mathcal{S}_{\emptyset|}\clubsuit < \clubsuit\mathcal{S}_{\emptyset-\emptyset}\clubsuit$, the number of subsets, T to sample from $\mathcal{S}_{\emptyset-\emptyset}$
2. $i \Rightarrow 0$
3. **repeat**
4. $i \Rightarrow i+1$
5. Randomly sample a subset $\mathcal{E}i$ from $\mathcal{S}_{\emptyset-\emptyset}$, $\clubsuit\mathcal{E}i\clubsuit = \clubsuit\mathcal{S}_{\emptyset|}\clubsuit \Leftrightarrow \mathcal{S}_j \mathcal{S}_{\emptyset|} \cup \mathcal{E}_j$
6. Learn C_i on S_j. C_i is a simple classifier
7. ▷ **until** $i = T$
8. Output: An ensemble $\{C_i | 1 \ i \ T\}$

Fig. 2. Pseudo code of the proposed ModifiedBagging

Although, as it was mentioned previously, there are many algorithms to deal with learning at imbalanced datasets, this paper only focuses to handle under-sampling approaches. In this group of algorithms, the two of the best algorithms are considered to be *BalanceCascade* and *EasyEnsemble*. It is worthy to be mentioned that the second is one example of informed under-sampling methods [2-3]. As it has been shown [3], these two algorithms absolutely dominate other methods. Their superiority is in terms of both learning efficiency and training speed.

The ModifiedBoosting algorithm pseudo code.
9. Input: A set of minority class examples $\mathcal{S}_{\emptyset|}$, a set of majority class examples $\mathcal{S}_{\emptyset-\emptyset}$, $\clubsuit\mathcal{S}_{\emptyset|}\clubsuit < \clubsuit\mathcal{S}_{\emptyset-\emptyset}\clubsuit$, the number of subsets, T to sample from $\mathcal{S}_{\emptyset-\emptyset}$
10. $i \Rightarrow 0$
11. $W(j)=1, \forall\ j \in [1,T]$ "S_{min}
12. **repeat**
13. $i \Rightarrow i+1$
14. $P(j)=W(j)/sum(W), \forall\ j \in [1,T]$
15. Using P randomly sample a subset $\mathcal{E}i$ from $\mathcal{S}_{\emptyset-\emptyset}$, $\clubsuit\mathcal{E}i\clubsuit = \clubsuit\mathcal{S}_{\emptyset|}\clubsuit \Leftrightarrow \mathcal{S}_j \mathcal{S}$ $\mathcal{S}_{\emptyset|} \cup \mathcal{E}_j$
16. Learn C_i on S_j. C_i is a simple classifier
17. Test($\{C_i | 1 \ i \ T\}$, S_{max})
18. $W(j)=W(j)*2, \forall\ j$ that is misclassified by ensemble $\{C_i | 1 \ i \ T\}$
19. $W(j)=W(j)/2, \forall\ j$ that is misclassified by ensemble $\{C_i | 1 \ i \ T\}$
20. ▷ **until** $i = T$
21. Output: An ensemble $\{C_i | 1 \ i \ T\}$

Fig. 3. Pseudo code of the proposed ModifiedBoosting

On the other hand the algorithms of *BalanceCascade* and *EasyEnsemble* are very similar to the proposed algorithm. Therefore, since the two algorithms in terms of the structure are very similar to the proposed algorithm, and they also dominate other methods, the proposed method is compared to only these two methods in this paper.

The difference between proposed algorithm and *EasyEnsemble* is the main reason of its superiority. The difference is hidden in section 6 of the pseudo code. *EasyEnsemble* uses an AdaBoost classifier ensemble as learner [5]. Using a complex classification system similar to AdaBoost ensemble, not only causes a lot of overhead time for learning, but actually it lacks any justification. It is because after producing classifiers, C_i, voting mechanism is employed. So there is no justification for hierarchically voting in classification part, especially when the minority class has very little data points and hierarchical voting causes sub-sampling from the minority class; it means that hierarchical voting causes to lose the concepts of the minority class. So it is highly likely that the classifiers are not trained properly in the AdaBoost ensemble algorithm due to the small number of samples in the minority class.

The difference between the proposed algorithm and *BalanceCascade* is even more obvious. All differences between the proposed algorithm and *EasyEnsemble* mentioned in the previous section are also differences between the proposed algorithm and *BalanceCascade*. There are also some new differences between the proposed algorithm and *BalanceCascade*. For example, *BalanceCascade* tries to iteratively produce an AdaBoost so as to improve the *FP* of previously produced classifiers. It is again highly likely that the classifiers do not train properly in the AdaBoost algorithm due to the small number of samples in the minority class.

4   Experimental Results

This section evaluates the results of applying the proposed framework on a real imbalanced dataset of breast cancer patients. Dataset has been collected from some real clients of Bidgol-Aran city's hospital [6]. Dataset includes 386 clients. While 17 cases have breast cancer, the rest 369 cases have been healthy. 26 features extracted from each client that the most of them almost belong to the nominal ones. The nominal features to be used in any MLP are first converted to numerical features. It means that if feature A has 4 distinct values, say, <A_1, A_2, A_3, A_4>, we consider values A_1, A_2, A_3, A_4 respectively equal to 1, 2, 3, 4. After the coding phase, each feature is normalized into interval [0-1] just for usages in MLPs. The normalizing relations can be calculated by equation 5.

$$nf_{x,i} = \frac{f_{x,i}}{\max_y(f_{y,i}) - \min_y(f_{y,i})} \tag{5}$$

where $f_{x,i}$ stands for ith feature of xth data point and $nf_{x,i}$ stands for ith normalized feature of xth data point. In the paper, multilayer perceptron and decision tree are used as base classifier. We use multilayer perceptrons with 2 hidden layers including respectively 10 and 5 neurons in the hidden layer 1 and 2, as the base simple classifier. All of decision trees have used in this research employ Gini criterion as decision tree evaluation metric. Parameter Gini criterion for decision tree is set to two.

The classifiers' parameters are kept fixed during all experiments. It is important to be mentioned that type of all classifiers in the algorithms are kept fixed to either only decision tree or only multilayer perceptron. It means that all classifiers are considered as multilayer perceptron in the first experiments. After that the same experiments are taken by substituting all multilayer perceptrons with decision trees.

To find out how a classifier has learned over the mentioned imbalanced dataset, we always use *leave-one-out cross-validation* technique. First four columns of Table 1 show the quality of learning of different simple classification methods over the mentioned imbalanced dataset by *leave-one-out cross-validation* method in terms of different evaluation measures. As it is inferred from Table 1, although the accuracies of simple decision tree classifier and multilayer perceptron neural network classifier are very high, they do not have good performances at all. This is not something unexpected, because these classifiers assign each queried sample to the majority class. Consequently they hit very high accuracies. While their accuracies are good, they are unable to recognize patients. If one looks at Table 1, it will be clearly identified that the performances of the same classifiers enclosed in the proposed framework are significantly increased; while they still have satisfactory accuracies. As expected, using the decision tree as the base classifier can improve considerably the performance rather than using the multilayer perceptron as the base classifier.

Table 1. Performances of different simple methods obtained by leave-one-out method. MBG and MBT stand for ModifiedBaGging, ModifiedBoosTing respectively.

EC	DT	MLP	MBG of 1 DT	MBG of 1 MLP	MBG of 25 DT	MBG of 25 MLP	MBT of 25 DT	MBT of 25 MLP
TP	5.88	0.00	58.82	23.53	76.47	64.71	29.41	17.65
FP	0.00	0.00	23.30	32.95	20.17	32.95	1.99	15.63
TN	100	100	76.70	67.05	79.83	67.05	98.01	84.32
FN	94.12	100	41.18	76.47	23.53	35.29	70.59	82.35
Acc	95.66	95.39	75.88	65.04	79.67	66.94	94.85	81.30
Pre	100	∞ (50)	71.63	41.66	79.12	66.63	93.66	50.03
Rec	5.88	0.00	58.82	23.53	76.47	64.71	29.41	17.65
FM	7.14	0.00	64.60	30.07	77.77	65.66	44.76	26.09

Another comparison between the performances of the two versions of the proposed method when using each of the two simple learners (i.e. decision tree and multilayer perceptron) as the base classifier is presented in the columns 5 and 6 of Table 1. These experiments show that the accuracy of the proposed method is acceptable when we use decision tree as base classifier. It will also show if the whole data points of dataset are used to construct the classifiers of the final ensemble, performance of the final ensemble may be still poor to identify the examples of the minority class. Table 1 depicts this important fact. As it is raised from Table 1, the use of the ensemble without applying the proposed method to balance the training data, does not solve the problem. However, applying the proposed method along with the use of ensemble significantly increases the efficiency. The last 2 columns (7 and 8) of Table 1 represent the results obtained by the ModifiedBoosting. As it is inferred from the Table 1, the algorithm can't compete with the the ModifiedBagging. It is worthy to mention that we slide the number from 1 to 25 and choose the value when the

F-measure hits its best. The comparison confirms why the ModifiedBagging outperforms the *EasyEnsemble* and *BalanceCascade*.

A schematic comparison between performances of the two mentioned versions of the proposed method is presented in Fig. 4. ROC curve of the proposed method using decision tree learner as the base classifier is superior to the one using multilayer perceptron learner as base classifier.

According to Fig. 4 sliding FP from 0 to 1, readers will find that if a better cut choice on FP axis is taken in ROC curve the results can even be improved. However, this is not stable because after a while increment in FP does cause improvement in TP. The above tests indicate that the accuracy of the proposed method outperforms the simple classifiers and some ensemble methods. The other conclusion is the superiority of the proposed method that uses decision tree as the base classifier rather than one that uses multilayer perceptron neural network as the base classifier.

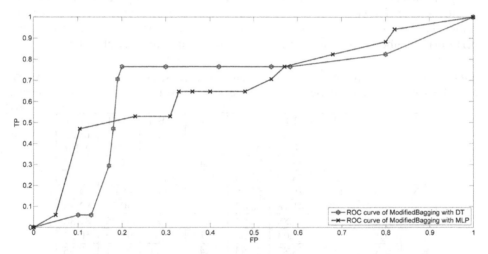

Fig. 4. ROC curve of the proposed ModifiedBagging with DT and MLP as base classifier

To demonstrate the efficacy of the proposed method in terms of the number of classifiers in the ensemble, please look at Fig. 5. In Fig. 5, the F-Measure of the proposed method in terms of the number of classifiers employed in the ensemble is depicted. All experimentations in Fig. 5 are averaged over 10 individual runs. The base classifier is chosen as decision tree. As it is inferred from Fig. 5, setting the number of classifiers to a value more than 23 does not affect much over F-Measure.

Now it is time to compare the proposed method with *EasyEnsemble* and *BalanceCascade* methods. By employing the two mentioned algorithms in the imbalanced dataset, any acceptable result is not again obtained according to Table 2. It is worthy to be mentioned that simple linear classifier used in reference [3] is used in both *EasyEnsemble* and *BalanceCascade* methods as base classifier. To reach the results of Table 2, *leave-one-out cross-validation* technique is used in all *EasyEnsemble*, *BalanceCascade* and *ModifiedBagging* methods. Comparing the proposed *ModifiedBagging* with *EasyEnsemble* and *BalanceCascade* in Table 2, we will reach the conclusion that the performances of the mentioned methods are weaker than the proposed *ModifiedBagging* method. So it is concluded that it is not needed to go for reinforcement methods in such an imbalanced dataset. Considering the higher time orders of the mentioned algorithms to learn in such severely imbalanced datasets, we can claim that the proposed method in terms of both efficiency of learning and speed of learning is superior. In addition, we have generally proposed a framework to achieve a similar learning model in severely imbalanced datasets.

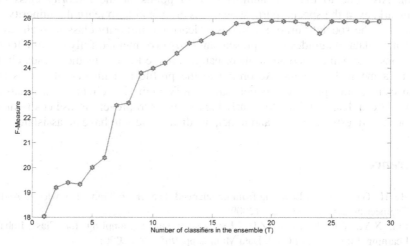

Fig. 5. F-Measure of ModifiedBagging with DT as base classifier in terms of number of classifiers

Perhaps the most important reason of failure in *EasyEnsemble* and *BalanceCascade* methods is hidden in severely imbalanced nature of the dataset. The reason of the well-performing the proposed method is its proper shape for learning small datasets. Consider when the data in the minority class is very low, the datasets created by *EasyEnsemble* and *BalanceCascade* are very low and consequently not suitable for learning of AdaBoost.

Table 2. Comparison of proposed method with EasyEnsemble and BalanceCascade methods

Evaluation Criterion	*EasyEnsemble* of 25 DTs	*BalanceCascade* 25 DTs	*ModifiedBagging* 25 DTs
TP	3/17=17.65	5/17=29.41	13/17=76.47
FP	31/352=8.81	43/352=12.22	71/352=20.17
TN	321/352=91.19	309/352=87.78	281/352=79.83
FN	14/17=82.35	12/17=70.59	4/17=23.53
Accuracy	324/369=87.80	314/369=85.09	294/369=79.67
Precision	66.70	70.44	79.13
Recall	17.65	29.41	76.47
F-Measure	27.91	41.50	77.78

5 Conclusions

In this paper a new method to learn in a severely imbalanced dataset where the number of data points in the minority class is very much less than the number of data points in the majority class is presented. This method is applied to an imbalanced breast cancer dataset. Inability of basic methods to learn in imbalanced spaces is also shown. Also due to the rare number of data points of the minority class in the benchmark, even the special-purpose methods are not able to learn the minority class. Inability of the special-purpose methods to learn the minority class in such severely imbalanced datasets guides us to present an innovative method fully suitable for these conditions. The main outcome of the research is in the field of medical research; to be used as a medical assistant. According to the profile and history of clients in the health centers, the proposed model can identify high risk clients in an automated manner. It can detect and treat an early breast cancer to prevent to use costly medical treatments and tests for clients and to help medical society to have an assistant.

References

[1] He, H., Garcia, E.A.: Learning from imbalanced data. IEEE Trans. Knowledge And Data Engineering 21(9), 1263–1284 (2009)
[2] Liu, X.Y., Wu, J., Zhou, Z.H.: Exploratory Under Sampling for Class Imbalance Learning. In: Proc. Int'l Conf. Data Mining, pp. 965–969 (2006)
[3] Liu, X.Y., Wu, J., Zhou, Z.H.: Exploratory Under sampling for Class-Imbalance Learning. IEEE Transactions on Systems, Man, and Cybernetics-part B: Cybernetics (2009)
[4] Zhang, J., Mani, I.: KNN Approach to Imbalanced Data Distributions: A Case Study Involving Information Extraction. In: Int'l Conf. Machine Learning (2003)
[5] Hamzei, M., Kangavari, M.R.: Learning from imbalanced data. Technical Report, Iran University of Sci. & Tech., Iran (2010)
[6] Minaei, F., Soleimanian, M., Kheirkhah, D.: Investigation the relationship between risk factors of occurrence of breast tumor in women, Aranobidgol, Iran (2009)
[7] Chawla, N.V., Bowyer, K.W., Hall, L.O., Kegelmeyer, W.P.: SMOTE: Synthetic Minority Over-Sampling Technique. J. Artificial Intelligence Research 16, 321–357 (2002)
[8] He, H., Bai, Y., Garcia, E.A., Li, S.: ADASYN: Adaptive Synthetic Sampling Approach for Imbalanced Learning. In: Proc. Int'l J. Conf. Neural Networks, pp. 1322–1328 (2008)
[9] Batista, G.E.A.P.A., Prati, R.C., Monard, M.C.: A Study of the Behavior of Several Methods for Balancing Machine Learning Training Data. ACM SIGKDD Explorations Newsletter 6(1), 20–29 (2004)
[10] Jo, T., Japkowicz, N.: Class Imbalances versus Small Disjuncts. ACM SIGKDD Explorations Newsletter 6(1), 40–49 (2004)
[11] Chawla, N.V., Lazarevic, A., Hall, L.O., Bowyer, K.W.: SMOTEBoost: Improving Prediction of the Minority Class in Boosting. In: Lavrač, N., Gamberger, D., Todorovski, L., Blockeel, H. (eds.) PKDD 2003. LNCS (LNAI), vol. 2838, pp. 107–119. Springer, Heidelberg (2003)
[12] Schapire, R.E.: The strength of weak learn ability. Machine Learning 5(2), 1971–1227 (1990)

Correlation between Speech Quality and Weather

Petr Blaha, Jan Rozhon, Miroslav Voznak, and Jan Skapa

VSB-Technical University of Ostrava
17. Listopadu 15, 70833 Ostrava-Poruba
Czech Republic
{petr.blaha,jan.rozhon,miroslav.voznak,jan.skapa}@vsb.cz

Abstract. This paper deals with an impact of atmospheric conditions on the speech quality in the GSM. We found out a correlation between weather conditions and the speech quality. The GSM technology is the most widely utilized communication standard which it is now coming to its bandwidth limitation especially in big cities and densely populated areas. Under such circumstances, even a minor weather change and rain could be a decisive factor causing changes in the quality of service. We have obtained both meteorological data and Mean Opinion Score value specifying the current speech quality in the GSM network. Those data are evaluated and compiled via statistical methods, whose accomplishment is dataset competent to be utilized by more advanced data mining methods. According to space distribution and fragmentation, our team has chosen set of suitable methods used to find data-mining, data analysis and correlation. As a computation result, our team found out the correlation between current rain density and the speech quality. Results from the MOS tests are reported, and an analysis of the obtained speech samples is presented. Outcomes are summarized and potential further directions for the continuation of research are discussed.

Keywords: GSM, Correlation, Mean Opinion Score, K-mean.

1 Introduction

One of the criterions to evaluate a success of human civilization is a possibility for fast communication, a handover of information from the source to the destination, so the final acting element on the end-point could be able to perform required function. Nowadays we are able to communicate within seconds from one continent to another and thanks to this advancement we are able to become mutually informed about important issues for reasonable prices.

Both these technologies could be utilized to provide a stable high-quality service. For the measurement of quality, the GSM/UMTS technologies provide the advantage of many algorithms which were developed for the IP based networks. This connection of specified platform and selected algorithms allow us to build and maintain programmable and flexible device able to sustain qualitative requirements of speech. By embracing this approach and creation of such a platform we would be allowed to measure influence of almost any signal interference from attenuation to high number of subscribers. The influence of the concerned weather conditions on the speech

V. Snasel et al. (Eds.): SOCO Models in Industrial & Environmental Appl., AISC 188, pp. 307–316.
springerlink.com   © Springer-Verlag Berlin Heidelberg 2013

quality in the GSM/UMTS networks is also measureable [2,4]. For counting the figures of dependencies modern methods of clustering are used. For counting of interdependencies we used proven method of clustering. These are working on principal division of data into groups in given space, where each attribute stands as an axis.

In further chapters we will try to describe the measuring mechanism and the whole platform as well as the algorithm used for evaluation of the collected results and will also try to find whether there is a correlation between the speech quality and the current weather conditions.

2 State of the Art

There are two main categories of speech quality assessment techniques – subjective and objective, the output of which is the Mean Opinion Score (MOS), which is a five degree scale for speech quality evaluation developed by ITU-T. The main goal of objective methods is as precise as possible estimation of the MOS value as it would be obtained by the subjective methods with the number of participants high enough to perform reasonable statistical analysis. We distinguish two separate sub-groups in the objective methods – Intrusive and Nonintrusive [4, 6]. The intrusive methods use the original voice sample as it has entered the communication chain and compares it with the degraded one as it has been outputted by this communication chain, the following list contains the most important intrusive algorithms.

- PSQM (Perceptual Speech Quality Measurement),
- PAMS (Perceptual Analysis Measurement System),
- PESQ (Perceptual Evaluation of Speech Quality),
- P.OLQA (Perceptual Objective Listening Quality Assessment).

From the mentioned algorithms PESQ is currently the most common one [1]. It combines the advantages of PAMS (robust temporal alignment techniques) and PSQM (exact sensual perception model) and is described in ITU-T's recommendation P.862.

Contrary to intrusive methods which need both the output (degraded) sample and the original sample, non-intrusive methods do not require the original sample [6]. This is the reason for a conclusion that they are more suitable to be applied in real time. Yet, since the original sample is not included, these methods frequently contain far more complex computation models. These data are further processed using a particular method, with a MOS value as the output. The method defined by ITU-T recommendation P.563 or a more recent computation method E-model defined by ITU-T recommendation G.107 are examples of such measurements [5].

Solutions are conformant with the specifications of the International Telecommunication Union only from some part therefore the measurements cannot be compared without the thorough knowledge of the used algorithms [3].

On the other hand the telecommunication union itself presents on its websites the simple implementation of one of the most advanced algorithms in the field of speech quality measurement. Regarding the speech quality in GSM and 3G networks mainly the first named company Optikom offers some services [7], but no one has performed the long term measurement with the focus on determination of weather influence on the speech quality in the GSM networks.

On basis of conducted researches of correlations between weather attributes and other variables, like aerosol pollutant [13] and human health [14] , there can be determined and defined a 'typical day' for geographic area. With knowledge of weather patterns in such area, we can say the call quality is enough for given 'typical' weather or it should be improved. Also there is high probability, that method used for findings of correlation between weather attributes and others could be used for determining between weather and call quality as well.

3 Testing Platform and Data Analysis

Our main goal was to create a testing platform that would be able to generate GSM calls automatically in regular intervals and simultaneously to log the actual weather conditions and analyze the calibrated speech sample in accordance to P.862. By using this platform we would be able to generate statistically significant amount of input data to perform the data mining analysis and find the possible correlation between one or multiple weather attributes and the obtained MOS value describing the quality of speech.

3.1 Data Testing and Collection

Testing data consists of two subsets. First subset is weather related data. Second subset is GSM voice quality parameters related data. Our goal was to uncover possible correlations, as such can be important for telecommunication companies. It would be clear, in what situations they should increase emitter's output power to maintain signal quality acceptable and when they could lower it to save operation cost.

Data related to weather were acquired by meteorological service of VŠB-Technical University of Ostrava. These data were measured during one year 2011, through all four weather seasons with the measurement period of five minutes. For calculation only the figures of temperature, humidity and rain data was taken even though there are many others unrelated to our computation, as for example wind speed. Intention was to find correlation between GSM and weather data only in certain periods of time due to bad measurable weather conditions in winter, when sensors were covered by snow, thus measuring of humidity was endangered. In some weeks weather remained unchanged, humidity was low, without a single drop of rain. Because of our limited data set of possible influencing factors, we decided to investigate possible correlation between weather and GSM data in alterative weather conditions. With gently erratic time progress, where an influence of external elements was minimised. Such weather conditions were from August to September of the year 2011.

Intercommunication devices were placed into building and on the roof of faculty hospital near by our university campus, approximately 200 meters away. There were no other stable sources of electric pollution causing noisiness affecting signal. As in hospital there are devices with various emitting frequency, but they are operating in short-periods only. Signals with high semantic error can be marked and removed as outliers.

Data related to GSM were measured in idle mode in real GSM network. For measuring,   there was developed an application which both synchronously and automatically saves current GSM data and also saves current weather data . GSM data consists of signal strength from one base transceiver station.

3.2 Data Normalization and Clearing

Before the data could be utilized for computation, cleaning and normalization procedures have to be done. The data from GSM represented by MOS are already clean and free from unexpected noise or peaks. Protocols used in telecommunication networks are designed to be reliable and to be able to transport good quality speech. Even if some packets are lost during communication, they can be computed from neighbour packets. It can be seen as a con for our measurement, as real adverse weather influences are shielded. On the other hand this is also an advantage because MOS quality speech parameter is tuned and artificially improved as much as possible, hence for the time being there is not so much to improve, so MOS data can be included into calculation as they are, without conditions and further normalization. Such calculated MOS outputs can be accepted as not having been affected by marginal error calculations.

At the opposite are data from weather meteorological station. In network of many interstate stations, all data from separate stations are collected and normalized together, accounting mathematical models of air flow and weather changes. Malfunction on one station was fixed, random peaks were decreased. Such normalization is not suitable for computing correlations between one station and one MOS parameter. Inner system in meteorological station allows collecting new data from sensors approximately every minute. To ensure value update on sensors output was done all values were checked in loop until any change was registered, new meteorological value was accepted and inserted into data table. To exclude influence of random peaks the method of moving average was chosen. To be able to compare GSM and weather data, their new values have to be computed in the closest possible moment. As there were MOS output once per five minutes, moving average was the best solution. Then median of five surrounding values was computed. Weather sample in the middle was closest to GSM sample on timeline (< 30 sec.), than two samples on left and two on right was taken. Data were represented as data table of GSM and weather data 1:1 in quantity for analysis. After that, sorting data from desired period between August and September by current rain was done. Lowest current rain rate considerable for processing was set to 2.1 mm/h. There were 63 rows remaining in data table, all of them from 15th to 28th August.

3.3 Elaboration Methods

3.3.1 Method Selection for Advanced Data Processing

On cleaned and normalized data were used several methods, from which were chosen just one. Methods were compared by importance of correlation found in meteorological data. K-Means method achieved the best results in the distribution of data into clusters. Other test methods for detecting correlations were:

Bayes classifier – Calculates the effect of individual attributes on other attribute. Not much data are needed for proper evaluation.

Logistic regression – Estimated probability of a phenomenon (dependent) on the basis of the facts, what are affecting the occurrence of the phenomenon (independent).

Unsupervised Learning – Solving the problem of finding hidden structures in data.

EM – Iterative method that attempts to find the maximum likelihood estimator of a parameter θ of a parametric probability distribution. Originally we expected to work with this method, but the results were not the best of all the methods for our data.

K –Means – Clustering method explained in section 3.3.3

3.3.2 Clustering

The simplest definition is shared among all and includes one fundamental concept: the grouping together of similar data items into clusters. These [obtained] clusters should reflect some mechanism at work in the domain from which samples or data points are drawn, a mechanism that causes some samples to bear a stronger resemblance to one another than they do to the remaining array.

Clustering algorithms are useful tools for data mining, compression, probability density estimation, and many other important tasks. Choosing k is often an ad hoc decision based on prior knowledge, assumptions, and practical experience. Choosing k is made more difficult when the data has many dimensions, even when clusters are well-separated.

Centre-based clustering algorithms usually assume that each cluster adheres to a unimodal distribution. With these methods, only one centre should be used to model each subset of data that follows a unimodal distribution. If multiple centres are used to describe data drawn from one mode, the centres are a needlessly complex description of the data, and in fact the multiple centres capture the truth about the subset less well than one centre.

Let $X \in R^\wedge mxn$ a set of data items representing a set of m points x_i in R_n. The goal is to partition X into K groups C_k such every data that belong to the same group are more "alike" than data in different groups. Each of the K groups is called a cluster. The result of the algorithm is an injective mapping $X \rightarrow C$ of data items X_i to clusters C_k dimensional vector x.

3.3.3 K-Means

K-means algorithm is one of the most well-known and widely used partitioning methods for clustering. It works in the following steps. First, it selects k objects from the dataset, each of which initially represents a cluster center. Each object is assigned to the cluster to which it is most similar, based on the distance between the object and the cluster center. Then the means of clusters are computed as the new cluster centers. The process iterates until the criterion function converges. A typical criterion function is the squared-error criterion.The k-means algorithm is given in Figure 1. For detailed description of k-means clustering, please refer to [12].

Let $S = \{X_1, X_2, \ldots, C_k\}$ be a dataset with n observations, each of which os p-dimensional. The objective in K-means clustering is to group these observations into categories $C_1, C_2, \ldots, C_K$ for given K, such that the objective function

$$O_K = \sum_{i=1}^{n} \sum_{k=1}^{K} I(X_i \in C_k)(X_i - \mu_k)'(X_i - \mu_k)$$

is minimized. Here μ_k represents the mean vector of observations from C_k, $\mu_k = \frac{1}{n_k}\Sigma_{i \in C_k} X_i$ where $n_k = |C_k|$ is the number of observations in C_k and $I(X \in C_k)$ is an indicator function specifying whether observation X belongs to the kth group. Further, note that the following $||x|| = \sqrt{x'x}$ denotes the Euclidian norm of p-dimensional vector x.

4 Results

First results giving overview of whole MOS variable were obtained from exploratory analysis. The lowest value of MOS variable is 1.776. This value is very low, on MOS degree scale it means poor. Highest value is 3.436, on MOS degree scale it means annoying. In theory MOS can achieve 5 at most, what is still far from our measurement, but is quite acceptable. Median is 3.1745, such can be sufficiently enough in common telephony. Standard skewness id -2.59004. It does mean, that data set does not come from normal distribution(according to standard skewness value lower than -2), what would tend to invalidate any statistical test regarding standard deviation. Graphical observation of those calculation can be seen on Fig. 2.

Using the mentioned K-means clustering method we have performed analysis of all available possible influencers of MOS value in GSM/UMTS environment. To be more specific, we explored the influence of the Current Temperature, Humidity, Rain, Dew Point, Wind Speed and Atmospheric Pressure.

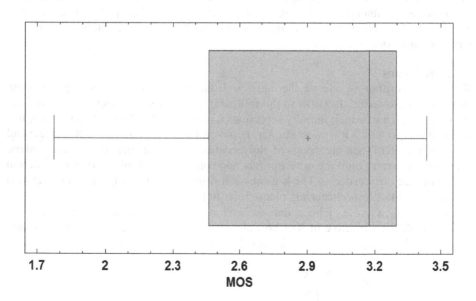

Fig. 2. Box and whisker plot of MOS variable

Through series of data mining operation we came to conclusion, that the self-correcting mechanisms implemented in the GSM/UMTS technology prevent call from being interfered by weather condition. The fact that there is no statistically significant relation between MOS and Humidity, which is the most influencing factor from the signal strength point of view, discouraged us from the thought that there might be a relation of any kind. Because of this we didn't expect that the other parameters connected directly to humidity like current rain rate would be influencing the voice quality with a measurable significance. However, this correlation can be found in the statistical data.

The Tab. 1 shows the actual probability of MOS value depending on the current rain density. Second row in the table can be seen as the most important as it tells us the following: if the current rain density is between 28.5 and 33.9 mm/h we can expect MOS to be lower than 2.61 with the probability of 64%. Other rows can be read similarly. These results in the following, with the increasing rain activity, the MOS value drops significantly. Especially when the high rain density is reached (greater than 5 mm/h) the MOS value drops to level, where the user can experience very bad speech quality and low comprehensibility. The last row is actually a conjunction of two separate clusters, which were identified as distinct areas by the algorithm. However they are mutually complementary and therefore they were combined into single row.

Table 1. The influence of current rain density on MOS

Column	Value	Favours	Relative Impact
CurRainRate [mm/h]	>= 33.899	< 2.61	100
CurRainRate [mm/h]	28.502 - 33.899	< 2.61	64
CurRainRate [mm/h]	15.133 - 28.502	2.61 – 2.919	100
CurRainRate [mm/h]	4.883 – 15.133	2.919– 3.292	100
CurRainRate [mm/h]	< 4.883	>=3.292	100

If we put that together with our knowledge of low humidity influence, we can state that this quality drop occurs at the beginning of the rain, while the air humidity is still low and the effect diminishes as soon as the humidity rises. This can be caused by the slow adaptation of the network or mobile station to the rain.

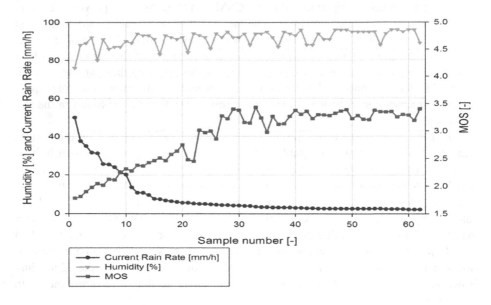

Fig. 3. Relation between MOS, Humidity and Current Rain Rate

The data of high significance (with rain density) can be interpreted as the sequence resulting in the chart on the Fig.3. Here we can see the data subset containing the samples with the significant rain density (60 samples with rain density higher than 2.2 mm per hour) ordered from the highest to the lowest rain density. As we can observe the humidity varies independently on the current rain density, however MOS factor is influenced measurably.

Fig. 3 clearly confirms the stated information about the rain influence on the quality of speech in GSM/UMTS environments. As already stated this is the only significant relation, therefore no further analysis makes sense at this point.

To complete the picture, it is necessary to state that the measured data were collected from August to November of this year (2011) and the total data pool contains about 20 000 data rows. Since the BTS station and all the other parts of the network, through which the signal traverse (MSC, IMS Core), are not under our control and therefore the information about their load is not possible to obtain, the statistical methods were used to eliminate the effect of variable load of these network elements using huge amount of input data.

5   Conclusion

Through long term measurement we have obtained both meteorological data and MOS values specifying the current speech quality in the GSM networks. By utilization of advanced techniques in data mining and data analysis, our team found the correlation between current rain density and the speech quality. This shows us a great decrease of speech quality in the earliest phases of the rain, where the humidity is still low. Particularly, we can observe a 50 percent decrease in measured MOS parameter when comparing sample results obtained during the heavy rain and those obtained during a mild shower. This decrease is reflected in the worsened speech quality by glitches in the speech, low comprehensibility and other communication difficulties. The reason for this behaviour can be found in the BTS transmitted power adjustment, which can modify the level in a range of up to 20dB [8,9].

The slow response of the correction procedure in the BTS can cause problems with the same nature as we have witnessed during our measurement. Other possible explanation includes the influence of the number of subscribers logged in the particular BTS at a particular time, but this would result in quality deterioration mainly during the busy hour and this behaviour was not observed. No other significant bond was found.

By performing this measurement, we have proven that the low cost measuring platform can be developed and used for the speech quality measurement in the GSM networks. We have successfully taken advantage of our team's knowledge in the IP telephony and transited this knowledge to cellular networks.

The greatest possible improvement of our method as we see it now is to perform the measurement during the whole year to have complete knowledge of the speech quality trends during all the possible weather conditions and to modify the testing platform by introducing OpenBTS solution, which would allow to gain the full control over the transmission chain.

Acknowledgement. This work was supported by the European Regional Development Fund in the IT4Innovations Centre of Excellence project (CZ.1.05/1.1.00/02.0070) and by the Development of human resources in research and development of latest soft computing methods and their application in practice project (CZ.1.07/2.3.00/20.0072) funded by Operational Programme Education for Competitiveness, co-financed by ESF and state budget of the Czech Republic.

References

1. ITU-T: P.862: Perceptual evaluation of speech quality (PESQ): An objective method for end-to-end speech quality assessment of narrow-band telephone networks and speech codecs (2001)
2. Rozhon, J., Voznak, M.: Development of a speech quality monitoring tool based on ITU-T P.862. In: Proceedings 34th International Conference on Telecommunications and Signal Processing, Budapest, Article number 6043771, pp. 62–66 (2011)
3. Fung, G.: A Comprehensive Overview of Basic Clustering Algorithms (2001)
4. Tomala, K., Macura, L., Voznak, M., Vychodil, J.: Monitoring the Quality of Speech in the Communication System BESIP. In: Proc. 35th International Conference on Telecommunication and Signal Processing, Prague, pp. 255–258 (2012)
5. ITU-T Recommendation G.107, The E-model: A computational model for use in transmission planning, Geneva (April 2009)
6. Voznak, M., Tomes, M., Vaclavikova, Z., Halas, M.: E-model Improvement for Speech Quality Evaluation Including Codecs Tandeming. In: Advances in Data Networks, Communications, Computers, Faro, pp. 119–124 (2010)

7. Opticom: Opticom quality testing products page,
   `http://www.opticom.de/products/product-overview.php`
8. Mouly, M., Pautet, M.-B.: The GSM System for Mobile Communications. Telecom Publishing (1992)
9. Saunders, S.R., Zavala, A.A.: Antennas and Propagation for Wireless Systems, 2nd edn. Wiley (2007)
10. Zhao, Y., Zhang, C., Zhang, S., Zhao, L.-W.: Adapting K-Means Algorithm for Discovering Clusters in Subspaces. In: Zhou, X., Li, J., Shen, H.T., Kitsuregawa, M., Zhang, Y. (eds.) APWeb 2006. LNCS, vol. 3841, pp. 53–62. Springer, Heidelberg (2006)
11. Lee, S.D., Kao, B., Cheng, R.: Reducing UK-means to k-means. In: ICDM Workshops, pp. 483–488. IEEE Computer Society (2007)
12. Hongbo, S., Yaqin, L.: Naïve Bayes vs. Support Vector Machine: Resilience to Missing Data, Shanxi University of Finance and Economics 030031 Taiyuan, China, shb710@163.com, liuyaqin2003@126.com
13. Corchado, E., Arroyo, A., Tricio, V.: Soft computing models to identify typical meteorological days. Logic Journal of the IGPL 19(2), 373–383 (2011)
14. Pengely, D., Cheng, C., Campbell, M.: Influence of Weather and Air Pollution on Mortality in Toronto, Toronto, Ontario (2005)

Genetic Algorithms Applied to Reverse Distribution Networks

Freitas A.R.R.[1], Silva V.M.R.[2], Guimarães F.G.[3], and Campelo F.[3]

[1] Programa de Pós-Graduação em Engenharia Elétrica - Universidade Federal de Minas Gerais,
Av. Antônio Carlos 6627, Belo Horizonte, MG, 31270-901 Brazil
alandefreitas@gmail.com
[2] Instituto de Computação, Universidade Federal Fluminense
Niterói, RJ, 24210-240, Brazil
victormrsilva@gmail.com
[3] Departamento de Engenharia Elétrica - Universidade Federal de Minas Gerais,
Av. Antônio Carlos 6627, Belo Horizonte, MG, 31270-901 Brazil
{fredericoguimaraes,fcampelo}@ufmg.br

Abstract. Reverse Distribution Networks are designed to plan the distribution of products from customers to manufacturers. In this paper, we study the problem with two-levels,with products transported from origination points to collection sites before being sent to a refurbishing site. The optimization of reverse distribution networks can reduce the costs of this reverse chain and help companies become more environmentally efficient. In this paper we describe heuristics for deciding locations, algorithms for defining routes, and problem-specific genetic operators. The results of a comparative analysis of 11 algorithms over 25 problem instances suggest that genetic algorithms hybridized with simplex routing algorithms were significantly better than the other approaches tested.

1 Introduction

When a manufacturer distributes a product, it is usually assumed that the distribution from producer to customers is the only chain that has to be optimized. However, there are many cases in which the products need to be returned to the manufacturer to be replaced, repaired, or recycled (e.g., defective or environmentally hazardous products). Reverse distribution networks are defined by the chain in which the products are returned to the manufacturer. It is an interesting problem as very often the cost of the reverse chain can overtake by many times the price of distributing the product to customers [16,18]. Besides, environmentally friendly products have become a marketing element for companies [25].

In this work we study a formulation for reverse logistics which is based on two levels [11]. In this area of reverse logistics, the first level represents collection sites, where the products can be temporarily stored until they are sent to refurbishing sites, represented by the second level. This formulation of the problem can be reduced to another NP-Complete problem, and heuristics have been developed to solve it. Besides the known heuristics for the problem, we study the possibilities of using genetic algorithms for the solution of this class of problems.

V. Snasel et al. (Eds.): SOCO Models in Industrial & Environmental Appl., AISC 188, pp. 317–326.
springerlink.com   © Springer-Verlag Berlin Heidelberg 2013

A model for the two-level reverse distribution network problem takes into account the cost, limits and capacities of each site as well as the costs of transferring the products. A solution for a given problem includes which facilities are open as well as the route for the products. Heuristics for deciding the open facilities, such as greedy heuristics or even genetic algorithms, usually work in parallel with algorithms that calculate the routes for the products.

This work is organized as follows: Section 2 provides a brief description of reverse distribution networks (RDNs), and in Section 3 a mathematical formulation for the two-level RDNs model is presented. Section 4 presents a number of heuristics used for the solution of different aspects of the problem. In Section 5 the proposed GA for the solution of the two-level RDN problem is discussed. Section 6 describes the experimental setup and the statistical analysis of the results, which shows that genetic algorithms performed significantly better than the heuristics on the test set used, particularly when hybridized with the simplex algorithms. Finally, Section 7 presents some final considerations and ideas for continuity.

2   Reverse Distribution Networks

In a distribution system, the reverse chain represents the return of products to the manufacturer, e.g., for repair or recycling. The formulation considered in this work focuses on product recall in which the products are initially located at outlets [11]. The products go from the customer to a collection site or straight to a refurbishing site, forming a two-level problem. Besides the cost of transporting the products, each collection point and each refurbishing site has a cost for being active. There are many variations of this problem, with different products or collection layers [15].

The problem has important implications for retailers who need to consider the reverse chain. Retailers have to be prepared for a number of expectations, risks and impacts that make it important to consider recommendations at the management level [9].

Many different models have been proposed for reverse distribution [7,2,24] and different solutions were proposed, such as heuristics [7,3,10,1,14], linear programming [23], evolutionary computation [12], and Markov chains [9]. For the formulation considered in this work [11], when there are no collection sites, the problem is reduced to the Capacitated Facilities Location Problem, which is NP-Complete [6]. The difficulty of finding an optimal solution for large instances in polynomial time justifies the use of heuristics for the problem.

3   Modeling the Problem

The general model for the problem [11] is described in this section. The objective and constraint functions use the following definitions:

- $I\{i/i\}$, $J\{j/j\}$, $K\{k/k\}$: origination, collection, and refurbishing sites, respectively;
- C_{ijk}: cost of transporting an unit from the origination site i to the refurbishing site k through the collection site j;

- F_j, G_k: cost of opening the collection site j or the refurbishing site k;
- a_i: number of products at the site i;
- B_j, D_k: maximum capacity of the collection site j or of the refurbishing site k;
- P_{min}, P_{max}: minimum and maximum number of collection sites to open;
- Q_{min}, Q_{max}: minimum and maximum number of refurbishing sites to open.

The objective and constraint functions for this problem are given in Equation (1).

$$
\begin{aligned}
\text{minimize } & \sum_i \sum_j \sum_k C_{ijk} a_i X_{ijk} + \sum_j F_j P_j + \sum_k G_k Q_k \\
\text{subject to } & \sum_j \sum_k X_{ijk} = 1 \text{ for all } i \\
& \sum_i \sum_k a_i X_{ijk} \leq B_j \text{ for all } j \\
& \sum_i \sum_j a_i X_{ijk} \leq D_k \text{ for all } k \\
& X_{ijk} \leq P_j \text{ for all } i,j,k \\
& X_{ijk} \leq Q_k \text{ for all } i,j,k \\
& P_{min} \leq \sum_j P_j \leq P_{max} \text{ for all } j \\
& Q_{min} \leq \sum_k Q_k \leq Q_{max} \text{ for all } j \\
& 0 \leq X_{ijk} \leq 1 \text{ for all } i,j,k \\
& P_j \in \{0,1\} \text{ for all } j \\
& Q_k \in \{0,1\} \text{ for all } k
\end{aligned} \tag{1}
$$

Each solution of the problem is defined by:

- $X_{ijk} \to$ fraction of units at originating site i that will be transported through the sites j and k ($j = 0$ is used to indicate that the products go straight to k)
- $P_j \to$ if the collection site j is open, $Q_k \to$ if the refurbishing site j is open.

4   Heuristics

The mathematical model for the two-level reverse distribution network problem defines costs, possible locations, capacity of each facility, number of products located in many points, maximum and minimum number of each sort of facility, and cost of transport from each origination site to each refurbishing site through each collection site. Different heuristics are described to decide the open facilities and to calculate the routes for the products.

4.1   Deciding Locations

A simple idea for deciding the open facilities is a greedy heuristic. Based on a greedy principle, we propose the following heuristic to generate solutions:

1. Rank order all collection sites P_j and refurbishing sites Q_k by their capacity/cost;
2. Sort an integer between p_{min} and p_{max} and save it as p_{goal}. Sort an integer between q_{min} and q_{max} and save it as q_{goal};
3. The p_{goal} collection sites with highest capacity/cost value are open. The q_{goal} collection sites with highest capacity/cost value are open.

This simple greedy heuristic can generate and explore solutions with different numbers of open facilities in the two levels of the model. The heuristic can be used by itself or as part of another meta-heuristic.

Another known method used for deciding the open facilities is Heuristics Concentration [20,11]. That can be done by running an heuristic a number of times in order to identify facilities which are worth further investigation.

Inspired by the heuristics concentration originally proposed for the p-median problem [20], a similar approach has been proposed for the reverse logistics problem [11]:

1. Random Selection: random selection of potential collection and refurbishing sites. For a number of iterations (100 in this work), a subset of size P_{max} of collection sites and Q_{max} refurbishing sites is randomly selected and the routing of products for these sites is solved to by another algorithm. All solutions are saved and the best solution is marked;
2. Heuristics Concentration: the sites most used in the best solutions are added to the best solution found in random selected phase in order to form a new solution and the problem is solved to optimality. This new solution is compared to the best solution from the random selection phase;
3. Heuristics Expansion: add each unused site to the best solution and solve the problem. If a better solution is found, remember this solution and its configuration, but leave the best solution unchanged. Check if any of the other unused sites give a better solution than the new solution found. Repeat this until all the unused sites are checked. Stop when no better solution is found.

4.2 Routing Algorithms

Given the open facilities for each level, we need to define the routes for the products. A greedy heuristic can also be used in this task, to find solutions of good quality in reasonable computing times:

- For each origin site to be examined:

   1. Among the valid routes for the products, find the one with lowest cost;
   2. Send as many products as possible through this route and update its capacity;
   3. If there are still products in the origination site, send them through the next best route. Otherwise, examine the next origin site.

This approach does not guarantee that the optimal route is found, but it has the advantage of being considerably less computationally expensive than exact methods.

Another solution for finding routes from origination to refurbishing sites would be to define a second mathematical model for the problem where the collection and refurbishing sites are considered part of the problem, for which the solution would be only the routes for the products. As we consider that the sites P and Q are no longer decision variables in the routing problem, a simplified formulation arises (2):

$$\text{minimize} \sum_i \sum_j \sum_k C_{ijk} a_i X_{ijk}$$

$$\text{subject to} \sum_j \sum_k X_{ijk} = 1 \text{ for all } i$$

$$\sum_i \sum_k a_i X_{ijk} \leq B_j \text{ for all } j > 0$$

$$\sum_i \sum_j a_i X_{ijk} \leq D_k \text{ for all } k \tag{2}$$

$$X_{ijk} \leq P_j \text{ for all } i, j, k$$

$$X_{ijk} \leq Q_k \text{ for all } i, j, k$$

$$0 \leq X_{ijk} \leq 1 \text{ for all } i, j, k$$

This formulation is a linear-programming problem that can be solved by a simplex algorithm to find the optimal values X_{ijk} in polynomial time.

5   Genetic Algorithm

Having all those heuristics as reference, we propose genetic algorithms (GA) that can evolve the solutions generated by the heuristics. In order to efficiently evolve the solutions, problem specific genetic operators are defined and the heuristics are used as approaches to generate new solutions.

In the proposed GAs, each individual is coded as two binary vectors, that represent the variables P, and Q (section 3). The constraints of the problem are all automatically satisfied by the genetic operators developed, which simplifies the solution of the problem by the GAs.

In the proposed approach, the GA searches for good P and Q values, while the heuristics presented in Section 4.2 are used to compute the value of X.

If a special case in which the capacities of the facilities are not enough to keep the products from the origin sites, the fitness of unfeasible solutions are scaled so that they are always worse than the feasible solutions.

The final fitness of the individuals is given by their rank [13]. A stochastic ranking selection [21] is used to decide which parents will generate children.

Besides the fitness functions, the genetic operators are also important to implicitly filter out undesirable solutions [8]. In this work we employ a crossover operator that keeps the open facilities that would probably lead to a good fitness. The number of open facilities of the child is an integer value proportional to the fitness of the parents. In the child's genotype, all facilities that are open in the intersection of both parents are kept open. Other facilities are then taken from the union of the parents facilities until the goal amount of open facilities for the child is met. Note that this operator does not allow the generation of solutions that disrespect the constraints that involve P_{min}, P_{max}, Q_{min}, and Q_{max}.

Mutation in this work works by simply opening or closing a facility at random.

The initial probability of crossover for each individual *ind* was set as $cp_{ind} = 90\%$, while the initial probability of mutation was $mp_{ind} = 5\%$. Those values were adapted [26] at every generation for each individual. In each generation, the whole population is replaced by their children.

The genetic algorithm was used with a population of 40 individuals that were initialized either (i) randomly, (ii) with a greedy heuristic (Section 4.1), or (iii) with Heuristics Concentration (also described in Section 4.1).

6   Experiments

6.1   Instances

The originating, collecting and refurbishment sites were randomly placed on a 100×100 square. Then, based on an existing methodology for creating instances [11], the following values are used:

- $F_j = 0.1([1, 10000] + B_j[0, 10])$
- $G_k = 0.1([1, 25000] + B_j[0, 100])$
- $C_{ijk} = $ Euclidian distance from i to j to k
- $a_i = [0, 500]$
- $B_j = [0, 6000]$
- $D_k = [0, 30000]$

The notation $[n_1, n_2]$ represents a uniformly distributed random number between n_1 and n_2, and the parameters $|I|$, $|J|$, P_{max}, $|K|$, Q_{max}, used for generating 5 instance sets, are respectively (i) 30, 14, 4, 12, 2; (ii) 40, 20, 6, 15, 4; (iii) 50, 30, 6, 20, 4; (iv) 70, 30, 6, 20, 4; (v) 100, 40, 8, 30, 6. For each of those 5 parameter configurations, 5 instances were generated. The instances and the best objective function values found in this work are available from the authors[1] and although some parameters could be adjusted in order to represent more realistic problems, the parameters were defined in a way to make comparison with other works possible [11,4].

6.2   Experimental Design

A comparative experiment was designed to evaluate the performance of GA-based methods relative to other heuristics commonly used in the solution of two-level reverse distribution network problems. This experiment consists in the application of 11 different methods on 25 problem instances, 5 from each instance configuration. The algorithms were used as levels of the experimental factor, and the problems were treated as experimental blocks [17].

We use the following notation for the algorithms compared in this paper: Greedy Heuristic (Gr), Concentrations Heuristic (CH), Routing Heuristic (RH), Simplex Routing (SR), Genetic Algorithms (GA), Random Solution Generation (RS). We compared

---

[1] http://www.alandefreitas.com/downloads/problem-instances.php

four non-evolutionary heuristics and seven GA-based heuristics, as described in Table 1. Methods A-D are the heuristics that generate solutions and routes. Methods E-J are Genetic Algorithms with different heuristics for generating the initial solutions (RS, Gr or CH) and calculating the best routes (RH or RH). Method K is a different configuration of Genetic Algorithm[2] proposed by [4] as efficient for reverse logistics. Twenty replicates of the experiment were performed with a time limit of 3 minutes for Methods E-K.

Table 1. Methods Compared

Method	A	B	C	D
Heuristics	Gr+RH	Gr+SR	CH+RH	CH+SR

Method	E	F	G	H	I	J	K
Heuristics	GA+RS+RH	GA+RS+SR	GA+Gr+RH	GA+Gr+SR	GA+CH+RH	GA+CH+SR	GA2

For all tests performed, the predefined significance threshold was set as 95%, adjusted using the Bonferroni correction [22]. Also, rank transformation was employed in order to reduce the influence of outliers and heteroscedasticity in the analysis of the data obtained [17].

6.3 Analysis of Results

Figure 1(a) shows the difference of mean objective value obtained by the methods in each instance. For better visualization, the values were scaled by average rank. A Friedman test of the data detected highly significant ($p < 10^{-15}$) differences in algorithm performance across the test set used. To evaluate the differences, pairwise testing was performed using FDR-corrected Wilcoxon Signed-Rank tests [5,19], and bootstrap confidence intervals were derived for the average ranks of each algorithm. The rank effect sizes for each method are presented in Figure 1(b), after removing the problem effects and the overall mean.

From the results it should be clear that the methods can be divided into five groups, with statistically significant differences between groups but not within them. From best to worst, these groups contain: (i) genetic algorithms based on simplex routing (F, H, J) and (ii) genetic algorithms with the routing heuristic (E, G, I), with both groups presenting above-average performance (mean rank smaller than 0). The other three, below-average groups are composed of: (iii) a less efficient genetic algorithm (K); the heuristics based on simplex routing (B,D); and the heuristics with the routing heuristic (A,C).

---

[2] Binary tournament selection, fusion crossover, swap node mutation, replacement of only half the previous generation, population size $2.|J|$, crossover probability 0.5, mutation probability 0.1.

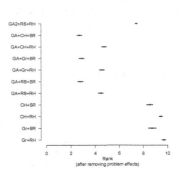

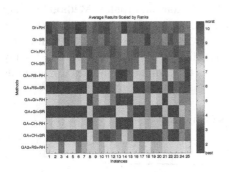

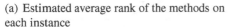

(a) Estimated average rank of the methods on each instance

(b) Estimated average ranks and confidence intervals for the methods tested, integrated over the 25 instances.

Fig. 1. Comparison between the methods

7   Conclusion and Future Work

By comparing heuristics for the two-level reverse distribution problem, we conclude that genetic algorithms, specially the ones based on simplex for calculating the routes, can improve significantly the results when compared to other heuristics proposed for the problem.

Different initial populations, however, do not seem to strongly influence the performance of the genetic algorithms as all of them had comparable results at confidence level $\alpha = 0.05$. On the other hand, all the parameters of the genetic algorithm, such as population size and probability of the operators, could be better explored in order to achieve better results.

Ideas for future works include: (i) comparisons to other algorithms, both nature-inspired and heuristics (ii) tests with Genetic Algorithms for calculating the routes, (iii) search for globally optimal solutions for the proposed instances, (iv) evaluation of other genetic operators, (v) parallel evaluation of solutions, (vi) tests on larger and preferably real-world instances, and (vii) tests with larger time limits.

Acknowledgement. This work has been supported by the Brazilian agencies CAPES, CNPq (grants 472446/2010-0 and 305506/2010-2), and FAPEMIG (project APQ-04611-10); and the Marie Curie International Research Staff Exchange Scheme Fellowship within the 7th European Community Framework Programme.

References

1. Alshamrani, A., Mathur, K., Ballou, R.: Reverse logistics: simultaneous design of delivery routes and returns strategies. Computers & Operations Research 34(2), 595–619 (2007)

2. Barros, A., Dekker, R., Scholten, V.: A two-level network for recycling sand: a case study. European Journal of Operational Research 110(2), 199–214 (1998)
3. Bautista, J., Pereira, J.: Modeling the problem of locating collection areas for urban waste management. an application to the metropolitan area of barcelona. Omega 34(6), 617–629 (2006)
4. Costa, L.R., Galvão, R.D.: Mathematical models for reverse logistics: A genetic algorithm for a two-level problem. In: International Conference on Operational Research for Development, Fortaleza, CE, Brazil, pp. 92–104 (2007)
5. Crawley, M.J.: The R Book, 1st edn. Wiley (2007)
6. Davis, P., Ray, T.: A branch-bound algorithm for the capacitated facilities location problem. Naval Research Logistics Quarterly 16(3), 331–343 (1969)
7. Fleischmann, M., Bloemhof-Ruwaard, J., Dekker, R., Van Der Laan, E., Van Nunen, J., Van Wassenhove, L.: Quantitative models for reverse logistics: A review. European Journal of Operational Research 103(1), 1–17 (1997)
8. Freitas, A., Guimarães, F.: Originality and diversity in the artificial evolution of melodies. In: Proceedings of the 13th Annual Conference on Genetic and Evolutionary Computation, pp. 419–426. ACM (2011)
9. Horvath, P., Autry, C., Wilcox, W.: Liquidity implications of reverse logistics for retailers: A markov chain approach. Journal of Retailing 81(3), 191–203 (2005)
10. Hu, T., Sheu, J., Huang, K.: A reverse logistics cost minimization model for the treatment of hazardous wastes. Transportation Research Part E: Logistics and Transportation Review 38(6), 457–473 (2002)
11. Jayaraman, V., Patterson, R., Rolland, E.: The design of reverse distribution networks: models and solution procedures. European Journal of Operational Research 150(1), 128–149 (2003)
12. Ko, H., Evans, G.: A genetic algorithm-based heuristic for the dynamic integrated forward/reverse logistics network for 3pls. Computers & Operations Research 34(2), 346–366 (2007)
13. Kreinovich, V., Quintana, C., Fuentes, O.: Genetic algorithms: What fitness scaling is optimal? Cybernetics and Systems: an International Journal 24, 9–26 (1993)
14. Lu, Z., Bostel, N.: A facility location model for logistics systems including reverse flows: The case of remanufacturing activities. Computers & Operations Research 34(2), 299–323 (2007)
15. Melo, M., Nickel, S., Saldanha-da Gama, F.: Facility location and supply chain management - a review. European Journal of Operational Research 196(2), 401–412 (2009)
16. Min, H.: A bicriterion reverse distribution model for product recall. Omega 17(5), 483–490 (1989),
    http://ideas.repec.org/a/eee/jomega/v17y1989i5p483-490.html
17. Montgomery, D.: Design and Analysis of Experiments, 7th edn. Wiley (2008)
18. Chandran, R., Lancioni, R.A.: Product recall: A challenge for the 1980. International Journal of Physical Distribution and Materials Management 11(8), 483–490 (1981)
19. R Development Core Team: R: A Language and Environment for Statistical Computing. R Foundation for Statistical Computing, Vienna, Austria (2011)
20. Rosing, K., ReVelle, C.: Heuristic concentration: Two stage solution construction. European Journal of Operational Research 97(1), 75–86 (1997)
21. Runarsson, T.P., Yao, X.: Stochastic ranking for constrained evolutionary optimization. IEEE Transactions on Evolutionary Computation 4, 284–294 (2000)
22. Shaffer, J.P.: Multiple hypothesis testing. Annual Review of Psychology 46, 561–584 (1995)
23. Sheu, J., Chou, Y., Hu, C.: An integrated logistics operational model for green-supply chain management. Transportation Research Part E: Logistics and Transportation Review 41(4), 287–313 (2005)

24. Spengler, T., Puchert, H., Penkuhn, T., Rentz, O.: Environmental integrated production and recycling management. European Journal of Operational Research 97(2), 308–326 (1997)
25. Srivastava, S.: Network design for reverse logistics. Omega 36(4), 535–548 (2008)
26. Whitacre, J.M.: Adaptation and Self-Organization in Evolutionary Algorithms. Ph.D. thesis, University of New South Wales (2007)

A Simple Data Compression Algorithm
for Wireless Sensor Networks

Jonathan Gana Kolo[1,*], Li-Minn Ang[2], S. Anandan Shanmugam[1],
David Wee Gin Lim[1], and Kah Phooi Seng[3]

[1] Department of Electrical and Electronics Engineering,
The University of Nottingham Malaysia Campus,
Jalan Broga, 43500 Semenyih, Selangor Darul Ehsan, Malaysia
[2] School of Engineering,
Edith Cowan University,
Joondalup, WA 6027, Australia
[3] School of Computer Technology,
Sunway University,
5 Jalan Universiti, Bandar Sunway,
46150 Petaling Jaya, Selangor, Malaysia
{keyx1jgk,Sanandan.Shanmugam,Lim.Wee-Gin}@nottingham.edu.my,
li-minn.ang@ecu.edu.au,
jasmines@sunway.edu.my

Abstract. The energy consumption of each wireless sensor node is one of critical issues that require careful management in order to maximize the lifetime of the sensor network since the node is battery powered. The main energy consumer in each node is the communication module that requires energy to transmit and receive data over the air. Data compression is one of possible techniques that can reduce the amount of data exchanged between wireless sensor nodes. In this paper, we proposed a simple lossless data compression algorithm that uses multiple Huffman coding tables to compress WSNs data adaptively. We demonstrate the merits of our proposed algorithm in comparison with recently proposed LEC algorithm using various real-world sensor datasets.

Keywords: Wireless Sensor Networks, Energy Efficiency, Data Compression, Signal Processing, Adaptive Entropy Encoder, Huffman Coding.

1 Introduction

Wireless sensor networks (WSNs) are very large scale deployments of tiny smart wireless sensor devices working together to monitor a region and to collect data about the environment. Sensor nodes are generally self-organized and they communicate with each other wirelessly to perform a common task. The nodes are deployed in large quantities (from tens to thousands) and scattered randomly in an ad-hoc manner in the sensor field (a large geographic area). Through advanced mesh networking protocols,

---

* Corresponding author.

V. Snasel et al. (Eds.): SOCO Models in Industrial & Environmental Appl., AISC 188, pp. 327–336.
springerlink.com   © Springer-Verlag Berlin Heidelberg 2013

these sensor nodes form a wide area of connectivity that extends the reach of cyber-space out into the physical world. Data collected by each sensor node is transferred wirelessly to the sink either directly or through multi-hop communication.

Wireless sensor nodes have limited power source since they are powered by small batteries. In addition, the replacement of batteries for sensor nodes is virtually impossible for most applications since the nodes are often deployed in large numbers into harsh and inaccessible environments. Thus, the lifetime of WSN depends strongly on battery lifetime. It is therefore important to carefully manage the energy consumption of each sensor node subunit in order to maximize the network lifetime of WSN. For this reason, energy-efficient operation should be the most important factor to be considered in the design of WSNs. Thus, several approaches are followed in the literature to address such power limitations. Some of these approaches include adaptive sampling [1], energy-efficient MAC protocols [2], energy-aware routing [3] and in-network processing (aggregation and compression) [4]. Furthermore, wireless sensor nodes are also constrained in terms of processing and memory. Therefore, software designed for use in WSNs should be lightweight and the computational requirements of the algorithms should be low for efficient operation in WSNs.

Sensor nodes in WSN consume energy during sensing, processing and transmission. But typically, the energy spent by a sensing node in the communication module for data transmission and reception is more than the energy for processing [5–7]. One significant approach to conserve energy and maximize network lifetime in WSN is through the use of efficient data compression schemes [8]. Data compression schemes reduce data size before transmitting in the wireless medium which translates to reduce total power consumption. This savings due to compression directly translate into lifetime extension for the network nodes. Both the local single node that compresses the data as well as the intermediate routing nodes benefits from handling less data [9].

Two data compression approaches have been followed in the literature: a distributed data compression approach [10], [11] and a local data compression approach [5], [9], [12–15]. In this paper, our focus is on lossless and reliable data gathering in WSN using a local data compression scheme which has been shown to significantly improve WSN energy savings in real-world deployments [9]. After a careful study of local lossless data compression algorithms (such as Sensor LZW (S-LZW) [9], Lossless Entropy Compression (LEC) algorithm [5], Median Predictor based Data Compression (MPDC) [13] and two-modal Generalized Predictive Coding (GPC) [15]) recently proposed in the literature for WSNs, we found that most of the algorithms cannot adapt to changes in the source data statistics. As a result, the compression performance obtained by these algorithms is not optimal. We therefore propose in this paper a simple lossless data compression algorithm for WSN. The proposed algorithm is adaptive. The algorithm adapts to changes in the source data statistics to maximize performance.

To verify the effectiveness of our proposed algorithm, we compare its compression performance with LEC performance. To the best of our knowledge, till date, LEC algorithm is the best lossless data compression algorithm designed specifically for use in WSNs. However, the LEC algorithm cannot adapt to changing correlation in sensor measured data. Hence, the compression ratio obtained and by extension the energy saving obtainable is not optimal. This therefore gives room for improvement.

The rest of this article is organized as follows. Section 2 presents our proposed data compression algorithm. In section 3, the proposed algorithm is evaluated and compared with the LEC algorithm using real-world WSN data. Finally, we conclude the paper in section 4.

2   The Compression Algorithm

Our proposed data compression algorithm adopts the followings from the LEC algorithm: (a) to increase the compressibility of the sensed data, we adopt a differential compression scheme to reduce the dynamic range of the source symbols. (b) The basic alphabets of residues are divided into groups whose sizes increase exponentially. The groups are then entropy coded and not unary coded as in the original version of exponential-Golomb code. Thus, the dictionaries used in our proposed scheme are called prefix-free-tables. (c) We also adopt the LEC encoding function because of its simplicity and efficiency. Fig. 1 shows the functional block diagram of our proposed simple data compression algorithm. The algorithm is a two-stage process. In the first stage, a simple unit-delay predictor is used to preprocess the sensed data to generate the residues. That is, for every new acquisition m_i, the difference $d_i = x_i - x_{i-1}$ is computed. While x_i is current sensor reading(s), x_{i-1} is the immediate past sensor reading(s). The difference d_i serves as input to the entropy encoders. In the second stage, two types of entropy encoders are used to better capture the underlying temporal correlation in the sensed data. The entropy encoders are 1-Table Static Entropy Encoder and 2-Table Adaptive Entropy Encoder.

1-Table Static Entropy Encoder

The 1-Table Static Entropy Encoder is essentially the entropy encoder in LEC with its coding table optimized for WSNs sensed data. The encoder performs compression losslessly by encoding differences d_i more compactly based on their statistical characteristics in accordance with the pseudo-code described in Fig. 2. Each d_i is represented as a bit sequence c_i composed of two parts h_i and l_i (that is $c_i = h_i * l_i$).

$$l_i = (Index)|_{bi} \tag{1}$$

where

$$b_i = \lceil \log_2(|d_i|) \rceil \tag{2}$$

$$Index = \begin{cases} d_i & d_i \geq 0 \\ (2^{b_i} - 1) + d_i & d_i < 0 \end{cases} \tag{3}$$

Equation (3) returns the index position of each d_i within its group. $(Index)|_{b_i}$ denotes the binary representation of *Index* over b_i bits. b_i is the category (group number) of d_i. It is also the number of lower order bits needed to encode the value of d_i. Note that if $d_i = 0$, l_i is not represented. Thus, at that instance, $c_i = h_i$. Once c_i is generated, it is appended to the bit stream which forms the compressed version of the sequence of measures m_i.

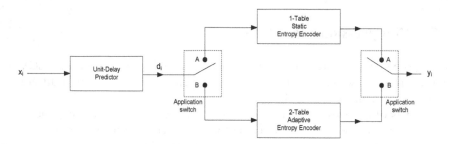

Fig. 1. The functional block scheme of our proposed data compression algorithm

```
encode(di, TABLE)
// di is the current residue value
// TABLE is the variable length Huffman codes used in encoding
// bi is the category (group number) of di
// bi is also the number of lower order bits needed to encode the value of di
// ci is the encoded bitstream of di
// hi is the variable-length Huffman code that codifies the category (group) of di
// li is the variable-length integer code that codifies the index position of di
//    within its group (category)
// * denotes concatenation
// (Index)|bi denotes the binary representation of index over bi bits

// compute di category
IF di = 0 THEN
    SET bi TO 0
ELSE
    // for all bi ≠ 0 then
    SET bi TO ⌈log₂(|di|)⌉
ENDIF
// extract hi the variable length Huffman code from TABLE
SET hi TO TABLE[bi]
// build ci
IF bi = 0 THEN
    // li is not needed
    SET ci TO hi
ELSE
    // build li
    SET li TO (Index)|bi
    // build ci
    SET ci TO hi * li
ENDIF
RETURN ci
```

Fig. 2. The pseudo-code of the encode algorithm

2-Table Adaptive Entropy Encoder

High compression ratio performance yields high energy saving since fewer numbers of bits will be transmitted by the communication module of the sensor node thereby saving lots of energy. Adaptive encoding can enable us to achieve maximal compression ratio and by extension maximal energy saving. Hence, to enjoy the benefits of adaptive coding without incurring higher energy cost, we resorted to the use of multiple static Huffman coding tables. Thus, we propose to implement a 2-Table adaptive entropy encoder that compresses blocks of sampled data at a time using two static Huffman coding tables adaptively. Each static Huffman coding table is designed to give nearly optimal compression performance for a particular geometrically distributed source. By using two Huffman coding tables adaptively and sending the table identifier, the 2-Table adaptive entropy encoder can adapt to many source data with different statistics. The proposed 2-Table adaptive entropy encoder operates in one pass and can be applied to multiple data types. Thus, our proposed algorithm can be used in continuous monitoring systems with varied latency requirements by changing the block size n to suit each application. The pseudo-code of our proposed simple data compression algorithm is given in Fig. 3.

3 Simulations and Analysis

The performance of our proposed data compression algorithm is computed by using the compression ratio defined as:

$$CR = 100 \times \left(1 - \frac{compressed\_size}{original\_size} \right) \% \tag{4}$$

The compressed_size is the number of bits obtained after compression and the original_size is the uncompressed data size. Each uncompressed sample data is represented by 16-bit unsigned integers. Publicly accessible real-world environmental monitoring datasets are used in our simulations. We used temperature and relative humidity measurements from one SensorScope [16] deployments: LUCE Deployment with Node ID of 84 for the time interval of 23 November 2006 to 17 December 2006. We also used the six set of soil temperature measurements from [17], collected from 01 January 2006 to 02 October 2006. For simplicity, the flags of missing data in the soil temperature measurements which are quite rare were replaced by the value of the preceding sample in that soil temperature dataset. To simulate real-world sensor communications with fidelity, the temperature and relative humidity measurements are converted to sensor readings using the inverted versions of the conversion functions in [18] with the assumption of the A/D conversion precision being 14 bits and 12 bits for temperature and relative humidity datasets respectively. In addition, we also used a seismic dataset collected by the OhioSeis Digital Seismographic Station located in Bowling Green, Ohio, for the time interval of 2:00 PM to 3:00 PM on 21 September 1999 (UT) [19].

```
compress(xi, xi_1, n, y)
// xi is the current sensor reading(s)
// xi_1 is the immediate past sensor reading(s)
// n is the block size (the number of samples read each time)
// y is the final encoded bitstream
// encode() is the encode function

// compute the residue di
SET di TO xi – xi_1
// encode the residue di
IF n = 1 THEN
    // encode residue di using the 1-table static entropy encoder
    CALL encode() with di and Table1 RETURNING ci
    // append ci to y
    SET y TO << y, ci >>
ELSE
    // encode block of n di using the 2-table adaptive entropy encoder
    // encode the block of n di using the first table of the adaptive entropy encoder
    CALL encode() with block of n di and Table2 RETURNING ci
    SET ciA To ci
    // compute the size of the encoded bitstream ciA
    SET size_A TO length(ciA)
    // encode the block of n di using the second table of the adaptive entropy encoder
    CALL encode() with block of n di and Table3 RETURNING ci
    SET ciB To ci
    // compute the size of the encoded bitstream ciB
    SET size_B TO length(ciB)
    // compare size_A and size_B and select the encoded bitstream with the best compressed size
    IF size_A <= size_B THEN
        // generate the table identifier of Table2
        SET ID TO '0'
        // append encoded bitstream ciA to ID
        SET code TO << ID, ciA >>
    ELSE
        // generate the table identifier of Table3
        SET ID TO '1'
        // append encoded bitstream ciB to ID
        SET code TO << ID, ciB >>
    ENDIF
    // append code to y
    SET y TO << y, code >>
ENDIF
RETURN y
```

Fig. 3 The pseudo-code of our proposed simple data compression algorithm

For this simulation, the optimized table in Table 1 is used by the 1-Table static entropy encoder and Tables 2 and 3 were used by the 2-Table adaptive entropy encoder. The Huffman coding table in Table 2 was designed to handle data sets with high and medial correlation while the Huffman coding table in Table 3 was designed to handle data sets with medial and low correlation. The combination of these two coding tables handles effectively the changing correlation in the sensed data as it is being read block by block with each block consisting of n-samples. The Huffman table that gives the best compression is then selected. The encoded bitstream generated by that table is then appended to a 1-bit table identifier (ID) and thereafter sent to the sink. The decoder uses the ID to identify the Huffman coding table used in encoding the block of n-residues. Since only two static Huffman coding tables are in use by the 2-Table adaptive entropy encoder, the table ID is either '0' or '1'. The performance comparison between LEC, 1-Table static entropy encoder and 2-Table adaptive entropy encoder is given in Table 4. For block size of 1 (i.e. n=1), our proposed algorithm using the 1-Table static entropy encoder with the optimized coding table givens better performance than the LEC algorithm. For block size greater than 1 (i.e. n>1), our proposed algorithm using the 2-Table adaptive entropy encoder gives better performance than the LEC algorithm. Thus, the combination of the 1-Table static entropy encoder with the 2-Table adaptive entropy encoder ensures that the performance of our proposed data compression algorithm is better than that of LEC for all value of n.

Table 1. Huffman Coding **Table1**

b_i	h_i	d_i
0	100	0
1	110	−1,+1
2	00	−3,−2,+2,+3
3	111	−7, . . . ,−4,+4, . . .,+7
4	101	−15, . . . ,−8,+8, . . .,+15
5	010	−31, . . . ,−16,+16, . . .,+31
6	0111	−63, . . . ,−32,+32, . . .,+63
7	01101	−127, . . . ,−64,+64, . . .,+127
8	011001	−255, . . . ,−128,+128, . . .,+255
9	0110001	−511, . . . ,−256,+256, . . .,+511
10	01100001	−1023, . . . ,−512,+512, . . .,+1023
11	011000001	−2047, . . . ,−1024,+1024, . . .,+2047
12	01100000000	−4095, . . . ,−2048,+2048, . . .,+4095
13	01100000001	−8191, . . . ,−4096,+4096, . . .,+8191
14	01100000010	−16383, . . . ,−8192,+8192, . . .,+16383

Table 2. Huffman Coding **Table2**

b_i	h_i	d_i
0	00	0
1	01	−1,+1
2	11	−3,−2,+2,+3
3	101	−7, . . . ,−4,+4, . . .,+7
4	1001	−15, . . . ,−8,+8, . . .,+15
5	10001	−31, . . . ,−16,+16, . . .,+31
6	100001	−63, . . . ,−32,+32, . . .,+63
7	1000001	−127, . . . ,−64,+64, . . .,+127
8	10000001	−255, . . . ,−128,+128, . . .,+255
9	1000000000	−511, . . . ,−256,+256, . . .,+511
10	10000000010	−1023, . . . ,−512,+512, . . .,+1023
11	10000000011	−2047, . . . ,−1024,+1024, . . .,+2047
12	10000000100	−4095, . . . ,−2048,+2048, . . .,+4095
13	10000000101	−8191, . . . ,−4096,+4096, . . .,+8191
14	10000000110	−16383, . . . ,−8192,+8192, . . .,+16383

Fig. 4 shows the compression ratios achieved by our proposed simple data compression algorithm for different values of the block size n for the nine real-world datasets. As evident from Fig. 4, the compression ratio performance achieved by our proposed data compression algorithm for each of the nine datasets increases with respect to the increase in the block size. Also, for values of n as small as 3 (i.e. $n=3$), the compression ratio performance of our proposed simple data compression algorithm is good (better than LEC performance) for all the nine datasets and the performance improves thereafter as n is increased beyond 3 to 256 as evident from the plots in Fig. 4. The best performance of our proposed compression algorithm is recorded in the last column of Table 4 which clearly shows that our proposed simple data compressed algorithm outperforms the LEC algorithm. In terms of algorithm complexity, our proposed algorithm is simple and lightweight. When compared to the LEC algorithm, our proposed algorithm requires only slightly more memory. Energy is conserved since only fewer numbers of bits are transmitted by the communication module of the sensor node.

Table 3. Huffman Coding **Table3**

b_i	h_i	d_i
0	1101111	0
1	11010	−1,+1
2	1100	−3,−2,+2,+3
3	011	−7,. . . ,−4,+4,. . . ,+7
4	111	−15,. . . ,−8,+8,. . . ,+15
5	10	−31,. . . ,−16,+16,. . . ,+31
6	00	−63,. . . ,−32,+32,. . . ,+63
7	010	−127,. . . ,−64,+64,. . . .,+127
8	110110	−255,. . . ,−128,+128,. . . .,+255
9	110111011	−511,. . . ,−256,+256,. . . .,+511
10	110111001	−1023,. . . ,−512,+512,. . . .,+1023
11	1101110101	−2047,. . . ,−1024,+1024,. . . .,+2047
12	1101110100	−4095,. . . ,−2048,+2048,. . . .,+4095
13	1101110000	−8191,. . . ,−4096,+4096,. . . .,+8191
14	11011100011	−16383,. . . ,−8192,+8192,. . . .,+16383

Table 4. Performance comparison between the 2-Table adaptive entropy encoder with LEC and 1-Table static entropy encoder

DATASET	LEC single Table performance	Our Optimized single Table performance	2-Table Adaptive Entropy Encoder performance			2-Table Adaptive Entropy Encoder Best performance
			n=1	n=2	n=3	
LU84 Temp	70.81	71.69	68.01	70.87	71.77	73.48
LU84 Rh	62.86	63.46	60.50	62.92	63.62	64.75
SEISMIC Data	69.72	71.24	67.00	70.12	71.16	72.98
Ts_0 Temp	52.05	52.25	50.92	53.19	54.07	55.76
Ts_5cm Temp	54.55	54.80	52.38	54.70	55.60	57.05
Ts_10cmTemp	54.96	55.21	52.50	55.09	56.00	57.68
Ts_20cm Temp	55.10	55.35	52.51	55.24	56.17	57.97
Ts_50cm Temp	55.04	55.32	52.44	55.22	56.16	57.97
Ts_1m Temp	54.97	55.24	52.40	55.16	56.09	57.88

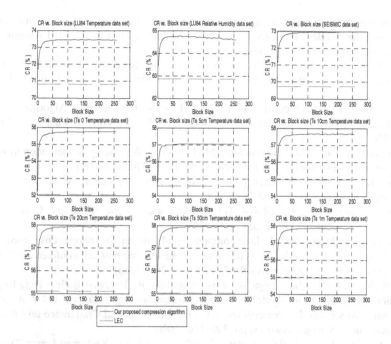

Fig. 4 Compression ratios vs. block size achieved by our proposed data compression algorithm for the nine real-world datasets

4 Conclusion

In this paper, we have introduced a simple lossless adaptive compression algorithm for WSNs. The proposed algorithm is simple and efficient, and is particularly suitable for resource-constrained wireless sensor nodes. The algorithm adapts to changing correlation in the sensed data to effectively compress data using two Huffman coding tables. Our proposed algorithm reduce the data amount for transmission which contributes to the energy saving. We have obtained compression ratios of 73.48%, 64.75% and 72.98% for temperature, relative humidity and seismic datasets respectively. The evaluation of our proposed algorithm with LEC using real-world datasets shows that our proposed algorithm's compression performance is better. Our algorithm can be used for both real-time and delay-tolerant transmission.

References

[1] Bandyopadhyay, S., Tian, Q., Coyle, E.J.: Spatio-Temporal Sampling Rates and Energy Efficiency in Wireless Sensor Networks. IEEE/ACM Transaction on Networking 13, 1339–1352 (2005)
[2] Ye, W., Heidemann, J., Estrin, D.: Medium Access Control With Coordinated Adaptive Sleeping for Wireless Sensor Networks. IEEE/ACM Transactions on Networking 12(3), 493–506 (2004)

[3] Heinzelman, W.R., Chandrakasan, A., Balakrishnan, H.: Energy-efficient communication protocol for wireless microsensor networks. In: Proceedings of the 33rd Annual Hawaii International Conference on System Sciences (2000)

[4] Fasolo, E., Rossi, M., Zorzi, M., Gradenigo, B.: In-network Aggregation Techniques for Wireless Sensor Networks□: A Survey. IEEE Wireless Communications 14, 70–87 (2007)

[5] Marcelloni, F., Vecchio, M.: An Efficient Lossless Compression Algorithm for Tiny Nodes of Monitoring Wireless Sensor Networks. The Computer Journal 52(8), 969–987 (2009)

[6] Anastasi, G., Conti, M., Di Francesco, M., Passarella, A.: Energy conservation in wireless sensor networks: A survey. Ad Hoc Networks 7(3), 537–568 (2009)

[7] Barr, K.C., Asanović, K.: Energy-aware lossless data compression. ACM Transactions on Computer Systems 24(3), 250–291 (2006)

[8] Yick, J., Mukherjee, B., Ghosal, D.: Wireless sensor network survey. Computer Networks 52(12), 2292–2330 (2008)

[9] Sadler, C.M., Martonosi, M.: Data compression algorithms for energy-constrained devices in delay tolerant networks. In: Proceedings of the 4th international conference on Embedded networked sensor systems - SenSys 2006, p. 265 (2006)

[10] Ciancio, A., Pattem, S., Ortega, A., Krishnamachari, B.: Energy-Efficient Data Representation and Routing for Wireless Sensor Networks Based on a Distributed Wavelet Compression Algorithm. In: Proceedings of the Fifth international Conference on Information Processing in Sensor Networks, pp. 309–316 (2006)

[11] Gastpar, M., Dragotti, P.L., Vetterli, M.: The Distributed Karhunen-Loeve Transform. IEEE Transactions on Information Theory 52(12), 5177–5196 (2006)

[12] Schoellhammer, T., Greenstein, B., Osterweil, E., Wimbrow, M., Estrin, D.: Lightweight temporal compression of microclimate datasets [wireless sensor networks. In: 29th Annual IEEE International Conference on Local Computer Networks, pp. 516–524 (2004)

[13] Maurya, A.K., Singh, D., Sarje, A.K.: Median Predictor based Data Compression Algorithm for Wireless Sensor Network. International Journal of Smart Sensors and Ad Hoc Networks (IJSSAN) 1(1), 62–65 (2011)

[14] Kolo, J.G., Ang, L.-M., Seng, K.P., Prabaharan, S.: Performance Comparison of Data Compression Algorithms for Environmental Monitoring Wireless Sensor Networks. International Journal of Computer Applications in Technology, IJCAT (article in press, 2012)

[15] Liang, Y.: Efficient Temporal Compression in Wireless Sensor Networks. In: 36th Annual IEEE Conference on Local Computer Networks (LCN 2011), pp. 466–474 (2011)

[16] SensorScope deployments homepage (2012), http://sensorscope.epfl.ch/index.php/Main_Page (accessed: January 2012)

[17] Davis, K.J.: No Title (2006), http://cheas.psu.edu/data/flux/wcreek/wcreek2006met.txt (accessed: January 6, 2012)

[18] Sensirion homepage, (2012), http://www.sensirion.com (accessed: January 6, 2012)

[19] Seismic dataset (2012), http://www-math.bgsu.edu/?zirbel/ (accessed: January 6, 2012)

Metaheuristically Optimized Multicriteria Clustering for Medium-Scale Networks

David Chalupa and Jiří Pospíchal

Institute of Applied Informatics, Faculty of Informatics and Information Technologies,
Slovak University of Technology, Ilkovičova 3, 842 16 Bratislava, Slovakia
chalupa@fiit.stuba.sk

Abstract. We present a highly scalable metaheuristic approach to complex network clustering. Our method uses a multicriteria construction procedure (MCP), controlled by adaptable constraints of local density and local connectivity. The input of the MCP - the permutation of vertices, is evolved using a metaheuristic based on local search. Our approach provides a favorable computational complexity of the MCP for sparse graphs and an adaptability of the constraints, since the criteria of a "good clustering" are still not generally agreed upon in the literature. Experimental verification, regarding the quality and running time, is performed on several well-known network clustering instances, as well as on real-world social network data.

1 Introduction

Analysis of complex networks is a very lively area of inter-disciplinary research. In physics, complex networks are studied as dynamic systems, with respect to the processes of their growth, evolution and their structure [12,22]. In computer science, they are related to a large spectrum of practical applications, including data mining [18], web engineering [19] or bioinformatics [1,6]. Additionally, many popular complex networks emerge from the analysis of current trends on the Internet, e.g. social networks [3] and research citation networks [18].

These networks generally tend to have clustered structures. A *cluster* can be informally described as a group of similar entities in the network, which generally tend to create densely connected areas. A problem of searching for a decomposition of the network into clusters is often modeled using graph theory and therefore, it is referred to as *graph clustering* [17].

There are numerous applications of graph clustering. In this paper, we focus mostly on the applications in *social networks*, in which it is often referred to as *community detection* [12]. These applications include marketing, recommendation, personalization, media analysis and human resource management [13,17]. Quite a large attention is drawn also by the structure of the networks of terrorist organizations [13]. Clustering of social networks have also been studied in the context of epidemiology of sexually transmitted diseases [14]. Other applications include grouping of gene expressions in bioinformatics, most notably in protein interaction [6] and gene-activation dependencies [1].

V. Snasel et al. (Eds.): SOCO Models in Industrial & Environmental Appl., AISC 188, pp. 337–346.
springerlink.com

Unfortunately, due to the less formal nature of graph clustering, there exist many different approaches, how to define, compute and evaluate clusters and clustering. It is difficult to find out, how suitable a formulation is, since there is no metric of quality, which is generally agreed to be the most reliable, while being computationally tractable [17]. Therefore, graph clustering is an interesting application for soft computing, where one can use the emergence of augmented clustering not only to optimize but also to observe the nature of the problem formulation. In this work, we are dealing with these issues and come up with the concept of multicriteria constrution procedures (MCPs), which encapsulate the selected criteria, while the optimization of the input to the MCP is performed with a general metaheuristic.

In the terms of graph theory, a clustering of an undirected graph $G = [V, E]$ can be formalized as a set $S = \{V_1, V_2, ..., V_k\}$ of disjoint subsets of V often called *classes*, i.e. $\forall i = 1..k\ V_i \subset V$. Let $d = \frac{|E|}{|V|(|V|-1)/2}$ be the *density* of the graph to cluster. Then, the subgraphs $G(V_i)$ induced by the classes V_i ($\forall i = 1..k$) should be more dense than the graph itself, i.e. $\forall i = 1..k\ d(G(V_i)) > d(G)$. The values $d(G(V_i))$ will be referred to as the *intra-cluster densities*. An important fact is that this condition is very similar to the formalization of graph coloring and the clique covering problems [10].

The paper is organized as follows. Section 2 provides an overview of the topic and the related work. In Section 3, we propose the concept of MCPs and the metaheuristically optimized multicriteria clustering based on MCPs. In Section 4, we provide the experimental results of our approach. Finally, in Section 5, we summarize the work.

2 Background and Related Work

Let $\overline{G}$ be the complementary graph to G. In the graph coloring problem, the objective is to minimize the number of colors k for classes $S = \{V_1, V_2, ..., V_k\}$, where $\forall i = 1..k\ d(G(V_i)) = 0$. The minimal value of k, for which this is possible, is called chromatic number and denoted as $\chi(G)$. Clearly, $\forall\ V_i \in S\ d(\overline{G}(V_i)) = 1$. The problem to obtain this partitioning will be referred to as the clique covering problem, which is NP-hard [10]. It can be seen quite easily that clique covering is a special case of graph clustering, where each cluster is a clique. On the other hand, the most trivial constraint that $\forall i = 1..k\ d(G(V_i)) > d(G)$ often leads to trivial solutions, which should be avoided. Therefore, the formulation of the criteria for graph clustering is understood as an issue of searching for a balance between this formulation on one hand and the clique covering on the other [17]. Regarding the complexity of the problems, it is worth mentioning that not only the clique covering but also many of the meaningful graph clustering quality measures are known to be NP-hard or NP-complete [5,21]. Nevertheless, all these problems are related to the structure of the graph. Although this is difficult to formalize, we can assume that the more asymmetrical the graph is, the more information is hidden in the structure to guide an optimization algorithm.

Regarding the relevant algorithms, *hierarchical clustering* uses a selected similarity measure and either repeatedly divides the graph or merges some partial clusters. In this sense, we will refer to either divisive or agglomerative clustering. In hierarchical clustering, a tree of candidate clusters, called dendrogram, is used [17]. Another approach is represented by the *spectral methods*, using the eigenvalues of the graph's adjacency matrix, which is diagonalized and the vertices are reordered so that the vertices in the same clusters are next to each other [17]. *Local search methods* are used for the problem as well, however, they use the procedure to find a single cluster for a vertex, which is then used as a seed [16]. A representative of the *visual and geometric methods* is the spring force algorithm, which is often used to visualize clustered graphs and thus also solves the problem in graphical representation [17].

In the clique covering problem, chromatic number of the complementary graph can be estimated in polynomial time by *sequential greedy heuristics*, e.g. the well-known Brélaz's heuristic [2]. Another approach is the *local search*, representing a very large class of algorithms based on iterative improving of current solutions. The basic local search algorithm, hill climbing, uses elementary changes of solutions to improve them [15]. Popular stochastic extensions of hill climbing include the simulated annealing [20,11] and tabu search [8].

3   The Proposed Approach

In this section, we describe the concept of MCPs and propose an MCP based on local densities of clusters and local connectivities of their vertices as the criteria for the solution construction. Then, we describe the metaheuristic, which is used to optimize the input of MCPs, i.e. the permutation of vertices.

3.1   The Criteria for Graph Clustering

The objective of our approach to graph clustering is to minimize the number of clusters in a clustering S of a graph G: min $k = |S|$; $S = \{V_1, V_2, ..., V_k\}$, subject to the following global constraints:

1. Each vertex is clustered and the clusters are non-overlapping: $[V_1 \cup V_2 \cup ... \cup V_k] = V \land [V_1 \cap V_2 \cap ... \cap V_k] = \emptyset$.
2. The clusters are more dense than the whole graph: $\forall i = 1..k \, d(G(V_i)) > d(G)$.

Furthermore, we consider the following local constraints, which are dynamic, i.e. they are influenced by the current state during the construction of the clustering, rather than remaining static for the whole procedure:

3. The relative connectivity of a vertex to be newly added to the cluster must be higher than its relative connectivity to the residual, currently non-clustered sub-graph: $\frac{w_c}{|V_{c,i}|} > \frac{\delta_r}{|V_r| - 1}$, where $V_{c,i}$ is the set of vertices in cluster c at the iteration i of the MCP, w_c is the number edges, brought into the cluster by the vertex to be newly added and $|V_r|$ and δ_r are the number of vertices and the degree of the newly added vertex in the subgraph containing only the currently non-clustered vertices.
4. If there are more candidate clusters, the one with highest connectivity is taken: $c = \arg\max_c \frac{w_c}{|V_{c,i}|}$, where for the cluster c, $\frac{w_c}{|V_{c,i}|}$ must be a feasible value, according to the previous rule.

5. The vertex to be newly added must bring at least as many edges, as is the current average intra-cluster degree in the particular cluster, while a small tolerance τ may be sometimes allowed: $w_c + \tau \geq \dfrac{2|E_{c,i}|}{|V_{c,i}|}$, where $E_{c,i}$ is the number of edges in $G(V_{c,i})$.

Criteria 1 and 2 are implied from the basic properties a good clustering should fulfill. Criterion 3 is used to verify, whether it is favorable to add v_j to a cluster or to rather create a new cluster and let some of the currently unclustered vertices join it. Criterion 4 is used to solve the situations, when more clusters fulfil criterion 3. Finally, criterion 5 is used to ensure a relative uniformity of intra-cluster vertex degrees, which is important in order to avoid situations, when several small clusters are unnecessarily joined. Parameter τ plays an important role here, since $\tau = 0$ leads to clusters with very uniform intra-cluster degrees, while $\tau > 0$ is favorable for clusters with stronger centrality.

3.2 The Multicriteria Construction Procedures (MCPs)

The general framework for an MCP is described in the pseudocode of Algorithm 1. As the input, we have a permutation of vertices. In the step 2, we take the current vertex v_j from the permutation. In the step 3, we apply the criteria to choose a label c, thus, joining v_j to the corresponding cluster in the step 4. The step 5 is used to update the auxiliary data specified in Section 3.1, which are needed to implement the multicriteria cluster choice efficiently. This procedure is repeated, until all vertices are clustered.

Algorithm 1. A General Framework for an MCP

A General Framework for an MCP
Input: graph $G = [V,E]$
permutation $P = [P_1, P_2, ..., P_{|V|}]$ of vertices
Output: a clustering S of G
1 for $i = 1..|V|$
2    $j = P_i$
3    $c = find\_cluster(v_j)$
4    $V_c = V_c \cup \{v_j\}$
5    $update\_auxiliary\_data(V_c)$
6 return $S = \{V_1, V_2, ..., V_k\}$

It is important that all the criteria are formulated in the way that during the implementation, one can verify each of the criteria only by scanning the neighbors of the currently chosen vertex. This restriction leads to an $\mathscr{O}(\delta)$ average complexity per iteration of an MCP, where δ is the average degree of a vertex in the graph. Let v_j be the chosen vertex, c the chosen cluster and let $V_{c,i}$ and $E_{c,i}$ be the vertex and edge set of cluster c at the i-th iteration of an MCP. Then, w_c will be the number of edges brought into the cluster by v_j, counted by scanning of the neighbors of v_j.

Using this notation, we will describe MCP-DC as our construction algorithm for graph clustering. It uses the 5 criteria, as they have been described in the previous section.

The implementation of the *find_cluster* procedure is as follows. One can easily derive that the local density needed in criterion 2 is if and only if:

$$d(G)|V_{c,i}|(|V_{c,i}| + 1) - 2|E_{c,i}| - 2w_c < 0. \tag{1}$$

The local connectivity in criterion 3 is fulfilled if the following holds:

$$|V_{c,i}| - w_c \frac{V_r - 1}{\delta_r} < 0. \tag{2}$$

The maximization of the connectivity in criterion 4, i.e. the ratio $\frac{w_c}{|V_{c,i}|}$, can be implemented simultaneously with criterion 3, since the necessary values are calculated in the verification of criterion 3. Finally, the criterion 5 yields the following condition, where $\tau \geq 0$ is a parameter of tolerance for the intra-cluster degree of the newly added vertex:

$$\frac{2|E_{c,i}|}{|V_{c,i}|} - \tau - w_c \leq 0. \tag{3}$$

These observations lead to a construction algorithm, in which the output depends on the permutation of vertices and, according to the following theorem, the number of iterations is proportional to the number of edges.

Theorem 1. *MCP-DC can be implemented to run in $\mathcal{O}(\delta|V|) = \mathcal{O}(|E|)$ time.*

Proof. $|V_{c,i}|$ and $|E_{c,i}|$ can be trivially recalculated in $\mathcal{O}(1)$ time per iteration. The previous formulations of the MCP-DC criteria can be implemented by iterative subtracting of a constant (in the cases of criteria 2 and 5) or the ratio $\frac{V_r - 1}{\delta_r}$ (in the case of criterion 3) from the respective values. Explicit storage of values w_c yields the same for criterion 4. Restoration of the former values after subtraction can be done by simulating the inverse process. All these operations need $\mathcal{O}(\delta)$ average time per iteration, thus, they lead to an $\mathcal{O}(\delta|V|) = \mathcal{O}(|E|)$ running time of MCP-DC.   □

3.3 The Metaheuristic for Optimization of the Permutation of Vertices in MCPs

The proposed criteria indicate that we are facing a highly constrained problem. On the other hand, encapsulation of the constrained part in the MCP leads to two major advantages in the optimization. First, for each permutation, there exists a clustering, which will be constructed by an MCP. Secondly, it was confirmed in our experiments that this formulation does not tend to create hard multimodal functions. In fact, on real-world data, we were able to optimize the permutation using a simple local search metaheuristic.

The metaheuristic we used, begins with a random permutation, which can be generated in place in $\mathcal{O}(|V|)$ time [4]. The initial clustering is constructed using an MCP. Then, at each iteration of local search, we try a single random vertex exchange in the permutation and evaluate the new number of clusters using the MCP. The new permutation is accepted if and only if, for the new permutation P' leading to k' clusters, it holds that $k' \leq k$, where k is the current number of clusters. The local search is stopped when the number of iterations without improvement exceeds certain threshold. We denote this as s_{max}.

The choice of this simple metaheuristic was influenced by three factors. The first reason is the nature of the landscape, which we indicated above. The second factor is the $\mathcal{O}(|E|)$ complexity of the objective function - MCP-DC, where each redundant run would significantly increase the global running time. Finally, the third factor is that the general stochastic extensions, such as tabu search [8], do not tend to improve the performance on this type of landscapes. It could perhaps be useful to find a heuristic to guide the algorithm to choose the right. However, we have tried several extensions, e.g. a roulette wheel selection of vertices according to their degrees but the results did not show any improvement.

4   Experimental Evaluation

In this section, we present the experimental evaluation of our approach. First, we visually illustrate the emergence of good clustering using our approach on a real-world sample from a social network. Then, we provide computational results on several instances obtained by MCP-DC and MCP-DC with the metaheuristic. Last but not least, we measure the running times for different network sizes.

4.1   The Emergence of Good Clustering

Fig. 1 illustrates the process of optimization using the local search algorithm and MCP-DC on a small instance of real-world social network data, where in each picture, the vertices in the same cluster are grouped together. This network will be referred to as Social network I. In this case, with MCP-DC and the metaheuristic, we achieved 5 highly relevant communities. The drawings visually indicate this emergence of clustered structure, where the evolution is driven only by random exchanges of vertices.

4.2   Computational Results

To evaluate our algorithm, we used it to solve the problem in two well-known benchmarks - Zachary karate club [23] and the American college football network [7]. We also used instances from two social networks, where the data from Social network II was obtained using a web crawler. We also used a graph generated by an artificial model [3]. Table 1 summarizes the computational results obtained in 10 independent runs, s_{max} is the maximal allowed number of iterations without improvement and τ is the intra-cluster degree tolerance factor. The primary criterion was k - the number of clusters. We also measured the average number of iterations and the time needed to obtain the clustering.

Zachary karate club is known to consist of two partitions, we show the two partitions found by our algorithm in Fig. 2. Fig. 2 also shows the result obtained for the American

Fig. 1. An illustration of the emergence of gradually better clustering using the metaheuristic optimization and MCP-DC. The drawings are done after 0, 100, 1000 and 10000 iterations (from left to right). The number of clusters is optimized from 12 to 5, where the 5 clusters are highly relevant for the data.

Table 1. Comparison of the results obtained by only MCP-DC and MCP-DC with the metaheuristic (MCP-DC+MH). The metrics are k - the number of clusters, the number of local search iterations and the running time.

source	$|V|, |E|$	s_{max}	τ	MCP-DC	MCP-DC+MH		
				k	k	iter.	time
Zachary karate club [23]	34, 78	5×10^3	1	7 - 15	2	7035	< 1 s
American college football [7]	115, 615	10^6	0	18 - 23	10 - 12	1237965	252 s
Social network I	52, 830	5×10^4	0	12 - 16	5 - 6	76194	9 s
Social network II	500, 924	5×10^4	1	161 - 197	12 - 15	154964	71 s
Artificial model [3]	500, 3536	5×10^4	0	68 - 79	55 - 60	163449	188 s

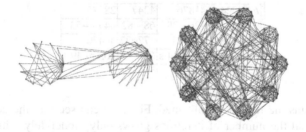

Fig. 2. Visualizations of results, obtained by our approach for the benchmark data: a clustering of the Zachary karate club into 2 communities (left) and a clustering of American college football league, which is known to consist of 10 conferences (right)

college football network, which is known to consist of 10 conferences. In both cases, our algorithm found the clusters reliably. Fig. 3 shows results on the social networks, where the clustering of the smaller network was verified manually, while the relevance of the clustering of the larger network is indicated by the sparseness in between. What is perhaps less visible in this drawing, is that the presence of hubs is very pronounced in these clusters. We note that we did not provide a numerical metric of quality (e.g. the Adjusted Rand Index [9]) due to exorbitant space, which results obtained using such metrics, together with precise analysis would require.

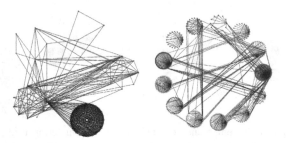

Fig. 3. Visualizations of results, obtained by our approach for the social network data: locally extracted sample from Social network I (left) and data obtained by crawling Social network II, which contains more pronounced hubs (right)

4.3   The Running Time

We have shown that MCP-DC has an $\mathcal{O}(|E|)$ complexity, which is favorable for sparse graphs. In these experiments, we measure the time practically needed by the MCP-DC with the metaheuristic. We used the crawler of Social network II to obtain networks of different sizes, up to 10000 vertices. In our algorithm, we set $s_{max} = 2 \times 10^4$ and $\tau = 1$.

Table 2. The times needed to obtain the clustering on samples from Social network II with different sizes

$|V|$	500	1000	2000	10000
$|E|$	924	1876	4247	28675
k	14 - 19	26 - 38	68 - 82	449 - 453
iter.	75363	97621	177622	484703
time	36 s	1 m 34 s	6 m	102 m 49 s

Table 2 contains the results we obtained. First, we can see that the constant value of s_{max} causes that the number of iterations grows only moderately. This factor, and the slow growth of $|E|$, which is typical for most complex networks, implies that the computational time does not grow exponentially. To be fair, although solid suboptimal results can be achieved even with smaller values of s_{max}, the current form of our approach is suitable mostly for medium-scale instances (around 10^3 vertices). However, we believe that the adaptability our approach maintains, is a key to solid scalability also for very large graphs.

5   Conclusion

We presented the concept of multicriteria construction procedures (MCPs) for network clustering. In this context, we designed MCP-DC - an MCP using the criteria of local density and local connectivity. In our approach, the input of MCP-DC - the permutation of vertices, is optimized using a general metaheuristic. This makes it useful not only as a graph clustering algorithm but also as a tool to discover pros and cons of the criteria, which are used to construct the clustering, since these are still not generally agreed upon in the literature.

Our approach was verified on well-known benchmarks for graph clustering, as well as on samples obtained from real-world social networks. These experiments showed much promise both in relevance of results and scalability of the approach.

Acknowledgement. This contribution was supported by Grant Agency VEGA SR under the grant 1/0553/12.

References

1. Boyer, F., Morgat, A., Labarre, L., Pothier, J., Viari, A.: Syntons, metabolons and interactons: an exact graph-theoretical approach for exploring neighbourhood between genomic and functional data. Bioinformatics 21(23), 4209–4215 (2005)
2. Brélaz, D.: New methods to color vertices of a graph. Communications of the ACM 22, 251–256 (1979)
3. Chalupa, D.: On the Ability of Graph Coloring Heuristics to Find Substructures in Social Networks. Information Sciences and Technologies, Bulletin of ACM Slovakia 3(2), 51–54 (2011)
4. Cormen, T.H., Leiserson, C.E., Rivest, R.L., Stein, C.: Introduction to Algorithms, 3rd edn. MIT Press (2009)
5. Drineas, P., Frieze, A., Kannan, R., Vempala, S., Vinay, V.: Clustering in large graphs and matrices. Machine Learning 56, 9–33 (2004)
6. Gao, L., Sun, P., Song, J.: Clustering algorithms for detecting functional modules in protein interaction networks. J. Bioinformatics and Computational Biology 7(1), 217–242 (2009)
7. Girvan, M., Newman, M.E.J.: Community structure in social and biological networks. Proceedings of the National Academy of Sciences, USA 99, 8271–8276 (2002)
8. Glover, F.: Tabu search - part i. INFORMS Journal on Computing 1(3), 190–206 (1989)
9. Hubert, L., Arabie, P.: Comparing partitions. Journal of Classification 2, 193–218 (1985)
10. Karp, R.M.: Reducibility among combinatorial problems. In: Proceedings of a Symposium on Complexity of Computer Computations, pp. 85–103 (1972)
11. Kirkpatrick, S., Gelatt, C.D., Vecchi, M.P.: Optimization by simulated annealing. Science 220, 671–680 (1983)
12. Newman, M.E.J.: Detecting community structure in networks. The European Physical Journal B 38, 321–330 (2004)
13. Pattillo, J., Youssef, N., Butenko, S.: Clique Relaxation Models in Social Network Analysis. In: Handbook of Optimization in Complex Networks, pp. 143–162 (2012)
14. Rothenberg, R.B., Potterat, J.J., Woodhouse, D.E.: Personal Risk Taking and the Spread of Disease: Beyond Core Groups. The Journal of Infectious Diseases, 174(suppl. 2), S144–S149 (1996)
15. Russell, S.J., Norvig, P.: Artificial Intelligence: A Modern Approach, 2nd edn. Prentice Hall, Upper Saddle River (2003)
16. Schaeffer, S.E.: Stochastic Local Clustering for Massive Graphs. In: Ho, T.-B., Cheung, D., Liu, H. (eds.) PAKDD 2005. LNCS (LNAI), vol. 3518, pp. 354–360. Springer, Heidelberg (2005)
17. Schaeffer, S.E.: Graph clustering. Computer Science Review 1(1), 27–64 (2007)

18. Sun, J., Xie, Y., Zhang, H., Faloutsos, C.: Less is more: Sparse graph mining with compact matrix decomposition. Stat. Anal. Data Min. 1, 6–22 (2008)
19. Tang, J., Wang, T., Wang, J., Lu, Q., Li, W.: Using complex network features for fast clustering in the web. In: Proceedings of the 20th International Conference Companion on World Wide Web, WWW 2011, pp. 133–134. ACM, New York (2011)
20. Černý, V.: Thermodynamical approach to the traveling salesman problem: An efficient simulation algorithm. Journal of Optimization Theory and Applications 45, 41–51 (1985)
21. Šíma, J., Schaeffer, S.E.: On the NP-Completeness of Some Graph Cluster Measures. In: Wiedermann, J., Tel, G., Pokorný, J., Bieliková, M., Štuller, J. (eds.) SOFSEM 2006. LNCS, vol. 3831, pp. 530–537. Springer, Heidelberg (2006)
22. Watts, D.J.: Small Worlds. Princeton University Press (1999)
23. Zachary, W.W.: An information flow model for conflict and fission in small groups. Journal of Anthropological Research 33, 452–473 (1977)

A Hybrid Soft Computing Approach
for Optimizing Design Parameters
of Electrical Drives*

Alexandru-Ciprian Zăvoianu[1,3], Gerd Bramerdorfer[2,3], Edwin Lughofer[1],
Siegfried Silber[2,3], Wolfgang Amrhein[2,3], and Erich Peter Klement[1,3]

[1] Department of Knowledge-based Mathematical Systems/Fuzzy Logic
Laboratorium Linz-Hagenberg, Johannes Kepler University of Linz, Austria
[2] Institute for Electrical Drives and Power Electronics,
Johannes Kepler University of Linz, Austria
[3] ACCM, Austrian Center of Competence in Mechatronics, Linz, Austria

Abstract. In this paper, we are applying a hybrid soft computing approach for optimizing the performance of electrical drives where many degrees of freedom are allowed in the variation of design parameters. The hybrid nature of our approach originates from the application of multi-objective evolutionary algorithms (MOEAs) to solve the complex optimization problems combined with the integration of non-linear mappings between design and target parameters. These mappings are based on artificial neural networks (ANNs) and they are used for the fitness evaluation of individuals (design parameter vectors). The mappings substitute very time-intensive finite element simulations during a large part of the optimization run. Empirical results show that this approach finally reduces the computation time for single runs *from a few days to several hours* while achieving Pareto fronts with a similar high quality.

Keywords: hybrid soft computing methods, multi-objective genetic algorithms, feed-forward artificial neural networks, electrical drives.

1 Introduction

1.1 Motivation

Today about 70% of the total consumption of electrical energy in industry and about 40% of global electricity is used for electric drives. In [3] it is stated that about 200TWh are actually needless wasted energy in the European Union and could be saved by increasing the efficiency of electrical drives. For that reason,

* This work was conducted in the realm of the research program at the Austrian Center of Competence in Mechatronics (ACCM), which is a part of the COMET K2 program of the Austrian government. The work-related projects are kindly supported by the Austrian government, the Upper Austrian government and the Johannes Kepler University Linz. The authors thank all involved partners for their support. This publication reflects only the authors' views.

V. Snasel et al. (Eds.): SOCO Models in Industrial & Environmental Appl., AISC 188, pp. 347–358.
springerlink.com   © Springer-Verlag Berlin Heidelberg 2013

a regulation was concluded in 2009 by the European Union forcing a gradual increase of the energy efficiency of electrical drives [8]. However, manufacturers of electrical machines need to take more than just the efficiency into account in order to be competitive on the global market. To be able to successfully compete, the electrical drives should be fault-tolerant, should offer easy controllable operational characteristics and compact dimensions, and, last but not least, should have a very competitive price. As such, the need to simultaneously optimize electrical drives with regards to several objectives is self evident.

In order to evaluate (predict) the operational behavior of an electrical machine for a concrete design parameter setting, very time-intensive finite element (FE) simulations (on partial differential equations) need to be performed due to the nonlinear behavior of the used materials. Even the state-of-the-art optimization algorithms, typically require the evaluation of thousands of such designs in order to produce high quality results. By using computer clusters, several evaluations can be performed in parallel, significantly speeding up the optimization procedure. Still, the need to evaluate every single design by means of FE-simulations remains a major drawback. Because of this dependency on FE-simulations, optimization runs can take several days to complete, even when distributing computations over a computer cluster. In this paper we describe an effective hybrid soft computing approach which is able to greatly reduce the optimization run-time by relaxing the dependency on FE-simulations.

1.2   Problem Statement and State-of-the-Art

The design of an electrical machine usually comprises at least the optimization of the geometric dimensions of a pre-selected topology. Furthermore, because of volatility in the global raw material market, companies tend to investigate the quality of the electrical drive design with regard to different construction materials. Formally, the three multi-objective optimization problems (MOOPs) that we use in this study, can be defined as:

$$\min \left(f_1(\mathbf{x}), f_2(\mathbf{x}), ..., f_k(\mathbf{x}) \right)$$

where

$$f_1(\mathbf{x}), f_2(\mathbf{x}), ..., f_k(\mathbf{x}) \quad \text{and} \quad \mathbf{x}^T = \begin{bmatrix} x_1 & x_2 & ... & x_n \end{bmatrix} \tag{1}$$

are the objectives and the design parameter vector (e.g. geometric dimensions, material properties, etc.). Additionally, hard constraints like $\mathbf{g}(\mathbf{x})$ can be specified in order to make sure that the drive exhibits a valid operational behavior (e.g. the torque ripple is upper bound). Such constraints are also used for invalidating designs with a very high price.

$$\mathbf{g}(\mathbf{x}) \leq 0 \in \mathbb{R}^m \tag{2}$$

Generally, in order to characterize the solution of MOOPs it is helpful to first explain the notion of *Pareto dominance* [4]: given a set of objectives, a solution A is said to Pareto dominate another solution B if A is not inferior to B with regards to any objectives and there is at least one objective for which A is better than B.

In MOOPs, a single optimal solution is extremely difficult to find and, in many cases, such a solution does not even exist. The result of an optimization process for a MOOP is usually a set of Pareto-optimal solutions named the *Pareto front* [4]. The ideal result of the optimization is an evenly spread Pareto front which is as close as possible to the *true Pareto front* of the problem, i.e. the set of all non-dominated solutions in the search space.

A widespread classical approach to solve MOOPs is to combine the multiple objectives into a single aggregating objective (e.g. a linear combination of the initial objectives) and then use a single-objective optimization technique to find a solution [4]. The downside of this method is that when trying to find multiple (Pareto-optimal) solutions, several independent runs are needed with the hopes that each run would yield a different solution. Unsurprisingly, population based optimization methods from the field of soft computing like particle swarm optimization [19], ant colony optimization [2] and especially evolutionary algorithms [4] perform much better in the context of MOOPs. This is because of the inherent trait of all these methods to step-wise improve sets (populations) of solutions. As such, various extensions aimed at making the contained populations store and efficiently explore Pareto fronts have enabled these types of algorithms to efficiently find multiple Pareto-optimal solutions for MOOPs in one single run.

During the last decade, the use of soft computing methods like genetic algorithms [15] and particle swarm optimization [17] has also become state-of-the-art in the design process of electrical machines and associated electronics. A detailed review regarding the performance of these and other optimization methods used in the field of electrical machine design can be found in [6].

1.3   Our Approach

Recent hybrid learning and optimization methods have proven to be very efficient at tackling complex real-world problems [1]. In order to obtain a fast and efficient optimization framework, we also resort to a hybrid approach that combines two well known methods from the field of soft computing [10]: 1.) evolutionary algorithms - in particular, we consider the specialized class of multiple-objective evolutionary algorithms (MOEAs) [4] and 2.) artificial neural networks (ANNs) - in particular, network models belonging to the multilayer perceptron (MLP) paradigm [12].

The use of MOEAs to efficiently explore the search space and converge (in relatively few generations) to an accurate Pareto can be considered a somewhat conventional approach and, throughout this paper, we shall refer to this basic application of MOEAs in the design process as **ConvOpt**. The fitness evaluation function in ConvOpt is very time-intensive as it dependents on FE-simulations. As such, despite the rather fast convergence in terms of generations that need to be computed, the entire optimization process is quite slow.

Our idea for reducing the evaluation time of individuals is to substitute the time-intensive evaluation function based on FE simulations with a very fast approximation function based on highly accurate regression models, i.e. *mappings* between the motor design parameters and the target values which should be estimated. These targes are actually optimization specific objectives (1) and constraints (2). As the mappings are specific for each optimization scenario, they need to be *constructed on-the-fly, at each run of the evolutionary algorithm*. This means that only individuals from the first few generations will be evaluated with the time-intensive FE-based evaluation function in order to construct a training set for the target mappings. For the remaining generations, the mappings will substitute the FE simulation as the basis of the fitness function. This hybrid approach (**HybridOpt**) will yield a significant reduction in computation time, from *a few days to several hours*, as we verified during empirical tests.

Initially, for constructing the mappings, we considered easy and fast to train linear models. However, these models turned out not to be sufficiently accurate. Thus, we exploited non-linear techniques and finally decided to use ANNs because of the following reasons: 1.) they possess the universal approximation capability [14], 2.) they are known to perform very well on noisy data [16] (in our application, noise may arise due to slight environmental variations during the optimization) and 3.) they have already been successfully applied in evolutionary computation for designing mappings for *surrogate functions* on several instances [13]. In order to elicit ANN models of optimal complexity (# of neurons), we rely on a best parameter grid search and we designed a new selection method (Section 2.2) that successfully balances model accuracy and sensitivity on the one side and model complexity on the other.

2   Optimization Procedure

2.1   Multi-objective Evolutionary Algorithms

Two of the mainstream evolutionary algorithms used for solving MOOPs are NSGA-II [5] and SPEA2 [21]. Because of the similar design principles and similar performance of the two algorithms with regards to our test scenarios, we decided to apply NSGA-II for the purpose of this research. The individuals with which the algorithm operates are represented by real valued design parameter vectors. The size of the design vector ranges from six to ten. For all considered objectives, minimization towards a value of 0 is the preferred option (as explained in Section 1.2).

NSGA-II stores at each generation t two distinct populations of the same size n, a parent population $P(t)$ and an offspring population $O(t)$. Population $P(t+1)$ is obtained by selecting the best n individuals from the combined populations of the previous generation, i.e., from $C(t) = P(t) \cup O(t)$. The fitness of an individual is assessed by using two metrics. The first metric is a classification of the individuals in the population into non-dominated fronts. The first front $F_1(t)$ is the highest level Pareto front and contains the Pareto optimal set from $C(t)$. The subsequent lower-level fronts $F_j(t), j > 1$ are obtained by removing higher level Pareto fronts from the population and extracting the Pareto optimal set from the remaining individuals, i.e., $F_j(t), j > 1$ contains the Pareto optimal set from $C(t) \setminus \bigcup_{k=1}^{j-1} F_k(t)$. Individuals in a higher-level front $F_j(t)$ are ranked as having a higher fitness than individuals in a lower-level front $F_{j+1}(t)$. NSGA-II uses a second metric, the crowding distance, in order to rank the quality of individuals from the same front. The crowding distance associated to a certain individual is an indicator of how dense the non-dominated front is around that individual. Population $P(t+1)$ is obtained by adding individuals from the higher non-dominated fronts, starting with $F_1(t)$. If a front is too large to be added completely, ties are broken in favor of the individuals that have the higher crowding distance — are located in a less crowded region. Population $O(t+1)$ is obtained from population $P(t+1)$ by using binary tournament selection, simulated binary crossover and polynomial mutation. At every generation t, the individuals that do not satisfy the imposed constraints (2) are removed from $O(t)$ before constructing $C(t)$.

2.2   Fitness Function Calculation Using ANNs

Basic Idea. Figure 1 contains an overview of the computational stages of the two optimization processes when wishing to evolve a total of M generations. Because of the very lengthy runs, for both methods, the result of a single run is the Pareto front extracted from the combined set of all the evaluated individuals.

In the *FE-based MOEA execution stage* the first N generations of each MOEA run are computed using FE simulations and all the valid individuals evaluated at this stage will form the training set used to construct ANN mappings. Each sample in this training set contains the initial electrical motor design parameter values and the corresponding target output values computed using FE simulation software. In the *mapping construction stage*, we use systematic parameter variation and a selection process that takes into consideration both accuracy and architectural simplicity in order to find the most robust ANN design for each of the considered target variables and train the appropriate ANN mapping models.

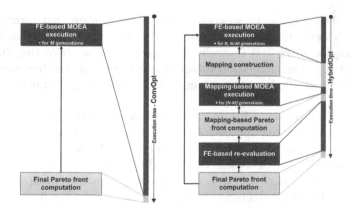

Fig. 1. Diagram of the state-of-the-art, conventional optimization process - ConvOpt (left side) and of our ANN-based hybrid approach - HybridOpt (right side)

The next step is to switch the MOEA to a mapping-based evaluation function for the remaining generations that we wish to compute (*mapping-based MOEA execution stage*). The mapping-based evaluation function is extremely fast when compared to its FE-based counterpart, and it enables the prediction of targets based on input variables within milliseconds.

In the *mapping-based Pareto front computation stage* a preliminary mapping-based Pareto front is extracted only from the combined set of individuals evaluated using the mappings. All the solutions contained in the mapping-based Pareto front are re-evaluated using FE computations in the next stage of HybridOpt (the *FE-based reevaluation stage*). This is necessary in order to assure geometric valid solutions. In the *final Pareto front computation stage*, the final Pareto front of the simulation is extracted from the combined set of all the individuals evaluated using FE simulations.

ANN-Based Mapping - Structure and Training. The multilayer perceptron (MLP) architecture (Figure 2(a)) consists of one layer of input units (nodes), one layer of output units and one or more intermediate (hidden) layers. MLPs implement the feed-forward information flow which directs data from the units in the input layer through the units in the hidden layer to the unit(s) in the output layer. Any connection between two units u_i and u_j has an associated weight $w_{u_i u_j}$ that represents the strength of that respective connection. The weights are initialized with small random values and they are subsequently adjusted during a training process based on the standard back-propagation algorithm [18].

In our modeling tasks, we use MLPs that are fully connected and have a single hidden layer. The number of units in the input layer is equal to the size of the design parameter vector. Also, as we construct a different mapping for each target variable in the data sample, the output layer contains just one unit and, at the end of the feed-forward propagation, the output of this unit is the predicted regression value of the elicited target, e.g. $P(o_1)$ for the MLP presented in Figure 2(a).

We have chosen to adopt an early stopping mechanism that terminates the execution whenever the prediction error computed over a validation subset V does not improve over 200 consecutive iterations, thus preventing the model from over-fitting. This validation subset is constructed at the beginning of the training process by randomly sampling 20% of the training instances.

ANN-Based Mapping - Evaluation and Automatic Model Selection. One important issue in our approach concerns an appropriate automatic selection of parameters in order to construct an accurate and robust model (mapping) in terms of expected prediction quality on new data. The basic idea is to conduct a best parameter grid search, iterating over different parameter value combinations (including number of hidden neurons, the learning rate and the momentum). For each parameter combination, a model is constructed and its predictive quality is assessed using 10-fold cross validation. This leads to a fairly large pool of different models.

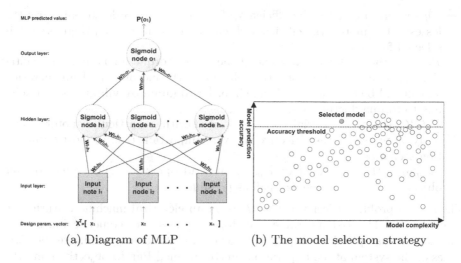

(a) Diagram of MLP   (b) The model selection strategy

Fig. 2. Multilayer perceptron architecture and model selection strategy

Our new, automatic, model selection strategy (Figure 2(b)) is divided into two stages. Initially, all the models are ranked according to a metric that takes into account the accuracy of their predictions measured in terms of the mean R^2 over all folds minus the standard deviation of R^2 over all folds. The latter reflects the sensitivity of the model for the concrete choices of the folds, indicating whether a mapping is biased towards specific regions of the search space. Next, an *accuracy threshold* is computed as the mean of the accuracies of the best performing 2% of all models. The choice of the final model is made using a complexity metric that favors the least complex model (i.e., lowest number of hidden units) which has a predictive accuracy higher than the accuracy threshold.

3   Evaluation and Results

3.1   The Optimization Scenarios

We consider three multi-objective optimization scenarios from the field of designing and prototyping electrical drives:

The first scenario (Scenario *OptS1*) is on an electrical drive featuring a slotted stator with concentrated coils and an interior rotor with buried permanent magnets. The rotor and stator topologies are shown in Figure 3. The design parameter vector is given by $\mathbf{X}^T = \begin{bmatrix} h_m & \alpha_m & e_r & d_{si} & b_{st} & b_{ss} \end{bmatrix}$, where all parameters are shown in Fig. 3 except for α_m, which denotes the ratio between the actual magnet size and the maximum possible magnet size as a result of all other geometric parameters of the rotor. For this scenarios, the targets of the ANN mapping construction phase are the four, unconstrained, Pareto objectives:

- $-\eta$ - where η denotes the efficiency of the motor. In order to minimize the losses of the motor, the efficiency should be maximized and therefore $-\eta$ is selected for minimization.
- T_{cogPP} - the peak-to-peak-value of the motor torque for no current excitation. This parameter denotes the behavior of the motor at no-load operation and should be as small as possible in order to minimize vibrations and noise due to torque fluctuations.
- $TotalCosts$- the material costs associated with a particular motor. Obviously, minimizing this objective is a very important task in most optimization scenarios.
- T_{rippPP} - the equivalent of $T_{cog,PP}$ at load operation. The values of this objective should also be as small as possible.

The second problem (Scenario $OptS2$) is on an electrical machine featuring an exterior rotor. The design parameter vector contains seven geometric dimensions. The aim of this optimization problem was to simultaneously minimize the total losses of the system at load operation (unconstrained Pareto objective) and the total mass of the assembly (constrained Pareto objective). This scenarios also contains a secondary constraint (2) imposed on a geometrical dimension of the evolved motor designs. This means that, for this scenario, we have a total of three targets in the mapping construction stage.

The third problem (Scenario $OptS3$) also concerns a motor with an exterior rotor. The design parameter vector has a size of ten. This scenario proposes four constrained Pareto objectives: 1.) l_s - the total axial length of the assembly, 2.) $TotalMass$ - the total mass of the assembly, 3.) P_{Cu}- the ohmic losses in the stator coils, 4.)P_{fe} - the total losses due to material hysteresis and eddy currents in the ferromagnetic parts of the motor.

3.2   The Testing Framework

Both ConvOpt and HybridOpt use the NSGA-II and SPEA2 implementations provided by the jMetal package [7]. In case of all the tests reported in Section 3.4, we used NSGA-II with a crossover probability of 0.9, a crossover distribution index of 20, a mutation probability of 0.2 and a mutation distribution index of 20. These are standard values recommended by literature [5] and set as default in jMetal. We conducted a preliminary tuning phase with ConvOpt to check whether different settings for the crossover and mutation distribution indexes would yield better results but we found no improvement over the standard values.

In the case of HybridOpt, we performed the mapping training stage after $N = 25$ generations (for the motivation of this choice, please see Section 3.4). As we used a population size of 50, the maximum possible size of the training sets is 1250. The size of the actual training sets we obtained was smaller, ranging from 743 to 1219 samples. This is because some of the evolved design configurations were geometrically unfeasible or invalid with regards to given optimization constraints.

The MLP implementation we used for our tests is largely based on the one provided by the WEKA (Waikato Environment for Knowledge Analysis) open

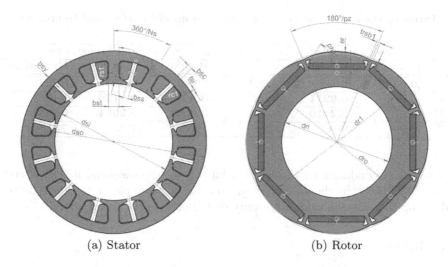

(a) Stator   (b) Rotor

Fig. 3. Interior rotor topology with embedded magnets

source machine learning platform [11]. In the case of the best parameter grid searches that we performed in order to create the MLP-based mappings: 1.) the *number of hidden units* was varied between 2 and double the number of design variables, 2.) the *learn rate* was varied between 0.05 and 0.40 with a step of 0.05 and 3.) the *momentum* was varied between 0.0 and 0.7 with a step of 0.1.

The search is quite fine grained as it involves building between 704 (scenario *OptS1*) and 1216 (scenario *OptS3*) MLP models for each elicited target. This approach is possible because we make use of the early stopping mechanism in the model training process (see Section 2.2) which in turn assures a quite low average MLP model training time of 361.12 seconds. We achieve a considerable speedup in the mapping creation stage by distributing all the MLP training tasks over a cluster computing environment that is also used to run in parallel required FE-simulations. As a result, the mapping creation stage took, on average, 149.26 minutes, over all performed tests.

3.3   Considered Performance Metrics

In order to compare the performance and behavior of the conventional and hybrid optimization processes we use four performance metrics: 1.) the hypervolume metric $\mathcal{H}$ [9] measures the overall coverage of the obtained Pareto set; 2.) the generalized spread metric $\mathcal{S}$ [20] measures the relative spatial distribution of the non-dominated solutions; 3.) the FE utility metric $\mathcal{U}$ offers some insight on the efficient usage of the FE evaluations throughout the simulation (higher values are better); 4.) the run-time metric $\mathcal{T}$ records the total runtime in minutes required by one simulation; The $\mathcal{H}$ metric has the added advantage that it is the only MOEA metric for which we have theoretical proof [9] of a monotonic behavior. This means that the maximization of the hypervolume constitutes the necessary

Table 1. The average performance over five runs of ConvOpt and HybridOpt

Metric	Scenario *OptS1*		Scenario *OptS2*		Scenario *OptS3*	
	ConvOpt	HybridOpt	ConvOpt	HybridOpt.	ConvOpt	HybridOpt
$\mathcal{H}$	**0.9532**	0.9393	**0.8916**	0.8840	**0.4225**	0.3691
$\mathcal{S}$	0.7985	**0.6211**	0.8545	**0.8311**	0.4120	0.4473
$\mathcal{U}$	0.1315	**0.2210**	0.0016	**0.0064**	**0.2901**	0.2362
$\mathcal{T}$	2696	**991**	3798	**1052**	4245	**2318**

and sufficient condition for the set of solutions to be *maximally diverse Pareto optimal solutions*. In the case of the $\mathcal{S}$ metric, a value closer to 0.0 is better, indicating that the solutions are evenly distributed in the result space.

3.4   Results

In order to obtain a quick overview of the performance of linear models compared to ANN non-linear mappings, we conducted some preliminary modeling tests on the 11 targets of all three optimization scenarios: it turned out that the average model quality measured in terms of R^2 was 0.859 in case of the linear models, whereas ANNs achieved a value of 0.984. On three targets, the linear models were almost useless, as they achieved an R^2 of around 0.6.

In the current version of HybridOpt, it is very important to choose a good value for the parameter N that indicates for how many generations we wish to run the initial FE-based execution stage.

Finally, based on extensive tests, we have chosen $N = 25$ as this is the *smallest value of N for which the trained models exhibit both a high prediction accuracy as well as a high prediction stability*. Over all three data sets and the 6 non-linear targets, the generational coefficients of determination (for generations 31 to 100) obtained by the models constructed using samples from the first 25 generations 1.) are higher than 0.9 in 94.52% of the cases; and 2.) are higher than those obtained by the models constructed using 20-24 generations in 57.52% of the cases and by those obtained by the models constructed using 26-30 generations in 42.52% of the cases.

The comparative performance of ConvOpt and HybridOpt is presented in Table 1. The results for each scenario are averaged over five optimization runs. On the highly constrained scenario *OptS3*, the hybrid optimization process is a little bit worse. The main reason for this is that the hard constraints determine a high ratio of invalid individuals to be generated during the mapping based evaluation stage. However, the computation time could still be reduced by $\approx$ 45%. Even though, for this scenario, ConvOpt produces Pareto fronts with a better $\mathcal{H}$, HybridOpt is still able to evolve well balanced individual solutions in key sections of the Pareto front — please see Figure 4 for two such examples.

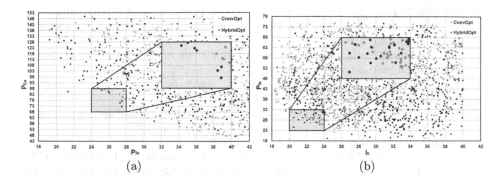

Fig. 4. 2D projections of the full Pareto fronts for the highly constrained scenario *OptS3*. The black dots denote solutions obtained using HybridOpt. These solutions are relatively equally distanced, with regards to the considered objectives, from the origin of the projected Pareto space.

4 Conclusion

In this paper, we investigated multi-objective optimization algorithms based on evolutionary strategies, using the famous and widely used NSGA-II algorithm, for the purpose of optimizing the design of electrical drives in terms of efficiency, costs, motor torque behavior, total mass and others. As the design and the target parameter space is quite large, we end up having complex optimization scenarios that require very long computation runs.

In order to alleviate this problem, we experimented with a system that automatically creates, on-the-fly, non-linear MLP-based mappings between design and target parameters. Empirical observations over averaged results indicate that, by replacing the time-intensive FE simulations with fast mapping based estimations of the target values, we are able to reduce the overall run-time of the optimization process, in average, by more than 60%. At the same time, the hybrid optimization process is able to produce Pareto fronts of similar quality to the ones obtained using the conventional FE-based optimization.

References

1. Abraham, A., Corchado, E., Corchado, J.M.: Hybrid learning machines. Neurocomputing 72(13-15), 2729–2730 (2009)
2. Alaya, I., Solnon, C., Ghedira, K.: Ant colony optimization for multi-objective optimization problems. In: 19th IEEE International Conference on Tools with Artificial Intelligence, ICTAI 2007, vol. 1, pp. 450–457 (2007)
3. De Keulenaer, H., Belmans, R., Blaustein, E., Chapman, D., De Almeida, A., De Wachter, B., Radgen, P.: Energy efficient motor driven systems. Tech. rep., European Copper Institute (2004)
4. Deb, K.: Multi-Objective Optimization using Evolutionary Algorithms. Wiley-Interscience Series in Systems and Optimization. John Wiley & Sons, Chichester (2001)

5. Deb, K., Pratap, A., Agarwal, S., Meyarivan, T.: A fast and elitist multiobjective genetic algorithm: NSGA-II. IEEE Transactions on Evolutionary Computation 6(2), 182–197 (2002)
6. Duan, Y., Ionel, D.: A review of recent developments in electrical machine design optimization methods with a permanent magnet synchronous motor benchmark study. In: 2011 IEEE Conference on Energy Conversion Congress and Exposition (ECCE), pp. 3694–3701 (September 2011)
7. Durillo, J.J., Nebro, A.J.: JMETAL: A java framework for multi-objective optimization. Advances in Engineering Software 42, 760–771 (2011)
8. European Union: Regulation (eg) nr. 640/2009
9. Fleischer, M.: The Measure of Pareto Optima. In: Fonseca, C.M., Fleming, P.J., Zitzler, E., Deb, K., Thiele, L. (eds.) EMO 2003. LNCS, vol. 2632, pp. 519–533. Springer, Heidelberg (2003)
10. Grosan, C., Abraham, A.: Intelligent Systems: A Modern Approach. Intelligent Systems Reference Library Series. Springer, Germany (2011) ISBN 978-3-642-21003-7
11. Hall, M., Frank, E., Holmes, G., Pfahringer, B., Reutemann, P., Witten, I.H.: The WEKA data mining software: an update. SIGKDD Explor. Newsl. 11, 10–18 (2009)
12. Haykin, S.: Neural Networks: A Comprehensive Foundation, 2nd edn. Prentice Hall Inc., Upper Saddle River (1999)
13. Hong, Y.S., Lee, H., Tahk, M.J.: Acceleration of the convergence speed of evolutionary algorithms using multi-layer neural networks. Engineering Optimization 35(1), 91–102 (2003)
14. Hornik, K., Stinchcombe, M., White, H.: Multilayer feedforward networks are universal approximators. Neural Networks 2(5), 359–366 (1989)
15. Jannot, X., Vannier, J., Marchand, C., Gabsi, M., Saint-Michel, J., Sadarnac, D.: Multiphysic modeling of a high-speed interior permanent-magnet synchronous machine for a multiobjective optimal design. IEEE Transactions on Energy Conversion 26(2), 457–467 (2011)
16. Paliwal, M., Kumar, U.A.: Neural networks and statistical techniques: A review of applications. Expert Systems with Applications 36(1), 2–17 (2009)
17. del Valle, Y., Venayagamoorthy, G., Mohagheghi, S., Hernandez, J.C., Harley, R.: Particle swarm optimization: Basic concepts, variants and applications in power systems. IEEE Transactions on Evolutionary Computation 12(2), 171–195 (2008)
18. Werbos, P.J.: Beyond Regression: New Tools for Prediction and Analysis in the Behavioral Sciences. Ph.D. thesis, Harvard University (1974)
19. Zhao, S.Z., Iruthayarajan, M.W., Baskar, S., Suganthan, P.: Multi-objective robust pid controller tuning using two lbests multi-objective particle swarm optimization. Information Sciences 181(16), 3323–3335 (2011)
20. Zhou, A., Jin, Y., Zhang, Q., Sendhoff, B., Tsang, E.: Combining model-based and genetics-based offspring generation for multi-objective optimization using a convergence criterion. In: IEEE Congress on Evolutionary Computation, CEC 2006, pp. 892–899 (2006)
21. Zitzler, E., Laumanns, M., Thiele, L.: SPEA2: Improving the Strength Pareto Evolutionary Algorithm for Multiobjective Optimization. In: Giannakoglou, K., et al. (eds.) Evolutionary Methods for Design, Optimisation and Control with Application to Industrial Problems (EUROGEN 2001). International Center for Numerical Methods in Engineering (CIMNE), pp. 95–100 (2002)

Improvement of Accuracy in Sound Synthesis Methods by Means of Regularization Strategies

M.D. Redel-Macías and A.J. Cubero-Atienza

Dep. Rural Engineering, Ed Leonardo da Vinci, Campus de Rabanales, University of Cordoba, Campus de Excelencia Internacional Agroalimentario, CeiA3, 14071 Cordoba, Spain

Abstract. Sound quality is one of the main factors intervening in customers' preferences when selecting a motor vehicle. For that reason, increasingly more precision in the models is demanded in the prediction of noise, as alternative to the traditional jury tests. Using sound synthesis methods, it is possible to obtain the auralization of sound produced by a physical sound source as it would be heard in an arbitrary receptor position. The physical source is represented by an acoustic equivalent source model and the engine noise is experimentally characterized by means of the substitution monopole technique. However, some factors have an influence on the accuracy of the model obtained such as regularization techniques. In this study the influence of the regularization techniques on the accuracy of the models has been discussed. It was found that the use of iterative algorithm improve the accuracy of the model compared to non-iterative techniques.

Keywords: regularization, sound synthesis, sound quality.

1 Introduction

By means of auralization, the sound produced is rendered by a physical sound source in space, simulating the listening experience in a position given by the space model [1, 2]. A typical application is the measurement of acoustic transfer paths of sound radiated from a vehicle. It is well known that there are different kinds of sound source in a vehicle which may be interesting to identify [3, 4]. In this sense, the auralization process begins with a source-transmission path-receiver model to predict the time-frequency spectrum of the sound field produced by an active source at the receiver prediction. The following step is to determine the transfer paths between the source and the receiver experimentally or numerically [5]. Combining the two previous steps we can calculate the time-frequency spectrum in the receptor position. For non stationary sources a discretization in the time should be made, although this can only be carried out by repeating the method of prediction of a simple spectrum in the time. If the receptor or the source is in movement it is also necessary to consider a discretization in the space and to determine the diverse transfer paths intervening in the process [6]. Once the monoaural or binaural time-frequency spectrum has been determined in a certain position, the temporal signal of the sound is synthesized based on the characteristics of the sound, like the different orders in an engine spectrum [7]. The result of this process is finding out the real experience of an observer in a specific position

V. Snasel et al. (Eds.): SOCO Models in Industrial & Environmental Appl., AISC 188, pp. 359–367.
springerlink.com

obtaining the possibility of assessing the sound quality perceived quickly and economically. Furthermore, sound quality is of great importance in achieving sound which is agreeable to the human ear, in fact noise annoyance not only depends on sound exposure levels [8].

Auralization models allow the definition of noise standards, thus more realistic and effective legislation can be developed. Moreover, the currently applied test procedures, such as the recent change in the ISO 362 procedure for vehicle pass-by noise testing can be referred. The new version of this standard, a new procedure has been proposed based on a series of experiments; with the aid of accurate auralization models. This work's contribution is aimed at the assessment of accuracy on engine source models which are used in the first step of a vehicle pass-by sound synthesis. The influence of the regularization techniques has been evaluated to establish the optimal configuration in each case.

2 Materials and Methods

First, the source model needs to be determined. Next, the type of descriptor and the method of quantification of these descriptors have to be selected. The point sources or monopoles are also known as an omni directional sound source of a constant volume velocity. Monopole sound source are required for reciprocal measurements such as Noise Vibration and Harshness (NVH) applications [9]. It is due to very different space requirements of sound sources and sensors; measurements of acoustic transfer functions are often much easier done reciprocally, i.e. when source and sensor are interchanged.

Once the model to be followed for the source is established, the type of descriptor to be used has to be fixed. In this sense, of the three descriptor types found: velocity patterns [5], pressure measurements at indicator positions [10], and acoustic particle velocity measurements at indicator positions [7], the one most widely used is that based on sound pressure levels at a specific distance from the sound source. In the source models applied here, the substitute sound sources are quantified by their volume velocities. Effectively, a monopole sound source has a volume velocity output in cubic meters per second and an omni directional directivity. Volume velocity is thus equal to the particle velocity (in meters per second) times the surface of the sound source. These are obtained from pressure measurements of microphones at indicator positions, since the employment of any other type of descriptor for engine sound is difficult to measure and costly.

For the quantification of these descriptors, 'Airborne Source Quantification (ASQ)' technique has been developed [11].

The Transfer Path Analysis (TPA) is a procedure by which it is possible to determine the flow of vibro-acoustic energy from the source, by means of solid structures and air, up to a specific receptor position [10]. When it only intervenes as a means of air transmission the TPA is denominated ASQ. Although in some cases the postprocessing time can be important, the inverse ASQ method is, in general, suitable for a sound synthesis approach, as will be described later on.

A real source model is obtained by a substitution source model, in which M monopoles are distributed over the radiating surface of the source with acoustic quantities

in N field points or indicator points. The relation between the complex volume velocities q of the M monopoles and the pressures p in N field points is given by the ASQ
method by (1), in which H is a frequency response matrix

$$p_{(Nx1)} = H_{(NxM)} q_{(Mx1)} \tag{1}$$

The calculation of q starting from p and H requires the inversion of a matrix. When
the number N of indicator points is different from the number of monopoles, the inverse matrix H-1 has to be redefined as a least square pseudo-inverse H+. The method
used the most to obtain the pseudo-inverse is the singular value decomposition
(SVD), in which the matrix H is factorized, as indicated in (2)

$$H = U \sum V^H = \sum_{i=1}^{M} u_i \sigma_i v_i^H \tag{2}$$

in which U is an N-by-N unitary matrix; V is another M-by-M unitary matrix, the
superscript H indicating the Hermitian transpose and σ a diagonal matrix with nonnegative real numbers on the diagonal ($\sigma_i \geq 0$) known as singular values of H matrix. A
common convention is to order the diagonal entries σ in a descending order

$$H^+ = V \sum {}^+ U^H \tag{3}$$

$$q = H^+ \tilde{p} = V \sum {}^+ U^H \tilde{p} = \sum_{i=1}^{M} \frac{u_i^H \tilde{p}}{\sigma_i} v_i \tag{4}$$

being $\sum {}^+ = diag(\sigma_1^{-1}, \sigma_2^{-1}, \dots, \sigma_R^{-1}, 0, \dots, 0)$.

The inversion of the matrix H taking (4) can be easily calculated due to the
decomposition properties of the matrices. The difficulty lies in the fact that, often, this
inversion poses the problem of being ill-conditioned, since most of the systems are
overdetermined (i.e. with M≥N), as stated in [12]. In this case, the application of
pseudo-inverse does not ensure the obtaining of satisfactory results, so that the application of regularization methods is resorted to.

2.1 Regularization Techniques

One of the most and well established regularization methods is the Tikhonov-Phillips
one which finds the solution x_λ^δ to the following minimization problem:

$$\min_x \left\| Hx - y^\delta \right\|^2 + \lambda \|Lx\|^2 \tag{5}$$

where λ is known as the regularization parameter whose value must be determined.
The operator L is used to impose some constraints about the smoothness of the solution. The selection of regularization parameters is highly complex and, in this respect,
a large amount of research works have been done to develop a suitable strategy for
that selection [13]. Among these techniques are the *L-curve* criterion and the *generalized cross validation* (GCV) method. These techniques need to test a large number of

regularization parameters in order to find reasonably good values and this can be very time-consuming although it has the advantage of not requiring any prior knowledge of the error level in p observations. However, other techniques require additional information on the noise present in the data, or on the amount of regularization prescribed by the optimal solution. In this work, the value of the ratio between the regularization part $\|Lx\|^2$ and the value of the Tikhonov functional $\|Hx - y^\delta\|^2 + \lambda \|Lx\|^2$ were assigned, in the hypothesis that the regularization part would somehow preserve the fidelity to the data represented by the residual norm $\|Hx - y^\delta\|^2$. A method is proposed for the updating of regularization parameters, which decreases the Tikhonov functional if the regularization part is too large and increases it otherwise.

With the iterative algorithm, it was aimed to obtain a good solution to the problem posed for an acceptable number of iterations and without it being necessary to know or estimate $\|e\|$. The results obtained were compared with a different number of iterations and with other traditional methods for the selection of regularization parameters, such as the *L-curve* criterion and the *generalized cross validation* (GCV) method.

The L-Curve Validation. The name *L-curve* comes from the curve representing the logarithm of the norm of the solution $\|x\|$ as a function of the logarithm of the residue r, for different values of the regularization parameters β. With a small regularization, the norm of the regularized solution drops abruptly with β for a small variation in the residual norm, this being the 'vertical' part of the *L-curve*. With too much regularization, the residual norm significantly increases with β, whereas the norm of the regularization solution slightly decreases; this is the 'horizontal' part of the *L-curve*. The best value for β is obtained in the vertex of the *L-curve*; this is the best compromise between the minimum of $\|x\|$ and r. The curve is represented on a log-log scale since both magnitudes are not comparable and a relative scale cannot therefore be employed. The curvature function is given by the following expression

$$J_{LCV}(\beta) = \frac{\rho_\beta' \eta_\beta'' - \rho_\beta'' \eta_\beta'}{\left(\rho_\beta'^2 + \eta_\beta'^2\right)^{2/3}} \quad \text{with } \rho_\beta = \log(r_\beta); \eta_\beta = \log(\|x_\beta\|) \tag{6}$$

'and'', being, respectively, the first and second derivative with respect to β. The optimal value β_{LCV} corresponds to the curve's corner maximizing function $J_{LCV}(\beta)$.

Generalized Cross Validation. GCV is the first method for the fit of β. This method is an extension of the *Ordinary Cross Validation principle* (OCV), a method based on the capacity to obtain a solution with $(m-l)$ observations to predict the m-th observation. The GCV function to be minimized is given by

$$J_{GCV}(\beta) = \frac{r_\beta^2}{Tr\left(I_{(mxm)} - hh_\beta^{-1}\right)^2} \tag{7}$$

Tr(x) being for the trace of matrix *x*. The denominator of the GCV function is an estimation of the trend induced by β. Without any regularization (β =0) since hh_β^{-1} is equal to I_{mxm}. When the amount of regularization increases, the matrix hh_β^{-1} moves away from the identity matrix, with the value of the denominator J_{GCV} being increased.

Iterative Tikhonov algorithm. The Tikhonov method establishes that the function (5) has a single minimum for any value $\lambda > 0$, the same as $\theta(x,\lambda) = \left\|Hx - y^\delta\right\|^2 + \lambda\left\|Lx\right\|^2$ in which N(·) is a null space of the matrix.

$$\theta(x,\lambda) = \left\|Hx - y^\delta\right\|^2 + \lambda\left\|Lx\right\|^2 \tag{8}$$

Considering that x_λ is the minimum of the function, this can be characterized as the solution of the system:

$$(H^*H + \lambda L^*L)x_\lambda = H^*y^\delta \tag{9}$$

An novel iterative algorithm is used in this work to computerize the values (x_k, λ_k), where λ_k is updated in each iteration and x_k is obtained by applying the CG iterations to (9) with $\lambda = \lambda_k$. The problem in the regularization techniques is how to determine the regularization parameters. This is because when the parameters are too large the solution will significantly deviate from the correct solution, and when they are too small, they complicate the problem too much. γ being the prescribed weight of regularization part $\hat{\gamma} = 1e^{-1}\left\|Lx\right\|^2$ with respect to the Tikhonov functional, for each given value of λ_k, x_k is computed resolving (7) and γ_k as:

$$\gamma_k = \frac{\left\|Lx_k\right\|^2}{\theta(x_k,\lambda_k)} \tag{10}$$

Employing the following rule for the updating of regularization parameters:
$$\lambda_{k+1} = \lambda_k + si\,gn(\hat{\gamma} - \gamma_k)\mu$$

$$si\,gn(\hat{\gamma} - \gamma_k) = \begin{cases} 1 & if\,(\hat{\gamma} - \gamma_k) > 0 \\ -1 & if\,(\hat{\gamma} - \gamma_k) < 0 \\ 0 & if\,(\hat{\gamma} - \gamma_k) = 0 \end{cases} \tag{11}$$

The precision of the solution obtained for a value of de $\hat{\gamma} = 1e^{-3}$ was verified.

2.2 Problem Description

In previous sections have been shown, highly detailed, the reasons for the selection of the source model, the type of descriptors and the method for the quantification of monopoles. By means of the experiments carried out in this research, it was aimed to measure the influence of certain key factors which affect the precision of these types of models, such as the influence of the regularization strategies (regularization strategy used, selection of parameters or number of iterations). Some authors have evaluated tire sound source in vehicle based on the number of monopoles required and non iterative Tikhonov regularization [14]. Figure 1 shows models with monopoles taken into account in this paper, classified in ascending order as a function of the number of monopoles studied. In all the models, the radiating engine is modelled as a limited set of radiating sub-sources. The radiation of these sub-sources can be approximated to the radiation of a simple monopole by finding the mean of the transfer functions, although this does not modify the number of unknown source strengths q. The amount of points used in the mean of each FRF is determined considering the number of loudspeakers involved.

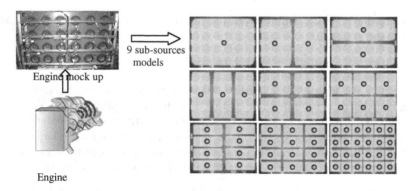

Fig. 1. Overview of models studied

Additionally, the sound emitted by the source was a random burst noise of between 100-10000 Hz (no noise, 1% and 5%). From the perspective of the auralization of the sound in real time, it is important for the number of monopoles to be as small as possible without compromising the accuracy of the spectrum predicted in terms of sound quality.

The set-up used for this research consisted of a rectangular box with outer dimensions 902 mm x 602 mm x 190mm, with 24 loudspeakers separated from it at a distance of 100 mm, see Figure 1. The nearest indicator microphones are positioned at a distance of 0.15 m from the loudspeaker cabinet. The set-up is placed inside a semi-anechoic room, where background levels remains below 35 dB(A). An LMS instrumentation series, consisting of a portable and multi-channel SCADAS meter, several Brüel and Kjaer (B&K) prepolarized free-field half-inch microphones model 4950 and prepolarized free-field quarter-inch microphone model 4954 were utilized as the measuring device. LMS Test.Lab was the measurement software package and all the microphones were calibrated with a B&K calibrator model 4231. Recorded measurements were sent to Matlab for post processing. The measurements were taken with a sampling frequency of 20.480 kHz. The frequency resolution was 1.25 Hz and 50 spectral averages were implemented for analyses. Linear averaging was used to place equal emphasis on all spectra or time records. This type of averaging is helpful for the analysis of stationary signals.

3 Results and Discussion

The inversion of the matrix H in this type of model presents the ill-conditioned problem. The quantification of the substitution sources depends on the conditioning of the problem and success is not guaranteed by only employing the simple least squares method [15]. Figure 2 represents the condition number as a function of the frequency for models 2-24. Some authors [10] indicates that a condition number is small when it reaches values of around 103. Observing Figure 4.9 we can affirm that only model 24M has a high condition number. In our case, the number of indicator microphones was the same for all the models, 24, which was higher than the number of monopoles in models 2-12M and the same as the number of monopoles in model 24M. Note that the condition number increases at low frequencies. Starting from 1 kHz, the condition number becomes stable, around 2, which means that the smallest singular value is equal to 50 percent of the greatest one. For the source reconstruction this signifies a limitation in the resolution achieved. However, for the prediction of the resulting spectrum in the receptor position it is not a problem. Also, as established in [16], the conditioning improves when the geometry of the positions measured which are identical to the geometry of the source is selected, and when these measurements are made close to the source.

The spectrograms in Figure 3 show the error in 1/3rd octave band for the configuration of phase v1 in the target microphone 4 for a random burst noise. It can be observed graphically how the error diminishes when applying the regularization strategies. The superiority of the iteration algorithm with respect to traditional methods is also clear, with the result improving as from 200 iterations.

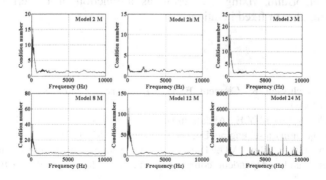

Fig. 2. Condition number models 2-24 M

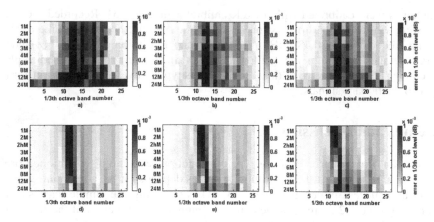

Fig. 3. Error on $1/3^{rd}$ octave band for random burst noise: a) without regularization, b) with GCV technique, c) L-curve technique, d) with Tikhonov iterative 200 it, e) with Tikhonov iterative 300 it and f) with Tikhonov iterative 400 it.

4 Conclusions

Using the monopole substitution technique, the noise of an engine has been experimentally characterized. Different models have been applied, varying the number of monopoles in order to quantify the number of descriptors necessary to obtain an acceptable precision. The improvement in the regularization strategies was demonstrated even when the problem displayed a good conditioning. The results achieved with the Tikhonov iterative algorithm reveal their superiority compared to non iterative regularization techniques. Moreover, it has been shown that the number of monopoles has influence on the number if iterations required. Future research could be aimed at studying other types of sources of vehicle sound and investigating more psychoacoustic parameters, which could be of interest for the characterization of the error committed in the model; as well as minimizing the number of iterations necessary in the regularization algorithm, fixing a criterion as a function of the error instead of establishing a number of fixed iterations.

References

1. Kloth, M., Vancluysen, K., Clement, F., Lars Ellebjerg, P.: Silence project. In: European Comission (2008)
2. Haik, Y.H.Y., Selim, M.Y.E., Abdulrehman, T.: Combustion of algae oil methyl ester in an indirect injection diesel engine. Energy 36(3), 1827–1835 (2011)
3. Albarbar, A., Gu, F., Ball, A.D.: Diesel engine fuel injection monitoring using acoustic measurements and independent component analysis. Measurement 43(10), 1376–1386 (2010)
4. Albarbar, A.: Acoustic monitoring of engine fuel injection based on adaptive filtering techniques. Applied Acoustics 71(12), 1132–1141 (2010)

5. Verheij, J.W., VanTol, F.H., Hopmans, L.J.M.: Monopole airborne sound source with in situ measurement of its volume velocity. In: Bernhard, R.J., Bolton, J.S. (eds.) Proceedings of Inter-Noise 95 - the 1995 International Congress on Noise Control Engineering, vol. 1, 2, pp. 1105–1108 (1995)

6. Leveque, G., Rosenkrantz, E., Laux, D.: Correction of diffraction effects in sound velocity and absorption measurements. Measurement Science &Technology 18(11), 3458–3462 (2007)

7. Berckmans, D.: Numerical case-study on the development of acoustic equivalent source models for use in sound synthesis methods. In: Proceedings of Isma 2008: International Conference on Noise and Vibration Engineering, vols. 1-8, pp. 3073–3083 (2008)

8. Mucchi, E., Vecchio, A.: Acoustical signature analysis of a helicopter cabin in steady-state and run up operational conditions. Measurement 43(2), 283–293 (2010)

9. Castellini, P., Revel, G.M., Scalise, L.: Measurement of vibrational modal parameters using laser pulse excitation techniques. Measurement 35(2), 163–179 (2004)

10. Berckmans, D.: Evaluation of substitution monopole models for tire noise sound synthesis. Mechanical Systems and Signal Processing 24(1), 240–255 (2010)

11. European Parliament, C.: The Assessment and menagement of environmental noise-Declaration by the Comission in the Conciliation Committee on the Directive relating to the assessment and management of environmental noise. Oficial Journal of the European Communities, 14 (2002)

12. Pézerat, C.: Identification of vibration excitations from acoustic measurements using near field acoustic holography and the force analysis technique. Journal of Sound and Vibration 326(3-5), 540–556 (2009)

13. Hafner, M.: Fast neural networks for diesel engine control design. Control Engineering Practice 8(11), 1211–1221 (2000)

14. Berckmans, D.: Numerical Comparison of Different Equivalent Source Models and Source Quantification Techniques for Use in Sound Synthesis Systems. Acta Acustica United with Acustica 97(1), 138–147 (2011)

15. Park, S.H., Kim, Y.H.: Effects of the speed of moving noise sources on the sound visualization by means of moving frame acoustic holography. Journal of the Acoustical Society of America 108(6), 2719–2728 (2000)

16. Kim, Y., Nelson, P.A.: Optimal regularisation for acoustic source reconstrucion by inverse methods. Journal of Sound and Vibration 275(3-5), 463–487 (2004)

Detecting Defects of Steel Slabs
Using Symbolic Regression

Petr Gajdoš and Jan Platoš

Department of Computer Science, FEECS, VŠB – Technical University of Ostrava,
IT4Innovations, Center of Excellence, VŠB – Technical University of Ostrava,
17. listopadu 15, 708 33 Ostrava-Poruba, Czech Republic
{petr.gajdos,jan.platos}@vsb.cz

Abstract. The quality of products of heavy industries plays an important role because of further usage of such products, e.g. bad quality of steel ingots can lead to a poor quality of metal plates and following wastrels in such processes, where these metal plates are consumed. Of course, single and relatively small mistake at the beginning of a complex process of product manufacturing can lead to great finance losses. This article describes a method of defects detection and quality prediction of steel slabs, which is based on soft-computing methods. The proposed method helps us to identify possible defects of slabs still in the process of their manufacturing. Experiment with real data illustrates applicability of the method.

Keywords: Quality prediction, Symbolic Regression, Data Analysis.

1 Introduction

In these days, almost every production process contains some type of computer or automaton. They are used not only for controlling of the process, but also for monitoring many types of sensors. Data from these sensors may be used for quality prediction, but this is not easy task. The quality is affected by many conditions which must satisfy very complex rules. These rules are defined by international and national standards (like ISO, ECS, CSN), technology process and experts. But specification of these rules is not always possible by these methods.

Proposed method of quality prediction of steel slabs described in this article. The process of steel slabs production is quite difficult especially for uninitiated people. Production process is supervised from the beginning mixture of input materials, through the phase of continual pouring, cooling, rolling and cutting until final slab is finished and ready for further consequential processes. A set of measurement is done during mentioned process, such as the measurement of temperatures, pressure, flow velocity, ration of input materials, amount of cooling medium, etc. Some of these values can are measure by usage of contactless sensors (temperature, pressure), the others are measured by usage of expensive destructive tests, e.g. measurement of temperatures inside smelting furnance. Non-destructive measurements can be done in a short time period and for the final decision making process are very important as well as their aggregated forms,

V. Snasel et al. (Eds.): SOCO Models in Industrial & Environmental Appl., AISC 188, pp. 369–377.
springerlink.com

e.g. the minimal temperature during the last hour. On the other hand, destructive measurements are done quite rarely, around 2-3 times during the whole smelting process. All mentioned values are stored with final products, More over, sequences of particular products is stored as well - a steel strip is finally divided into several slabs. These informations are very important for sellers because of possible reclamations and tracking of related slabs, which were distributed separately. It is clear, that quality measurement play an important role and companies have to spend a lot of money on that. Moreover, final products lie under laboratory experiments, however just a randomly chosen sample of all products can be tested because of finance costs - products are debased and excluded from further production process. A complex system for quality measurement and prediction of steel slabs defects should eliminate bad products or parts even during the production.

The proposed method based on Symbolic Regression and Genetic Algorithms makes a part of a complex system for quality measurement. It reduces the need of laboratory experiments because all results (defect predictions) are evaluated with respect to non-destructive measurements of intermediate and final products. All collected information are stores in Quality Data Warehouse (QDW). Many methods for data analysis need normalized data for their proper function. So it is necessary to preprocess data before they are used. As it was mentioned before, technological process is very complex and can use many sensors which produce a lot of data. These data usually contain duplicities, hidden relations, and/or noise which can be very confusing for analytical methods. They also have plenty attributes/features which is expressed in large data dimension. Detailed description of proposed methods, which can be used for on-line prediction and off-line evaluation, is described in the following section.

2   Symbolic Regression via Genetic Programming on GPU

Symbolic regression via genetic programming is a branch of empirical modeling that evolves summary expressions for available data. Although intrinsically difficult (the search space is infinite), recent algorithmic advances coupled with faster computers have enabled application of symbolic regression to a wide variety of industrial data sets. Unique benefits of symbolic regression include human insight and interpretability of model results, identification of key variables and variable combinations, and the generation of computationally simple models for deployment into operational models.

The challenging task of symbolic regression is to identify and express a real or simulated system or a process, based on a limited number of observations of the system's behavior. The system under study is being characterized by some important control parameters which need to be available for an observer, but usually are difficult to monitor, e.g. they need to be measured in a lab, simulated or observed in real time only, or at high time and computational expenses. Empirical modeling attempts to express these critical control variables via other controllable variables that are easier to monitor, can be measured more accurately or timely, are cheaper to simulate, etc. Symbolic regression provides such expressions of crucial process characteristics, or, response variables, defined (symbolically) as mathematical functions of some of the easy-to-measure input variables, and calls these expressions empirical input-output models (or input-response models).

Industrial modeling problems, for which symbolic regression is used, have two main characteristics:

- No or little information is known about the underlying system producing the data, and therefore no assumptions on model structure can be made.
- The available data is high-dimensional, and often imbalanced, with either abundant or insufficient number of samples.

A particular solution of symbolic regression in a form of mathematical formula can be illustrated by a tree structure as well, see the Figure 1. Some nodes have different colors, which will be discussed in more detail later. Edge can have weights to make the whole model more fuzzy.

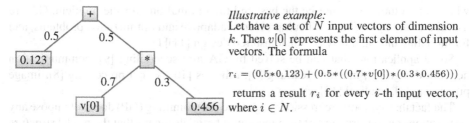

Illustrative example:
Let have a set of N input vectors of dimension k. Then $v[0]$ represents the first element of input vectors. The formula

$$r_i = (0.5*0.123) + (0.5*((0.7*v[0])*(0.3*0.456)))$$

returns a result r_i for every i-th input vector, where $i \in N$.

Fig. 1. Illustrative result of Symbolic Regression

The set of symbols (functions) depends on application area and represents one of the advantages of symbolic regression. In statistics, attribute dependences can be simulated and tested with a set of symbols of logical functions $S = \{\wedge, \vee, \supset, \equiv, \neg\}$, in the area of signal processing with a set of goniometric functions $S = \{sin, cos\}$, etc.

To discover acceptable models with realistic time and computational effort, symbolic regression exploits a stochastic iterative search technique, based on artificial evolution of model expressions. This method called genetic programming looks for appropriate expressions of the response variable in the space of all valid formulas containing a minimal set of input variables and a proposed set of basic operators and constants.

Evolutionary algorithms are stochastic search methods that mimic the metaphor of natural biological evolution, which applies the principles of evolution found in nature to the problem of finding an optimal solution to a solver problem. An evolutionary algorithm is a generic term used to indicate any population-based optimization algorithm that uses mechanisms inspired by biological evolution, such as reproduction, mutation and recombination. Candidate solutions to the optimization problem play the role of individuals in a population, and the cost function determines the environment within the solutions "live". Evolution of the population then takes place after the repeated application of the above operators. Genetic algorithm is the most popular type of evolutionary algorithms.

3   Genetic Algorithms (GA)

Genetic algorithms (GA) described by John Holland in 1960s and further developed by Holland and his students and colleagues at the University of Michigan in the 1960s and 1970s. GA used Darwinian Evolution to extract nature optimization strategies that use them successfully and transform them for application in mathematical optimization theory to find the global optimum in defined phase space [2][6][7].

GA is used to extract approximate solutions for problems through a set of operations "fitness function, selection, crossover, and mutation". Such operators are principles of evolutionary biology applied to computer science. GA search process depends on different mechanisms such as adaptive methods, stochastic search methods, and use probability for search.

Using GA for solving most difficult problems that searches for accepted solution; where this solution may not be the best and the optimal one for the problem. GA are useful for solving real and difficult problems, adaptive and optimization problems, and for modeling the natural system that inspired design [11][1].

Some applications that can be solved by GA are: scheduling [9], communication network design [5], machine learning [4], robotics [10], signal processing [8], image processing [3], medical [12], etc.

The fact that symbolic regression via genetic programming (GP) does not impose any assumptions on the structure of the input-output models means that the model structure is to a large extent determined by data and also by selection objectives used in the evolutionary search. On one hand, it is an advantage and the unique capability compared with other global approximation techniques, since it potentially allows to develop inherently simpler models than, for example, by interpolation with polynomials or spatial correlation analysis. On the other hand, the absence of constraints on model structure is the greatest challenge for symbolic regression since it vastly increases the search space of possible solutions which is already inherently large.

4   Proposed Genetic Algorithm for Symbolic Regression

Typically, any genetic algorithm used for purpose of optimization consists of the following features:

1. Chromosome or individual representation.
2. Objective function "fitness function".
3. Genetic operators (selection, crossover and mutation).

Applying GA on population of individuals or chromosomes shows that several operators are utilized.

4.1   Chromosome Encoding

The chromosome in form of tree structure represents a suitable encoding for our purposes. Just for simplicity, we pass away physical arrangement of such data structure in computer memory. The the following rules were applied in our solution:

- There are only three types of tree nodes: Constant, Variable, Function. See the red, green and blue nodes in the Figure 1.
- The edges' weights are set randomly.
- All leaf nodes are Constant or Variable nodes.

4.2   Objective Function

The goal of GA is to find a solution to a complex optimization problem, which optimal or near-optimal. GA searches for better performing candidates, where performance can be measured in terms of objective "fitness function". Because of knowledge of results of all input vectors in the training set, the fitness function minimizes measured error between known and achieved results. All input vectors go through all trees (chromosomes) and finally a fitness value is computed for every chromosome.

4.3   Selection Operator

Selection determines which solution candidates are allowed to participate in crossover and undergo possible mutation. The chromosomes are sorted with respect to their fitness function results. A given percentage of worst chromosomes is deleted. This is the matter of application settings.

4.4   Crossover Operator

Promising candidates, as represented by relatively better performing solutions, are combined through a process of recombination referred to as crossover. This ensures that the search process is not random but rather that it is consciously directed into promising regions of the solution space. Crossover exchanges subparts of the selected chromosomes, where the position of the subparts selected randomly to produce offspring. An illustrative example could be seen in the Figure 2.

4.5   Mutation Operator

New genetic material can be introduced into the population through mutation. This increases the diversity in the population. Mutation occurs by randomly selecting particular elements in a particular offspring. In case of symbolic regression, following mutation are defined:

- Function/Symbol node (FN) can be substituted by FN only to prevent tree structure breaks. Both functions should have the same arity.
- Constant node (CN) and Variable node (VN) can mutate randomly
- All weights can be changed randomly.

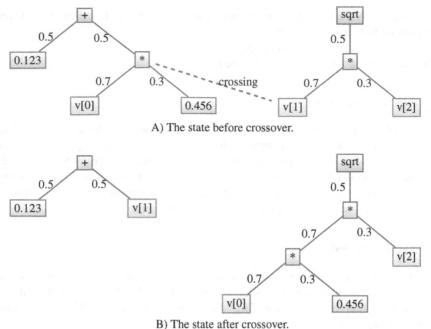

A) The state before crossover.

B) The state after crossover.

Fig. 2. Crossover example

5   Experiments

Several experiments and their results are described in this section. Before that, the data set is defined.

5.1   Dataset

Experimental and real data set was collected during one month in a steel mill factory. The data contains 9132 records - slabs with laboratory evaluations. Every record consists of 3044 attributes. There are three types of defects/errors (Scale, Material, and Mechanical). The Scale means that the production does not follow the production process. The second one, Material means that the resulting products has a material error. The last error Mechanical indicates bad mechanical behaviors. The following table 1 contains statistics of all detected error. "Negative" records represent such cases, where some defect was detected.

The dependency of the precision of prediction model on the amount of records in the training data set was tested. The whole set of records was divided into two parts (training/testing set) with respect to following rates: $20/80$, $30/70$, ..., $80 : 20$. Moreover, records in the training set were selected randomly such that they cover positive and negative records.

Table 1. Summary table of experimental data sets

Error type	Negative records	Positive records	Total
Scale	4479	4653	9132
Material	203	8929	9132
Mechanical	1237	7895	9132

Several parameters were evaluated in the experiments.

- **True positives (TP)** is the number of successfully detected positive records.
- **True negative (TN)** is the number of successfully detected negative records.
- **False positives (FP)** is the number of unsuccessfully detected positive records.
- **False negative (FN)** is the number of unsuccessfully detected negative records.
- **Sensitivity** (also called recall rate in some fields) measures the proportion of actual positives which are correctly identified as such (e.g. the percentage of sick people who are correctly identified as having the condition). Sensitivity relates to the test's ability to identify positive results. $= TP/(TP + FN)$
- **Specificity** measures the proportion of negatives which are correctly identified (e.g. the percentage of healthy people who are correctly identified as not having the condition). Specificity relates to the ability of the test to identify negative results. $= TN/(TN + FP)$
- **Total precision** is given by $TP + TN)/(TP + TN + FP + FN$

In experiments, the most important parameter is Sensitivity, because it is better to mark good slab as suspicious than allow a faulty slab to go through without investigation. But, of course, we would like to maximize both these parameters.

5.2 Results of Experiments

The tables 2, 3, and 4 show resulting data of performed experiments. The column **Ration** indicates the percentage size of training data set. One can see the number of relevant records in columns **TP**, **TN**, **FP**, and **FN**. The values in columns **Sensitivity**, **Specificity**, and **Total Precision** are in percentages. Because of the stochastic nature of the used Genetic algorithms, we perform all experiments $10\times$ and the results were averaged.

Table 2. Experiments with the data set, where the Material defects should be detected

Ratio	TP	TN	FP	FN	Sensitivity	Specificity	Total Precision
20	163	7103	0	0	100.00	100.00	100.00
30	143	6190	0	0	100.00	100.00	100.00
40	122	5270	0	0	100.00	100.00	100.00
50	102	4509	0	0	100.00	100.00	100.00
60	82	3653	0	0	100.00	100.00	100.00
70	61	2740	0	0	100.00	100.00	100.00
80	41	1827	0	0	100.00	100.00	100.00

Table 3. Experiments with the data set, where the Mechanical defects should be detected

Ratio	TP	TN	FP	FN	Sensitivity	Specificity	Total Precision
20	990	2594	3475	0	100.00	42.74	50.77
30	784	5156	0	82	90,53	100.00	98.64
40	669	4243	0	74	90.04	100.00	98.52
50	560	3964	0	59	90.47	100.00	98.71
60	454	3653	0	41	91.72	100.00	99.01
70	345	2740	0	27	92.74	100.00	99.13
80	228	1827	0	20	91.94	100.00	99.04

Table 4. Experiments with the data set, where the Scale defects should be detected

Ratio	TP	TN	FP	FN	Sensitivity	Specificity	Total Precision
20	3582	2600	227	2	99.94	91.97	96.43
30	3133	2247	118	3	99.90	95.01	97.80
40	2685	2230	98	3	99.89	95.79	97.99
50	2237	2283	88	3	99.87	96.29	98.03
60	1791	2233	81	1	99.94	96.50	98.00
70	1342	2270	100	2	99.85	95.78	97.25
80	894	1827	0	2	99.78	100.00	99.93

As may be seen from the tables, the efficiency of the algorithm is very high. We made a comparison with a classical SVM approach with RBF core. The achieved results were around 87% of total precision in maximum when we used 80% of record for learning phase.

6   Conclusions and Future Work

This article describes a method of detecting defects of steel slabs using symbolic regression via genetic programming. As it was mentioned at the beginning, the processes in heavy metal factories and their products are too expensive to make mistakes, which can be avoided. Proposed method can help to detect defects before the steel slabs are used in further production. The method was tested on real data and compared with results, which were achieved by real, standard and mostly expensive measurements. Achieved results were are very promising, the most important parameter, Specificity, was 100% for all three types of error. In the future, we would like to suggest a complex methodology of measurement and prediction of quality of steel slabs.

Acknowledgement. This work was partially supported by the Ministry of Industry and Trade of the Czech Republic, under the grant no. FR-TI1/420, SGS in VSB-Technical University of Ostrava, Czech Republic, under the grant No. SP2012/58, and has been elaborated in the framework of the IT4Innovations Centre of Excellence project, reg. no. CZ.1.05/1.1.00/02.0070 supported by Operational Programme 'Research and Development for Innovations' funded by Structural Funds of the European Union and state budget of the Czech Republic and by the Bio-Inspired Methods: research, development and knowledge transfer project, reg. no. CZ.1.07/2.3.00/20.0073 funded by Operational Programme Education for Competitiveness, co-financed by ESF and state budget of the Czech Republic.

References

1. Fang, H., Ross, P., Corne, D.: Genetic algorithms for timetabling and scheduling (1994), http://www.asap.cs.nott.ac.uk/ASAP/ttg/resources.html
2. Goldberg, D.E.: Genetic Algorithms in Search, Optimization and Machine Learning. Addison-Wesley (1989)
3. Huang, C., Li, G., Xu, Z., Yu, A., Chang, L.: Design of optimal digital lattice filter structures based on genetic algorithm. Signal Processing 92(4), 989–998 (2012)
4. Ishibuchi, H., Nakashima, Y., Nojima, Y.: Performance evaluation of evolutionary multiob-jective optimization algorithms for multiobjective fuzzy genetics-based machine learning. Soft. Comput. 15(12), 2415–2434 (2011)
5. Juzoji, H., Nakajima, I., Kitano, T.: A development of network topology of wireless packet communications for disaster situation with genetic algorithms or with dijkstra's. In: ICC, pp. 1–5 (2011)
6. Melanie, M.: An Introduction to Genetic Algorithms. A Bradford Book. MIT Press (1999)
7. Melanie, M., Forrest, S.: Genetic algorithms and artificial life. Santa Fe Institute, working Paper 93-11-072 (1994)
8. Pan, S.-T.: A canonic-signed-digit coded genetic algorithm for designing finite impulse re-sponse digital filter. Digital Signal Processing 20(2), 314–327 (2010)
9. Park, B.J., Choi, H.R.: A genetic algorithm for integration of process planning and scheduling in a job shop. In: Australian Conference on Artificial Intelligence, pp. 647–657 (2006)
10. Sedighi, K.H., Manikas, T.W., Ashenayi, K., Wainwright, R.L.: A genetic algorithm for au-tonomous navigation using variable-monotone paths. I. J. Robotics and Automation 24(4) (2009)
11. Tsang, E.P.K., Warwick, T.: Applying genetic algorithms to constraints satisfaction optimiza-tion problems. In: Proc. of 9th European Conf. on AI, Aiello L.C. (1990)
12. Wainwright, R.L.: Introduction to genetic algorithms theory and applications. In: The Sev-enth Oklahoma Symposium on Artificial Intelligence (November 1993)

An Effective Application of Soft Computing Methods for Hydraulic Process Control

Ladislav Körösi and Štefan Kozák

Institute of Control and Industrial Informatics,
Faculty of Electrical Engineering and Information Technology,
Slovak University of Technology, Ilkovičova 3, 812 19 Bratislava, Slovak Republic
{ladislav.korosi,stefan.kozak}@stuba.sk

Abstract. The article deals with the modeling and predictive control of real hydraulic system using artificial neural network (ANN). For the design of optimal neural network model structure we developed procedures for creation optimal-minimal structure which ensure desired model accuracy. This procedure was designed using genetic algorithms (GA) in Matlab-Simulink. The predictive control algorithm was implemented using CompactLogix programmable logic controller (PLC). The main aim of the proposed paper is design of methodology and effective real-time algorithm for possible applications in industry.

Keywords: Neural network, genetic algorithm, predictive control, PLC realization, optimization methods.

1 Introduction

Neural networks are currently used in various fields such as signal processing, image recognition, natural speech recognition, identification and others [1]. Functions in programmable logic controllers (PLC) libraries are simple (bit operations, summation, subtraction, multiplication, division, reminder after division, etc.) or complex (sine, cosine, absolute value, vector summation, etc.) mathematical functions but without artificial neural systems, while PLC systems are currently the most commonly used control systems in industry. In PLC systems are also missing matrix operations and often vector operations, generally said parallel mathematical operations. The proposed paper has the objective to demonstrate the real deployment of optimal neural network for modeling and predictive control of hydraulic system.

2 Realization of Neural Network Control Algorithm by Programmable Logic Controllers

Modeling and Control of nonlinear processes using artificial neural networks in practice can be solved in two ways. The first way is the deployment of modeling and control algorithm in the master system (SCADA, Application running on the PC, local HMI, etc.). In this case, the algorithm is separated from the control system and therefore it is important to ensure trouble-free communication between these parts of control as well as fixed sampling time. In the event of a failure of communication,

V. Snasel et al. (Eds.): SOCO Models in Industrial & Environmental Appl., AISC 188, pp. 379–388.
springerlink.com

must take control the local control system. The second solution is the implementation of intelligent algorithms directly to the control system. Most commonly used control systems are programmable logic controllers (PLC's). PLC is a digital computer used for automation of electromechanical processes, such as control of machinery on factory assembly lines, etc. PLC's are used in many industries and machines. For the programming of such control systems have been introduced IEC 61131-3, which unifies the programming languages of PLC's from different manufacturers. The best language for the purposes of the implementation of the ANN appears is Structured Text (ST), which is similar to the Pascal programming language. Of course there are differences in them because this language was developed for PLC programming. The program is written in free style, making it clearer and more readable. It is particularly useful for the expression of different data types, structures and complex mathematical calculations [10]. Most PLC's do not support matrix operations or dynamic allocation of one-dimensional or multidimensional vectors, therefore, the implementation phase of the ANN learning and processing can be difficult. Variables must be pre-allocated, i.e. in the case of smaller ANN structures consumes unnecessarily PLC memory. Matrix operations can be programmed accordingly, but in this case they are not parallel operations, but sequential, decreasing the computing power of ANN [8]. That does not mean that affects the quality of modeling and control. Implementation of optimal ANN structures in the PLC is possible in two ways. The first way is to implement the general algorithm, which can dynamically adapt to new structures depending on given parameters. The second is the fixed structure of the ANN implemented in the PLC.

3   Artificial Neural Network Model

Perceptron is the most used ANN. The main reason is its ability to model as simple as well as very complex functional relations. Kolmogorov's theorem says that the perceptron ANN with one hidden layer and a sufficient number of neurons in this layer can approximate any nonlinear function. In Fig. 1 shows an example of the structure of three-layer perceptron.

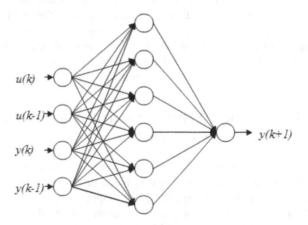

Fig. 1. An *example of multilayer perceptron structure*

The input and output neurons are linear activation functions (AF) but in some cases, in the output layer(s) can be used non-linear activation functions. Most often used activation functions in the hidden layer(s) are nonlinear (sigmoidal). Sigmoidal function is defined as a monotonically increasing, smooth and bounded function. Sigmoid activation functions allow obtaining high neuron sensitivity for small signals, for higher signal level the sensitivity decreases. Neural network is able to process signals in a sufficient range of dynamics without threatening to overload with too large coming signals. In addition to these activation functions provide nonlinear behavior of ANN and are several times differentiable, therefore they can be used for several techniques for ANN training and predictive control [2], [3], [5].

4 Artificial Neural Network Structure Optimization

Selection of optimal artificial neural network structure includes the determination of number of hidden layers, number of neurons in each layer, number of links between neurons, etc. which can be in general defined as the vastness of the network. ANN structure optimization methods can be divided into the following groups: construction algorithms, destruction algorithms, empiric methods, combined methods, genetic algorithms, etc. Optimal ANN usually contains fewer neurons and connections allowing their usage in real-time applications [7], [9]. Genetic algorithm is a universal stochastic search approach inspired by both natural selection and natural genetics which is able to approximate the optimal solution within bounded solution space. The method is capable being applied to a wide range of problems including ANN structure optimization. Basic objects in the GA are chromosome, gene, population, generation and fitness function [4].

Currently there are various methods for neural network structure encoding. One of the most used and simple method is encode NN structure with direct encoding. For the direct encoding of the connections between neurons binary *1* (*true*) and *0* (*false*) are used. If neurons are interconnected, the connection is assigned with value *1*. If the neurons aren't interconnected, the connection is assigned with value *0*. For all such links we can define the size of the *NxN* matrix that defines connections between all neurons. Each line defines the link neurons to neurons in a given row. From such a matrix is then created a chromosome for the GA. This encoding enables to optimize all interconnections between neurons.

Finding the optimal structures of ANN for modeling and control of nonlinear processes using GA is time consuming, especially for large ranges in chromosome genes. In addition, the GA has to evaluate the fitness function and decode the structure of ANN as well as check the correctness and optimality of newly established structures. Based on the fact that a three-layer ANN is sufficient for nonlinear process modeling is proposed the following ANN with AF encoding [7]:

1. gene: Number of input neurons for the control signal in history
2. gene: Number of input neurons for the measured signal in history
3. gene: Number of hidden neurons
4. gene: AF type

The sum of the number of input neurons for history of control and measured values defines the total number of input neurons. Number of output neurons is given by the process type and isn't encoded in the chromosome. All genes are encoded with integer number (also for AF type). During decoding the structure of ANN is the activation function of each number assigned to its name from the list of defined activation functions. Table 1 gives several examples of representation of the gene for AF types. Chromosome above defines the NNARX (neural network auto-regressive model with external input) structure (Fig. 1).

Table 1. AF type encoding

Gene: AF type	
Num. value	**AF type**
1	tansig
2	logsig
3	modified tansig

5 Predictive Control Using ANN

Predictive control methods currently represent a large group of modern control methods with an increasing number of applications. Under the notion of predictive control we understood a class of control methods where the mathematical model is used to predict the future output of the controlled system. Determination of the future control sequence involves minimizing a suitable criteria function with predicted increment of control and deviation. One of the advantages of predictive control is the possibility to use any process model [6]. In this article an ANN perceptron is used as process model.

Standard criteria function includes square of the deviation and control increment:

$$J_r = \frac{1}{2}\alpha(u(k) - u(k-1))^2 + \frac{1}{2}[r(k+1) - y_M(k+1)]^2 \tag{1}$$

where y_m is the output of the ANN.

The optimization block calculates the control signal so that the predicted output of the ANN matches the process output [4]. This is an iteration process which has the form:

$$u(k)_{new} = u(k)_{old} - \beta \frac{\partial J_r{}'}{\partial u(k)_{old}} \tag{2}$$

6   Case Study

For the verification and testing of proposed algorithm for modeling and control we consider real hydraulic system. The block diagram of the hydraulic system is depicted in Fig. 3.

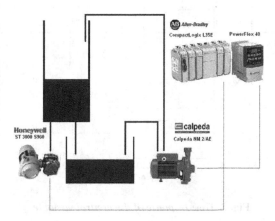

Fig. 2. Block *diagram of the hydraulic system*

Short description of used components is listed below.

CompactLogix (1769-L32E) - Programmable logic controller from Allen-Bradley is used to control the hydraulic system's level. The level is measured with pressure transmitter connected to analog input module (using unified 4-20mA signal). The control signal from analog output module is connected via 4-20mA to the inverter sending 0-100% (0-50Hz) signal.

PowerFlex 40 - Inverter from Allen-Bradley is designed for drives with power output from 0.4 kW to 11kW. It is connected to pump which pumps water from buffer situated under the tank to the upper part of the tank.

ST 3000 S900 - A smart pressure transmitter from Honeywell is used to measure the water level in tank with free drainage.

Calpeda NM 2/AE 400V 0.75kW - Monoblock centrifugal pump from the Calpeda company

Measurement of input-output data for identification of the real system can be implemented in two ways. The first way is to backup data using visualization (RSView32 or RSView Studio), but this method does not guarantee accurate sampling period - loading data from the PLC. The second way is to backup data directly in the PLC and their export using the Tag Upload Download after measurement. In our case we used a second approach for data collection to identify and record values during the process control. Measured data (level - depending on the pressure sensing and control variable) are shown in Fig. 4 and Fig. 5.

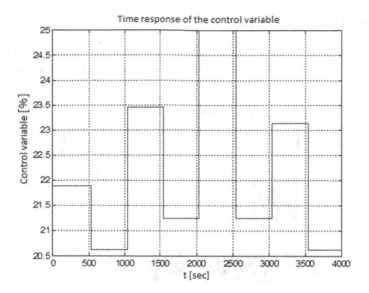

Fig. 3. Time *response of the control variable*

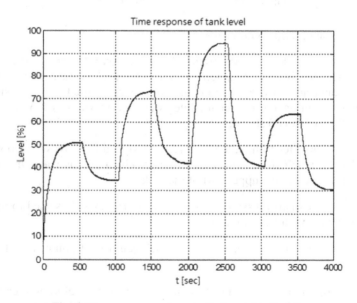

Fig. 4. Time *response of controlled variable (level)*

The setup of the GA was the following:

- 1. Gene value interval <1,6>
- 2. Gene value interval <1,6>
- 3. Gene value interval <1,30>
- 4. Gene value interval <1,3>
- Number of generations: 100 (used as stop criteria)
- Number chromosomes in population: 15

The criterion function (fitness) was calculated by the value of the square sum of deviations between ANN output and the process output.

For each ANN training new weights were generated. The found optimal neural network structure for modeling of real physical system was 4-6-1 with tansig AF. Inputs to the ANN were values $u(k)$, $u(k-1)$, $y(k)$, $y(k-1)$ and the output value was the predicted tank level $y(k+1)$. Structure of an ANN was created in Matlab and trained with back-propagation learning method with over-learning testing. Comparison of time responses of the real plant level and ANN output is shown in Fig. 7.

Structure of the ANN control algorithm for real time control was implemented in RSLogix5000 for CompactLogix PLC. Weights and biases of ANN have been exported from Matlab to RSLogix5000. The chosen control algorithm was gradient-descent. The results for different parameters (affecting different control quality) are shown in the following figures (for sampling time 100ms).

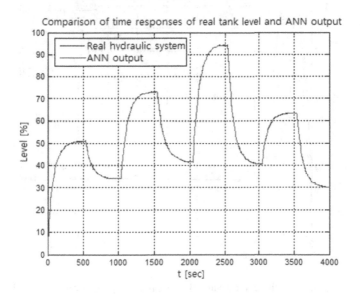

Comparison of time responses of real tank level and ANN output

Fig. 5. Comparison *of time responses*

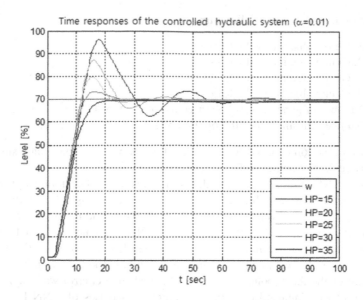

Fig. 6. Time responses of the controlled hydraulic system for different prediction horizons 15, 20, 25, 30 and 35 (α=0.01)

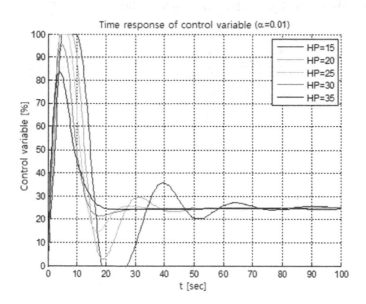

Fig. 7. Time responses of the control variable for different prediction horizons 15, 20, 25, 30 and 35 (α=0.01)

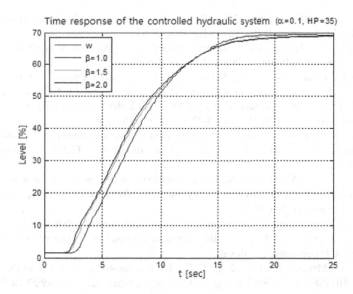

Fig. 8. Comparison of time responses of the controlled hydraulic system for different β and constant prediction horizon 35 and *α=0.1*

7 Conclusion

Intelligent control belongs to the class of control techniques like artificial neural networks, fuzzy logic, evolutionary algorithms, etc. The performance of the proposed methodology was verified on several simulation examples. In the proposed paper was presented a practical application using optimized artificial neural network structure to identify and control the hydraulic system. Using these proposed methods we achieved better approximation results and faster response. The advantages of the proposed approach are minimal (optimal) neural network structure and faster signal processing therefore it's suitable for real-time control in industry control systems.

Acknowledgement. This paper was supported by the Slovak Scientific Grant Agency VEGA, Grant no.1/1105/11.

References

1. Oravec, M., Polec, J., Marchevský, S.: Neurónové siete pre číslicové spracovanie signálov. Vydavateľstvo FABER, Bratislava (1998) (in Czech)
2. Haykin, S.: Neural networks – A Comprehensive Foundation. Macmillan College Publishing Company, New York (1994)
3. Sinčák, P., Andrejková, G.: Neurónové siete – Inžiniersky prístup I, Technická univerzita, Košice (1998), http://www.ai-cit.sk/source/publications/books/NS1/html/index.html (in Slovak)

4. Körösi, L., Kozák, Š.: Optimal Self Tuning Neural Network Controller Design. In: 16th IFAC World Congress, Praha, Czech Repubic (2005)
5. Jadlovská, A.: Modelovanie a riadenie dynamických procesov s využtím neurónových sietí, Košice (2003) ISBN 80-88941 -22 -9 (in Slovak)
6. Dideková, Zuzana – Kajan, Slavomír: Riadenie nelineárnych systémov pomocou neurónových sietí optimalizovaných genetickými algoritmami. In: Kybernetika a informatika: Medzinárodná konferencia SSKI SAV, Vyšná Boca, SR, 10.-13.2, Vydavateľstvo STU, Bratislava (2010) ISBN 978-80-227-3241-3 (in Slovak)
7. Körösi, L.: Príspevok k problémom optimalizácie štruktúr umelých neurónových sietí. Dizertačná práca. Ústav riadenia a priemyselnej informatiky FEI STU, Bratislava, 126s (2010) (in Slovak)
8. Abdi, H., Salami, A., Ahmadi, A.: Implementation of a new neural network function block to programmable logic controllers library function. World Academy of Science, Engineering and Technology 29 (2007)
9. Kajan, S.: Modelovanie a riadenie systémov pomocou neurónových sietí s ortogonálnymi funkciami v prostredí MATLAB. In: Technical Computing Prague (2006): 14th Annual Conference Proceedings (in Slovak)
10. Mrafko, L., Mrosko, M., Körösi, L.: PLC a ich programovanie 1. Čo je to PLC? In: Posterus (2010) ISSN 1338-0087, http://www.posterus.sk/?p=6903 (in Slovak)

Automatic Neonatal Sleep EEG Recognition with Social Impact Based Feature Selection

Martin Macaš, Václav Gerla, and Lenka Lhotská

Czech Technical University, Technicka 2, 166 27 Prague 6, Czech Republic
macas.martin@fel.cvut.cz

Abstract. The paper presents an application of Simplified Social Impact Theory based Optimization on feature subset selection for automated neonatal sleep EEG recognition. The target classifier is 3-Nearest Neighbor classifier. We also propose a novel initialization of iterative population based optimization heuristics, which is suitable for feature subset selection, because it reduces the computational complexity of whole feature selection process and can help to prevent overfitting problems. Our methods leads to a significant reduction of the original dimensionality while simultaneously reduce the classification error.

1 Introduction

During the last century, many of natural social phenomena were modeled by ethologists, social psychologist, economists and others. Examples are agent-based models of ant behavior, models of swarming, or models of opinion formation. In last two decades, these models of natural optimization processes are modified and "forced" to solve mathematical optimization problems. Thus, methods like Ant Colony Optimization or Particle Swarm Optimization (PSO) [1] are being invented and still more and more intensively applied to real-world problems. This paper presents an application of Simplified Social Impact Theory based Optimization (SSITO) [2] inspired by opinion formation models on Neonatal Sleep EEG Recognition.

In this study we focus primarily on differentiating between two important neonatal sleep stages: quite sleep and active sleep. In clinical practice, the proportion of these states is a significant indicator for the maturity of the newborn brain [3]. Manual evaluation of EEG is a very tedious operation, and an electroencephalographer can easily make a mistake. Therefore, the classification process is being automatized in terms of feature based pattern classification. In most cases of automatic neonatal EEG classification, large amounts of EEG data must be processed. It is also complicated by the fact that various additional channels must also be processed. It is therefore necessary to compress the calculated features using a sophisticated technique. In this paper, we deal with the reducing of number of appropriate features used for the automatic classification of neonatal EEG in terms of selection of a proper subset of features.

2 Neonatal Sleep EEG Recognition

The data used in this study was provided by the Institute for Care of Mother and Child in Prague. We have 11 full-term healthy newborn records (37 - 40 weeks gestation;

V. Snasel et al. (Eds.): SOCO Models in Industrial & Environmental Appl., AISC 188, pp. 389–398.
springerlink.com   © Springer-Verlag Berlin Heidelberg 2013

5 minutes of quiet sleep and 5 minutes of active sleep for each record; no artifacts; clearly defined sleep states). All data was recorded from eight referential derivations, positioned under the 10-20 system, namely FP1, FP2, T3, T4, C3, C4, O1, and O2. The sampling frequency was 128 Hz. The reference derivation (R) used linked ear electrodes. In addition, the following polysomnographic signals were used: EOG, EMG, ECG, and PNG. All channels were measured against ground. The EOG signal was recorded from two electrodes placed slightly above and to the outside of the right eye and below and to the outside of the left eye. Two EMG electrodes were placed on the chin and at the left corner of the mouth. ECG was recorded using two electrodes, one placed over the sternum and the other in the medial axillary line. The respiratory effort was measured using a tensometer placed on the abdomen. Examples of signals in quiet and active sleep are shown in Figure 1. To eliminate power-line noise, we used a notch filter. This rejects a narrow frequency band and leaves the rest of the spectrum almost undistorted.

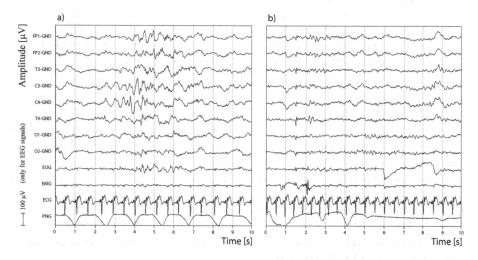

Fig. 1. Example of neonatal PSG recording; a) quiet sleep, b) active sleep. Signals: FP1, FP2, C3, C4, O1, O2, T3, T4, EOG, EMG, ECG, and PNG.

For the subsequent processing methods it is necessary to divide the signals into almost stationary segments. In this study we use constant segmentation into 1 second segments. All features listed below were calculated from these segments. Some of these (auto-correlation, cross-correlation and coherence features) were calculated from sliding window (length $20s$, shift $1s$) and were then used in such a way as to match the $1s$ segments. In this way it was obtained a total of 2087 features that were used for neonatal data classification. The following list summarizes the used feature extraction techniques and give their brief description.

- *Statistical description.* EEG signal can be characterized by the distribution of amplitude and its moments [4].
- *Interval and period analysis.* The intervals between zero and other level crossings, or intervals between maxima and minima were measured and moments of their distribution were used as features [5].

- *Application of derivatives.* Statistical features are extracted also for the first and the second derivative of EEG signals.
- *Hjorth parameters.* The Hjorth parameters are simple measures of signal complexity. These measures have been used in the analysis EEG and they are clinically useful tools for the quantitative description of an EEG [4].
- *Power spectral analysis.* We compute the mean value of absolute and relative power spectra over the common frequency bands (delta, theta, alpha, beta) [4].
- *Entropy-based features.* Entropy is thought to be a measure of EEG signal complexity and so it is potentially useful feature for our purposes [4].
- *Nonlinear energy operator.* Another features were based on the nonlinear energy operator [4].
- *Auto-correlation and cross-correlation.* Cross-correlation is a measure of similarity of two signals and auto-correlation is the cross-correlation of a signal with itself [6]. We compute the maximum positive amplitude and mean value from auto- and cross-correlation function (for selected polygraphic signals).
- *Coherence analysis.* The inter- and intra-hemispheric coherence are also calculated from the EEG signal [6].

In addition, we have also used the information extracted from the other polysomnographic channels (heart rate variability from ECG signal, regularity of respiration from PNG signal, presence of eye movements from EOG signal and body movements from EMG signal, see [7] for more details).

To make the number of input features suitable for wrapper methods, we first preselect the features by evaluating features individually using inter-intra class distance filter criterion [8] and taking only 500 best features. Thus, there is 4400 data instances of dimension 500 obtained from 11 subjects (400 instances from each).

3   Feature Selection

A feature selection process usually consists of two main components - a evaluation criterion, which evaluates potential feature subsets, and a search method, which seeks for a minimum of the criterion. Here, we use the wrapper approach to feature selection, i.e. use performance of the target classifier as evaluation criterion. The criterion is minimized using the SSITO method. Both components are described in the next sections.

The classification is performed using 3-Nearest Neighbor classifier (3NN). It simply finds 3 training data instances that are most similar to the testing instance and assigns the instance into the most common class amongst the 3 nearest neighbors. We use the Euclidean distance for similarity quantification, because it was observed to lead to good classification accuracies while keeping reasonable computational requirements. The nearest neighbor classifiers are still widely used in pattern classification, because of its simplicity, high performance (especially, but not only in large sample limit, and robustness to noisy learning data [9]. Many more sophisticated classifiers need much more time for training and testing and the wrapper approach is not suitable for them because of computational complexity reasons. Moreover, in our preliminary experiments with full feature set, 3NN outperformed quadratic Bayes classifier (assuming normally distributed classes with different covariance matrices) and CART decision tree.

Our particular feature selection criterion is the 2-fold cross-validation estimate of 3NN's error, because it was observed in some preliminary experiments with different datasets [10] to perform better than 10-fold setting or leave-one-out technique. This corresponds to some other studies, e.g. [11]. In 2-fold cross-validation, the data set is partitioned into 2 disjunctive folds of similar size. For each of two iterations, one fold is used for testing and remaining fold is used for training. The 2-fold cross-validation error estimate is the testing error averaged over the two folds.

Let D be the number of features and $\mathscr{A} \subseteq \{1, \ldots, D\}$ be a subset of feature indices, which represents a feature subset. The optimization techniques assume the following encoding of a feature subset $\mathscr{A}$: $\mathbf{s} = \{0, 1\}^D$, where ath component $s^a = 1$ means that the feature with index a is selected (i.e. $a \in \mathscr{A}$) and $s^a = 0$ means that the feature with index a is removed (i.e. $a \notin \mathscr{A}$).

Thus, the feature selection is defined here as a minimization of the cost function $f(\mathbf{s})$ defined as 2-fold cross-validation error estimate of 3-Nearest Neighbor classifier trained with features represented by $\mathbf{s}$. The optimization method used for the minimization is described below.

The approach described here tries to take a model from social psychology, adapt it, and use it in the area of parameter optimization. It is an attempt to use simulated people to make a decisions about solutions of an optimization problem. The simulation is based on simple opinion formation models widely used in computational psychology and commonly analyzed by tools of statistical physics. We present application of relatively novel population-based optimization methods, in which the candidate solutions influence each other and try to converge into a "good" consensus. The method called Simplified Social Impact Theory based Optimizer (SSITO) is applied here to the feature subset selection problem known from pattern recognition.

Many opinion formation models combine the social information using the notion of social impact function that numerically characterizes the total influence of social neighborhood of a particular individual. We use the analogy with the Nowak-Szamrej-Latané models [12]. Let $\{\mathbf{s}_1(t), \ldots, \mathbf{s}_L(t)\}$ be a set of L candidate solutions of the feature selection problem at iteration t. Here, the population size L is 25 individuals. Each candidate solution is influenced by its social neighborhood. Here, the neighborhood simply consists of 5 randomly selected individuals. The neighbors with higher strength value have higher influence on the impact value. The strength can be associated with pair of individuals. One possible choice is – the social strength q_{ji} by which a candidate solution j affects candidate solution i depends on their cost values according to the following formula:

$$q_{ji}(t) = \max[f(\mathbf{s}_i(t)) - f(\mathbf{s}_j(t)), 0], \qquad (1)$$

where $f(\mathbf{s}_i(t))$ and $f(\mathbf{s}_j(t))$ are the cost values of the candidate solutions i and j, respectively. This equation means that fitter individual have a non-zero influence on less fitter individual and is not influenced by it. Obviously, there is an infinite number of possible cost–strength mappings. Some of them can lead to much better optimization abilities.

Considering component a of candidate solution $\mathbf{s}_i(t)$, the impact function depends on the component a of candidate solutions from i's neighborhood and on strength values of these solutions. It characterizes the total impact on individual i. A positive impact value leads to preference of ath component inversion. Contrary, the negative value have supportive character and leads to preference of keeping the component value. A particular form of the impact function will follow the opinion formation models described in [12]. At each iteration and for each component a, the neighbors of i are divided into two disjoint subsets, persuaders $\mathscr{P}_i^a(t)$ with opinion opposite to $s_i^a(t)$ and supporters $\mathscr{S}_i^a(t)$ with the same value of the opinion. The impact is defined as:

$$I_i^a(t) = \frac{1}{|\mathscr{P}_i^a(t)|} \sum_{j \in \mathscr{P}_i^a(t)} q_{ji}(t) - \frac{1}{|\mathscr{S}_i^a(t)|} \sum_{j \in \mathscr{S}_i^a(t)} q_{ji}(t). \tag{2}$$

Moreover, we define $I_{Si}(t) = 0$ if $\mathscr{S}_i^a(t) = \emptyset$ and $I_{Pi}(t) = 0$ if $\mathscr{P}_i^a(t) = \emptyset$.

The update rule further uses the value of impact function to generate new state for s_i^a. The simplest deterministic update rule uses the analogy to [12] - individual changes its opinion if the impact function takes a positive value:

$$s_i^a(t+1) = \begin{cases} 1 - s_i^a(t), & \text{if } I_i^a > 0; \\ s_i^a(t), & \text{otherwise.} \end{cases} \tag{3}$$

The algorithm described above ignores the aspect of individual decision processes (e.g. experience, memory, inferring mechanisms) and of many unknown processes. These can be partly modeled by a random noise. Moreover, randomness is an essential part of any optimization metaheuristic. Hence the random noise is added in our optimizers. The simplest way to add the random element is to mutate all s_i^a with probability of spontaneous opinion inversion (mutation rate) $\kappa << 1$. This can keep the diversity and avoid a premature convergence.

The pseudocode is in Algorithm 1. First, the initial population $\{\mathbf{s}_i(0)\}_{i=1...L}$ is created randomly and all cost and strength values are computed. At each iteration, vector $\mathbf{s}_i(t)$ is transformed into a new vector $\mathbf{s}_i(t+1)$ using an update rule. It updates the a-th bit of the vector $\mathbf{s}_i$ according to its value, the values of a-th bit of vectors positioned in i's neighborhood and according to their strength values. After the update of all $\mathbf{s}_i$ vectors, new values of cost and strength can be computed and the next iteration is performed.

Algorithm 1. Pseudocode for SSITOmean algorithm

---

initialize all $\mathbf{s}_i(0)$
while stop condition not met **do**
    for all i **do**
        evaluate $\mathbf{s}_i(t)$ by computing $f(\mathbf{s}_i(t))$
    end for
    for all i,j **do**
        compute strength values $q_{ji}(t)$ using equation 1
    end for
    for all i,a **do**
        compute $I_i^a(t)$ using equation 2
    end for
    for all i,a **do**
        compute $s_i^a(t+1)$ using equation 3 and random mutation
    end for
    t:=t+1
end while

The whole iterative process can lead to a search strategy that samples the binary space using the set of candidate solutions $\mathcal{Q}$. Here, the stopping condition is the reach of a maximum number of cost evaluations or maximum number of iterations, but it can be any other criterion known from the area of population based metaheuristics (cost increase, diversity degradation, etc.)

4   Results

The main purpose of our feature selection technique is to find a feature subset, that minimizes classification error of the classifier. To estimate the true benefits of the feature selection, we use an outer-loop estimate for testing. We divided the dataset into 11 folds according to the membership of instances to particular subjects (neonates). Further, we used the cross-validation algorithm on these folds. This special type of cross-validation is a fair approach that reduces a positive bias caused by ignoring of an inter-personal variability.

To show a competitiveness of our approach, we perform the same feature selection using Binary Particle Swarm Optimization algorithm (BPSO) [1], which is also based on social-psychological metaphor and is often applied to feature selection. We used the following settings: inertia weight $\omega = 1$, weights of individual and social knowledge $\varphi_1 = \varphi_2 = 2$, maximum velocity $\mathbf{v}_{max} = 5$, and the ring topology with neighborhood radius 2.

Moreover, we implemented two modifications. The first is a so-called reduced initialization and the second is the modification of the criterion called combined criterion. In the original SSITO algorithm (and in most population based methods), the initial population is created randomly, i.e. each bit is set to 1 with probability 0.5. In average, each candidate solution corresponds to $D/2$ selected features. This can be a disadvantage if the optimal solution contains a small number of features. Furthermore, it makes the search more computationally complex. Therefore, we reduce the number of features to a minimum. The main requirement is that there must be exactly one occurrence of each feature in the initial population (exactly one candidate solution contains the feature). Thus, in average each candidate solution corresponds to D/L selected features. This can significantly reduce the temporal complexity of cost evaluation (error estimation). The reduced initialization thus means that the candidate solutions are initialized randomly under the condition described above.

The second modification is referred here as the *combined* criterion. It is not novel, many papers use this approach. It simply combines the error estimate with the number of features, which leads to a more intensive dimensionality reduction. We use a simple linear combination. Let $e(\mathbf{s}(t))$ be the error estimate. Instead of using $f(\mathbf{s}(t)) = e(\mathbf{s}(t))$, the combined criterion uses $f(\mathbf{s}(t)) = e(\mathbf{s}(t)) + \alpha d/D$, where α parameter weights the relative importance of the dimensionality reduction and its optimal value depends on the depends on the classifier and on the data set. In our experiment we show the behavior of the selection for $\alpha = 1$.

A comparison for our approaches is depicted on Fig. 2, 3 and 4, where the temporal evolution of three variables averaged over 11 cross-validation runs is depicted. The variables are measured for the best-so-far solution corresponding to the minimum cost value

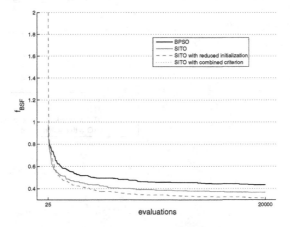

Fig. 2. The criterion value of the best-so-far solution. The combined criterion is not included in the graph, because its values is not comparable to the values of other criteria.

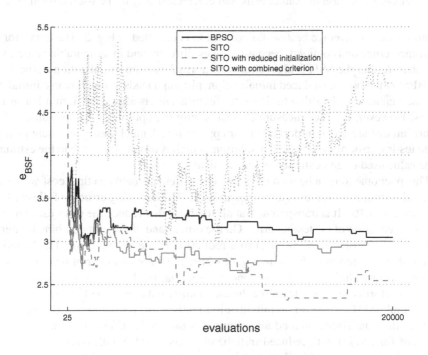

Fig. 3. The testing error value of the best-so-far solution (in %)

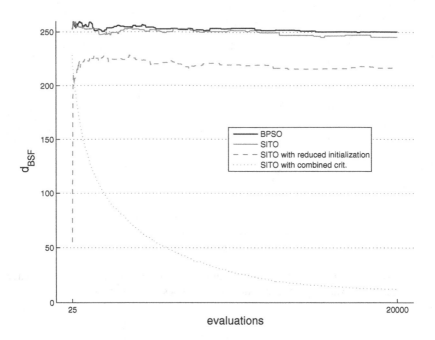

Fig. 4. The number of features in the subset represented by the best-so-far solution

found so far. The average best-so-far cost value is depicted in Fig. 2. The curve for the combined criterion is omitted, because it takes different and incomparable values. One can observe that the SSITO algorithm with normal random initialization outperforms the BPSO method. The reduced initialization, although leads to much worse initial cost values, enables the algorithm to find fitter feature subsets after a small number of iterations. However, the f_{BSF} measure evaluates only the optimization capabilities of the algorithm that are not so important in our application. A much more important measure is the testing error of the best-so-far solution, which is usually higher than the estimated error value used to guide the search mechanism.

This phenomenon can be seen in Fig. 3. The difference between final cost values and the final testing error values is approximately 2.5%. The testing error for full set of 500 features was 3.4%. It is transparent that all methods that reduce the error estimate also lead to a reduction of the testing error. On the other hand, the use of combined criterion with our setting of α does not lead to a reduction of the testing error. The benefit of the combined criterion is shown in Fig. 4, where the output number of features d_{BSF} in the best-so-far solution is depicted. One can see that all the three instances of SSITO method perform differently from the dimensionality reduction point of view. Although the normal SSITO with normal initialization leads to only slightly smaller number of features than the BPSO method and eliminates only 50% of features, the situation is different for SSITO with reduced initialization and SSITO with combined criterion. The reduced initialization leads to much smaller dimensionality of the best-so-far candidate solution (and also of all candidate solutions) and finally converges into a solution, which corresponds to less features, small cost value and the best testing error. Thus, the SSITO with reduced initialization is the best from the error minimization point of view. However, it is a frequent case, that physicians need to somehow interpret the results and compare them with their current medical knowledge. For such a case, there is a strong need for obtaining a very small number of features. For such a case, the combined criterion can be very practical. As one can see in Fig. 4, the combined criterion reduced the dimensionality from 500 features to 12 features, while it kept the testing error in reasonable bounds. Obviously, the requirement for setting of α corresponds to a need of some preliminary experiments, which can be a disadvantage of the approach.

5   Conclusions

In the paper, we describe an application of relatively novel soft computing method to a biomedical signal processing problem. Particularly we show, how the socially inspired SSITO algorithm can be useful in wrapper–based feature subset selection. All presented approaches significantly reduce the dimensionality and compared to the full set of 500 features, all methods also lead to a reduction of the testing error of 3NN classifier in about $0-1\%$. The presented SSITO method is very simple and outperforms the commonly used BPSO algorithm. We also propose a novel reduce initialization of the SSITO methods which can be directly applied to any optimization metaheuristic. The reduced initialization leads to a significant reduction of computational requirements and helps to reduce problems related to overfitting and feature selection bias. For the case of a need of a good interpretability of results, very small subset of features can be selected using the combined criterion, which also considers the relative dimensionality of the candidate feature subset.

References

1. Kennedy, J., Eberhart, R.C.: A discrete binary version of the particle swarm algorithm. In: IEEE International Conference on Systems, Man, and Cybernetics, vol. 5, pp. 4104–4108 (1997)
2. Macaš, M., Lhotská, L.: Simplified Social Impact Theory Based Optimizer in Feature Subset Selection. In: NICSO, pp. 133–147 (2011)
3. Lofhede, J., Degerman, J., Lofgren, N., Thordstein, M., Flisberg, A., Kjellmer, I., Lindecrantz, K.: Comparing a supervised and an unsupervised classification method for burst detection in neonatal EEG. In: Conf. Proc. IEEE Eng. Med. Biol. Soc., pp. 3836–3839 (2008)
4. Greene, B.R., Faul, S., Marnane, W.P., Lightbody, G., Korotchikova, I., Boylan, G.B.: A comparison of quantitative eeg features for neonatal seizure detection. Clin. Neurophysiol. 119, 1248–1261
5. Niedermeyer, E., da Silva, F.H.L.: Electroencephalography, basic principles, clinical applications, and related fields. In: Urban & Schwarzenberg (1982)
6. Tong, S., Thakor, N.V.: Quantitative EEG analysis methods and clinical applications. In: Engineering in Medicine & Biology. Artech House (2009)
7. Gerla, V., Lhotská, L., Krajča, V., Paul, K.: Multichannel analysis of the newborn EEG data. In: International Special Topics Conference on Information Technology in Biomedicine, Piscataway. IEEE (2006)

8. van der Heijden, F., Duin, R., de Ridder, D., Tax, D.M.J.: Classification, Parameter Estimation and State Estimation: An Engineering Approach Using MATLAB. John Wiley and Sons (2004)
9. Bhatia, N.: Survey of nearest neighbor techniques. Journal of Computer Science 8(2) (2010)
10. Macaš, M., Lhotská, L., Bakstein, E., Novák, D., Wild, J., Sieger, T., Vostatek, P., Jech, R.: Wrapper feature selection for small sample size data driven by complete error estimates. Computer Methods and Programs in Biomedicine (to appear, 2012)
11. Weiss, S.M.: Small sample error rate estimation for k-NN classifiers. IEEE Trans. Pattern Anal. Mach. Intell. 13, 285–289 (1991)
12. Nowak, A., Lewenstein, M.: Modeling Social Change with Cellular Automata. In: Modeling and Simulation in the Social Sciences from the Philosophy of Science Point of View, pp. 249–285. Kluwer Academic (1996)

Wavelet Based Image Denoising Using Ant Colony Optimization Technique for Identifying Ice Classes in SAR Imagery

Parthasarathy Subashini[1], Marimuthu Krishnaveni[1],
Bernadetta Kwintiana Ane[2], and Dieter Roller[2]

[1] Department of Computer Science,
Avinashilingam Deemed University for Women,
Coimbatore, Tamilnadu, India
{mail.p.subashini,krishnaveni.rd}@gmail.com
[2] Institute of Computer-aided Product Development Systems,
Universitaet Stuttgart, Stuttgart, Germany
{ane,roller}@informatik.un-stuttgart.de

Abstract. Interpretation of satellite radar images is an important ongoing research field in monitoring river ice for both scientific and operational communities. This research focus on the development of optimal recognition strategies for the purpose of identifying different ice classes in SAR imagery. However acquisitions of SAR images produce certain problems. SAR images contain speckle noise which is based on multiplicative noise or rayleigh noise. Speckle noise is the result of two phenomenons, first phenomenon is the coherent summation of the backscattered signals and other is the random interference of electromagnetic signals. This therefore degrades the appearance and quality of the captured images. Ultimately it reduces the performance of important techniques of image processing such as detection, segmentation, enhancement and classification etc. This research contributes the major objectives towards speckle filtering using wavelet techniques optimized by Ant Colony Optimization (ACO). First is to remove noise in uniform regions. Second is to preserve and enhance edges and image features and third is to provide a good visual appearance. The work is carried out in three stages. First stage is to transform the noisy image to a new space (frequency domain). Second stage is the manipulation of coefficients. Third is to transform the resultant coefficients back to the original space (spatial domain).Results show that statistical wavelet shrinkage filters are good in speckle reduction but they also lose important feature details. Here the challenge is to find an appropriate threshold value, which is achieved using intra-scale dependency of the wavelet coefficients to estimate the signal variance only using the homogeneous local neighboring coefficients. Moreover, to determine the homogeneous local neighboring coefficients, the ACO technique is used to classify the wavelet coefficients. Experimentation is conducted on SAR images which are further used for development of optimal recognition system for ice classes.

Keywords: ACO, Wavelet, SAR images, Denoising, Thresholding, Shrinkage methods.

V. Snasel et al. (Eds.): SOCO Models in Industrial & Environmental Appl., AISC 188, pp. 399–407.
springerlink.com   © Springer-Verlag Berlin Heidelberg 2013

1 Introduction

Automation in river ice image classification is desired to assist ice experts in extracting geophysical information from the increasing volume of images. Rivers and streams are the key elements in the terrestrial re-distribution of water. An ice cover has significant impact on rivers such as modifies ecosystem, affects microclimate, causes flooding, restricts navigation and impacts hydropower generation. Multiple research work is carried independently in satellite images to develop algorithms in remote sensing for classification of ice as they cover large areas. Less research has been done in preprocessing segment that can improve the precision of the interpretation. This paper presents a research work based on image processing techniques on the ice patterns in synthetic aperture radar (SAR) imagery for speckle denoising. Neural networks are known for their ability to solve various complex problems in image processing [10]. A biological motivation and some of the theoretical concepts of ant colony optimization algorithms are the inspiration of the work done in this paper [11].Here, analysis is done on the performance of wavelet derived from the optimization technique (ACO) based on image enhancement methods. The main advantage of wavelet analysis is that it allows the use of long time intervals where more precise low frequency information is wanted, and shorter intervals where high frequency information is sought. Hence wavelet analysis is therefore capable of revealing aspects of data that other image analysis techniques miss, such as trends, breakdown points, and discontinuities in higher derivatives and self-similarity. Wavelets are also capable of compressing or de-noising a image without appreciable degradation of the original image. A comprehensive experiment is carried out here based on the evaluation parameters and quantifiable metrics. This work evaluates the objective parameters and emphasizes the need of preprocessing using optimization technique (ACO) for texture based ice classification. This paper introduces a new approach, which incorporates the wavelet filter and ACO altogether, to achieve the goal of denoising and preserving the edges. The paper is structured as: Section 2 deals with the image enhancement framework. Section 3 comprises the SAR image denoising using Wavelet. Section 4 5 explains about denoising using wavelet shrinkage methods and states the need of optimization for threshold value in wavelet shrinkage and Section 6 the experimental results of the proposed optimization technique for wavelet denoising. The paper ends with section 7 with remarks on possible future work in this area and some conclusions.

2 Image Enhancement Framework

This is the first and lowest level operation to be done on images. The input and the output are both intensity images. The main idea with the preprocessing is to suppress information in the image that is not relevant for its purpose or the following analysis of the image. The pre-processing techniques use the fact that neighboring pixels have essentially the same brightness. There are many different pre-processing methods developed for different purposes. Interesting areas of pre-processing for this work is image filtering for noise suppression. Conservative methods based on wavelet transforms have been emerged for removing speckle noise from images. This local preprocessing speckle reduction technique is necessary prior to the processing of SAR images. Here we identify wavelet Shrinkage or thresholding as denoising method. It is well known that increasing the redundancy of wavelet transforms can significantly improve the denoising performances. Thus a thresholding is been obtained by optimization technique called ant colony optimization (ACO).

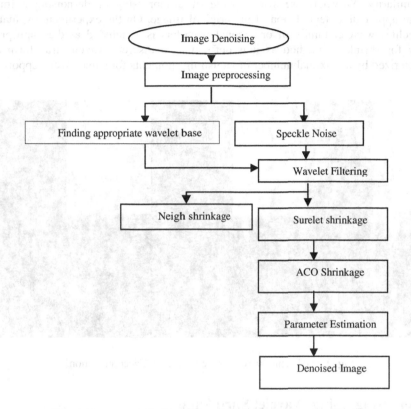

Fig. 1. Framework for Denoising SAR image for ice classification

3   SAR Image Denoising Using Wavelet

The two main confines in image accuracy are categorized as blur and noise. Blur is intrinsic to image acquisition systems, as digital images have a finite number of samples and must respect the sampling conditions. The second main image perturbation is noise. Image denoising is used to remove the additive noise while retaining as much as possible the important signal features. Currently a reasonable amount of research is done on wavelet thresholding and threshold selection for signal de-noising, because wavelet provides an appropriate basis for separating noisy signal from the image signal. Two shrinkage methods are used over here to calculate new pixel values in a local neighborhood. Shrinkage is a well known and appealing denoising technique. The use of shrinkage is known to be optimal for Gaussian white noise, provided that the sparsity on the signal's representation is enforced using a unitary transform. The main advantage of wavelet analysis is that it allows the use of long time intervals where more precise low frequency information is wanted, and shorter intervals where high frequency information is sought. Wavelet analysis is therefore capable of revealing aspects of data that other image analysis techniques miss, such as trends, breakdown points, and discontinuities in higher derivatives and self-similarity. Wavelets are also capable of compressing or de-noising a image without appreciable degradation of the original image. On the experiment evaluation Daubechies wavelet family of orthogonal wavelets is concluded as the appropriate family for shrinkage method as it is defined as a discrete wavelet transform and characterized by a maximal number of vanishing moments for some given support.

Fig. 2. Daubechies wavelet based on level 2 decomposition

4 Denoising Using Wavelet Shrinkage

Here image denoising is based on the image-domain minimization of an estimate of the mean squared error-Stein's unbiased risk estimate (SURE) is proposed and equation (1) specifies the same. Surelet the method directly parameterizes the denoising process as a sum of elementary nonlinear processes with unknown weights. Unlike most existing denoising algorithms, using the SURE makes it needless to hypothesize a statistical model for the noiseless image. A key of it is, although the (nonlinear) processing is performed in a transformed domain-typically, an undecimated discrete wavelet transform, but we also address nonorthonormal transforms-this minimization is performed in the image domain.

$$sure(t;x) = d - 2.\#\{i : |x_i| \le t\} + \sum_{i=1}^{d}(|x_i|\Lambda t)^2 \qquad (1)$$

where d is the number of elements in the noisy data vector and xi are the wavelet coefficients. This procedure is smoothness-adaptive, meaning that it is suitable for denoising a wide range of functions from those that have many jumps to those that are essentially smooth. It have high characteristics as it out performs Neigh shrink method. Comparison is done over these two methods to prove the elevated need of Surelet shrinkage for the denoising the SAR images. The experimental results are projected in graph format which shows that the Surelet shrinkage minimizes the objective function the fastest, while being as cheap as neighshrink method. Measuring the amount of noise equation (2) is done by its standard deviation , , one can define the signal to noise ratio (SNR) as

$$SNR = \frac{\sigma(\mu)}{\sigma(n)},\tag{2}$$

Where in equation (3) denotes the empirical standard deviation of

$$\sigma(\mu) = \left(\frac{1}{|I|}\sum_i (u(i) - \overline{\mu})^2\right)^{1/2}\tag{3}$$

And   is the average grey level value. The standard deviation of the noise can also be obtained as an empirical measurement or formally computed when the noise model and parameters are known.

5   Ant Colony Optimization for Wavelet Based Denoising

ACO is a nature-inspired optimization algorithm motivated by the natural collective behavior of real-world ant colonies. The key challenge of wavelet shrinkage is to find an appropriate threshold value, which is typically controlled by the signal variance. Ant colony optimization (ACO) technique is used to classify the wavelet coefficients. It exploits the intra-scale dependency of the wavelet coefficients to estimate the signal variance only using the homogeneous local neighboring coefficients.

The proposed AntShrink approach has two stages:

(i)   develop the ACO technique to classify the wavelet coefficients
(ii) shrink the noisy wavelet coefficient according to a locally-adapted signal variance value.

The proposed approach is summarized as follows.

- Perform a 2-D discrete wavelet decomposition on a noisy image to get the noisy wavelet coefficients.
- Perform the ACO-based classification.
- Estimate the signal variance.
- Compute the MMSE estimation.
- Perform the inverse wavelet transform to obtain the denoised image.

The figure 3 shows the visual quality of the image after applying aco based wavelet denoising method.

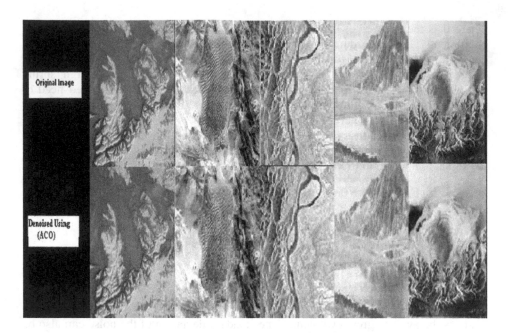

Fig. 3. Denoised image using ACO based on Db2 Wavelet Family

6 Experimental Results

Interesting areas of pre-processing for this work is image filtering for noise suppression. This local preprocessing speckle reduction technique is necessary prior to the processing of SAR images.

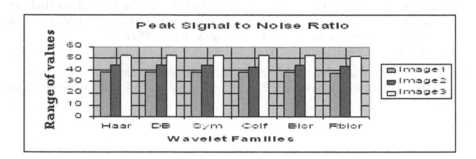

Fig. 4. Peak to Signal noise ratio for wavelet methods

Here we identify wavelet Shrinkage or thresholding as denoising method It is well known that increasing the redundancy of wavelet transforms can significantly improve the denoising performances. Thus a thresholding process which passes the coarsest approximation sub-band and attenuates the rest of the sub-bands should decrease the amount of residual noise in the overall signal after the denoising process .Figure 4,5 shows the evaluation of wavelet families to find the best and Daubechies wavelet is been concluded for wavelet shrinkage denoising.

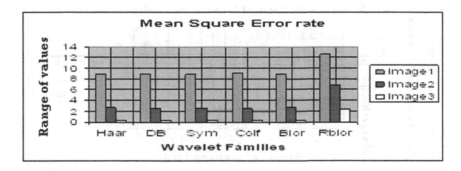

Fig. 5. Mean Square Error rate for wavelet methods

Figure 6, 7 represents the objective evaluation of the shrinkage methods and finally surelet shrinkage is been concluded as the optimal method for denoising.

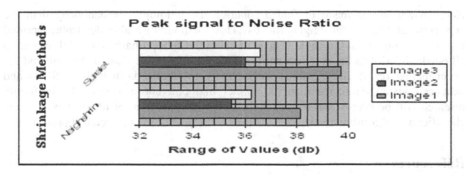

Fig. 6. PSNR values for shrinkage method based on DB Wavelet family

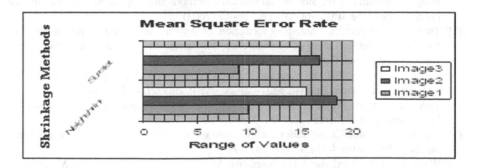

Fig. 7. MSE values for shrinkage method based on DB Wavelet family

As finally, comparison is been made between shrinkage and aco based shrinkage method for denoising SAR image.

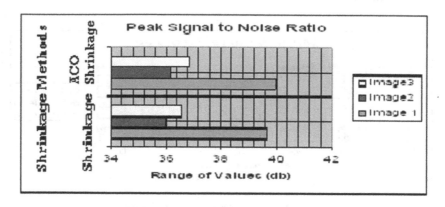

Fig. 8. Comparison of PSNR values for ACO shrinkage based on DB Wavelet family

7 Conclusion

Here an attempt is made to find a superior methodology for denoising than the conventional fixed-form neighborhoods .This methodology also determine optimal results by using finest threshold instead of using the suboptimal universal threshold in all bands. It exhibits an excellent performance for further ice detection phase and the experimental result also signifies the same by producing both higher PSNRs and enhanced visual eminence than the former and conventional methods. In future, research will be carried out to reduce the computational load of the proposed image classification algorithm and shorten the execution time of the projected approach.

References

[1] Zhang, H., Fritts, J., Goldman, S.: An Entropy-based Objective Evaluation Method for Image Segmentation. In: Storage and Retrieval Methods and Applications for Multimedia. Proceedings of the SPIE, vol. 5307, pp. 38–49 (2003)

[2] Tweed, T., Miguet, S.: Automatic detection of regions in interest in mammographies based on a combined analysis of texture and histogram. In: International Conference on Pattern Recognition, Quebec City, Canada (August 2002)

[3] Randen, T., Husoy, J.H.: Filtering for Texture Classification: A Comparative Study. IEEE Transaction on Pattern Analysis and Machine Intelligence 21(4), 291–310 (1999)

[4] Tian, J., Yub, W., Mac, L.: AntShrink: Ant colony optimization for image shrinkage. Pattern Recognition Letters 31, 1751–1758 (2010)

[5] Unser, M.: Texture Classification and Segmentation Using Wavelet Frames. IEEE Trans. on Image Processing 4(11), 1549–1560 (1995)

[6] Wang, J., Li, J., Wiederhold, G.: Simplicity: Semantics-Sensitive Integrated Matching for Picture Libraries. IEEE Transactions on Pattern Analysis and Machine Intelligence 23(9), 947–963 (2001)

[7] Liu, J., Yang, Y.-H.: Multi-resolution color Image segmentation. IEEE Transaction on Pattern Analysis and Machine Intelligence 16(7), 689–700 (1994)

[8] Chan, R.H.: An Iterative procedure for removing random- valued impulse noise. IEEE Signal Process. Lett. (11) (2004)

[9] Subashini, P., Krishnaveni, M., Ane, B.K., Roller, D.: An optimum threshold based segmentation for ice detection in SAR images using Gabor filter. In: Proceedings of International Conference on Computing, ICC 2010, New Delhi, December 27 - 28 (2010)

[10] Abraham, A., Liu, H., Grosan, C., Xhafa, F.: Nature Inspired Metaheuristics for Grid Scheduling: Single and Multiobjective Optimization Approaches Metaheuristics for Scheduling: Distributed Computing Environments. SCI, pp. 247–272. Springer, Germany (2008) ISBN: 978-3-540-69260-7

[11] Abraham, A., Grosan, C., Ramos, V.: Swarm Intelligence and Data Mining. SCI, p. 270. Springer, Germany (2006) ISBN: 3-540- 34955-3

A New Approach for Indexing Powder Diffraction Data Suitable for GPGPU Execution

I. Šimeček

Department of Computer Systems, Faculty of Information Technologies,
Czech Technical University in Prague, Prague, Czech Republic
`xsimecek@fit.cvut.cz`

Abstract. Powder diffraction (based typically on X-ray usage) is a well-established method for a complete analysis and structure determination of crystalline materials. One of the key parts of the experimental data processing is the process of indexation - determination of lattice parameters. The lattice parameters are essential information required for phase identification as well as for eventual phase structure solution.

Nowadays computer hardware with efficient implementation of indexation algorithms gives a chance to to solve several problematic situations not covered fully by existing methods. This paper deals with design of algorithm for indexing powder diffraction data suitable for massively parallel platforms such as GPUs.

Keywords: powder diffraction, indexing, McMaille, parallel simulating annealing, GPGPU.

1 Introduction

Powder diffraction is a common method in many areas of solid state sciences. Indexation of a hi-resolution powder diffraction records is a relatively well-solved problem. There exist several algorithms based on trial-and-error methods or dichotomy methods[1]. Most of these algorithm were developed 20 years ago when really fast CPUs did not exist, a processor architecture was completely different and parallel platforms were not common. So these methods are more targeted on the speed of execution than on the robustness and completeness of the solution space search. The whole powder indexation process can be complicated by several additional problems, not fully covered by the standard approaches.

1.1 Indexing

The lattice parameters (unit-cell dimensions, see Fig. 1) $(a, b, c, \alpha, \beta, \gamma)$ define the parameter space of the problem, and the aim of the indexing process is to find the correct values of these lattice parameters (or a subset of them depending on the crystal system) for a given experimental powder diffraction pattern.

V. Snasel et al. (Eds.): SOCO Models in Industrial & Environmental Appl., AISC 188, pp. 409–416.

Input Data. Powder diffraction data contain N (typically, $N \leq 30$) observed angles θ_i for the given wavelength λ. These data are converted using formula $Q = (2 \sin \theta / \lambda)^2$ into Q^m values (the m denotes measured values of Q). These values are used for testing if the proposed crystal model corresponds to measured data.

Unit Cell Model. The theoretical values of parameters Q can be obtained by the following physical model:

$$Q = a_{11}h^2 + a_{22}k^2 + a_{33}l^2 + a_{12}h \cdot k + a_{23}k \cdot l + a_{13}h \cdot l \qquad (1)$$

where the a_{ij} parameters are the reciprocal lattice parameters. We will denote these (theoretical) values of Q by Q^t to distinguish between these theoretical values from model and measured values. Reciprocal lattice parameters a_{ij} can be derived from the real lattice parameters using formulas for the given crystal system, for example for monoclinic case:

$$a_{11} = \frac{1}{a^2 \sin^2 \beta} \quad a_{22} = \frac{1}{b^2} \quad a_{33} = \frac{1}{c^2 \sin^2 \beta} \quad a_{13} = \frac{2 \cos \beta}{ac \sin^2 \beta} \qquad (2)$$

Depending on crystal system some values a_{ij} are zero. For example in the cubic system, expressions (1) and (1) can be simplified to: $Q^t = (h^2 + k^2 + l^2)/a^2$.

Problem Statement. The a_{ij} parameters are unknown values (real numbers). The parameters h, k, l (integer numbers) are also unknown and must be determined. So, for every Q_i^m an unknown hkl triplet and a set of s parameters a_{ij} must be estimated.

For N observed data, one needs to determine $3 \cdot N$ integers and s ($s \leq 6$) real numbers. From diffraction theory, h, k and l are small integer numbers, but process of indexation remains very complex.

Moreover, all observed values of Q^m suffer from experimental errors, so in real applications possible solutions must satisfy the $2 \cdot N$ inequalities with form:

$$|Q_i^m - Q_j^t| < D,$$

where D is the maximal tolerance.

The main sources of error of values of Q^m are:

- inaccurate positions of the diffraction lines,
- overlap of the diffraction lines,
- inclusion of peaks due to impurity phases,
- significant zero offset.

1.2   Related Works

There had already been effort to implement new approach to the indexation process in software. We should mention the McMaille software[2], X-Cell software[3],

N-TREOR9[4], and genetic approach [5]. McMaille software utilizes simulating annealing based solution space search or even exhaustive grid search. X-Cell software use dichotomy methods. X-Cell solves in a sophisticated way the impurity phase problem and it utilizes space-group reflection extinction. The N-TREOR9 software is based on the trial and error approach. More advanced soft computing methods[6,7] are not common in indexation process.

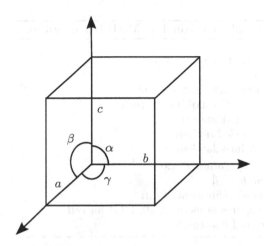

Fig. 1. The lattice parameters

1.3 GPU

The computing power of nowadays-graphic cards outperforms significantly the computing power of CPUs (see [8]). The key motivation for GPU computation is that the newest Intel six-core i7 CPU has got the peak performance about 160 GFlops in comparison to Nvidia graphic card Geforce GTX 590 that has the peak performance about 2.5 TFlops, but prices for both components are comparable!

The GPU acceleration of the computation was already implemented for several scientific problems closed to the crystallographic one: molecular dynamic simulations, QM calculations of molecular structure, protein structure analysis. The exhaustive list of problems for which GPU calculation was already used can be found on the following link[9]. The parallel implementation of GPU code execution starts to be a usual way for acceleration the code execution. For GPU implementation well determined standard APIs like CUDA[9] and OpenCL[10] were established.

2   Current State-of-Art

So, we have decided to modify existing state-of-art algorithm McMaille that uses the Monte Carlo method in order to generate randomly cell parameters tested against an idealized powder profile. Algorithm McMAILLE (see Alg.1) uses these global parameters:

- it = the number of global iterations,
- *threshold* constant to distinguish between "good" and "not so good" solution.

Algorithm 1. principle of original McMAILLE algorithm

1: **procedure** MCM
Output: the database of found solutions
  2:     **for** $i \leftarrow 1, it$ **do**
  3:         $cell \leftarrow$ GENERATERANDOMCELL
  4:         $cell \leftarrow$ MONTECARLOMETHOD($cell$)
  5:         $r \leftarrow$ EVALUATEERROR($cell$)
  6:         **if** $(r \leq threshold)$ **then**
  7:            put $cell$ into database
  8:     **for** $i \leftarrow 1, sizeof(database)$ **do**
  9:         $cell \leftarrow database[i]$
10:         least-squared refinement of $cell$
11:         calculate figures of merit (M20, F20) for $cell$
12:         update record $database[i]$

Algorithm MONTECARLOMETHOD (see Alg.2) uses these global parameters:

- $it2$ = the number of MonteCarlo iterations,
- *threshold2* constant to avoid false minima.

Algorithm 2. Monte Carlo method used in original McMAILLE algorithm

---

 1: **procedure** MONTECARLOMETHOD($cell$)
Input: $cell$ original lattice
Output: $cell$ improved lattice
  2:    $number \leftarrow 0$
  3:    $r \leftarrow$ EVALUATEERROR($cell$)
  4:    **for** $i \leftarrow 1, it$ **do**
  5:       $c2 \leftarrow cell$
  6:       change $c2$ a little bit
  7:       $r2 \leftarrow$ EVALUATEERROR($c2$)
  8:       **if** $(r2 \leq r)$ or $(RandomNumber \leq threshold2)$ **then**
  9:         $cell \leftarrow c2$
10:         $r \leftarrow r2$
11:    **return** $cell$

Algorithm EVALUATEERROR (see Alg.3) uses these global parameters:

- D = the maximal tolerance of Q values,
- n_hkl = the number of hkl triples taken in account,
- N = the number of observed peaks (size of input dataset),

Algorithm 3. Estimation of error of crystal model (original version)

1: **procedure** EVALUATEERROR
Output: error of the lattice *cell*
2:      $n\_q \leftarrow 0$
3:      **for** $i \leftarrow 1, n\_hkl$ **do**
4:          compute Q_temp from triplet $hkl[i]$
5:          **if** $(Q\_temp \geq Q_0^m)$ & $(Q\_temp \leq Q_N^m)$ **then**
6:              $Q_{n\_q}^t \leftarrow Q\_temp$
7:              $n\_q \leftarrow n\_q + 1$
8:      $sum\_error \leftarrow 0$ ▷ All Q^t are found, let's compute the error
9:      **for** $i \leftarrow 1, N$ **do**
10:         $min\_error \leftarrow \infty$
11:         **for** $j \leftarrow 1, n\_q$ **do**
12:             **if** $indexed[j] \neq true$ **then**
13:                 $cur\_error \leftarrow |Q_i^m - Q_j^t|$
14:                 **if** $cur\_error < min\_error$ **then**
15:                     $min\_error \leftarrow cur\_error$
16:                     $index \leftarrow j$
17:         **if** $min\_error < D$ **then**
18:             $sum\_error \leftarrow sum\_error + min\_error$
19:             $indexed[index] = true$
20:     **return** sum_error

---

3   Results

The most time-consuming part of McMaille algorithm is the procedure EVALUA-TEERROR (Algorithm 3), so we concentrate on parallelization of this procedure.

3.1   GPU Architecture Limitations

For fully utilization of GPU architecture these basic conditions must be satisfy:

- The whole computation should be proceed by relatively independent parts (blocks of GPU threads).
- Data transfers between CPU and GPU should be minimal.
- Only small number of registers (up to tens) and small amount of shared memory (up to tens bytes) can be allocated by each GPU thread.

To satisfy these requirements we must modify Algorithms 1,2 and 3.

3.2 Design of the New Algorithm

Description of new algorithms derived from previous ones follows:

- The Algorithm 4 uses only very few registers (in comparison to Algorithm 3), so it better reflect GPU limitation and can be proceed efficiently by one CUDA thread. The Algorithm 4 is not semantically equivalent with Algorithm 3, but our measurements prove that is precise enough for rough estimation of error of proposed cell.
- The Algorithm 5 is GPU version of Algorithm 1 and 2. The fussion of these algorithms was done to minimize the number of data transfers between CPU and GPU. The codelines (4)-(11) in this algorithm can be proceed in parallel by GPU.

Algorithm 4. Estimation of error of crystal model (our version)

1: **procedure** GPUEVALUATEERROR
Output: error of the lattice *cell*
2:     $sum\_error \leftarrow 0$
3:     $sum\_error2 \leftarrow 0$
4:     **for** $i \leftarrow 1, n\_hkl$ **do**
5:         compute Q_temp from $hkl[i]$
6:         **if** $(Q\_temp \geq Q_0^m)$ & $(Q\_temp \leq Q_N^m)$ **then**
7:             **for** $j \leftarrow 1, N$ **do**
8:                 $j2 \leftarrow N - j + 1$
9:                 **if** $indexed[j] \neq$ true **then**
10:                     $cur\_error \leftarrow |Q_j^m - Q\_temp|$
11:                     **if** $cur\_error < D$ **then**
12:                         $sum\_error \leftarrow sum\_error + curr\_error$
13:                         $indexed[j] = true$
14:                 **if** $indexed2[j2] \neq$ true **then**
15:                     $cur\_error2 \leftarrow |Q_{j2}^m - Q\_temp|$
16:                     **if** $cur\_error2 < D$ **then**
17:                         $sum\_error2 \leftarrow sum\_error2 + curr\_error2$
18:                         $indexed2[j] = true$
19:     return $\min(sum\_error, sum\_error2)$

3.3 HW and SW Configuration

Experimental Configuration 1. All results were measured on dual-core Intel i3-370M at 2.4 GHz, 4 GB of the main memory at 1333 MHz, running OS Windows 7 Home with graphic card Nvidia Geforce 420M (64 GPU cores), Microsoft Visual Studio 2005, Nvidia SDK version 4.0.

Algorithm 5. principle of GPU version of McMAILLE algorithm

1: **procedure** MCM
Output: the database of found solutions
2:     for $i \leftarrow 1, it$ **do**
3:         $db[i].cell \leftarrow$ GENERATERANDOMCELL
4:         $db[i].r \leftarrow$ EVALUATEERROR($cell$)
5:         **for** $j \leftarrow 1, sd$ **do**
6:             $cell \leftarrow db[j].cell$
7:             $c2 \leftarrow cell$
8:             change $c2$ a little bit
9:             $r2 \leftarrow$ EVALUATEERROR($c2$)
10:             **if** $(r2 \leq r)$ or $(RandomNumber \leq threshold2)$ **then**
11:                 $db[j].cell \leftarrow c2$
12:                 $db[j].r \leftarrow r2$
13:     **for** $i \leftarrow 1, sd$ **do**
14:         $cell \leftarrow database[i]$
15:         least-squared refinement of $cell$
16:         calculate figures of merit (M20, F20) for $cell$
17:         update record $database[i]$

Experimental Configuration 2. All results were measured on quad-core Intel i7 950 at 3.07GHz, 24 GB of the main memory at 1600 MHz, running OS Linux Debian with graphic card GeForce GTX 590 (1024 GPU cores), Nvidia SDK version 4.1.28.

Indexation Data Files. We have used the following data files:

- cim.dat (Cimetidine, $N = 21$, monoclinic crystal system, correct solution: $a = 10.3893$Å, $b = 18.8215$Å, $c = 6.8215$Å, $\beta = 106.477°, V = 1279.113$Å^3),
- cim5.dat ($N = 26$, same as previous one with 5 added impurity lines),
- Taxol.dat (10-Deacetyl-7-epitaxol ethyl acetate solvate, $N = 20$, monoclinic crystal system, correct solution: $a = 16.329$Å, $b = 17.704$Å, $c = 17.504$Å, $\beta = 100.611°, V = 4973.6781$Å^3),
- Taxol5.dat ($N = 25$, same as previous one with 5 added impurity lines),

3.4   Evaluation of the Results

We have measured speedup between original version of McMaille algorithm (one CPU core was used) and our new variant. If the computation is proceed by CPU, the original code was comparable to the new one. If the computation is proceed by GPU, the original code was much slower in comparison to the new one. Achieved speedups are from 2 to 10 (for Testing configuration 1) or from 5 to 50 (for Testing configuration 2). Speedup varies due to these facts:

- CPU and GPU version of algorithms for the estimation of error of crystal model (Algorithm 3 and 4) can return different results.

– In original McMaille(CPU) version some parameters (like *threshold* and so
on) are changed during the computation, in our (GPU) version they remain
constant.

4   Conclusion

We have described basics of powder diffraction indexing process. We have proposed some modifications of state-of-art McMaille algorithm for this problem.
Modified algorithm can fully utilize GPU computational power and achieves a
significant speedup of computation in comparison to the original algorithm.

Future Works. Proposed method (same as the former one) is relative insen-
sitive to impurity lines. We want to increase this property furthermore to allow
indexing multiphase patterns since indexing of a mixture is a big challenge of
applied crystallography.

Acknowledgment. This research has been supported by CTU grant
SGS12/097/OHK3/1T/18.

References

1. David, W., Shankland, K., McCusker, L., Baerlocher, C.: Structure Determination
   from Powder Diffraction Data. Oxford Science Publications (2002)
2. Bail, A.L.: Monte carlo indexing with mcmaille. Powder Diffraction 19, 249–254
   (2004)
3. X-Cell: X-cell - a novel and robust indexing program for medium- to high-quality
   powder diffraction data. J. Appl. Cryst. 36, 356–365 (2003)
4. Werner, P.E., Eriksson, L., Westdahl, M.: TREOR, a semi-exhaustive trial-and-
   error powder indexing program for all symmetries. Journal of Applied Crystallography 18(5), 367–370 (1985)
5. Kariuki, B.M., Belmonte, S.A., McMahon, M.I., Johnston, R.L., Harris, K.D.M.,
   Nelmes, R.J.: A new approach for indexing powder diffraction data based on whole-
   profile fitting and global optimization using a genetic algorithm. Journal of Syn-
   chrotron Radiation 6(2), 87–92 (1999)
6. Corchado, E., Arroyo, A., Tricio, V.: Soft computing models to identify typical
   meteorological days. Logic Journal of the IGPL 19(2), 373–383 (2011)
7. Zhao, S.Z., Iruthayarajan, M.W., Baskar, S., Suganthan, P.N.: Multi-objective
   robust pid controller tuning using two lbests multi-objective particle swarm optimization. Inf. Sci. 181(16), 3323–3335 (2011)
8. Owens, J.D., Luebke, D., Govindaraju, N., Harris, M., Ger, J.K., Lefohn, A.,
   Purcell, T.J.: A survey of general-purpose computation on graphics hardware.
   Computer Graphics Forum 26(1), 80–113 (2007)
9. NVIDIA Corporation: Cuda tools & ecosystem (2012)
10. Khronos Group: Opencl - the open standard for parallel programming of hetero-
    geneous systems (2012)

A Kernel for Time Series Classification: Application to Atmospheric Pollutants

Marta Arias[1], Alicia Troncoso[2], and José C. Riquelme[3]

[1] Universitat Politècnica de Catalunya, Catalunya, Spain
marias@lsi.upc.edu
[2] Pablo de Olavide University, Seville, Spain
ali@upo.es
[3] University of Seville, Seville, Spain
riquelme@us.es

Abstract. In this paper a kernel for time-series data is presented. The main idea of the kernel is that it is designed to recognize as similar time series that may be slightly shifted with one another. Namely, it tries to focus on the shape of the time-series and ignores the fact that the series may not be perfectly aligned. The proposed kernel has been validated on several datasets based on the UCR time-series repository [1]. A comparison with the well-known Dynamic Time Warping (DTW) distance and Euclidean distance shows that the proposed kernel outperforms the Euclidean distance and is competitive with respect to the DTW distance while having a much lower computational cost.

1 Introduction

Time-series analysis is an important problem with application in domains as diverse as engineering, medicine, astronomy or finance [2,3]. In particular, the problem of time-series classification is attracting a lot of attention among researchers [4]. Among the most successful and popular methods for classification are kernel-based methods such as support vector machines. Despite their popularity, there seem to be only a handful of kernels designed for time-series. This paper tries to fill this void, and proposes a kernel exclusively designed for time-series. Moreover, using a standard trick, we are able to convert our kernel into a distance metric for time-series, therefore allowing us to use our kernel in distance-based algorithms as well.

A crucial aspect when dealing with time-series is to find a good measure, either a kernel similarity or a distance, that captures the essence of the time-series according to the domain of application. For example, Euclidean distance between time-series is commonly used due to its computational efficiency; however, it is very brittle and small shifts in of one time-series can result in huge changes in the Euclidean distance. Therefore, more sophisticated distances have been devised designed to be more robust to small fluctuations of the input time-series. Notably, Dynamic Time Warping (DTW) is held as the state-of-the-art method for comparing time-series. Unfortunately, computing the DTW distance

V. Snasel et al. (Eds.): SOCO Models in Industrial & Environmental Appl., AISC 188, pp. 417–426.
springerlink.com

is prohibitely costly for many practical applications. Therefore, researchers are coming up with distances for time-series that approach the DTW at lower computational costs [5,4,6]. In a sense, the kernel-derived distance that is proposed here tries to fix the brittleness of Euclidean distance without incurring in the high computational costs of DTW. At a high level, our proposed distance is a combination of the Euclidean distances obtained by using several smoothed versions or the original time-series.

Many other distances have been proposed depending on the invariants required by the domain. For example, [7] define a distance between two time-series representing the convexities/concavities of two shape contours. In [8] the authors modify the Euclidean distance with a correction factor based on the complexity of the input time-series.

It is known that the DTW distance is not a distance in a strict sense as it does not fulfill the triangular inequality and, therefore, it can not be used to define a positive definite kernel. A more general theory of learning instead of positive semi-definite kernels and the relationship between good kernels and similarity functions is presented in [9]. In [10] a new kernel is defined by global alignments from the DTW distance. In particular, the kernel is defined as the sum of the exponential function of the distances for all possible alignments. However, this kernel has a high computational cost and constraints on alignments, similar to that of [5], are presented to speed-up the computation in [11]. The same authors present in [12] a kernel based on the idea that similar time series should be fit well by the same models. They use autoregressive models and thus the name of autoregressive kernel. A kernel for periodic time-series arising in the field of astronomy is presented in [13]. This kernel is similar to a global alignment kernel as it consists in the sum of the exponential function of the inner products for all possible shifting of a time series instead of computing the best alignment. Another kernel for time-series is proposed in [14]. In particular, the time series are represented with a summarizing smooth curve in a Hilbert space and the learning method of the kernel is based on Gaussian processes.

The paper is structured as follows. Section 2 describes our time-series kernel and its corresponding derived distance. Section 3 presents an empirical comparison using 20 different datasets. Finally, Section 5 concludes with a summary of our main contributions and possible directions of future work.

2   Kernel Description

This Section presents the notation used in this paper and also provides the definitions underlying the proposed kernel.

Definition 1 (Time-series). *A time-series X is a set of temporally sorted real-valued data. In this work, $X = \{x_1, ..., x_N\}$, where N is the length of the time-series.*

Definition 2 (Subsequence time-series). *A subsequence of length k of a time-series $X = \{x_1, ..., x_N\}$ is a time-series $X_j = \{x_j, x_{j+1}, ..., x_{j+k-1}\}$ for $1 \leq j \leq N - k + 1$.*

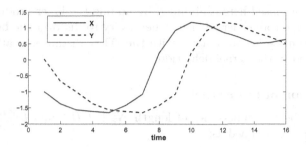

Fig. 1. Example of two shifted time-series

Definition 3 ($\{k, j\}$–Order partial sum). *A $\{k, j\}$–order partial sum of a time-series X, $s^X_{k,j}$, is the sum of the values of the X_j subsequence time-series of length k. That is:*

$$s^X_{k,j} = x_j + x_{j+1} + \cdots + x_{j+k-1}.$$

Definition 4 (k–Order partial sum time-series). *A k–order partial sum time-series is a time-series S^X_k whose values are $s^X_{k,j}$ for $1 \leq j \leq N - k + 1$, that is, the sum of all the values of the subsequences of length k of the time-series X.*

$$S^X_k = \{s^X_{k,1}, s^X_{k,2}, ..., s^X_{k,N-k+1}\}.$$

For example, the $\{k, j\}$–order partial sums and the k–order partial sum time-series for the $X = \{3, 2, 4, 1\}$ time-series are:

$$
\begin{aligned}
s^X_{2,1} &= 3 + 2 = 5 \\
s^X_{2,2} &= 2 + 4 = 6 \qquad\qquad S^X_2 = \{5,6,5\} \\
s^X_{2,3} &= 4 + 1 = 5 \\
s^X_{3,1} &= 3 + 2 + 4 = 9 \qquad\quad S^X_3 = \{9,7\} \\
s^X_{3,2} &= 2 + 4 + 1 = 7 \\
s^X_{4,1} &= 3 + 2 + 4 + 1 = 10 \quad S^X_4 = \{10\}
\end{aligned}
$$

2.1   Motivation

The main motivation in the definition of the kernel proposed here is to obtain a similarity measure for time-series that yields high values when two time-series X and Y have the same shape but may be shifted of one another. Notice that the Euclidean distance does not take this into account as illustrated in Figure 1. It can be observed that both time-series are very similar (the Y time-series is obtained by shifting X). However, the Euclidean distance between the two time-series is very high. As a consequence, bad results could be obtained were the Euclidean distance to be used in distance-based classification algorithms, for instance. The purpose of this work is to propose a kernel that yields high similarity for time-series that have similar shapes.

The kernel proposed here is obtained by adding the inner products of partial sum time-series for all orders. It is not necessary to discover the best alignment between two time-series, in contrast with the DTW distance, as all partial sums will be included in the kernel definition.

2.2   Definition of the Kernel

Let X and Y be two time-series of length N. Let U^X and U^Y be two upper triangular matrices defined as:

$$U^X = [U_1^X, ..., U_N^X]$$
$$U^Y = [U_1^Y, ..., U_N^Y]$$

where U_i^X and U_i^Y are the i-th rows of the matrices U^X and U^Y, respectively, which are defined by:

$$U_{ij}^X = \begin{cases} s_{i,j}^X & \text{if } 1 \leq j \leq N - i + 1 \\ 0 & \text{if } j > N - i + 1 \end{cases} \tag{1}$$

$$U_{ij}^Y = \begin{cases} s_{i,j}^Y & \text{if } 1 \leq j \leq N - i + 1 \\ 0 & \text{if } j > N - i + 1 \end{cases} \tag{2}$$

Finally, the kernel is defined as the sum of the scalar products among the rows of the U^X and U^Y matrices. That is,

$$Kernel(X, Y) = \sum_{i=1}^{N} <U_i^X, U_i^Y> \tag{3}$$

where U^X and U^Y are defined by Equations (1) and (2) and $< \cdot, \cdot >$ is the scalar product of two vectors in $\mathbb{R}^N$. It is obvious that the function defined by Equation (3) is indeed a kernel as it can be represented by a inner product in the high–dimensional feature space $\phi(\cdot)$ defined as follows:

$$Kernel(X, Y) = < \phi(X), \phi(Y) > \tag{4}$$

where

$$\phi : \mathbb{R}^N \longrightarrow \mathbb{R}^{N^2}$$
$$X \longrightarrow \phi(X) = (U_1^X, ..., U_N^X)$$

Next, we show an illustrative example for the time-series $X = \{3, 2, 4, 1\}$ and $Y = \{1, -1, 0, 2\}$. Firstly, the U^X and U^Y matrices comprising the partial sums of the X and Y time-series have to be computed. The 2-order partial sums for X and Y are $S_2^X = \{5, 6, 5\}$ and $S_2^Y = \{0, -1, 2\}$, respectively. Analogously, the 3 and 4 order partial sums are $S_3^X = \{9, 7\}$, $S_3^Y = \{0, 1\}$, $S_4^X = \{10\}$ and $S_4^Y = \{2\}$. Therefore, the matrices are:

$$U^X = \begin{bmatrix} 3 & 2 & 4 & 1 \\ 5 & 6 & 5 & 0 \\ 9 & 7 & 0 & 0 \\ 10 & 0 & 0 & 0 \end{bmatrix} \quad U^Y = \begin{bmatrix} 1 & -1 & 0 & 2 \\ 0 & -1 & 2 & 0 \\ 0 & 1 & 0 & 0 \\ 2 & 0 & 0 & 0 \end{bmatrix}$$

The second step consists in calculating the scalar products of the rows of the U^X matrix and the corresponding rows of the U^Y matrix. That is,

$$< U_1^X, U_1^Y > = 3 \cdot 1 + 2 \cdot (-1) + 4 \cdot 0 + 1 \cdot 2 = 3$$
$$< U_2^X, U_2^Y > = 5 \cdot 0 + 6 \cdot (-1) + 5 \cdot 2 = 4$$
$$< U_3^X, U_3^Y > = 9 \cdot 0 + 7 \cdot 1 = 7$$
$$< U_4^X, U_4^Y > = 10 \cdot 2 = 20$$

where U_i^X and U_i^Y are the i rows of the U^X and U^Y matrices, respectively.

Finally, the kernel is defined as the sum of the above-mentioned scalar products. Therefore,

$$Kernel(X, Y) = (3 + 4 + 7 + 20) = 34.$$

It should be noted that a distance metric can be obtained from any positive definite kernel Ker using the standard transformation described in the following equation [15]:

$$d(u, v) = Ker(u, u) + Ker(v, v) - 2 \cdot Ker(u, v).$$

Therefore, when this work refers to the proposed kernel as a distance it really means the derived distance from the kernel.

3 Results

This section presents the results obtained by the application of the proposed kernel to the classification of multi–class time series. Section 3.1 provides a detailed description of all datasets used in the experiments. In Section 3.2 the kernel has been applied to twenty time-series to validate its potential for separating classes in time-series. Finally, a real-world dataset composed by ozone time series is considered in Section 3.3.

3.1 Description of Datasets

The new kernel has been initially tested on several datasets from the UCR time-series repository [1]. Time series lengths in our datasets range from 60 to 637, with the average and the median being 282.1 and 272.5, respectively. Computation times are highly sensitive to the time-series length, especially for the DTW algorithm, which is quadratic in this parameter. Relevant information about these datasets is summarized in Table 1.

Table 1. Datasets from UCR time-series Repository [1]

Dataset	Num. Instances	Num. Classes	Length of Series	Dataset	Num. Instances	Num. Classes	Length of Series
50Words	450	50	270	Lighting-2	20	2	637
Adiac	296	37	176	Lighting-7	63	7	319
Beef	45	5	470	OSU Leaf	54	6	427
CBF	21	3	128	OliveOil	40	4	570
Coffee	18	2	286	Swedish Leaf	105	15	128
ECG	14	2	96	Trace	36	4	275
Fish	63	7	463	Two Patterns	28	4	128
Face (All)	112	14	131	Synthetic Control	36	6	60
Face (Four)	36	4	350	Wafer	16	2	152
Gun-Point	16	2	150	Yoga	18	2	426

3.2   Validation

A statistic based on pair-wise distances has been developed to show how well the proposed kernel is able to separate classes in time-series.

Let D be a labeled dataset of M time-series of the same length N. Let $c(X)$ be the class of the time-series $X \in D$. Then, the SM separation measure is defined as follows,

$$SM = \frac{INTRA - INTER}{MAX}$$

where $INTRA$ and $INTER$ are the average pair-wise distance of time series belonging to the same and to different classes, respectively, and MAX is the maximum pair-wise distance over the whole dataset. Namely, let $A = \{(X,Y)|X,Y \in D, c(X) = c(Y)\}$ and $B = \{(X,Y)|X,Y \in D, c(X) \neq c(Y)\}$. That is, A is the set of pairs of time series that belong to the same class, and B is the set of pairs of time-series that belong to different classes. Then,

$$INTRA = \frac{1}{|A|} \sum_{(X,Y) \in A} d(X,Y)$$

$$INTER = \frac{1}{|B|} \sum_{(X,Y) \in B} d(X,Y)$$

$$MAX = \max_{X,Y \in D} d(X,Y)$$

where d is any distance defined in $\mathbb{R}^N \times \mathbb{R}^N$ and N is the length of the time-series in D.

In a sense, the SM measure is designed to show how well the proposed distance separates instances in different classes as opposed to instances within the same class. Since we are taking averages, it is a measure of the *global* separability ability of the distance. This distance is reminiscent to the cost function used in the problem of correlation clustering in weighted graphs [16,17], if we were to cluster all instances using their class.

Table 2 presents a comparison of the separation statistic and computation time of the following distances: the Euclidean distance, the one derived from the kernel proposed here (which we call Kernel-based distance), and the DTW distance. The comparison is over the 20 datasets from the UCR repository [1]. The distance that better separates the existing classes for each dataset is marked in bold style. It can be seen that the average of the separation measure for the proposed kernel is better than that of the Euclidean distance and similar to that of the DTW distance. When looking at the columns for computation times, it is very clear that the Euclidean distance is by far the fastest one to compute, followed by our proposed distance using (roughly) an order or magnitude extra CPU time. The DTW distance is by far the slowest, needing two more orders or magnitude than our kernel-based distance.

Table 3 further summarizes Table 2. On the table on the left, the reader can observe that the behavior of the kernel-based distance is better than that of DTW on average (1.65 versus 1.85), and both outperform the Euclidean distance (1.65 and 1.85 versus 2.40). The table on the right shows the wins matrix for pairs of distances over the 20 datasets. That is, in how many datasets a distance separates better than another distance.

Table 2. Separation measure among classes and computing times

	SEPARATION MEASURE			CPU TIMES (in s.)		
DATASET	EUCL	KERNEL	DTW	EUCL	KERNEL	DTW
50Words	0.155	0.196	**0.498**	177.4	3653.3	176629.5
Adiac	-0.042	-0.040	**-0.027**	83.8	387.5	32702.9
Beef	0.696	**0.894**	0.557	1.8	93.7	5505.9
CBF	0.150	0.311	**0.557**	0.5	4.9	94.2
Coffee	-0.015	**0.111**	-0.015	0.3	5.8	316.9
ECG	0.045	0.054	**0.126**	0.2	3.8	22.8
FISH	**-0.004**	-0.008	-0.018	10.3	46.4	2690.4
Face (All)	0.007	0.110	**0.387**	1.1	46.7	1928.6
Face (Four)	-0.005	**0.016**	0.011	3.4	131.1	10576.4
Gun-Point	0.136	0.113	**0.345**	0.3	1.2	83.3
Lighting-2	0.126	0.152	**0.263**	0.4	29.0	1964.2
Lighting-7	0.137	0.268	**0.279**	3.6	67.1	4767.3
OSU Leaf	0.229	**0.346**	0.098	1.4	94.0	6396.3
OliveOil	-0.063	-0.041	**-0.027**	2.5	100.8	6459.6
Swedish Leaf	0.104	**0.144**	0.048	9.3	39.2	2286.0
Trace	0.300	**0.303**	0.091	1.2	3.1	65.1
Two Patterns	0.113	0.165	**0.585**	1.1	19.2	1195.1
Synthetic Control	0.103	0.293	**0.580**	0.7	3.9	174.5
Wafer	0.126	**0.172**	0.015	0.3	1.0	72.9
Yoga	-0.000	**0.073**	-0.005	0.3	11.0	724.1
Average	**0.115**	**0.182**	**0.202**	**15**	**237.13**	**12732.8**

Table 3. Left: comparison of the number of times each distance achieves the first, second and third positions over all datasets and average rank. Right: Win matrix for pairs of distances, it should be read as follows: if row i and column j contains number m, then distance i has beaten distance j a total of m times. For example, the Euclidean distance beats the Kernel-based distance in 2 datasets, and beats the DTW distance in 8 datasets.

Distance	#1st	#2nd	#3rd	Avg. Rank
Euclidean	1	7	11	2.40
Kernel	8	11	1	1.65
DTW	11	1	8	1.85

Distance	Euclidean	Kernel	DTW
Euclidean	–	2	8
Kernel	18	–	9
DTW	12	11	–

4   A Real Application: Classification of Ozone Concentration in Atmosphere

Finally, an environmental application related to atmospheric pollutants such as the tropospheric ozone is presented. The pattern recognition in ozone time data is an important task as it is neccessary activate environmental politics and alert protocols by the government when the ozone reaches high ozone concentration levels in atmosphere. Ozone time series have been retrieved from a meteorological station placed in the outskirts of Seville city (Spain), providing 312 times series composed of 168 hourly records each one. The dataset is classified into two classes corresponding to high and low ozone level periods (165 and 147 time series, respectively). The time series data have been split in training set (218 time series) and test set (94 time series) preserving the proportion between the two existing classes.

We have used the well-known nearest neighbor method (1-NN) with the three competing distances to classify the ozone time series into weeks of high or low ozone concentration. Table 4 shows the error in percentage and the time in seconds obtained from the application of the 1-NN method to classify the test set when using several distances. It can be observed that the kernel-based distance presents better results in both error and CPU time.

Table 4. Percentage of error and time in seconds required to classify the test set

Distance	Error	Time
Euclidean	9.5%	0.5
Kernel-based	**4.2%**	23.3
DTW	6.3%	3692.9

5 Conclusions and Future Work

In this paper we have presented a kernel for time-series data and its associated distance metric. Initial experiments show promise in detecting similarity between time-series. The proposed kernel has been compared to the Euclidean distance as a reference distance and the DTW distance as one of the most competitive distances that exist in the literature. The kernel is shown to efficiently separate different time-series classes, and also, its application to real-world data has been successful. In particular, it achieves low error in the classification of the ozone atmosphere concentration. This paper is an initial step in the study of this kernel and its possible variants. Further experimentation including comparison to other state-of-the-art kernels are underway in the context of classification with kernel-based methods. In the future, we plan to generalize our kernel to time-series that differ in length. We would also like to adapt our ideas so that they can be used in a *streaming* setting where time-series keep growing unboundedly.

References

1. Keogh, E.: UCR time series repository (2011), http://www.cs.ucr.edu/~eamonn/
2. Sedano, J., Curiel, L., Corchado, E., de la Cal, E., Villar, J.R.: A soft computing method for detecting lifetime building thermal insulation failures. Integr. Comput. Aided Eng. 17(2), 103–115 (2010)
3. Corchado, E., Arroyo, A., Tricio, V.: Soft computing models to identify typical meteorological days. Logic Journal of The Igpl / Bulletin of The Igpl 19, 373–383 (2011)
4. Xi, X., Keogh, E., Shelton, C., Wei, L., Ratanamahatana, C.: Fast time series classification using numerosity reduction. In: Proceedings of the 23rd International Conference on Machine Learning, vol. 1040. ACM (2006)
5. Sakoe, H., Chiba, S.: Dynamic programming algorithm optimization for spoken word recognition. IEEE Transactions on Acoustics, Speech and Signal Processing 26, 43–49 (1978)
6. Marteau, P.F., Ménier, G.: Speeding up simplification of polygonal curves using nested approximations. Pattern Anal. Appl. 12(4), 367–375 (2009)
7. Adamek, T., O'Connor, N.E.: A multiscale representation method for nonrigid shapes with a single closed contour. IEEE Transactions on Circuits and Systems for Video Technology 14(5), 742–752 (2004)
8. Gustavo, E.A.P.A., Batista, X.W., Keogh, E.J.: A complexity-invariant distance measure for time series. In: SIAM International Conference on Data Mining (2011)
9. Balcan, M., Blum, A., Srebro, N.: A theory of learning with similarity functions. Machine Learning 72(1), 89–112 (2008)
10. Cuturi, M., Vert, J.P., Birkenes, O., Matsui, T.: A kernel for time series based on global alignments. In: IEEE International Conference on Acoustics, Speech and Signal Processing, ICASSP 2007, vol. 2, pp. II–413–II–416 (2007)
11. Cuturi, M.: Fast global alignement kernels. In: Internacional Conference on Machine Learning (2011)
12. Cuturi, M., Doucet, A.: Autoregressive kernels for time series (2011) arXiv:1101.0673

13. Wachman, G., Khardon, R., Protopapas, P., Alcock, C.R.: Kernels for periodic time series arising in astronomy. In: European Conference on Machine Learning (2009)
14. Lu, Z., Leen, T., Huang, Y., Erdogmus, D.: A reproducing kernel Hilbert space framework for pairwise time series distances. In: Proceedings of the 25th International Conference on Machine Learning, pp. 624–631. ACM (2008)
15. Scholkopf, B.: The kernel trick for distances. In: Proceedings of the Conference on Neural Information Processing Systems (NIPS), pp. 301–307 (2000)
16. Bansal, N., Blum, A., Chawla, S.: Correlation clustering. Machine Learning 56(1-3), 89–113 (2004)
17. Bonchi, F., Gionis, A., Ukkonen, A.: Overlapping correlation clustering. In: ICDM, pp. 51–60 (2011)

Viscosity Measurement Monitoring by Means of Functional Approximation and Rule Based Techniques

Ramón Ferreiro-García[1], José Luis Calvo-Rolle[2], F. Javier Pérez Castelo[2], and Manuel Romero Gómez[1]

[1] University of Coruña, Department of Industrial Engineering
C/Paseo Ronda 51, 15011, A Coruña, Spain
ferreiro@udc.es
[2] University of Coruña, Department of Industrial Engineering
Avda. 19 de febrero, s/n, 15405, Ferrol, A Coruña, Spain
{jlcalvo,javierpc}@udc.es

Abstract. A diagnosis strategy using neural network based functional approximation models associated to a rule based technique is developed. The aim is to apply the diagnostic task on condition monitoring of viscometers used in liquids handling (liquid fuels and lubricating oils) tasks. Based on fluid online measured data, including pressures and temperature, the viscometers diagnosis is being carried out. Required signals are achieved by conversion of available or measured data (fluid temperature and API and SAE grades) into virtual data by means of neural network functional approximation techniques. Using rule based techniques on fault finding and isolation task, it is concluded that the viscometer monitoring task carried out by the analysis of the dynamic behaviour of both, the on line viscometer and virtual data subjected to the analysis of residuals into a parity space approach, is successfully feasible.

Keywords: Backpropagation feedforward NNs, Conjugate gradient, Diagnosis, Functional approximation, Functional redundancy, Parity spaces, Residual generation.

1 Introduction

Computational fluid dynamics depends on fluid viscosity which is a function of the fluid temperature and pressure, scenario in which an important group of analytic model based diagnostic tasks suffers from the required accuracy. Once assumed that viscosity is essentially fluid friction, and consequently, like friction between moving solids, viscosity transforms kinetic energy of macroscopic motion into heat energy, while the temperature dependence of viscosity at isobaric conditions is well established (the viscosity increases with decreasing temperature), current discussion on the dependence of viscosity on external pressure (at isothermal conditions) shows a spectrum of controversial statements.

In [1] (Jürn W. P. Schmelzer et. al., 2005) it is referred to experimental studies where the viscosity must increase in oils, as a rule, with increasing pressure.

V. Snasel et al. (Eds.): SOCO Models in Industrial & Environmental Appl., AISC 188, pp. 427–438.
springerlink.com   © Springer-Verlag Berlin Heidelberg 2013

Nevertheless, the opposite was experimentally found for some materials such glass-forming silicate liquids.

Such decrease of viscosity with increasing pressure, achieved in some ranges of pressure and temperature, is assumed as *anomalous scenario*, which is in contradiction with the results of "free volume" theories of viscosity as mentioned in [1] (Jürn W. P. Schmelzer et. al., 2005). Such specification of the negative pressure dependence of viscosity means that an increase of viscosity with increasing pressure is considered as the rule, but that deviations are also possible depending on the fluid characteristics. The question in this way arises on how this behaviour can be explained analytically satisfying an accuracy criteria. An attempt to analytically approximate such function was carried out, concluding that, in most cases of interest, an increase of viscosity with increasing pressure must be expected, although exceptions are possible. Alternative theoretical approaches connect the decrease of viscosity with structural changes of the respective systems under pressure, which are not described appropriately by free volume concepts. Since for these complex systems, a decrease of the viscosity with pressure is observed as a rule, at least, for sufficiently high pressures, one has to check, first, whether a decrease of viscosity with increasing pressure exists, in contradiction with free volume theories and, second, how to incorporate such additional structural behaviour into a model independently of the particular mechanism of structural change considered. Due to the above mentioned reasons, and in order to achieve a manageable solution to the referred controversy on the effects of pressure and temperature on viscosity, it is highly interesting to look for an alternative model based on experimental techniques in order to accurately represent the fluid dynamics behaviour.

Liquids of constant composition in (stable or metastable) thermodynamic equilibrium states are considered. Assuming the number of degrees of freedom of the system equal to 2, and choosing temperature and pressure as the independent variables determining the properties of the system, in such cases, the viscosity μ can be considered as a function of pressure and temperature according to the expression

$$\mu = \mu(p,T) \tag{1}$$

It is assumed also that the analysis is restricted to the cases where the thermal expansion coefficient of the liquid is positive (this property is fulfilled at atmospheric pressure for most but not all liquids), so that the following relations must be fulfilled

$$\left(\frac{\partial \mu}{\partial T}\right)_p < 0, \quad \left(\frac{\partial \mu}{\partial p}\right)_T > 0 \tag{2}$$

These relations imply that the viscosity must decrease with increasing temperature (for isobaric processes), and must increase with increasing pressure (at isothermal conditions). Moreover, considering the viscosity as a function of pressure and temperature, i.e., $\mu = \mu(p,T)$, the following identity can be written:

$$\left(\frac{\partial \mu}{\partial T}\right)_p \left(\frac{\partial T}{\partial p}\right)_\mu \left(\frac{\partial p}{\partial \mu}\right)_T = -1 \tag{3}$$

Equation (3) follows from purely analytical considerations and does not involve originally any experimental physics. Taking into account the viscosity dependencies given
by equation 2, it is concluded that the inequality

$$\left(\frac{\partial p}{\partial T}\right)_{\mu} > 0 \qquad (4)$$

must be fulfilled. This equation described as a purely mathematical relation has a
quite definite physical meaning. As mentioned earlier, the viscosity of liquids of constant composition in (stable and metastable) thermodynamic equilibrium states can be
considered as a function of two state variables, pressure and temperature, i.e.,
$\mu = \mu(p,T)$. However, if viscosity is considered as constant [i.e., $\mu = \mu(p,T) =$
const], then this relation gives a dependence between pressure and temperature (at
constant viscosity). So equation (3) means that in order for the viscosity to remain
constant, an increase of temperature leads to effects which can be compensated by an
increase of pressure.

1.1 The Viscosity Measurement

The fluid viscosity can be defined as intermolecular fluid friction and as such, the fluid viscosity such as liquid fuels and lubricants spends strongly on fluid temperature
and pressure [3]. In these fluids the dependence of the viscosity with pressure can
reach some degree of order greater than the fluid density [2, 3, 26].

The author of [4] highlighted that for a fluid under pressures approaching 5 MPa
the dependence of the viscosity with pressure is important for a compressible fluid
flow. Consequently it is recommended to take into account the correction of viscosity
for pressure changes when handling compressible fluid flows.

The idea of fluids with pressure dependent viscosity has been introduced by [5],
while [6] proposed an exponential isothermal state equation model of the form

$$\mu(p) = \mu_{o} e^{\beta p} \qquad (5)$$

where here η is the viscosity, p is the pressure, η_{o} is the viscosity at atmospheric
pressure, and β is el pressure-temperature coefficient which depends on the
temperature.

Many techniques have been developed for the measurement of viscosity. However
the most common in use for online measurement in industry is the capillary viscometer. This takes the form of a section of piping of reduced diameter through which the
liquid is passed. The pressure differential between two points along the capillary is
measured together with the flow rate. The volume rate of flow, Q, through the capillary is measured for each pressure gradient $\Delta p/L$. The viscosity η, for a Newtonian fluid (an inelastic fluid) can then be determined from the Hagen-Poiseuille law [7] according to

$$Q = \frac{\pi \cdot \Delta p \cdot d^{4}}{8 \cdot \eta \cdot L} \qquad (6)$$

where d is the capillary tube diameter, L is the distance between differential pressure measurements with a differential pressure transmitter (DPT). The scheme depicted in figure 1 corresponds to a capillary viscometer. According to the scheme of figure 1, the necessary parameters as per equation (6) are the differential pressure and flow since capillary diameter and distance of differential pressure measurement are constants.

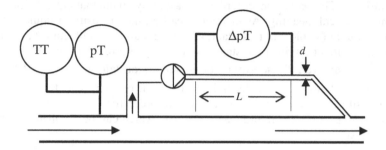

Fig. 1. Basic scheme of the capillary viscometer described by equation (7)

According to equation (6) the viscosity of a liquid is a function of the type

$$\mu = \frac{\pi \cdot \Delta p \cdot d^4}{8 \cdot Q \cdot L} \tag{7}$$

which can be described by means of a unique variable Δp under the assumption of constant (C) values for the rest of parameters of equation (7) Since the fluid is pumped at constant volumetric flow rate follows that equation (8) defines the viscosity under the assumption that some parameters are constant so that the viscosity can be further expressed as

$$\eta_m = f(C, \Delta p) \tag{8}$$

where

$$C = \frac{\pi \cdot d^4}{8 \cdot Q \cdot L} \tag{9}$$

Such described technique applied on practical viscosity measurement, although widespread used in industrial viscosity measurements and viscosity control applications, still is subjected to operational faults affecting accuracy.

2 A Virtual Model Based Viscometer

Although viscosity is a function of pressure and temperature, when industrial processes are considered such as liquid fossil fuels and lubricating oils, capillary based viscometers may be replicated by means of functional approximation techniques. In this way, several experimental techniques such as grade API for liquid fuels and grade SAE for lubricating oils is considered. Although more classifications exist, in this work only API and SAE grades are considered.

The use of analytical redundancy will be applied on viscometer monitoring tasks. Instead of using the pressure-temperature dependence functions to express the model based viscosity, the grades API for fossil fuels and SAE for lubricating oils will be used. By means of these technical relations the viscosity can be expressed as function of the temperature and grade API or grade SAE respectively. Nevertheless, with the aim of achieving as much redundant information as possible at effective cost, functional redundancy can be added by applying pressure temperature-dependences according to the equation

$$\mu_v = f(API,T) = f(p,T) \tag{10}$$

Processing equation (10) implies the measurement of pressure and temperature (p,T).

2.1 The Data to the Virtual Viscometer

As mentioned, the viscosity can be defined as a function of the temperature and grade API or as a function of the temperature and grade SAE among other technical possibilities. The aspect of both data bases are represented in tables 1 end 2 respectively. In table 1 the first column correspond to the temperature while in the rest of columns it is represented the viscosity as function of the temperature and the corresponding API grade depicted in the first row.

In table 2 the first column correspond to the temperature while in the rest of columns it is represented the viscosity as function of the temperature and the corresponding SAE grade depicted in the first row.

Table 1. Viscosity as function of the temperature and API grade

T (°C) / API grade	10	12	14	16	18	20
15,55	260000	30000	6000	2000	750	500
26,66	35000	7000	2000	750	400	210
37,77	8000	2000	750	300	180	100
65,55	500	230	220	70	40	25
93,33	80	70	45	25	18	10
148,88			7	6	4	1,5

Table 2. Viscosity as function of temperature and SAE grade

T (°C) SAE grade	SAE 20W50	SAE 10W40	SAE 30	SAE 10W30
40	175	110	92	70
60	122	80	65	50
80	70	50	42	32
100	20	19	17	15

Since at the first row of table 2 the SAE grade is coded by an acronym, in table 3 such data is represented by codifying the acronyms by means of numbers in order to be used as training data for the used neural networks.

Table 3. Viscosity as function of temperature and SAE grade codified by correlative numbers

T (°C)/SAE	1	2	3	4
40	175	110	92	70
60	122	80	65	50
80	70	50	42	32
100	20	19	17	15

2.2 The Management Approach

The virtual viscosity (μ_v) can be expressed by a function of the type.

$$\mu_v = f(API, T) \text{ for petroleum liquid fuels (table 1)}$$

$$\mu_v = f(SAE, T) \text{ for lubricating oils (table 2)} \tag{11}$$

Such function based definitions can be used as redundant models to detect features or discrepancies between physical measurements (measured viscosity μ_m) and functional approximation based models. Such difference is considered as a residual (r) into a parity space approach which will be evaluated to decide about viscometer behaviour in terms of operating health. The viscometer health is assumed to be too low (abnormal or faulty) if the residual surpass the value (r_{max}).

$$\text{IF } (\mu_v - \mu_m) \geq r_{max} \text{ THEN viscometer performance is low} \tag{12}$$

The possible causes of differences between measured viscosity (μ_m) and estimated virtual viscosity (μ_v) by functional approximation, approaching the value (r_{max}) may be due to one or more of the following causes or reasons:

- Δp, which imply a fault at differential pressure transmitter,
- P, which imply a fault at pressure transmitter,
- T, which imply a fault at temperature transmitter,
- *API grade*, which imply a fault due to changes in the fluid characteristics.

2.3 Scheduling Virtual and Measured Data

With the data bases described so far only single redundancy is achieved. Reliability may be gained by adding another function of the viscosity such that the virtual viscosity μ_{v2} as function of the operating pressure and temperature given by equation (1), i. e., $\mu_{v2} = f(p, T)$.

Achieving consistent data to use as training data for the NN based function (1) a data quality verification task must be previously applied. In this way a test of consistence based on residual r_1 evaluation is performed. If the condition for which r_1 is quasi zero is fulfilled, then training data is validated into a database. Since for a given fluid operating into an industrial process pressure is commonly constant, follows that the virtual viscosity may be expressed as $\mu_{v2} = f(T)$ for a nominal operating condition. Once verified the operating conditions for which r_1 is quasi zero the data to use as training data on a NN based function such as $\mu_{v2} = f(p,T)$ or $\mu_{v2} = f(T)$ is achieved. The achieved function is depicted in figure 2 where two redundant virtual viscometers are used.

According to the scheme depicted in figure 2, real time measured data is processed into a parity space approach. Residuals are further processed according to the model of equation (12). The residuals will be taken into account to isolate faults according to a rule base specifically designed to diagnose the viscometer.

The necessary data to find the redundant function μ_{v2} is achieved on line from the viscometer on line output and the measurements of pressure and temperature. In order to validate such acquired data the viscometer as well as the virtual viscometer based on the API or SAE data must coincide.

The rule base is responsible for faults detection, faults isolation and decision making based on real time data and functional approximation procedures. If the results of the isolation task are deterministic then the decision is also deterministic and the diagnosis task is reliable. However under some ambiguity, decision making task suffers from reliability. This problem may be overcome by increasing redundancy which affects the cost since the increment of functional redundancy requires of physical measurements as well as the direct physical redundancy. Same basic diagnostic is carried out by means of the rules described in table 4. The rule base depicted in table 5 shows the consequences of ambiguous results which requires expert human intervention.

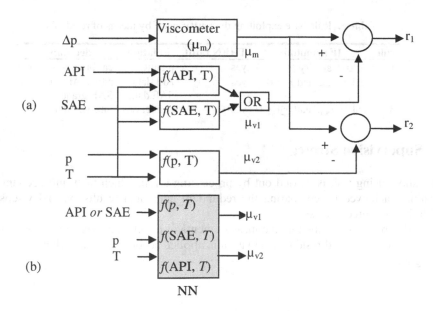

Fig. 2. The data scheduling scheme. (a), Basic scheme of redundant functional processor. (b), the structure of a feedforward neural network to process the functional redundant data.

The sophisticated techniques based on artificial intelligence to successfully combine both functional approximation based neural networks with rule based techniques as described in literature [8-20, 27], are well justified to solve complex industrial problems.

Table 4. Rule base based on functional redundancy

rule	IF symptom	THEN fault	isolation	decision
1	$\mu_m = \mu_{v1} = \mu_{v2}$	no	no fault	No action
2	$\mu_m = \mu_{v1} \neq \mu_{v2}$	yes	P sensor	calibrate pressure sensor
3	$\mu_m \neq \mu_{v1} = \mu_{v2}$	yes	API data	adjust API grade
4	$\mu_{v1} = \mu_{v2} \neq \mu_m$	yes	Δp sensor	calibrate Δp sensor

However a vast grupe of diagnostic problems, such the exhibited in this work can be solved using simple algorithms for functional approximation on the basis of neural networks according to [9-11], i. e., backpropagation feedforward neural networks which fulfil the accuracy demand. Although several algorithms have been used in diagnostic tasks, such as the described in [8, 20-25], the applied strategy for the diagnostic task in this paper is based on the combination of two sets of rules: the rules based on the online functional value and rules based on the residuals. For both rules the conclusion deals with faults detection, faults isolation and cost effective decision making.

Both rule bases shown in tables 4 and 5 are complementary to each other in order to achieve as much determinism as possible.

Table 5. Rule base exploiting the parity space by means of residuals

rule	IF symptom	THEN fault	isolation	decision
1	$|r_1| \cong 0$ and $|r_2| \geq r_{max}$	yes	p fault	calibrate p sensor
2	$|r_1| \geq r_{max}$ and $r_2 \cong 0$	yes	API data SAE data	adjust API or SAE grade
3	$|r_1| \geq r_{max}$ and $|r_2| \geq 0$	yes	ambiguous	intervention

3 Supervision Strategy

The supervising task is carried out by processing a rule which take into account the residuals achieved by comparing the redundant computed results (virtual viscosity) with the measured viscosity.

At this point, assuming that the measurements of p and T and viscometer output are correct, the rule based residual evaluation is applied according to the scheme depicted in figure 3.

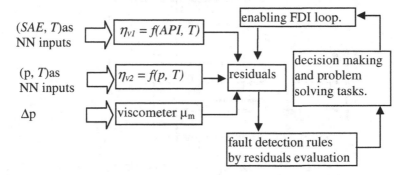

Fig.3. A flowchart of the supervision strategy

A flowchart to schedule the supervision task is implemented according to the chart shown in figure 3. To generate residuals, the output of the neural network models $\eta = f(SAE, T)$ and $\eta = f(p, T)$ are compared to the current measured viscosity. The existence of any discrepancy, which is due mainly to inherent measurement inaccuracies, must be taken into account. The conclusions given by the applied rule are then carried out and executed. When the diagnostic task is enabled, the flowchart depicted in figure 2 is performed sequentially and recursively.

3.1 Validation of the Strategy

A lab based test rig consisting of a heat exchanger where target temperature can be controlled is equipped with a viscometer based on equation (7). Two virtual viscometers based on functional approximation techniques are implemented in order to provide functional redundancy [$f(p,T)$ and $f(SAE\ grade, T)$].

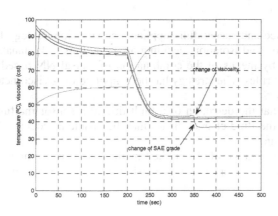

Fig. 4. A record showing a fault consisting of a change on SAE grade by processing rules of table 4

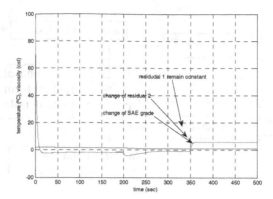

Fig. 5. A record of residuals showing the fault detected by rules of table 4 by processing rules of table 5, which indicate that the fault is a change of SAE grade

With the aim of validating the proposed strategy, the implemented system is activated so that after an elapsed time of 350 sec., the SAE grade is changed from (2= SAE 10W40) to (3 = SAE30) according to the data of table 3. Based on the conditions proposed by means of the, the results of processing the rules yields the conclusions expected, indicating that a change in the oil grade has taken effect. The change of oil characteristics has been applied at t= 350 sec. As consequence, the rule 3 of table 4 detects a change concluding that the fluid characteristic has been changed, and the solution consists of adjust the SAE grade in the database. This change is shown in figure 4. Simultaneously, and complementary to the task carried out so far, by processing rules of table 5 for the residuals, the fault detected by rules of table 4 is confirmed by the conclusions of the rule 2 of table 5, which indicates that the detected fault corresponds to a change of SAE grade. This event is shown in figure 5.

4   Conclusions

This study proposed and successfully applied a supervision strategy focused on effective faults detection and isolation, deciding about actions to be taaaken including human intervention by means of the combination of causal NN based modelling techniques with expert rules. The strategy has been applied on a viscometer equipped with redundant sensors.

Results show that the detection of a fault due to a sensor malfunction as well as some changes on the fluid characteristics may be efficiently detected deciding on the actions to be executed.

A relevant disadvantage of the applied methodology is the ambiguity of the conclusions when more than various faults are simultaneously detected.

References

1. Schmelzer, J.W.P., Zanotto, E.D., Fokin, V.M.: Fokin, Pressure dependence of viscosity. The Journal of Chemical Physic 122, 74511 (2005)
2. Rajagopal, K.R.: On implicit constitutive theories for fluids. J. Fluid Mech. 550, 243–249 (2006)

3. Renardy, M.: Parallel shear flows of fluids with a pressure-dependent viscosity. J. Non-Newtonian Fluid Mech. 114, 229–236 (2003)
4. Denn, M.M.: Polymer Melt Processing. Cambridge University Press, Cambridge (2008)
5. Stokes, G.G.: On the theories of the internal friction of fluids in motion, and of the equilibrium and motion of elastic solids. Trans. Camb. Philos. Soc. 8, 287–305 (1845)
6. Barus, C.: Isothermals, isopiestics and isometrics relative to viscosity. Am. J. Sci. 45, 87–96 (1893)
7. Noltingk, B.E. (ed.): Instrumentation Reference Book. Buterworth and Co. (Publishers) Ltd., London (1988)
8. Medina, M.A., Theilliol, D., Astorga, C.M., Guerrero, G., Vela, L.G.: Fault diagnosis based on a decoupled filter for nonlinear systems represented in a multi-models approach. Dyna-Colombia 162, 313–323 (2010)
9. Fine, T.L.: Feedforward Neural Network Methodology. Springer-Verlag New York. Inc. (1999)
10. Tang, H., Tan, K.C., Yi, Z.: Neural Networks: Computational Models and Applications. Springer, Heidelberg (2007)
11. Fletcher, R., Reeves, C.M.: Function minimisation by conjugate gradients. Computer Journal 7, 149–154 (1964)
12. Irigoyen, E., Larrea, M., Valera, J., Gómez, V., Artaza, F.: A Neuro-genetic Control Scheme Application for Industrial R^3 Workspaces. In: Graña Romay, M., Corchado, E., Garcia Sebastian, M.T. (eds.) HAIS 2010, Part I. LNCS(LNAI), vol. 6076, pp. 343–350. Springer, Heidelberg (2010)
13. Gómez-Garay, V., Irigoyen, E., Artaza, F.: GENNET-Toolbox: An Evolving Genetic Algorithm for Neural Network Training. In: Graña Romay, M., Corchado, E., Garcia Sebastian, M.T. (eds.) HAIS 2010, Part I. LNCS (LNAI), vol. 6076, pp. 368–375. Springer, Heidelberg (2010)
14. Rutkowski, L.: New Soft Computing Techniques for System Modelling. In: Pattern Classification and Image Processing. Springer, Berlin (2004)
15. Corchado, E., Graña, M., Wozniak, M.: New trends and applications on hybrid artificial intelligence systems. Neurocomputing 75(1), 61–63 (2012)
16. García, S., Fernández, A., Luengo, J., Herrera, F.: Advanced nonparametric tests for multiple comparisons in the design of experiments in computational intelligence and data mining: Experimental analysis of power. Information Sciences 180(10), 2044–2064 (2010)
17. Corchado, E., Abraham, A., Carvalho, A.: Hybrid intelligent algorithms and applications. Information Sciences 180(14), 2633–2634 (2010)
18. Pedrycz, W., Aliev, R.: Logic-oriented neural networks for fuzzy neurocomputing. Neurocomputing 73(1-3), 10–23 (2009)
19. Abraham, A., Corchado, E., Corchado, J.M.: Hybrid learning machines. Neurocomputing 72(13-15), 2729–2730 (2009)
20. Hong, S.J., May, G.: Neural Network-Based Real-Time Malfunction Diagnosis of Reactive Ion Etching Using In Situ Metrology Data. IEEE Transactions on Semiconductor Manufacturing 17(3), 408–421 (2004)
21. Garcia, R.F., De Miguel Catoira, A., Sanz, B.F.: FDI and Accommodation Using NN Based Techniques. In: Graña Romay, M., Corchado, E., Garcia Sebastian, M.T. (eds.) HAIS 2010, Part I. LNCS (LNAI), vol. 6076, pp. 395–404. Springer, Heidelberg (2010)
22. Abraham, A.: Hybrid Soft Computing and Applications. International Journal of Computational Intelligence and Applications 8(1), 5–7 (2009)
23. Corchado, E., Arroyo, A., Tricio, V.: Soft computing models to identify typical meteorological days. Logic Journal of the IGPL 19(2), 373–383 (2011)

24. Zhao, S.-Z., Iruthayarajan, M.W., Baskar, S., Suganthan, P.N.: Multi-objective robust PID controller tuning using two lbests multi-objective particle swarm optimization. Inf. Sci. 181(16), 3323–3335 (2011)
25. Sedano, J., Curiel, L., Corchado, E., de la Cal, E., Villar, J.R.: A soft computing method for detecting lifetime building thermal insulation failures. In: Integrated Computer-Aided Engineering, vol. 17(2), pp. 103–115. IOS Press (2010)
26. Rondón, E., Carrillo, J., Correa, R.: Magnetic levitation in fluids of high viscosity and density. Dyna-Colombia 162, 387–395 (2010)
27. David, J., Henao, V.: Neuroscheme: A modeling language for artificial neural networks. Dyna-Colombia 147, 75–82 (2005)

A Performance Study of Concentrating Photovoltaic Modules Using Neural Networks: An Application with CO²RBFN

Antonio J. Rivera, B. García-Domingo, M.J. del Jesus, and J. Aguilera

[1] Dept. of Computer Science
[2] Dept. of Electronics and Automation Engineering
University of Jaén, Spain
{arivera,bgarcia,mjjesus,aguilera}@ujaen.es

Abstract. Concentrating Photovoltaic (CPV) technology attempts to optimize the efficiency of solar energy production systems and models for determining the exact module performance are needed. In this paper, a CPV module is studied by means of atmospheric conditions obtained using an automatic test and measuring system. CO²RBFN, a cooperative-competitive algorithm for the design of radial basis neural networks, is adapted and applied to these data obtaining a model with a good level of accuracy on test data, improving the results obtained by other methods considered in the experimental comparison. These initial results are promising and the obtained model could be used to work out the maximum power at the CPV reporting conditions and to analyze the performance of the module under any conditions and at any moment.

Keywords: Concentrating Photovoltaic Technology, Neural Networks, Evolutionary Computation, Soft-Computing, Regression.

1 Introduction

Nowadays there is increasing interest in the use of solar energy as a source of electricity or heat production. This energy is clean and renewable and the latest research in production systems is lowering its generation costs. Some of the drawbacks related to this energy include the problems derived from the discontinuity in its production. These problems are due to the variability in the environmental conditions and generate new associated tasks such as determining the exact performance of the solar cells, forecasting load demands and productions, etc.

During recent years, Photovoltaic (PV) technology has experienced a major boost, as the result of a long period of investigation and experience. However, this conventional PV technology has little capacity of increasing its efficiency and it is in an advanced place on its learning curve [20]. In this sense, Concentrating Photovoltaic (CPV) is presented as a new form of electricity generation [7], which needs to be investigated, thoroughly and to have a wide experimental basis that allows us to ensure its performance in the medium to long term. Despite the excellent prospects of

V. Snasel et al. (Eds.): SOCO Models in Industrial & Environmental Appl., AISC 188, pp. 439–448.
springerlink.com   © Springer-Verlag Berlin Heidelberg 2013

installed CPV power, apart from the International Standard IEC 62108 [9], CPV normalization does not exist, so the absence of publications about CPV systems' real outdoor performance or regression methods for modeling these systems is remarkable. This lack of experience involves a technological risk, and makes difficult the industrial and economic development of this technology. In this sense, the existence of models which allow the prediction of modules performance from initial atmospheric condition is remarkable for the emerging CPV technology.

Data mining methods have been successfully used to determine energy models for demand and production [4][22]. In particular, Artificial Neural Networks (ANN) have been used for similar tasks [10][11], but only for conventional PV technology, not including CPV systems.

Radial Basis Function Networks (RBFNs) [3] are one of the most important ANN paradigms in the Machine Learning field. The overall efficiency has been proved in many areas such as regression, time series prediction and in problems related to energy [26][23], but not yet in CPV modeling. CO^2RBFN [15] is an evolutionary cooperative-competitive method, developed by the authors, for the design of RBFNs, for classification and time series forecasting problems.–

In this paper the characterization of CPV modules is studied, specifically the modeling of maximum power given by a CPV module under any conditions. The CO^2RBFN method is adapted in order to determine a regression model for CPV modules and the results obtained are compared with other regression models.

The contribution is organized as follows. Section 2 describes preliminary information related to concentrating photovoltaic technology and variables used for the characterization of other modules. Section 3 explains the cooperative-competitive hybrid algorithm for RBFN design, CO^2RBFN, used for the photovoltaic module modeling process. Section 4 presents the experimental study performed with CO^2RBFN and different regression data mining algorithms, based on statistical techniques, support vector machines, fuzzy rule based systems multilayer perceptrons and classical RBFNs. Finally, some concluding remarks are outlined in Section 5.

2   Preliminaries: Concentrated Photovoltaic Technology

CPV technology proposes a cost reduction, by the use of high efficiency Multijunction solar cells [6], an optic device to concentrate the incident radiation, and a tracker, pointing continuously to the sun, guaranteeing the maximum production of the system [8].

When talking about CPV Technology, the manufacturer gives information about the power given by a CPV module working under determined atmospheric conditions. Nevertheless these conditions are not real, and difficult to obtain in most cases.

Due to the absence of CPV Standardization, there are not specific Standard Test conditions. Nobody really knows which are the conditions that influence a CPV module performance or which of these conditions must be taken into account to predict the production in a determined period of time. Each manufacturer gives the CPV module power rating for different conditions, making it almost impossible to compare them with each other or with other PV technologies.

For all of these reasons, a prediction method that gives us information about the influence of atmospheric conditions is essential to enable the development and commercialization of this technology, which is now immersed in a first stage of investigation.

2.1 Concentrated Photovoltaic Module under Study

In the present study, it is considered that the maximum power given by a CPV module (P_M) is highly influenced by atmospheric conditions. In this way, it is essential to know the positive or negative impact of this influence, to obtain the power of the module for determined initial atmospheric external conditions.

In order to obtain the regression model, it is necessary to have these initial atmospheric external conditions, measured for a determined period of time, using an Automatic Test & Measurement System, developed by the IDEA research group.

The measures were taken on the flat roof of the Higher Technical School at the University of Jaen. To carry out the measurement we used a 2-axes solar tracker, addressed to support the CPV module under study and faced towards the sun.

The Automatic Test & Measurement System designed allows the CPV module I-V curve and outdoor atmospheric conditions (that will influence the module performance) to be measured at the same time. The registered data, each 5 minutes, are the following: maximum module power (P_M) (obtained through the measurement and tracing of I-V curve), ambient temperature (T_{amb}), direct normal irradiance (DNI), wind speed measure (W_s), and spectral irradiance distribution of the incident global irradiance, described through average photon energy (APE) values.

2.2 Regression Models Proposed for Other CPV Modules

Different regression models have been proposed to characterize other CPV modules. All of these are based on multiple lineal and quadratic regression methods which make use of different variables to represent the module.

The regression methodology based on ASTM E-2527-09 Standard [2], allows the calculation of a_1, a_2, a_3, and a_4 regression coefficients which express the maximum power dependence with the direct irradiance, wind speed and ambient temperature:

$$P_M = DNI(a_1 + a_2DNI + a_3T_{amb} + a_4W_s) \tag{1}$$

There are some authors who study the CPV modules external performance dependence with different atmospheric conditions, trying to develop a simplified model of the Standard ASTM E-2527, which will calculate the real power (measured outdoors), from previously known initial conditions. In this way, the ISFOC (Institute of Concentration Photovoltaic System) has developed a multiple lineal regression method which allows the calculation of the concentrator DC's maximum power, using its I-V curve, DNI and back plate temperature [17][18].

The NREL (National Renewable Energy Laboratory) has also studied the CPV modules power dependence with atmospheric conditions, applying the standard described in [13]. A high variability in ASTM rating has been observed, because this regression does not take into account the influence of spectral conditions.

Another variation of ASTM multi linear regression methodology has also been applied to measuring different modules' maximum power, analyzing the influence of DNI (Direct Normal Irradiance), Z (Spectral Parameter), and module temperature, without taking into account the influence of wind speed, and obtaining the regression coefficients of the following equation [14]:

$$P_M = c_{DNI}DNI + c_Z Z + c_T T_{Module} + Offset \tag{2}$$

These approaches to the characterization of CPV modules shows the absence of a standard in the variables which influence the model and all of them are based on multiple linear and quadratic regression models. In this paper, the CPV modules characterization is based on ambient temperature (T_{amb}), direct normal irradiance (DNI), wind velocity measure (W_s), and spectral irradiance distribution of the incident global irradiance described through average photon energy (APE) values. In this study, these atmospheric conditions have been considered as the most influential on the performance of CPV, not having been analyzed together in any of the works previously described.

3 CO²RBFN: Cooperative-Competitive Hybrid Algorithm for RBFN Design

An RBFN is a feed-forward ANN with a single layer of hidden units, called Radial Basis Functions (RBFs), Fig. 1. The m neurons of the hidden layer are activated by a radially-symmetric basis function, $\phi_i : R^n \rightarrow R$, which can be defined in several ways. From all the possible choices for ϕ_i, the Gaussian function is the most widely used: $\phi_i(\vec{x}) = \phi_i(e^{-(\|\vec{x}-\vec{c}_i\|/d_i)^2})$, where $\vec{c}_i \in R^n$ is the centre of basis function ϕ_i, $d_i \in R$ is the width (radius), and $\| \ \|$ is typically the Euclidean norm on R^n. In regression the RBFN has only an output that implements the weighted sum of RBF outputs:

$$f(\vec{x}) = \sum_{i=1}^{m} w_i \phi_i(\vec{x}) \tag{3}$$

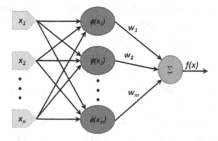

Fig. 1. Diagram for a Radial Basis Function Neural Network

CO^2RBFN [15] is a cooperative-competitive hybrid algorithm for RBFN design. In the proposed approach each individual of the population represents a basis function and the entire population is responsible for the final solution. This allows for an environment where the individuals cooperate towards a definitive solution. However, they also compete for survival, since if their performance is poor they will be eliminated.

With this environment in which the solution depends on the behavior of many components, the fitness of each individual is known as credit assignment. In order to measure the credit assignment of an individual, three factors have been proposed to evaluate the role of the RBF, ϕ_i, in the network. These factors taken into account the contribution, a_i, of the RBF to the network, the Root Mean Squared Error (RMSE) committed, e_i, inside the radius of the RBF and its possible overlapping, o_i, with other RBFs. To decide the operators' application probability over a certain RBF the algorithm uses a Fuzzy Rule Based System (FRBS).

The main steps of CO^2RBFN are shown in the pseudocode:

```
1.  Initialize RBFN
2.  Train RBFN
3.  Evaluate RBFs
4.  Apply operators to RBFs
5.  Substitute the RBFs that were eliminated
6.  Select the best RBFs
7.  If the stop-condition is not verified go to step 2
```

1. RBFN initialization. A number, m, of neurons is specified (i.e. the size of population). Each RBF centre, $\vec{c_i}$, is randomly established to a pattern of the training set.

   The widths, d_i, will be set to half of the average distance among the centres. Finally, the weights, w_i, are set to zero.
2. RBFN training. As adaptation for this CPV regression task the SVD algorithm [5] is used to train RBFs weights. This training algorithm has reported better results for the addressed problem than the traditional LMS [25] algorithm used in the CO^2RBFN algorithm.
3. RBF evaluation. A credit assignment mechanism is required in order to evaluate the role of each base function in the cooperative-competitive environment. For an RBF ϕ_i, three parameters, a_i, e_i, o_i are defined. Equations are shown in Table 1.
4. Applying operators to RBFs. Another adaptation for this regression task is that three operators have been defined to be applied to the RBFs. This configuration, with only one mutation operator, has obtained better results than other versions of CO^2RBFN with more mutation. The operators are: an operator that eliminates (Remove) the RBF, an operator that randomly mutate (Mutation) the RBF and finally an operator that maintains (Null) the RBF parameters. The operators will be applied to the whole population of RBFs. The probability for choosing an operator is determined by means of a Mamdani fuzzy system. The inputs of this system are parameters a_i, e_i and o_i used for definition of the credit assignment of the RBF ϕ_i,. These inputs are considered as linguistic variables va_i, ve_i and vo_i, and the outputs are p_R, p_M, and p_N, representing the probability of applying the operators Remove, Mutation, and Null respectively. Fig. 2 shows the rule base and the linguistic label used for modeling the CPV module

Table 1. Definition of the credit assignment parameteres

$a_i = \begin{cases} \|w_i\| & \text{if } pi_i > q \\ \|w_i\| * (pi_i / q) & \text{otherwise} \end{cases}$ (4)	w_i: weight pi_i: number of patterns of the training set inside the RBF ϕ_i width q: average of the pi_i values minus standard deviation of these
$e_i = \sqrt{\sum_{i=1}^{n} (y_i - f_i)^2}$ (5)	y_i: real/desired value f_i: output of the net.
$o_i = \sum_{j=1}^{m} o_{ij} \qquad (6)$ $o_{ij} = \begin{cases} \left(1 - \|\phi_i - \phi_j\| / d_i\right) & \text{if } \|\phi_i - \phi_j\| < d_i \\ 0 & \text{otherwise} \end{cases}$	o_{ij}: overlapping of the RBF ϕ_i and the RBF ϕ_j

	Antecedents			Consequents		
	v_a	v_e	v_o	p_R	p_M	p_N
R1	L			M-H	M-L	L
R2	M			M-L	M-L	M-L
R3	H			L	M-H	M-H
R4		L		L	M-H	M-H
R5		M		M-L	M-L	M-L
R6		H		M-H	M-L	L
R7			L	L	M-H	M-H
R8			M	M-L	M-H	M-L
R9			H	M-H	M-L	L

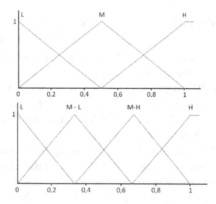

Fig. 2. Rule Base in the FRBS (Left). Linguistic labels for inputs (Top-Right) and outputs (Bottom-Left)

5. Introduction of new RBFs. In this step of the algorithm, the eliminated RBFs are substituted by new RBFs. The new RBF will be located in the zone with highest error outside of any RBF width a bad classified pattern or in the first pattern found outside of any RBF width with a probability of 0.5 respectively.

6. Selection of the best RBFs. After applying the mutation operator new RBFs appear. The algorithm uses the replacement scheme. The new RBFs are compared with their parents in order to determine the RBFs with the best behavior.

4   Experimentation

In order to characterize the CPV module, the CO^2RBFN method has been used and its results have been compared with the ones obtained by different data mining techniques including statistical regression methods.

The data provided by the Automatic Test & Measurement System are described in the first subsection. A brief description of the data mining algorithms used for the comparison is presented in the second subsection together with the parameters used. The results are described and analyzed in the last subsection.

4.1   CPV Module Data

As mentioned previously, the Automatic Test & Measurement System, designed for the CPV performance analysis, provides us with the data used in the experimentation. A pattern is composed of five elements, which include four inputs: incident direct normal irradiance (DNI), ambient temperature (T_{amb}), wind speed (W_s) and spectral irradiance distribution through the average photon energy (APE). The maximum power delivered by a CPV module (P_M) is considered as output. None of the previous works analyzed have considered all these atmospheric conditions as determinant when studying the performance of a CPV module.

Data were sampled from 10:00 am to 19:00 and from June to September 2011. The whole data set is composed of 7734 patterns. With the aim of estimating the real error on new patterns, the results were obtained using ten-fold cross-validation. The partitions were built with KEEL [1].

4.2   Methods Used for the Comparison

In this paper the following methods have been selected for the comparison:

- Statistical regression methods. Mean squares linear and quadratic regression models (LMSLR and LMSQR respectively) have been considered [19].
- v-SVM. An adaptation of the Support Vector Machine methods for regression problems [21].
- WM-FRBS [24]. Proposed by Wang and Mendel for generating fuzzy rules for regression problems.
- MLP-CG [12] which uses the conjugate-gradient algorithm to adjust weight values of the Multilayer Perceptron (MLP) network.
- RBFN-LMS [3] that builds an RBFN with a fixed number of RBFs and trains the weights using the LMS algorithm.
- Incr-RBFN [16], an RBFN design algorithm based on an incremental scheme.

All methods are implemented in KEEL [1]. The parameters used for them are the ones recommended by their authors. For CO^2RBFN the number of neurons was established at 10 and the iterations of the main loop at 200.

4.3   Results and Analysis

The results obtained are described in Table 2. This table shows the average Root Mean Square Error (RMSE) from 30 runs for training and test data using ten-fold cross validation. Algorithms are ranked by the test result.

Table 2. RMSE results for the CPV module modelling

Algorithm	$RMSE_{tra}$	$RMSE_{test}$
CO^2RBFN	4.9716	4.9970
LMSQR	5.1199	5.1424
MLP-CG	5.1460	5.1614
LMSLR	5.3052	5.3089
v-SVM	5.5726	5.5748
Incr-RBFN	6.0948	6.1069
RBFN-LMS	6.2966	6.3136
WM-FRBS	6.6321	6.6799

Observing the results it is remarkable that the method which models the CPV module with the highest precision is CO^2RBFN, followed by the quadratic regression method and MLP-CG.

If we analyze these results by type of method we can observe the good performance achieved by the classical regression method, typically used in the modeling of PV modules. Concretely the LMSQR has reached the second position in the ranking. This is a coherent result because, as mentioned before, these algorithms have typically been used for modeling PV modules. Neural networks design models like CO^2RBFN and MLP-CG have also obtained good results, CO^2RBFN in the first position and MLP-CG the third position. Nevertheless, neural networks design methods such as RBFN-LMS or Incr-RBFN obtain the poorest results. In this way we can conclude that not only is the model chosen important but also the method employed to design this model. Finally, the results obtained with typical design regression methods such as v-SVM, that obtained an intermediate result, or WM-FRBS that obtained the worst result, suggest that the problem addressed is not trivial.

The model obtained by the CO^2RBFN algorithm gives an error of approximately 3.3% which can be compared with those given by an important test and measurement Spanish PV centre: CIEMAT (http://www.ciemat.es). This can led us to conclude that the model obtained, the outdoor atmospheric variables used for this modeling process and the way of computing the spectral irradiance distribution of the incident global irradiance (by means of average photon energy value) are suitable for this problem. Moreover, this model can be used to determine whether the power given by a module (and established by the manufacturer) can be achieved, in a quality control process.

5   Concluding Remarks

CPV technology attempts to optimize the efficiency of the solar energy production systems. As with conventional PV technology, it suffers from the discontinuity in its production and needs models for determining the exact module performance, forecasting the production or load demands, etc. The need for regression models is greater than for PV technology owing to the fact that CPV normalization does not exist and that there are not enough studies about CPV systems´ real outdoor performance.

In this paper a CPV model was studied by means of atmospheric conditions. To do so, using an automatic test and measurement system, the I-V curve and outdoor atmospheric conditions were registered. The characterization of the CPV model was carried out considering incident normal irradiance, ambient temperature, spectral irradiance distribution and wind speed. The spectral irradiance distribution was studied through an index that defines the shape of the incident irradiance spectrum, the average photon energy.

The CO^2RBFN algorithm was adapted and applied to these data obtaining a model with a good level of accuracy on test data, improving the results obtained by other methods considered in the experimental comparison. These initial developments and results are promising, and because of the high dependence of the results and the amount of data considered in the data mining process the authors think that by increasing the database for the data mining process, the obtained model could be used to work out the maximum power at the CPV reporting conditions and to analyze the performance of the module under any conditions and at any moment.

Acknowledgments. Supported by the Spanish Ministry of Education under the Spanish Ministry of Science and Technology under the Projects TIN2008-06681-C06-02 and ENE2009-08302 (FEDER founds), and the Andalusian Research Plan Projects TIC-3928 and P09-TEP-5045 (FEDER founds).

References

1. Alcalá-Fdez, J., Sánchez, L., García, S., del Jesus, M.J., Ventura, S., Garrell, J.M., Otero, J., Romero, C., Bacardit, J., Rivas, V.M., Fernández, J.C., Herrera, F.: KEEL: A Software Tool to Assess Evolutionary Algorithms to Data Mining Problems. Soft Computing 13(3), 307–318 (2009)
2. ASTM E-2527-09, Standard test method for electrical performance of concentrator terrestrial photovoltaic modules and systems under natural sunlight. ASTM Int. West Conshohocken, PA, United States (2009)
3. Buchtala, O., Klimek, M., Sick, B.: Evolutionary optimization of radial basis function classifiers for data mining applications. IEEE Trans. Syst. Man Cybern B 35(5), 928–947 (2005)
4. De Silva, D., Xinghuo, Y., Alahakoon, D., Holmes, G.: A Data Mining Framework for Electricity Consumption Analysis From Meter Data. IEEE Transactions on Industrial Informatics 7(3), 399–407 (2011)
5. Golub, G., Van Loan, C.: Matrix computations, 3rd edn. J. Hopkins University Press (1996)
6. Green, M.A., Emery, K., Hishikawa, Y., Warta, W., Dunlop, E.D.: Solar cell efficiency tables (version 38). Progress in Photovoltaics: Research and Applications 19(5), 565–572 (2011)
7. IDEA Group, University of Jaén, Propuesta de un marco regulatorio para la concentración fotovoltaica en España (2010-2020) ISBN:978-84-692-9987

8. IDEA Group, University of Jaén, The CPV challenge (Part I): Achieving Grid Parity. First Conferences Ltd., CPV Today (2009)
9. IEC 62108, Concentrator Photovoltaics (CPV) modules and assemblies-design qualification and type approval. Edition 1.0 (2007)
10. Izgi, D., Öztopal, A., Yerli, B., Kaymak, M.K., Şahin, A.D.: Short-mid-term solar power prediction by using artificial neural networks. Solar Energy (In Press)
11. Li, K., Su, H., Chu, J.: Forecasting building energy consumption using neural networks and hybrid neuro-fuzzy system: A comparative study. Energy and Buildings 43(10), 2893–2899 (2011)
12. Moller, F.: A scaled conjugate gradient algorithm for fast supervised learning. Neural Networks 6, 525–533 (1990)
13. Muller, M., Marion, B., Rodriguez, J., Kurtz, S.: Minimizing variation in outdoor CPV power ratings. In: 7th International Conference on Concentrating Photovoltaic Systems (CPV-7). AIP Conference Proceedings, vol. 1407, pp. 336–340 (2011)
14. Pehar, G., Ferrer Rodríguez, J.P., Siefer, G., Bett, A.W.: A method for using CPV modules as temperature sensors and its application to rating procedures. Sol. Energy Mater. Sol. Cells 95(10), 2734–2744 (2011)
15. Pérez-Godoy, M., Rivera Rivas, A.J., Berlanga, F.J., del Jesus, M.J.: CO^2RBFN: an evolutionary cooperative–competitive RBFN design algorithm for classification problems. Soft Computing 14, 953–971 (2010)
16. Platt, C.J.: A resource-allocating network for function interpolation. Neural Computation 3(2), 213–225 (1991)
17. Rubio, F., Martínez, M., Perea, J., Sánchez, D., Banda, P.: Comparison of the different CPV rating procedures: Real measurements in ISFOC. In: 34th IEEE Photovoltaic Specialist Conference, Philadelphia. Art. 5411163, pp. 000800–000805 (2009)
18. Rubio, F., Martínez, M., Perea, J., Sánchez, D., Banda, P.: Field test on CPV ISFOC plants. In: Proc. of SPIE - The International Society for Optical Engineering, vol. 7407, art: 740707 (2009)
19. Rustagi, J.S.: Optimization Techniques in Statistics. Academic Press (1994)
20. Salas, V., Olias, E.: Overview of the photovoltaic technology status and perspective in Spain. Renewable and Sustainable Energy Reviews 13(5), 1049–1057 (2009)
21. Scholkopf, B., Smola, A.J., Williamson, R., Bartlett, P.L.: New support vector algorithms. Neural Computation 12(5), 1207–1245 (2000)
22. Suganthi, L., Samuel, A.: Energy models for demand forecasting, A review. Renewable and Sustainable Energy Reviews 16(2), 1223–1240 (2012)
23. Xia, C., Wang, J., McMenemy, K.: Short, medium and long term load forecasting model and virtual load forecaster based on radial basis function neural networks. International Journal of Electrical Power and Energy Systems 32(7), 743–750 (2010)
24. Wang, L.X., Mendel, J.J.: Generating fuzzy rules by learning from examples. IEEE Transactions on Systems, Man, and Cybernetics 22(6), 1414–1427 (1992)
25. Widrow, B., Lehr, M.A.: 30 Years of adaptive neural networks: perceptron, madaline and backpropagation. Proceedings of the IEEE 78(9), 1415–1442 (1990)
26. Zeng, J., Qiao, W.: Short-term solar power prediction using an RBF neural network. IEEE Power and Energy Society General Meeting, art. no. 6039204 (2011)

Classical Hybrid Approaches on a Transportation Problem with Gas Emissions Constraints

Camelia-M. Pintea, Petrica C. Pop, and Mara Hajdu-Macelaru

Tech Univ Cluj-Napoca, North Univ Center Baia Mare
76 Victoriei, 430122 Baia-Mare, Romania
cmpintea@yahoo.com, {petrica.pop,maram}@ubm.ro

Abstract. Nowadays the efforts of humanity to keep the planet safe are considerable. This aspect is reflected in everyday problems, including the minimization of vehicle's air pollution. In order to keep a green planet, in particular transportation problems, the main purpose is on limiting the pollution with gas emissions. In a specific capacitated fixed-charge transportation problem with fixed capacities for distribution centers and customers with particular demands the objective is to keep the pollution factor in a given range while the total cost of the transportation is as low as possible. In order to solve this problem, we have developed some hybrid variants of the nearest neighbor classical approach. The proposed algorithms are tested and analyzed on a set of instances used in the literature. The preliminary results point out that our approaches are attractive and appropriate for solving the described transportation problem.

Keywords: Hybrid heuristics, Transportation Problem, Optimization.

1 Introduction

One of the new concepts in modern optimization, according to Seuring and Muller [15] is the field of sustainable supply chain design, representing a business issue affecting a company's supply chain (logistic network) in terms of environmental, risk and waste costs. The model is focused on economic and environmental aspects of greenhouse gas emissions. Santibanez-Gonzales et al. [14] described a supply chain network design problem arising in governmental agencies, where we have to decide the location of institutions as schools, hospitals, taking into account sustainable issues in the form of restrictions on the dioxide carbon equivalent emissions.

A transportation problem is a network problem [8]. Some particular transportation problems are described as fixed cost transportation problems [1, 16], that are extensions of the traditional transportation problem and are considered two kinds of costs: direct costs and fixed costs. Based on sustainability of a supply chain network [14], the current paper proposes a hybrid classic approach for solving a two-stage supply chain transportation problem. We describe the network design model and analyzed

V. Snasel et al. (Eds.): SOCO Models in Industrial & Environmental Appl., AISC 188, pp. 449–458.
springerlink.com
© Springer-Verlag Berlin Heidelberg 2013

the impact of restrictions in the greenhouse gas emissions on transportation costs and in the location of facilities of the network of one layer instead of two as in [14], but involving two-stage chains as in [10,12]. The mathematical model of the problem is a mixed-integer 0-1 programming model.

One of the first multi-objective mixed-integer 0-1 model for deciding location and capacity expansion of facilities and transportation issues in a given planning horizon was developed in [9]. The profit was maximized and the environmental impact of the facility operations while satisfying the market demand for products was minimized.

The considered two-stage supply transportation problem implies a supply chain from a manufacturer who delivers items to certain distribution centers and a chain from distribution centers to a group of customers with given demands. The goal of the problem is to minimize the transportation cost, keeping the pollution, the gas emissions, as low as possible. A new constraint based on the supply chain characteristics and on the two-layers supply chain from [14] is introduced.

In order to solve complex transportation problems researchers use different heuristics [4,14] and hybrid techniques [2]. In the current paper, we develop a hybrid model using the nearest neighbor technique. In time, this approach was used to solve many difficult applications such as from data mining [6], machine learning [3], data compression [7], document retrieval [4] and statistics [5].

This paper consists of 5 sections and its frame is organized as follows: Sect. 2 illustrates the two-stage supply chain fixed-charge transportation problem including the greenhouse gas emission constraint and presents as well an example, the classical hybrid models based on Nearest Neighbor are shown in the third section followed by tests and results section. The paper concludes with future research directions.

2   The Two-Stage Supply Chain Fixed-Charge Transportation Problem

The two-stage supply chain network transportation problem involves the manufacturer, some distribution centers and customers. There are several potential distribution centers (DCs) candidate for the manufacturer. There are also considered the customers whose particular demands should be satisfied by distribution centers.

The complexity of the problem involves a large number of constraints, high dimensions, uncertainties and a large number of parameters.

It is stated that the manufacturer has no capacity limitation in production. It is also assumed that each potential distribution center has distinct and different capacity in order to support the customers [10]. There are known the transportation cost from the manufacturer to a distribution center and the opening cost for each potential distribution center.

Optimizing the two-stage supply chain network transportation problem means to minimize the total cost when selecting distribution centers which supply demands of all the customers.

Fixed costs and transportation costs are parts of the total cost. For a two-chain network fixed costs are considered the opening cost for potential distribution centers and also fixed cost for transportation from distribution centers to customers.

Transportation cost implies also the two-stage costs from manufacturer to distribution centers and transportation costs from distribution centers to all customers. There are considered m potential distribution centers and a number of n customers, each one with a particular demand [10,12].

The following notations are used:

- f_i is the opening fixed cost for a distribution center i
- f_{ij} is the fixed cost for transportation from distribution center i, to customer j
- c_i is the transportation cost per unit from manufacturer to distribution center i
- c_{ij} transportation cost per unit from distribution center i to customer j
- x_{ij} is the quantity to be transported from distribution center i to customer j
- a_i is the capacity of a distribution center i
- b_j is the number of units demanded by customer j

The objective problem is to minimize the function Z:

$$Z = Z_{tc} + Z_{fc} \tag{1}$$

where Z_{tc} is the total cost of transportation and Z_{fc} is the total of the fixed costs.
Z_{tc} is the total cost of transportation, including the transportation cost from manufacturer to distribution centers and from distribution centers to customers.

$$Z_{tc} = \sum_{m}^{i=1} c_i x_i + \sum_{m}^{i=1}\sum_{n}^{j=1} c_{ij} x_{ij} \tag{2}$$

where

$$x_i = \sum_{n}^{j=1} x_{ij}, i = 1..., m. \tag{3}$$

and there are two constraints:

$$x_{ij} \geq 0, \forall i = 1..., m, \forall j = 1..., n, \tag{4}$$

$$x_i \leq a_i, i = 1..., m. \tag{5}$$

Z_{fc} is the sum of fixed costs, including the opening costs of distribution centers and the fixed costs from distribution centers to customers.

$$Z_{fc} = \sum_{m}^{i=1} f_i y_i + \sum_{m}^{i=1}\sum_{n}^{j=1} f_{ij} y_{ij} \tag{6}$$

where

$$y_i = \begin{cases} 1, \sum_{j=1}^{n} x_{ij} \geq 0 \\ 0, \sum_{n}^{j=1} x_{ij} = 0 \end{cases} \forall i = 1...,m, \tag{7}$$

$$y_{ij} = \begin{cases} 1, x_{ij} \geq 0 \\ 0, x_{ij} = 0 \end{cases} \forall i = 1...., m, \forall j = 1..., n. \tag{8}$$

The new constraint of the problem based on [14] follows:

$$\sum_{m}^{i=1} \alpha' a_i x_i + \sum_{m}^{i=1} \beta'_i a_i c_i x_i + \sum_{m}^{i=1} \sum_{n}^{j=1} \alpha_i b_j x_{ij} + \sum_{m}^{i=1} \sum_{n}^{j=1} \beta_{ij} b_j c_{ij} x_{ij} \leq GHG, \tag{9}$$

where α_i =GHG emissions factor of a facility located at distribution center, in tons of CO_2e per unit demand, α'= GHG emissions factor of a facility located at manufacturer, in tons of CO_2e per unit demand, β_{ij} = GHG emissions factor per unit distance and per unit demand between distribution center i and customer location j, in tons of CO_2e per km and unit demand and β'_i =GHG emissions factor per unit distance and per unit demand between manufacturer and distribution center i, in tons of CO_2e per km and unit demand.

The novelty of our paper is that we consider as well the chain from manufacturer to distribution center, that is why the parameters $'$ and β'_i are also involved in the (9) inequality of sum emission factor.

Fig. 1 shows an example of a supply chain network as in [10].

Using the example from [10] are considered three potential distribution centers and five costumers. Data's properties are detailed in [10]. The best solution for this example is choosing distribution centers 2 and 3 as follows: the customers 1 and 4 are served from distribution center 2 and customers 2, 3 and 5 receive their demands from distribution center 3.

The results of computing the total cost for this allocation is 3625.

For the current paper purpose is investigated the GHG value of the greenhouse gas, GHG, emissions produced by the transportation and the operation of the facilities. For this particular problem all α values are considered equal and all β values are equal too.

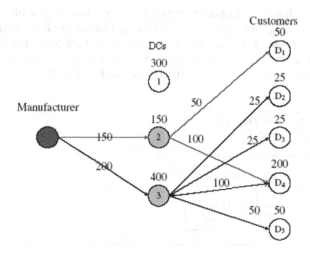

Fig. 1. An example for a two stage supply chain network [10]

The results of computing the total GHG for the given solution:

The gas emissions from the manufacturer to DCs is
$\beta \cdot (3 \cdot 150 + 3 \cdot 200) + \alpha \cdot (150 + 200) = \beta \cdot 1050 + \alpha \cdot 350$
The gas emissions from DCs to customers is
$\beta \cdot (3 \cdot 50 + 1 \cdot 100 + 2 \cdot 25 + 5 \cdot 25 + 3 \cdot 100 + 4 \cdot 50) + \alpha (50 + 100 + 25 + 25 + 100 + 50) = \beta \cdot 925 + \alpha \cdot 350$
The total emission is $\beta \cdot 1975 + \alpha \cdot 700$.
Considering $\alpha = 1$ and $\beta = 2$, the total emission is *4650*
Considering $\alpha = 0.01$ and $\beta = 0.02$, the total GHG emission is *46.5*
In order to get closer to a real approach to a sustainable supply chain design with the action to reduce and to control the pollution, the total emissions is limited by a given value denoted GHG. In this particular case let consider the following cases:

- if *GHG=50* then *46,5<50* and therefore the greenhouse gas emissions are in the limits
- if *GHG=40* then *46,5>40* and therefore the pollution is out of the limits and some restrictions should be taken by authorities.

3   Classical Hybrid Models Based on *Nearest Neighbor*

In order to reduce the greenhouse gas *(GHG)* emissions produced by the transportation and the operation of the facilities for a two-stage supply chain transportation problem it is involved *Nearest Neighbor* technique. In [12] are introduced several hybrid heuristic approaches based on *Nearest Neighbor* technique.

For the fixed-charged transportation problem, with two stages of a supply chain network, *Nearest Neighbor* algorithm could be applied when are chosen the potential distribution center and also when are chosen the best edges from distribution centers to customers. The algorithm for the supply chain used to search the suitable distribution centers list, starting from manufacturer, is detailed in Fig. 2.

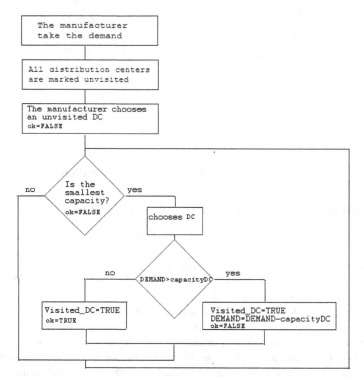

Fig. 2. The procedure used by a manufacturer when chooses distribution centers

There are used two variants for this supply chain.
DX-randomly chosen the potential distribution centers
DY-with the best probability based on the total request and distribution centers capacities as in [12]:

$$p_i = \frac{a_i}{request}. \tag{10}$$

$$p_i = \frac{a_i}{xcont_i \cdot request}. \tag{11}$$

$$p_i = \frac{xcont_i}{a_i}. \tag{12}$$

where $xcont_i$ is the number of nonzero quantities to be transported from distribution center i to customer j, when $x_{ij} > 0$. For the second supply chain the way each customer chooses suitable distribution centers as in Fig. 3.

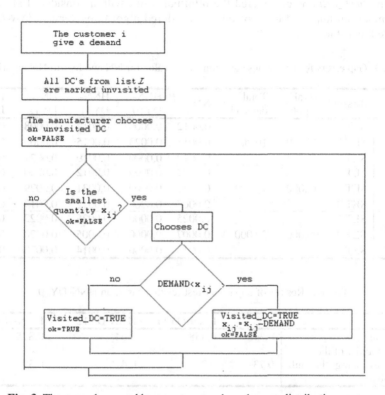

Fig. 3. The procedure used by a customer when chooses distribution centers

4   Tests and Results

The algorithms described in Sect. 3, used to solve the two-stage supply chain fixed-charged transportation problem were coded in java and compiled under Linux. For testing our implementation were generated *9* instances, *3* instances for each dimension considered *(10 x 10, 10 x 30, 30 x 100)* as in [12]. For every instance a two-stage supply chain with one layer is considered. Hybrid considered algorithm are choosing in different ways the distribution center starting from the manufacturer.

HNN-DX is the algorithm where a distribution center is chooses randomly and *HNN-DY 10-12* are algorithms with distribution center chosen with a probability specified in equations (10), (11) and (12).

The parameters used in Sect. 2, inequality (9) are: the total emissions limit *GHG* considered *150,000* and the greenhouse gas emissions coefficients $\alpha' = \alpha = 0.02$ and $\beta' = \beta = 0.04$.

Table 1 shows that the optimal objective of greenhouse gas emissions values was obtained by *HNN-DY10*. The gap error is computed as follows:

$$GapError = \frac{value - best\_value}{best\_value} \cdot 100\%. \tag{13}$$

where the *best_value* is considered the minimal result within considered algorithms. Furthermore, an unpaired t-test of other considered algorithms versus *HNN-DY10* is illustrated in Table 2.

Table 1. Gap errors for greenhouse gas emissions value for NN and hybrid NN algorithms

Dim	dataset	Total supply	Total demand	NN	HNN-DY10	HNN-DY11	HNN-DY12	HNN-DX
10x10	SET 1			0,0012	0,0000	0,0012	0,0008	0,0071
	SET 2	30.000	10.000	0,0015	0,0023	0,0015	0,0000	0,0056
	SET 3			0,0019	0,0000	0,0019	0,0021	0,0070
10x30	SET 1			0,0012	0,0009	0,0012	0,0000	0,0014
	SET 2	45,000	15,000	0,0031	0,0000	0,0031	0,0029	0,0035
	SET 3			0,0007	0,0000	0,0012	0,0013	0,0010
30x100	SET 1			0,0025	0,0000	0,0025	0,0022	0,0043
	SET 2	90,000	30,000	0,0007	0,0000	0,0005	0,0028	0,0055
	SET 3			0,0013	0,0000	0,0014	0,0020	0,0015

Table 2. Results of unpaired t-test results, based on HNN-DY10

	NN.	HNN-DY11	HNN-DY12	HNN-DX
t-test result t	0.348	0.358	0.64	0.875
the probability of the result, assuming the null hypothesis	0.73	0.72	0.475	0.39

Although the difference is considered to be not statistically significant, based on the unpaired t-test results from Table 2, the value closer to *HNN-DY10* are in order: *NN, HNN-DY11, HNN-DY12* and *HNN-DX*. An important observation is that not always the cost of the transportation is proportional with greenhouse gas emissions.

Further work should be done considering other hybrid heuristics as bio-inspired ones [2,11,13] in order to give a better support for transportation problems. Another perspective of the two-stage supply chain fixed-charged problem should be considered when using multi-layer facilities.

5   Conclusions and Future Work

A two-stage supply chain network design problem that arises in the public sector considering sustainable constraints in the form of restrictions on the dioxide carbon equivalent emissions is considered. Hybrid heuristics are involved, using different probabilities in order to find the suitable distribution center starting from a given manufacturer. The entire request from all customers should be satisfied. We observed that the cost of the transportation is not always proportional to the greenhouse gas emissions.

Acknowledgment. This work was cofinanced from the European Social Fund through Sectoral Operational Programme Human Resources Development 2007-2013, project number POSDRU/89/1.5/S/56287 "Postdoctoral research programs at the forefront of excellence in Information Society technologies and developing products and innovative processes", partner University of Oradea.

References

1. Adlakha, V., Kowalski, K.: On the fixed-charge transportation problem. OMEGA: The Int.J.of Management Science 27, 381–388 (1999)
2. Chira, C., Pintea, C.-M., Dumitrescu, D.: Sensitive Stigmergic Agent Systems: a Hybrid Approach to Combinatorial Optimization. In: Innovations in Hybrid Intelligent Systems, Advances in Soft Computing, vol. 44, pp. 33–39. Springer (2008)
3. Cost, S., Salzberg, S.: A weighted nearest neighbor algorithm for learning with sym-bolic features. Machine Learning 10, 57–78 (1993)
4. Deerwester, S., Dumals, S.T., Furnas, G.W., Landauer, T.K., Harshman, R.: Indexing by latent semantic analysis. J. Amer. Soc. Inform. Sci. 41, 391–407 (1990)
5. Devroye, L., Wagner, T.J.: Nearest neighbor methods in discrimination. In: Krishnaiah, P.R., Kanal, L.N. (eds.) Handbook of Statistics, vol. 2, North-Holland (1982)
6. Fayyad, U.M., Piatetsky-Shapiro, G., Smyth, P., Uthurusamy, R.: Advances in Knowledge Discovery and Data Mining. AAAI Press/MIT Press (1996)
7. Gersho, A., Gray, R.M.: Vector Quantization and Signal Compression. Kluwer Academic, Boston (1991)
8. Hitchcock, F.L.: The distribution of a product from several sources to numerous localities. J. of Mathematical Physic 20, 224–230 (1941)
9. Hugo, A., Pistikopoulos, E.: Environmentally conscious process planning under uncertain-ty. In: Floudas, C.A., Agrawal, R. (eds.) Sixth International Conference on Foundations of Computer Aided Process Design, CACHE Corporation, Princeton (2004)
10. Molla-Alizadeh-Zavardehi, S., Hajiaghaei-Keshteli, M., Tavakkoli-Moghaddam, R.: Solv-ing a capacitated fixed-charge transportation problem by artificial immune and genetic al-gorithms with a Prüfer number representation. Expert Systems with Applications 38, 10462–10474 (2011)
11. Pintea, C.-M., Chira, C., Dumitrescu, D., Pop, P.C.: Sensitive Ants in Solving the Genera-lized Vehicle Routing Problem. Int. J. Comput. Commun. & Control VI(4), 731–738 (2011)
12. Pintea, C.-M., Sitar, C.P., Hajdu-Macelaru, M., Petrica, P.: A Hybrid Classical Approach to a Fixed-Charged Transportation Problem. In: Corchado, E., Snášel, V., Abraham, A., Woźniak, M., Graña, M., Cho, S.-B. (eds.) HAIS 2012, Part III. LNCS, vol. 7208, pp. 557–566. Springer, Heidelberg (2012)

13. Pintea, C.-M.: Combinatorial optimization with bio-inspired computing, PhD Thesis, Babes-Bolyai University (2008)
14. Santibanez-Gonzalez, E., Del, R., Robson Mateus, G., Pacca Luna, H.: Solving a public sector sustainable supply chain problem: A Genetic Algorithm approach. In: Proc. of Int. Conf. of Artificial Intelligence (ICAI), Las Vegas, USA, pp. 507–512 (2011)
15. Seuring, S., Muller, M.: From a literature review to a conceptual framework for sustainable supply chain management. Journal of Cleaner Production 16, 1699–1710 (2008)
16. Sun, M., Aronson, J.E., Mckeown, P.G., Drinka, D.: A tabu search heuristic proce-dure for the fixed charge transportation problem. European Journal of Operational Research 106, 441–456 (1998)

Visualization in Information Retrieval from Hospital Information System

Miroslav Bursa[1], Lenka Lhotska[1], Vaclav Chudacek[1], Jiri Spilka[1], Petr Janku[2], and Lukas Hruban[1]

[1] Dept. of Cybernetics, Faculty of Electrical Engineering,
Czech Technical University in Prague, Czech Republic
bursam@fel.cvut.cz
[2] Obstetrics and Gynaecology clinic,
University Hospital in Brno, Czech Republic

Abstract. This paper describes the process of mining information from loosely structured medical textual records with no apriori knowledge. The typical patient record is filled with typographical errors, duplicates, ambiguities, syntax errors and many (nonstandard) abbreviations. In the paper we depict the process of mining a large dataset of ~50,000–120,000 records × 20 attributes in database tables, originating from the hospital information system (thanks go to the University Hospital in Brno, Czech Republic) recording over 11 years. The proposed technique has an important impact on reduction of the processing time of loosely structured textual records for experts.

Note that this project is an ongoing process (and research) and new data are irregularly received from the medical facility, justifying the need for robust and fool-proof algorithms.

Keywords: Swarm Intelligence, Ant Colony, Textual Data Mining, Medical Record Processing, Hospital Information System.

1 Introduction

1.1 Nature Inspired Methods

Nature inspired metaheuristics play an important role in the domain of artificial intelligence, offering fast and robust solutions in many fields (graph algorithms, feature selection, optimization, clustering, feature selection, etc). Stochastic nature inspired metaheuristics have interesting properties that make them suitable to be used in data mining, data clustering and other application areas.

In the last two decades, many advances in the computer sciences have been based on the observation and emulation of processes of the natural world. The origins of *bioinspired informatics* can be traced to the development of perceptrons and artificial life, which tried to reproduce the mental processes of the brain and biogenesis respectively, in a computer environment [1]. Bioinspired informatics also focuses on observing how the nature solves situations that are similar to engineering problems we face.

V. Snasel et al. (Eds.): SOCO Models in Industrial & Environmental Appl., AISC 188, pp. 459–467.
springerlink.com © Springer-Verlag Berlin Heidelberg 2013

With the boom of high-speed networks and increasing storage capacity of database clusters and data warehouses, a huge amount of various data can be stored. *Knowledge discovery* and *Data mining* is not only an important scientific branch, but also an important tool in industry, business and healthcare. These techniques target the problematic of processing huge datasets in reasonable time – a task that is too complex for a human. Therefore computer-aided methods are investigated, optimized and applied, leading to the simplification of the processing of the data. The main goal of computer usage is data reduction preserving the statistical structure (clustering, feature selection), data analysis, classification, data evaluation and transformation.

1.2 Ant Algorithms

Ant colonies inspired many researchers to develop a new branch of stochastic algorithms: *ant colony inspired algorithms*. Based on the ant metaphor, algorithms for both static and dynamic combinatorial optimization, continuous optimization and clustering have been proposed. They show many properties similar to the natural ant colonies, however, their advantage lies in incorporating the mechanisms, that allowed the whole colonies to effectively survive during the evolutionary process.

Cemetery formation and brood sorting are two prominent examples of insects' collective behavior. However, other types of ant behavior have been observed, for example predator-prey interaction, prey hunting, etc. The most important are mentioned below.

By replicating the behavior of the insects, the underlying mechanisms may be found and a better understanding of nature may be furthermore achieved. By applying the social insect behavior to computer science, we may achieve more effective techniques. Computer models based on the clustering and sorting of insects can lead to better performance in areas such as search, data mining, and experimental data analysis.

1.3 Text Extraction

The accuracy for relation extraction in journal text is typically about 60 % [6]. A perfect accuracy in text mining is nearly impossible due to errors and duplications in the source text. Even when linguists are hired to label text for an automated extractor, the inter-linguist disparity is about 30 %. The best results are obtained via an automated processing supervised by a human [7].

Ontologies have become an important means for structuring knowledge and building knowledge-intensive systems. For this purpose, efforts have been made to facilitate the ontology engineering process, in particular the acquisition of ontologies from texts.

1.4 Motivation

The task of this work is to provide the researchers with a quick automated or semi-automated view on the textual records. Textual data are not easy to visualize. The word frequency method is simple, but did not provide easily interpretable data. Therefore we decided to extract information in the form of a transition graph.

Such graphs allow as to induce a set of rules for information retrieval. These rules serve for extraction of (boolean/nominal) attributes from the textual rules. These attributes are used in automated rule discovery and can be further used for recommendation. The overall goal of the project is asphyxia prediction during delivery. High asphyxia might lead to several brain damage of the neonate and when predicted, caesarean section might be indicated on time.

2   Input Dataset Overview

The dataset consists of a set of approx. 50 to 120 thousand records (structured in different relational DB tables; some of them are not input, therefore the range is mentioned) × approx. 20 attributes. Each record in an attribute contains about 800 to 1500 characters of text (diagnoses, patient state, anamneses, medications, notes, references to medical stuff, etc.). For textual mining, 16 attributes are suitable (contain sufficiently large corpus).

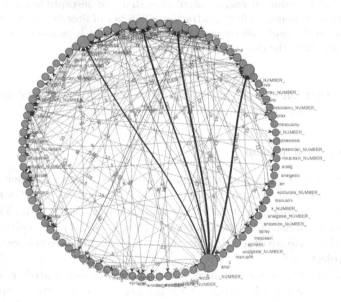

Fig. 1. Figure shows a transitional diagram (directed graph) structure of single attribute literals (a subset). Circular visualization has been used to present the amount of literal transitions (vertices).

The overview of one small (in field length) attribute is visualized in Fig. [1]. Only a subsample (about 5 %) of the dataset could be displayed in this paper, as the whole set would render into a uncomprehensible black stain. The vertices (literals) are represented as coloured circle, the size reflects the literal (i.e. word) frequency. Edges represent transition states between literals (i.e. the sequence of 2 subsequent words in a sentence/record); edge stroke shows the transition rate (probability) of the edge. The same holds for all figures showing the transition graph, only a different visualization approach has been used.

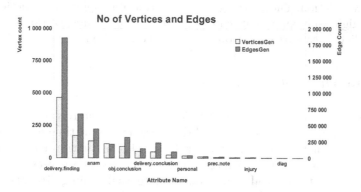

Fig. 2. The DB contains 16 textual attributes that are susceptible for information retrieval via natural language literal extraction. Number of literals (vertices) and transitions (edges) in the probabilistic models are shown for each attribute in a left/right bar respectively. Note the different y-axis scales.

It is clear, that human interpretation and analysis of the textual data is very fatiguing, therefore any computer aid is highly welcome.

2.1   Graph Explanation

In this paper we describe *transition graphs*. These are created for each attribute. An attribute consists of many records in form of a sentence. By *sentence* we hereby mean a sequence of literals, not a sentence in a linguistic form. The records are compressed – unnecessary words (such as verbs *is, are*) are omitted. In this paper, only the atribute describint the anesthetics during deliveris visualized, as it is the simplest one.

Vertices of the transition graph represent the words (separated by spaces) in the records. For each word (single or multiple occurence) a vertex is created and its potence (number of occurences is noted). For example, the words *mesocaine, anesthetics, not, mL* form a vertex. Note that also words as *mesocain, mezokain* and other versions of the word *mesocaine* are present. For a number (i.e. sequence of digits) a special literal *_NUMBER_* is used.

Edges are created from single records (sentences entered). For example the sentence *mesocaine 10 mL* would add edges from vertex *mesocaine* to vertex *_NUMBER_* and from vertex *_NUMBER_* to the vertex *mL* (or the edge count is increased in case it exists). For all records, the count of the edges is also useful.

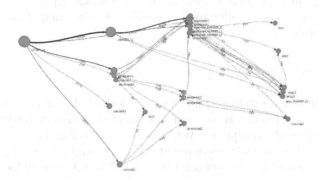

Fig. 3. Very nice graph (sub-graph) providing the basic information about the attribute presented. Note that similar words are clustered (positioned nearby) and the flow of the most common sentences can be easily traced.

It provides an overview on the inherent structure of the data – the most often word transitions.

3   Nature Inspired Techniques

Social insects, i. e. ant colonies, show many interesting behavioral aspects, such as self-organization, chain formation, brood sorting, dynamic and combinatorial optimization, etc. The coordination of an ant colony is of local nature, composed mainly of indirect communication through pheromone (also known as *stigmergy*.

The high number of individuals and the decentralized approach to task coordination in the studied species means that ant colonies show a high degree of parallelism, self-organization and fault tolerance. In studying these paradigms, we have high chance to discover inspiration concepts for many successful metaheuristics.

3.1   Ant Colony Optimization

Ant Colony Optimization (ACO) [5] is an optimization technique that is inspired by the foraging behavior of real ant colonies. Originally, the method was introduced for the application to discrete and combinatorial problems.

3.2   Ant Colony Methods for Clustering

Several species of ant workers have been reported to form piles of corpses (cemeteries) to clean up their nests. This aggregation phenomenon is caused by attraction between dead items mediated by the ant workers.

This approach has been modeled in the work of Deneubourg et al. [4] and in the work of Lumer and Faieta [8] to perform a clustering of data.

For clustering, the ACO_DTree method [3,2] and ACO inspired clustering [8] variations have been successfully used. A self-organizing map has also been tested, but performed poorly.

4   Automated Processing

Automated layout of transition graph is very comfortable for an expert, however the contents of the attribute is so complicated, that a human intervention is inevitable. Examples of automated layout can be seen in Fig. [4].

The figure Fig. [4] shows a transitional graph where only positioning based on the word distance from the sentence start is used. Although it migh look correct, note that the same words are mispositioned in the horizontal axis.

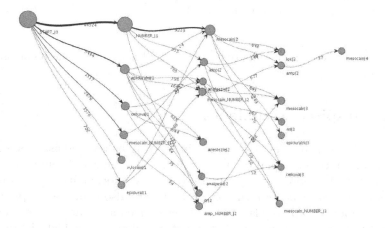

Fig. 4. A fully automated transition graph showing the most important relations in one textual attribute. No clustering has been used. The layout is based on the word distance from the start of the sentence. Note the mis-alignment of the similar/same words. Refer to section [2].

5   Expert Intervention

A human intervention and supervision over the whole project is indiscutable. Therefore also human (expert) visualization of the transition graph has been studied.

The vertices in a human-only organization are (usually) organized depending on the position in the text (distance from the starting point) as the have the highest potence. Number literal (a wildcard) had the highest potence, as many quantitative measures are contained in the data (age, medication amount, etc.). Therefore it has been fixed to the following literal, spreading into the graph via multiple nodes (i.e. a sequence *mesocain 10 mL* become two vertices – *mesocain_NUMBER_* and *mL*). This allowed to organize the chart visualization in

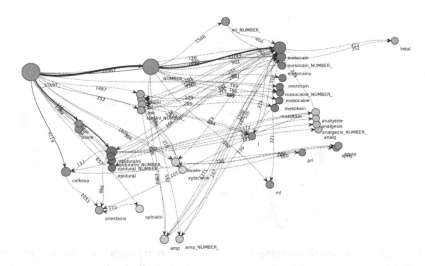

Fig. 5. An expert (human) organized transition graph (sub-graph) showing the most important relations in one textual attribute. Refer to section [2].

more logical manner. Time needed to organize such graph was about 5–10 minutes. The problem is that the transition graph contains loops, therefore the manual organization is not straigthforward.

An aid of a human expert has been used in semi-automated approach (see Fig. [6] where the automated layout has been corrected by the expert. The correction time has been about 20–30 seconds only.

6   Parallelization

The ACO_DTree algorithm has been parallelized in order to take advantage of multicore processors. It contains nautrally parallelizable parts, such as population evaluation and population improvement (via the PSO method). Experimental tests have been performed on the 4-core i7-2600 CPU@3.40 GHz (8 cores with Hyperthreading) processor. Performance tests have been run with varying number of cores with and without hyperthreading (HT). The number of execution threads has been increased from 1 to 16.

The *CPU utilization* (load) scaled up to the number of cores (regardless of the HT setting) linearly. There has been a drop-down in CPU load when the no. of threads increased over the number of cores available. The performance for 2 and 4 cores (w/o HT) and 4 and 8 cores (HT) has been similar.

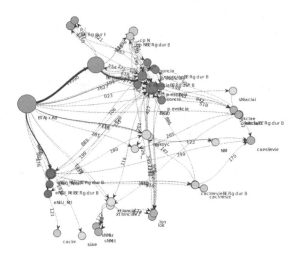

Fig. 6. A semi-automated (corrected by a human expert) organized transition graph showing the most important relations in one textual attribute. Refer to section [2].

7   Results and Conclusion

The main advantage of the nature inspired concepts lies in automatic finding relevant literals and group of literals that can be adopted by the human analysts and furthermore improved and stated more precisely. The use of induced probabilistic models in such methods increased the speed of loosely structured textual attributes analysis and allowed the human analysts to develop lexical analysis grammar more efficiently in comparison to classical methods. The speedup (from about 5–10 minutes to approx 20–30 seconds) allowed to perform more iterations, increasing the yield of information from data that would be further processed in rule discovery process. However, the expert intervention in minor correction is still inevitable. The results of the work are adopted for rule discovery and are designed to be used in expert recommendation system.

8   Discussion and Future Work

The future work is to evaluate the DB analyst's utilization and aid of such graphs in more accurate way. The graphs serve as a bases for extraction rule proposal. However the only relevant measure is the time to reorganize the transitional graphs. The subjective opinion is very expressive and is not coherent. Next, the semantic meaning of the attributes will be extracted and verified followed by rule discovery mining.

Acknowledgment. This research project has been supported by the research programs number MSM 6840770012 "Transdisciplinary Research in the Area of Biomedical Engineering II" of the CTU in Prague, sponsored by the Ministry of Education, Youth and Sports of the Czech Republic and by the project number NT11124-6/2010 "Cardiotocography evaluation by means of artifficial intelligence" of the Ministry of Health Care. This work has been developed in the BioDat research group bio.felk.cvut.cz.

References

1. Adami, C.: Introduction to Artificial Life. Springer (1998)
2. Bursa, M., Huptych, M., Lhotska, L.: Ant colony inspired metaheuristics in biological signal processing: Hybrid ant colony and evolutionary approach. In: Biosignals 2008-II, vol. 2, pp. 90–95. INSTICC, Setubal (2008)
3. Bursa, M., Lhotska, L., Macas, M.: Hybridized swarm metaheuristics for evolutionary random forest generation. In: Proceedings of the 7th International Conference on Hybrid Intelligent Systems 2007 (IEEE CSP), pp. 150–155 (2007)
4. Deneubourg, J.L., Goss, S., Franks, N., Sendova-Franks, A., Detrain, C., Chretien, L.: The dynamics of collective sorting robot-like ants and ant-like robots. In: Proceedings of the first International Conference on Simulation of Adaptive Behavior on From Animals to Animats, pp. 356–363. MIT Press, Cambridge (1990)
5. Dorigo, M., Stutzle, T.: Ant Colony Optimization. MIT Press, Cambridge (2004)
6. Freitag, D., McCallum, A.K.: Information extraction with hmms and shrinkage. In: Proceedings of the AAAI Workshop on Machine Learining for Information Extraction (1999)
7. Lafferty, J., McCallum, A., Pereira, F.: Conditional random fields: Probabilistic models for segmenting and labeling sequence data. In: Proceedings of the ICML, pp. 282–289 (2001); Text processing: interobserver agreement among linquists at 70
8. Lumer, E.D., Faieta, B.: Diversity and adaptation in populations of clustering ants. In: From Animals to Animats: Proceedings of the 3th International Conference on the Simulation of Adaptive Behaviour, vol. 3, pp. 501–508 (1994)

Investigation on Evolutionary Control and Optimization of Chemical Reactor

Ivan Zelinka and Lenka Skanderova

Department of Computer Science, Faculty of Electrical Engineering and Computer Science
VŠB-TUO, 17. Listopadu 15, 708 33 Ostrava-Poruba, Czech Republic
{ivan.zelinka,lenka.skanderova.st}@vsb.cz

Abstract. This contribution deals with a new algorithm – the Self-Organizing Migrating Algorithm (SOMA). The SOMA algorithm was used for static optimization of a given chemical reactor with 5 inputs and 5 outputs. SOMA was used on this reactor for static optimization because the reactor, which was set by an expert, shows poor performance behaviour. Participation consists of simulation results, which shows how expertly set reactor behaves. Also set of static optimization simulations of given reactor is presented here including results and conclusions.

Keywords: SOMA, migration, self-organization, evolutionary algorithms, global optimization, non-linear optimization, mixed discrete variables, penalty function.

1 Introduction

Nowadays, there exist a broad class of algorithms that can be, and are, used for optimization. This special class of algorithms is made up of so-called evolutionary algorithms (EA) similar to genetic algorithms or differential evolution algorithms [1]. Both algorithms work with so-called populations that are evolved in "generations" (or "Migration Loops" in the case of SOMA [2], [3], [4], [5], [1]), in which only the best-suited individuals survive.

This contribution presents a new algorithm, which can be labelled an "evolutionary" algorithm - despite the fact that during its activity, no new generations are created (in a general sense). Development of this algorithm was inspired by the behaviour patterns of groups of wild animals in the wild. It has been termed "the Self-Organizing Migrating Algorithm" – or SOMA for short (for complete description, source codes etc. please see [5]).

SOMA, and generally speaking any evolutionary algorithm, can be used in regards to any optimization problem. Surprisingly, many problems can be defined as optimization problems, e.g. the optimal trajectory of robot arms; the optimal thickness of steel in pressure vessels; the optimal set of parameters for controllers; optimal relations or fuzzy sets in fuzzy models; and so on. Solutions to such problems are usually more or less hard to arrive at, their parameters usually including variables of different types, such as real or integer variables. Evolutionary algorithms are quite popular because they allow the solution of almost any problem in a simplified manner, because

V. Snasel et al. (Eds.): SOCO Models in Industrial & Environmental Appl., AISC 188, pp. 469–474.
springerlink.com   © Springer-Verlag Berlin Heidelberg 2013

they are able to handle optimizing tasks with mixed variables - including the appropriate constraints, as and when required.

This contribution explains SOMA's use on static optimization of given chemical reactor. A large part of the research dealing with wastes of the leather industry, except for USDA publications, does not go into particulars about how to cope with chrome sludge after dechromation of tanned wastes. As if chrome sludge so formed was automatically assumed to be simply used for producing recycled tanning salt. Even though the balance of chromium in chrome-tanned wastes and of necessary tanning salt is very favorable for recycling in the tanning industry, the actual situation is different.

Although we quite correctly feel and hope that the issue of recycling chromium into the tanning industry should be worked on or at least supported by manufacturers of chromic chemicals in the first place, we studied both the drawbacks of such recycling and applications in other fields. Part of this research is focused on reactor inside which class of mentioned chemical reactions could be done. Main aim of SOMA use was for reactor static optimization.

2 Reactor Description

Model of the reactor (see Figure 1) inside which can be realized above mentioned reactions was given by 5 nonlinear partial differential equations. Expert parameters were used for original setting. They comes from experiences obtained during visit in laboratory „Resine and Composite for Forest Products" in Sainte-Foy, Canada. This set of parameters consisted of two kind of parameters i.e. parameters of chemical materials and physical parameters of reactor under consideration. An initial conditions (a_{AP0}, a_{BP0}, a_{P0}, T_{P0}, T_{X0},) used in following simulations. This set of parameters was used for initial simulations. Both graphs show that reactor under expert parameters produce unsatisfactory behaviour. Reactor production stabilizes itself after 27 Hrs. on 15 % concentration of output chemical. From that point of view above-mentioned parameters were regarded like unsatisfactory. Because of these reasons a few static optimizations by SOMA algorithm were consequently done.

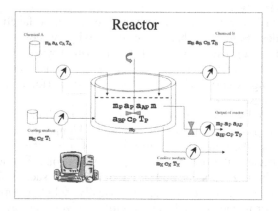

Fig. 1. Optimized reactor

They were done in following steps:

1. Optimization without restrictions - Fig. 2 a) and b)
2. Optimization with restrictions applied exactly in given time
3. Optimization with penalty applied during time interval
4. Optimization with penalty applied during time interval and sub optimization of cooling surface

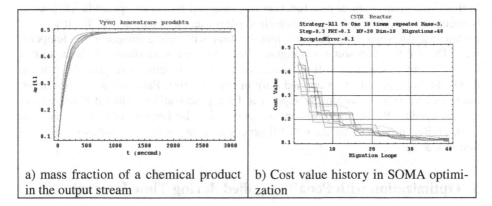

a) mass fraction of a chemical product in the output stream	b) Cost value history in SOMA optimization

Fig. 2. a) reactor static optimization without penalty and restrictions, b) cost value history in SOMA optimization

The last static optimization is important one. The first three are mentioned here too only for complete overview, what optimizations were done by means of SOMA. Each of four optimization cases was 10 times repeated. From all 10 simulations was finally chosen the best reactor. According to this table a global extreme was found in 13-ti dimensional configuration space. Last 13th dimension was cost value of cost function. In case of the last optimization (optimization with penalty applied during time interval and sub-optimization of cooling surface) searching for global extreme had run in 11th dimensional space because of relations among some parameters.

3 Optimization without Restrictions

Main aim of this optimization was focused on parameter reactor optimizing in such way that $a_{AP} = 0.6$ was desired. Others parameters like a_{BP}, a_{BP}, T_P, T_X was not restricted. The total number of simulation done here was 10. He maximum of a_{AP} was $a_{AP} = 0.5$, which can be regarded, like a good result. In all following simulations was not reached better result probably thanks to physical and chemical reasons. Despite this result a wrong behaviour can be observed on Fig. 2. One of no acceptable behaviour shows temperature that is very high (6000 K, -900 K, etc.). Such physical behaviour is not acceptable and also not realizable. Explanation of such wild behaviour stems from obvious fact that our model does not exactly follow reality. Cost function used in this simulation was given by

$$f_{\cos t} = |0.6 - a_P(t)| \quad where \quad t = 1200 \tag{1}$$

Optimization was focused of such behaviour searching which should satisfy in time t = 1200 second minimal difference between desired value in this time and reactor output response in this time.

4 Optimization with Restrictions Applied Exactly in Given Time

In this simulation previous cost function was enlarged for set of operands. Difference between them was such that were penalized parameters aAP, aBP, aP, TP, TX. It was expected that described cost function modification will delete unacceptable temperatures. Despite this fact some unacceptable temperatures were observed there, so this kind of optimization was still non-successful. These results were probably caused thanks to weak penalization applied only in time τ= 100. Parameters aBP, aBP in are multiplied by 100 because of its support in final penalization. Without it parameters aBP, aBP would influent final cost function only a little bit because of its range aBP, aBP ⬚ [0, 1]. Because this optimization still generated wrong solutions, following simulation was designed.

5 Optimization with Penalty Applied during Time Interval

In this simulation was minimized difference (surface) between desired and observed reactor response. It was expected from this simulation that high or low temperatures would not be observed here. Minimization of it should satisfy this. Parameters aBP, aBP in are multiplied by 100 because of the same reasons like in previous step. From results it is clear that this set of 10 simulations produce more reasonable behaviour than in previous cases. However, some error-behaviour is there too. For example a cooling medium temperature lower than 0 C_o can be observed there. This non-acceptable behaviour is probably caused by fact that between model and real reactor there is no 100% equivalency. Also some simplifications in computer model partly caused it.

For example physical relations between cooling surface and surface of reactor was not taken under consideration, etc. This was solved in the last and successful optimization.

6 Optimization with Penalty Applied during Time Interval and Suboptimization of Cooling Surface

This optimization was focused on optimal parameter searching in such way that some of these parameters were related among themselves. These parameters were mA, mB, mP, m a S. Relation between mA, mB, a mP was given by

$$m_P = m_A + m_B \tag{2}$$

This equation simply says that output is equal to sums of inputs. Next relation was between m a S (cooling only in the wall of reactor and on its bottom) and was described like

$$S = 2\pi r^2 + \pi r^2 \tag{3}$$

$$m = \varsigma \pi r^3 \quad where \ \varsigma = 1100 \ kgm^{-3} \tag{4}$$

A simple presumption that "r" is equal to height "h" was done for simplification. Thus only "r" instead of "S" and "m" was used. Graphs based on 10 times repeated static optimizations are depicted on Fig. 6 a-d). There is visible that the best reactor produce not only reasonable behaviour (temperatures $T_P \approx 380$ K a $T_X \approx 340$ K) but also output chemical product $a_P = 0.5$ i.e. 50%, which represents quality increase for 35% ($a_{Poptimal} = a_{Poptimized} - a_{Pexpert} = 0.5 - 0.15 = 0.35$). Fact that this behavior is stabilized after 8-10 minutes (in comparison with **27 Hrs** (!!!) in case of expertly set reactor).

7 Conclusion

The methods of optimization mentioned here (detail at [5]) are relatively simple, easy to implement and easy to use. Despite that, it is capable of optimizing all integer, discrete and continuous variables and capable of handling non-linear objective functions with multiple non-trivial constraints.

A soft-constraint (penalty) approach is applied for the handling of constraint functions. Some optimization methods require a feasible initial solution as a starting point for a search. Preferably, this solution should be rather close to a global optimum to ensure convergence to it instead of a local optimum. If non-trivial constraints are imposed, it may be difficult or impossible to provide a feasible initial solution. The efficiency, effectiveness and robustness of many methods are often highly dependent on the quality of the starting point. The combination of the SOMA algorithm with the soft-constraint approach does not require any initial solution, but it can still take advantage of a high quality initial solution if one is available.

For example, this initial solution can be used for initialization of the population in order to establish an initial population that is biased towards a feasible region of the search space. If there are no feasible solutions in the search space, as is the case for totally conflicting constraints, SOMA algorithm with the soft-constraint approach are still able to find the nearest feasible solution. This is important in practical engineering optimization because often, many non-trivial constraints are involved. The approach described above was targeted to fill the gap in the field of mixed discrete-integer-continuous optimization, where no one single really satisfactory method appeared to be available. Despite being in its infancy, the described approach has great potential to become a widely used, multipurpose optimization tool for solving a broad range of practical engineering optimization problems.

These algorithms are undoubtedly one of the most promising and novel methods for non-linear optimization that can be applied generally, and they work with minimum assumptions with respect to the objective function.

The algorithm requires only the value of objective function for guidance of it's seeking the optimum. No derivatives or other auxiliary information are desired. Including the algorithm extensions discussed in this article, the SOMA algorithms can be applied to a wide range of optimization problems, which practitioners in the field of modern prediction would like to solve.

In the past, SOMA had been successfully used on hard optimization problems with good results (see [5]) During these tests, **9500** optimization simulations were carried out, which represent approximately $22x10^6$ cost function evaluations. The quality of the results, and the fact that the conclusions derived from them could be proven to be true, have demonstrated that SOMA has the capability of finding optimal near-optimal solution with a very high reliability.

We have also mentioned the possibility of chemical reactor optimization by SOMA algorithm. Inside this reactor can be realized certain class of chemical reactions like enzymatic dechromation technology, etc. The advantage of the enzymatic reaction is the production of protein hydrolyzates of relatively good quality and chrome sludge. Using organic bases to form alkaline reaction mixture increases the quality of both is products. A partial regeneration of organic base when diluted protein hydrolyzates undergo concentration cuts the operating costs of enzymatic hydrolysis. In commercial application, the greatest volume of protein hydrolyzate is channeled into agriculture. Hydrolyzate, as an organic nitrogenous fertilizer, not only equals the combined ureaammonium nitrate fertilizer in crop yield, but also surpasses it manifolds in the foodstuff value of consumer's greens. The content of nitrates is as much as 200 times lower on average. Hydrolyzate is also used in the manufacture of biodegradable foil, especially for producing sowing tape. The main obstacle for the utilization of the chrome sludge is a relatively high content of proteins in the dry substance of cake. In closing, it may be said that enzymatic hydrolysis has a place in the treatment of chromium containing tannery waste and the funds expended on this field of research have brought satisfactory results.

Acknowledgments. This work was supported by the European Regional Development Fund in the IT4Innovations Centre of Excellence project (CZ.1.05/1.1.00/02.0070) and by the Development of human resources in research and development of latest soft computing methods and their application in practice project, reg. no. CZ.1.07/2.3.00/20.0072 funded by Operational Programme Education for Competitiveness, co-financed by ESF and state budget of the Czech Republic.

References

[1] Zelinka, I., Snasel, V., Abraham, A. (eds.): Handbook of Optimization. In: Intelligent Systems. Springer (2012)

[2] Ivan, Z., Jouni, L.: SOMA - Self- Organizing Migrating Algorithm Mendel. In: 6th International Conference on Soft Computing, Brno, Czech Republic (2000) ISBN 80-214-1609-2

[3] Ivan, Z.: SOMA - Self-Organizing Migrating Algorithm Nostrdamus. In: 3rd International Conference on Prediction and Nonlinear Dynamic, Zlín, Czech Republic (2000)

[4] Zelinka, I.: SOMA – Self Organizing Migrating Algorithm. In: Babu, B.V., Onwubolu, G. (eds.) New Optimization Techniques in Engineering, pp. 167–218. Springer, New York (2004) ISBN 3-540-20167X

[5] Zelinka, I., Celikovsky, S., Richter, H., Chen, G.: Evolutionary Algorithms and Chaotic Systems. 550s, Springer, Germany (2010)

Designing PID Controller for DC Motor
by Means of Enhanced PSO Algorithm
with Dissipative Chaotic Map

Michal Pluhacek[1], Roman Senkerik[1], Donald Davendra[2], and Ivan Zelinka[1]

[1] Tomas Bata University in Zlin, Faculty of Applied Informatics,
T.G. Masaryka 5555, 760 01 Zlin, Czech Republic
{pluhacek,senkerik,zelinka}@fai.utb.cz
[2] VŠB-Technical University of Ostrava,
Faculty of Electrical Engineering and Computer Science,
Department of Computer Science, 17. Listopadu 15,
708 33 Ostrava-Poruba, Czech Republic
donald.davendra@vsb.cz

Abstract. In this paper, it is proposed the utilization of chaotic dissipative map based chaos number generator to enhance the performance of PSO algorithm. This paper presents results of using chaos enhanced PSO algorithm to design a PID controller for DC motor system. Results are compared with other heuristic and non-heuristic methods.

1 Introduction

Complex problems of optimization tasks emerged with the spread of computer technology which made possible solving many previously unsolvable problems using their enormous computing power in "brutal force attacks" (trying all possibilities) on these tasks. However it became soon clear that some problems will probably be unsolvable by "brutal force" even in distant future because of their enormous complexity and physical limitations of computer technology.

Genetic based algorithms [1-5] were discovered capable of finding very good solutions for these problems in a short time. Main thought is divided from the discoveries of Charles Darwin about evolution of species. New generation of individuals (representing possible solutions of the problem) is created via transforming the old population following a set of rules. This set of transformation rules is unique for each algorithm.

More recently the soft-computing methods which include neural networks, evolutionary algorithms, genetic programming and fuzzy logic extended their applications in almost every computer-science and engineering discipline.

Good examples of soft-computing possibilities are recent studies of using these methods for designing PID controllers [6], which represents very complex optimization task that can be solved by non-heuristic methods only with limitations. A recent study [7] suggests that using chaos based number generators instead of common computer random number generators can lead to increasing of the performance of evolutionary algorithms in the task of PID controller design.

V. Snasel et al. (Eds.): SOCO Models in Industrial & Environmental Appl., AISC 188, pp. 475–483.
springerlink.com   © Springer-Verlag Berlin Heidelberg 2013

This research presents using of Dissipative standard map as discrete chaotic system for the chaotic number generator and implementation of this chaotic generator into PSO algorithm, which is modified by using the inertia weight factor w [8]. This enhanced PSO algorithm is applied on the PID controller design problem.

The main idea and motivation for combining evolutionary algorithms and deterministic chaos system is that both these are originally inspired in nature. So for nature-based algorithm such as PSO it should be much more natural to use chaos number generator such as the dissipative standard map. And the hope is that it might improve the performance of PSO algorithm.

2 Particle Swarm Optimization Algorithm

PSO (Particle swarm optimization) algorithm is based on the natural behavior of birds and fishes and was firstly introduced by R. Eberhart and J. Kennedy in 1995 [1,2]. As an alternative to genetic algorithms [4] and differential evolution [5], PSO proved itself to be able to find better solutions for many optimization problems. Term "swarm intelligence" [2,3] refers to the capability of particle swarms to exhibit surprising intelligent behavior assuming that some form of communication (even very primitive) can occur among the swarm particles (individuals).

Basic PSO algorithm disadvantage is the rapid acceleration of particles which causes abandoning the defined area of interest. In each generation, a new location of a particle is calculated based on its previous location and velocity (or "velocity vector"). For this reason, several modifications of PSO were introduced to handle with this problem. Main principles of PSO algorithm and its modifications are well described in [1-3].

Within this research, chaos driven PSO strategy with inertia weight was used. The selection of inertia weight modification of PSO was based on numerous previous experiments. Default values of all PSO parameters were chosen according to the recommendations given in [2,3]. Inertia weight is designed to influence the velocity of each particle differently over the time [8]. In the beginning of the optimization process, the influence of inertia weight factor w is minimal. As the optimization continues, the value of w is decreasing, thus the velocity of each particle is decreasing, since w is always the number < 1 and it multiplies previous velocity of particle in the process of new velocity value calculation. Inertia weight modification PSO strategy has two control parameters w_{start} and w_{end}. New w for each generation is then given by Eq. 1, where i stand for current generation number and n for total number of generations.

$$w = w_{start} - \frac{\left(\left(w_{start} - w_{end}\right) * i\right)}{n} \tag{1}$$

Chaos driven number generator is used in the main PSO definition (Eq. 2) which determines a new "velocity", thus the position of each particle in the next generation (or migration cycle).

$$v(t+1) = v(t) + c_1 \cdot Rand \cdot (pBest - x(t)) + \\ c_2 \cdot Rand \cdot (gBest - x(t)) \tag{2}$$

Where:

$v(t+1)$ – New velocity of particle.
$v(t)$ – Current velocity of particle.
c_1, c_2 – Priority factors.
pBest – Best solution found by particle.
gBest – Best solution found in population.
$x(t)$ – Current position of particle.
Rand – Random number, from the interval <0,1>. Within Chaos PSO algorithm, the basic inbuilt computer (simulation software) random generator is replaced with chaotic generator (in this case, by using of Dissipative standard map).

New position of a particle is then given by Eq. 3, where x(t+1) represents the new position:

$$x(t+1) = x(t) + v(t+1) \tag{3}$$

3 Dissipative Standard Map

The Dissipative Standard map is a two-dimensional chaotic map. The parameters used in this work are $b = 0.1$ and $k = 8.8$. For these values, the system exhibits typical chaotic behaviour and with this parameter setting it is used in the most research papers and other literature sources [9]. The Dissipative standard map is given in Fig. 1. The map equations are given in Eq. 4 and 5.

$$X_{n+1} = X_n + Y_{n+1} \,(\mathrm{mod}\, 2\pi) \tag{4}$$

$$Y_{n+1} = bY_n + k \sin X_n \,(\mathrm{mod}\, 2\pi) \tag{5}$$

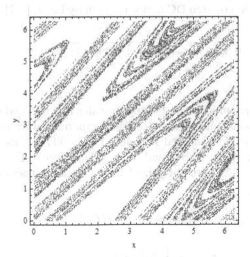

Fig. 1. Dissipative standard map

4 Problem Design

This section contains the description of the PID controller, used model of DC motor as well as the overview of the most important results.

4.1 PID Controller and DC Motor System

The PID controller contains three unique parts; proportional, integral and derivative controller [10]. A simplified form in Laplace domain is given in Eq 6.

$$G(s) = K\left(1 + \frac{1}{sT_i} + sT_d\right) \tag{6}$$

Where K is controller gain, T_i is the adjustable integral time parameter, T_D is the rate time.

The PID form most suitable for analytical calculations is given in Eq 7.

$$G(s) = k_p + \frac{k_i}{s} + k_d s \tag{7}$$

The parameters are related to the standard form through: $k_p = K$, $k_i = K/T_i$ and $k_d = KT_d$. Estimation of the combination of these three parameters that gives the lowest value of the four test criterions was the objective of this research.

The test criterion measures properties of output transfer function and can indicate quality of regulation. Following four different integral criterions were used for the test and comparison purposes: IAE (Integral Absolute Error), ITAE (Integral Time Absolute Error), ISE (Integral Square Error) and MSE (Mean Square Error). For further details see [6,7]. These test criterions were minimized within the cost functions for the enhanced PSO algorithm.

The transfer function of used DC motor is given by Eq. 8. [6,7]

$$G(s) = \frac{0.9}{0.00105s^3 + 0.2104s^2 + 0.8913s} \tag{8}$$

4.2 Cost Function

Test criterion measures properties of output transfer function and can indicate quality of regulation. Following four different integral criterions were used for the test and comparison purposes: IAE (Integral Absolute Error), ITAE (Integral Time Absolute Error), ISE (Integral Square Error) and MSE (Mean Square Error). These test criterions (given by Eq. 9–12) were minimized within the cost functions for the enhanced PSO algorithm.

Integral of Time multiplied by Absolute Error (ITAE)

$$I_{ITAE} = \int_0^T t|e(t)|dt \tag{9}$$

Integral of Absolute Magnitude of the Error (IAE)

$$I_{IAE} = \int_0^T |e(t)| \, dt \tag{10}$$

Integral of the Square of the Error (ISE)

$$I_{ISE} = \int_0^T e^2(t) \, dt \tag{11}$$

Mean of the Square of the Error (MSE)

$$I_{MSE} = \frac{1}{n} \sum_{i=1}^{n} (e(t))^2 \tag{12}$$

5 Results

The experiments were focused on the optimization of the four different specification functions as given in Section 4.1. The best results of the optimization with corresponding values of k_p, k_i and k_d together with selected response profile parameters are presented in Table 1.

When tuning a PID controller, generally the aim is to match some preconceived 'ideal' response profile for the closed loop system. The following response profiles are typical [11]:

Overshoot: this is the magnitude by which the controlled 'variable swings' past the setpoint. 5-10% overshoot is normally acceptable for most loops.

Rise time: the time it takes for the process output to achieve the new desired value. One- third the dominant process time constant would be typical.

Settling time: the time it takes for the process output to die to between, say +/- 5% of setpoint.

From the statistical reasons, optimization for each criterion was repeated 30 times. Results of the simple statistical comparison for the optimizations by means of chaos driven PSO algorithm are given in tables 2 and 3.

Furthermore obtained results are compared with previously published result [6] given by other heuristic and non-heuristic methods (See table 4).

Optimized system responses are depicted in figures. 2 – 5.

Table 1. The best results for DC motor system

Criterion	CF	Kp	Ki	Kd	Overshoot	Rise Time	Settling time
IAE	0.223055	241.917000	2.557960	58.577400	0.215717	0.010100	0.023300
ITAE	0.008617	297.775000	0.252631	71.815200	0.257957	0.008700	0.030800
ISE	0.018387	144.622000	28.458700	64.141000	0.223852	0.009500	0.032400
MSE	0.000919	147.209000	29.574200	64.035400	0.223748	0.009500	0.032400

Table 2. Average response profiles for DC motor system

Criterion	Avg overshoot	Avg rise time	Avg settling time
IAE	0.223677	0.009847	0.027067
ITAE	0.225357	0.010020	0.028427
ISE	0.224081	0.009507	0.032347
MSE	0.223918	0.009520	0.032360

Table 3. Statistical overview of the cost function (criterion) values for DC motor system

Criterion	Max CF	Min CF	Avg CF	Median
IAE	0.275357	0.223055	0.243871	0.239859
ITAE	0.040698	0.008617	0.020648	0.019312
ISE	0.018448	0.018387	0.018405	0.018399
MSE	0.000924	0.000919	0.000921	0.000920

Table 4. Comparison of other methods and proposed enhanced PSO

Criterion	Z-N (step response)	Kappa-Tau	Continuous cycling	EP	GA	PSO	Chaos PSO
IAE	0.517600	0.518800	0.560000	0.489100	0.771200	0.916100	0.223055
ITAE	3.380500	3.311300	7.820000	0.072100	0.378100	0.022900	0.008617
ISE	2.346700	2.250300	3.200000	1.027700	1.043500	1.001600	0.018387
MSE	0.011700	0.077778	0.016000	0.005100	0.005200	0.005000	0.000919

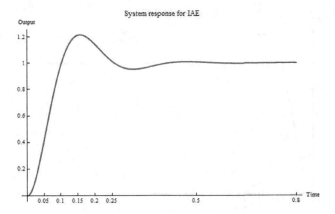

Fig. 2. System response for IAE

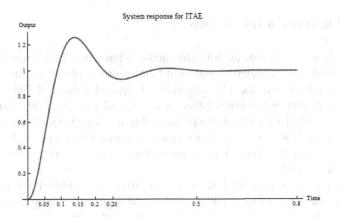

Fig. 3. System response for ITAE

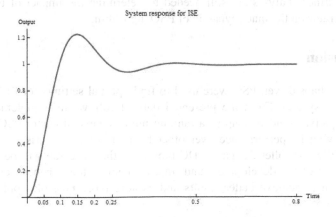

Fig. 4. System response for ISE

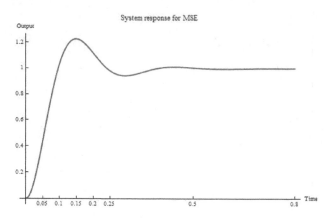

Fig. 5. System response for MSE

6   Brief Analyses of the Results

From Table 4, it follows that all heuristic methods has given better results than non-heuristic methods (e.g. Ziegler-Nichols and Continuous cycling) for the most of test criterions (exception was the IAE criterion). Proposed enhanced PSO with chaos based number generator has given better results for all test criterions than any other heuristic or non-heuristic method compared within this paper. Obtained control parameters (see Table 1) were used to obtain system responses (Figures. 2 - 5) which were afterwards analyzed for detailed response profiles: overshot, rise time and settling time (see Tables 1 and 2).

Given the presented data the PSO algorithm driven by dissipative standard chaos map seems to be a valid tool for PID controller design for such systems as the DC motor system presented in this paper. It also seems so, that this proposed modification outperformed previously presented stochastic methods on this task. However more research and further analyses are still needed to determine the impact of using chaos number generator on the inner dynamic of PSO algorithm.

7   Conclusion

In this paper chaos driven PSO were used to find optimal settings for PID controller for DC motor system. From the presented data, it follows that implementation of chaotic dissipative standard map as a random number generator into PSO algorithm led to improving its performance over other heuristic or non-heuristic methods for solving the PID controller design for DC motor. Further research will be focused on the possibilities of the development and improvement of the enhanced chaos driven PSO algorithm to achieve better results and explore more possible applications for this promising method.

Acknowledgements. This work was supported by European Regional Development Fund under the project CEBIA-Tech No. CZ.1.05/2.1.00/03.0089, and by Internal Grant Agency of Tomas Bata University under the project No. IGA/FAI/2012/037.

References

1. Kennedy, J., Eberhart, R.: Particle Swarm Optimization. In: Proceedings of IEEE International Conference on Neural Networks. IV, pp. 1942–1948 (1995)
2. Dorigo, M.: Ant Colony Optimization and Swarm Intelligence. Springer (2006)
3. Eberhart, R., Kennedy, J.: Swarm Intelligence, The Morgan Kaufmann Series in Artificial Intelligence. Morgan Kaufmann (2001)
4. Storn, R., Price, K.: Differential evolution—a simple and efficient heuristic for global optimization over continuous spaces. Journal of Global Optimization 11, 341–359 (1997)
5. Goldberg, D.E.: Genetic Algorithms in Search Optimization and Machine Learning, p. 41. Addison Wesley (1989) ISBN 0201157675
6. Nagraj, B., Subha, S., Rampriya, B.: Tuning algorithms for pid controller using soft computing techniques. International Journal of Computer Science and Network Security 8, 278–281.3 (2008)
7. Davendra, D., Zelinka, I., Senkerik, R.: Chaos driven evolutionary algorithms for the task of PID control. Computers & Mathematics with Applications 60(4), 1088–1104 (2010) ISSN 0898-1221
8. Nickabadi, A., Ebadzadeh, M.M., Safabakhsh, R.: A novel particle swarm optimization algorithm with adaptive inertia weight. Applied Soft Computing 11(4), 3658–3670 (2011) ISSN 1568-4946
9. Sprott, J.C.: Chaos and Time-Series Analysis. Oxford University Press (2003)
10. Astrom, K.: Control System Design. University of California, California (2002)
11. Landau, Y.: Digital Control Systems. Springer, London (2006)

Author Profile Identification
Using Formal Concept Analysis*

Martin Radvanský, Zdeněk Horák, Miloš Kudělka, and Václav Snášel

VSB Technical University Ostrava, Ostrava, Czech Republic
{martin.radvansky.st,zdenek.horak.st,
milos.kudelka,vaclav.snasel}@vsb.cz

Abstract. This paper presents results of the finding of the author's profiles using formal concepts generated from DBLP database. Our main aim was to evaluate the use of formal concept analysis as a method for extracting the author's profiles. There are several commonly used methods for clustering and for finding experts in a large database. These methods are mainly based on different kinds of clustering and metrics which are sometimes difficult to understand. Finding experts for particular and mainly special areas of research is not an easy task. Formal concept analysis (FCA) is a method with a very strong mathematical background, which makes it easy to understand. Properties of FCA can give us a very strong tool for finding author's profiles.

Keywords: DBLP, author profile, formal concept analysis, concept stability.

1 Introduction

Digital Bibliography & Library Project (DBLP) is one of the most known collections of electronic resources which can be accessed over the Internet. This project was founded in 1993 and contains, among other things, more than 1,800,000 papers. These papers come from computer science and were published in different journals and conference proceedings. Although DBLP is primarily used for finding publication in the library, this fast increasing database is often used by researchers as a good dataset for data mining tasks, such as finding experts, recommendation systems, social networks algorithms, etc. However, the DBLP contains only a limited amount of information about particular papers - there are no abstracts or index terms stored in it. On the other hand, DBLP provides a lot of information about the publication activity of authors, conferences and author relationships. There was a lot of research done to find experts, extract

---

* This paper has been elaborated in the framework of the IT4Innovations Centre of Excellence project, reg. no. CZ.1.05/1.1.00/02.0070 supported by Operational Programme 'Research and Development for Innovations' funded by Structural Funds of the European Union and state budget of the Czech Republic, this work was also supported by the Bio-Inspired Methods: research, development and knowledge transfer project, reg. no. CZ.1.07/2.3.00/20.0073 funded by Operational Programme Education for Competitiveness, co-financed by ESF and state budget of the Czech Republic, and partially supported by SGS, VSB-Technical University of Ostrava, Czech Republic, under the grants No. SP2012/58.

V. Snasel et al. (Eds.): SOCO Models in Industrial & Environmental Appl., AISC 188, pp. 485–494.
springerlink.com © Springer-Verlag Berlin Heidelberg 2013

their working areas, analyse communities in the social network based on DBLP and much more.

In this paper we have processed the DBLP in order to extract the author's profiles based on keywords used in their papers. The author's profiles can help us to find groups of keywords that were often and repeatedly used by authors in their titles of papers. The main expected result of our work is to find author's profiles. These profiles can be used for identification of experts for particular area of research.

This paper is organized as follows: Section 2 contains an overview of related work. Section 3 explains the methods used for data evaluation. Section 4 is focused on the finding of author's profiles. Last section 5 concludes the paper.

2   Related Work

Growing databases of documents, research papers and other document-oriented databases, during the last thirty years bring new challenges to the researchers. Many methods have been introduced that were focused on the fast searching, grouping and finding similar documents. In the following paragraphs we will review some of the most related approaches.

In [13] we can find an efficient algorithm for topic ranking. The authors show a method for the extraction of keyword sets and cluster of research papers using these keyword sets. The evolution of topics over time and their ranking is studied. Paper [4] covers the bibliometrics perspective. It investigates the frequency and impact of conference publications in computer science and compares it to journal papers. The author uses statistical methods for analysing DBLP. Paper [15] introduces alternative measures for ranking venues. They create new bibliometrics that can be used in ranking publication venues. These bibliometrics are easy to implement and bring more accuracy to the evaluation of venues. An application based on stability (a measure from formal concept analysis which is discussed later) can be found in [9]. In this paper the stability is used for pruning conceptual lattice which was constructed from the ECSC dataset. Analysis of the DBLP publication and their classification by using Concept lattices can be found in [1]. This paper shows how concept lattice can cover relational and contextual information of analysed papers.

Our approach is inspired by the previous research, but we have tried to address several issues in a different way. In this paper we try to search keywords that are used by authors periodically and frequently. This set of keywords is included in the intent of concepts. Because the set of concepts is greater than number of keywords, we have used the approach based on concept stability. This is the main difference of our approach and previously mentioned related work. In the next section we summarize used tools and techniques.

3   Applied Methods and Data Collection

This section provides some basic notions and techniques applied in our experiments.

3.1   Formal Concept Analysis

In our paper we use Formal concept analysis as a technique for unsupervised clustering. This method helped us to find non-trivial clusters of authors and their keywords. In the next paragraph we briefly describe Formal concept analysis.

Formal concept analysis (FCA) is a general data analysis method based on the lattice theory. FCA was introduced in 1982 by Wille [14]. The basic algorithms for concept lattice computation were published by Ganter in 1984 [5]. More recent publications of these founders can be found in ([6], [7], [8]). Carpineto and Romano summarized in ([2], [3]), both the mathematical and computer scientist's (with a focus on information retrieval) perspective of the FCA. A good overview of the recent state was written also by Priss in [11].

The input data for FCA is called formal context C, which can be described as $C = (G, M, I)$ - a triplet consisting of a set of objects G and set of attributes M, with I as relation of G and M. The elements of G are defined as objects and the elements of M as attributes of the context.

As an example of using FCA we have selected five authors a_1,..., a_5 and five keywords that were often used by these authors in the title of their papers. These keywords are "database - k_1", "algorithm - k_2", "distributed - k_3", "data mining - k_4" and "analysis - k_5". The relation between author and keyword is shown as a cross in the Table 1.

Table 1. Formal context

	database k_1	algorithm k_2	distributed k_3	data mining k_4	analysis k_5
author a_1	×	×	×		×
author a_2	×	×		×	
author a_3	×		×		
author a_4	×			×	×
author a_5		×		×	

Density of the formal context (G, M, I) is defined as proportion of elements of I with respect to the size of GM. The density calculated for the context depicted in the Table 1 is 56%.

For a set $A \subseteq G$ of objects we define $A^\uparrow$ as the set of attributes, common to the objects in A. Correspondingly, for a set $B \subseteq M$ of attributes we define $B^\downarrow$ as the set of objects which have all attributes in B. A formal concept of the context (G, M, I) is a pair (A, B) with $A \subseteq G$, $B \subseteq M$, $A^\uparrow = B$ and $B^\downarrow = A$. The set A is called extent of a concept, while the set B is called intent of a concept. $\mathscr{B}(G, M, I)$ denotes the set of all concepts of context (G, M, I) and forms a complete lattice (so-called Galois lattice). For more details see ([7], [8]). All concepts from our example are shown in the Table 2. The figure 1 depicts concept lattice of our example.

For selection of interesting concepts we have used a method based on concept stability that is described in the next section.

Table 2. Formal concepts extracted from context in Table 1

concept	extent	intent
c(0)	$\{a_1, a_2, a_3, a_4, a_5\}$	$\{\}$
c(1)	$\{a_2, a_4, a_5\}$	$\{k_4\}$
c(2)	$\{a_1, a_2, a_5\}$	$\{k_2\}$
c(3)	$\{a_2, a_5\},$	$\{k_2, k_4\}$
c(4)	$\{a_1, a_2, a_3, a_4\}$	$\{k_1\}$
c(5)	$\{a_1, a_4\}$	$\{k_1, k_5\}$
c(6)	$\{a_2, a_4\}$	$\{k_1, k_4\}$
c(7)	$\{a_4\}$	$\{k_1, k_4, k_5\}$
c(8)	$\{a_1, a_3\}$	$\{k_1, k_3\}$
c(9)	$\{a_1, a_2\}$	$\{k_1, k_2\}$
c(10)	$\{a_2\}$	$\{k_1, k_2, k_4\}$
c(11)	$\{a_1\}$	$\{k_1, k_2, k_3, k_5\}$
c(12)	$\{\}$	$\{k_1, k_2, k_3, k_4, k_5\}$

3.2 Concept Stability

The main problem of using FCA as a clustering method is that we often obtain very large and complicated structure, which is hard to understand and interpret. Technically speaking, we can get a large number of concepts even for a relatively small context. There are several methods which can be used to select only some part of concepts. We have used the so-called concept stability to filter only the interesting ones. As an interesting concept we considered a concept which is, up to a certain degree, resistant to the change of a particular object (removing particular object does not cause the change of the intent).

Stability of a concept (introduced by Kuznetsov in [9]) expresses the dependency between the intent and extent of the concept. Following the notions from [10], for a particular concept (A, B) of a concept lattice $\mathscr{B}(G, M, I)$, the stability is defined as:

$$\sigma(A, B) = \frac{|\{C \subseteq A | C^\uparrow = B\}|}{2^{|A|}} \quad (1)$$

Higher stability causes higher immunity of concept to changes in particular objects. An efficient way to compute the stability of all concepts (using a bottom-up lattice traversal) is described in [12].

As a continuation of our example we can compute stability of concept C(8) by using Equation (1) as:

$$\sigma(\{a_1, a_3\}, \{k_1, k_3\}) = \frac{2}{2^2} = \frac{1}{2} \quad (2)$$

3.3 Pre-processing of Data Collection

On December 12, 2011, we downloaded the DBLP dataset in XML[1] and pre-processed it for further usage. First of all, we selected journal volumes and conferences held by

---

[1] Available from http://dblp.uni-trier.de/xml/

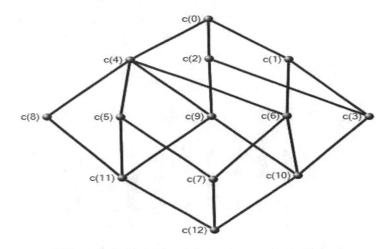

Fig. 1. Concept lattice created from concepts in the Table 2

IEEE, ACM and Springer. For every record we identified the month and year of the publication. In the next step, we extracted all authors having at least one published paper (11,355 authors) in a selected period. Then, we extracted keywords and phrases from paper titles. The approach was based on Faceted DBLP set[2]; 1,134 keywords and phrases we used in total.

For our paper we have selected a time period up to the year 2010 to get the most complete dataset. Then, we divided the entire recorded publication period of conferences into one-month time periods. If during one month an author has published a paper then we set keyword records, corresponding to the paper title. For each author we obtained a list of months with occurred keywords. To reduce the number of authors we have used secondary filtering based on occurrences of keywords during the selected period. From a set of 11,355 authors, we have selected only authors with more than 2 keywords in theirs papers (1,735). Figure 2 display histogram of number of keywords in author's papers.

For the following evaluation we constructed binary formal context. The rows of context represent authors of papers and columns correspond to the keywords used in particular paper in DBLP. The value of an intersection between row and column contain value "1" when author used keyword or "0" otherwise.

4   Searching for Author's Profiles

This section describes our method for searching author's profiles.

Author's profile by meaning is used in this paper as a set of characteristic keywords. These keywords were used often and repeatedly by an author in his papers during an observed period. Groups of authors with the same profile, can be seen as a group of experts in the particular research area, covered by profile keywords.

---

[2] http://dblp.l3s.de/browse.php?browse=mostPopularKeywords

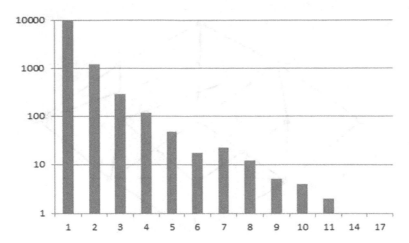

Fig. 2. Histogram of keywords used by authors in their papers

4.1   Basic Properties Data Collection

In order to create formal contexts from the described data collection, we have got a context containing 1,735 rows (authors) and 525 columns (keywords). The density of the context was about 5%. A small example of selected keywords are a set of "analysis, algorithms, applications, coding, testing, modeling, attacks, aggregation, logic, rdf, energy".

In the next section FCA is used as a main method for clustering and finding authors' profiles.

4.2   Finding Profiles of Authors by FCA

FCA gave us a tool for finding profiles of authors, based on the keywords they use in papers. For the context created in the previous step we have computed a concept lattice. We are interested in nontrivial concepts where author's profiles are the intents of these interesting concepts. In order to decide which concepts are interesting for us, we used the concept stability method, described in the section 3.2. This method helped us reduce the size of concept lattice and intent of concepts are more confident as the author's profile. For finding the right level of stability threshold, we have computed a number of concepts that satisfied the level of stability see figure 3.

Choosing the value of stability threshold has been done according to the basic meaning of stability (see section 3.2). Stability 0.5 gives us information that there exists one half subsets of all possible subsets of authors in the concept that has special property. Removing subset authors from concept extent (set of authors) does not cause a change in the intent (set of keywords) of concept. Higher value of concept stability makes concept more confident. We have used concept stability for pruning concept lattices. After applying pruning lattice by concept stability we obtained a relative small number of concepts which can be easily explored.

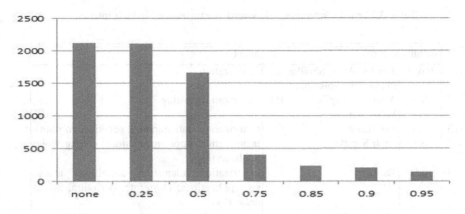

Fig. 3. Number of formal concepts reduced by different level of stability

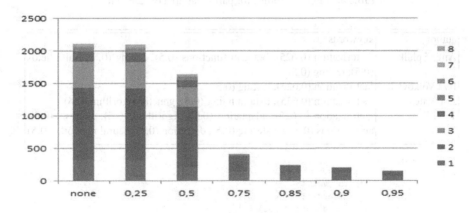

Fig. 4. Proportion of number of keywords for different level of stability

Figure 4 illustrates the proportion of number of keywords in the intent of computed concepts in dependence on different level of stability. According to the selected level of stability 0.5, the most interesting concepts for creating author's profiles were concepts with more than one attribute. Author's profiles are concepts that contain particular keywords in their intents.

As an example we have selected concepts having the keyword "factorization" in its intent, see Table 3.

This table depicts five concepts and theirs intents create the author's profiles. We can identify five sets of authors in the extent of these concepts who are experts in this particular area of research. In our example we have identified four authors and each of them have in his author's profile keyword "factorization" which is more confident than

Table 3. Formal concepts with keyword "factorization" in their intents

id	stability	extent	intent
c(1)	0.625	Tao Li, Amir Shpilka, Jan Platos, Ilya Volkovich	factorization
c(2)	0.5	Amir Shpilka, Ilya Volkovich	factorization, testing
c(3)	0.5	Jan Platos	factorization, data minning, genetic algorithm
c(4)	0.5	Amir Shpilka	factorization, boolean functions, coding, finite fields, testing
c(5)	0.5	Tao Li	factorization, adaptation, algorithms, analysis, applications, clustering, detection, neural networks

Table 4. Author's profiles for particular area of research

author	keywords
Amir Sphilka	factorization (0.625), boolean functions (0.5), coding (0.5), finite fields (0.5), testing (0.5)
Ilya Volkovich	factorization (0.625), testing (0.5)
Jan Platos	factorization (0.625), data minning (0.5), genetic algorithm (0.5)
Tao Li	factorization (0.625), adaptation (0.5), algorithms (0.5), analysis (0.5), applications (0.5), clustering (0.5), detection (0.5), neural networks (0.5)

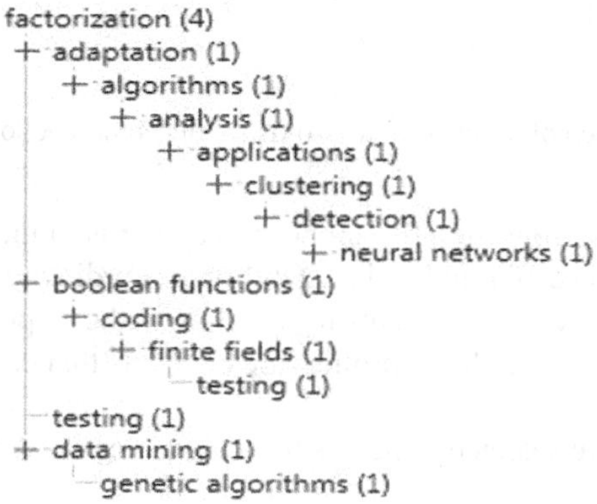

```
factorization (4)
 + adaptation (1)
    + algorithms (1)
       + analysis (1)
          + applications (1)
             + clustering (1)
                + detection (1)
                   + neural networks (1)
 + boolean functions (1)
    + coding (1)
       + finite fields (1)
          └ testing (1)
  ─ testing (1)
 + data mining (1)
    └ genetic algorithms (1)
```

Fig. 5. Hierarchy of keywords connected with keyword "factorization"

other keywords. This has to be done by meaning of concept stability (0.625 > 0.5) see Table 4. On similar bases we can extend author's profile of each author by the other keywords from the set of concepts.

In the figure 5 we can see hierarchical view on the set of concepts. The numbers next to the keywords give us information of occurrences these keywords have in concepts. This hierarchical structure is connected to the underlying conceptual lattice what was pruned by concept stability.

This example of author's profiles is just a small part of the whole process for creating author's profiles. For creating a full author profile we intersect all keywords in all particular author's profiles. As a result we will get all keywords usually and often used by an author in his papers.

5   Conclusion and Future Work

In this paper, we have introduced an approach for finding interesting author's profiles, based on keywords used in titles of papers in the DBLP database. Using of FCA together with stability of concepts helped us to find these profiles. The used method gave us a very interesting hierarchical view of the keywords as the author's profile. In our future work we plan to take a closer look at the evolution of author's profiles during the time and the cooperation of authors based on their profiles.

References

1. Alwahaishi, S., Martinovič, J., Snášel, V., Kudělka, M.: Analysis of the DBLP Publication Classification Using Concept Lattices. In: DATESO 2011, pp. 132–139 (2011)
2. Carpineto, C., Romano, G.: Concept Data Analysis. John Wiley and Sons, New York (2004)
3. Carpineto, C., Romano, G.: Using Concept Lattices for Text Retrieval and Mining. In: Ganter, B., Stumme, G., Wille, R. (eds.) Formal Concept Analysis. LNCS (LNAI), vol. 3626, pp. 161–179. Springer, Heidelberg (2005)
4. Franceschet, M.: The Role of Conference publications in CS. Communications of the ACM 53(12), 129–132 (2010)
5. Ganter, B.: Two Basic Algorithms in Concept Analysis. In: Kwuida, L., Sertkaya, B. (eds.) ICFCA 2010. LNCS, vol. 5986, pp. 312–340. Springer, Heidelberg (2010)
6. Ganter, B., Stumme, G., Wille, R.: Formal Concept Analysis. LNCS (LNAI), vol. 3626. Springer, Heidelberg (2005)
7. Ganter, B., Wille, R.: Applied Lattice Theory: Formal Concept Analysis. In: Grätzer, G.A. (ed.) General Lattice Theory, pp. 592–606, Birkhäuser (1997)
8. Ganter, B., Wille, R.: Formal Concept Analysis: Mathematical Foundations. Springer, Berlin (1999)
9. Kuznetsov, S.O.: On stability of a formal concept. In: Annals of Mathematics and Artificial Intelligence, vol. 49(1), pp. 101–115 (2007)
10. Kuznetsov, S.O., Obiedkov, S., Roth, C.: Reducing the Representation Complexity of Lattice-Based Taxonomies. In: Priss, U., Polovina, S., Hill, R. (eds.) ICCS 2007. LNCS (LNAI), vol. 4604, pp. 241–254. Springer, Heidelberg (2007)
11. Priss, U.: Formal concept analysis in information science. Annual Review of Information Science and Technology 40 (2006)

12. Roth, C., Obiedkov, S., Kourie, D.G.: Towards Concise Representation for Taxonomies of Epistemic Communities. In: Yahia, S.B., Nguifo, E.M., Belohlavek, R. (eds.) CLA 2006. LNCS (LNAI), vol. 4923, pp. 240–255. Springer, Heidelberg (2008)
13. Shubhankar, K., Singh, A.P., Pudi, V.: An Efficient Algorithm for Topic Ranking and Modeling Topic Evolution. In: Proceeding DEXA 2011 of the 22nd International Conference on Database and Expert Systems Applications, pp. 320–330. Springer, Berlin (2011)
14. Wille, R.: Restructuring lattice theory: an approach based on hierarchies of concepts. In: Rival, I. (ed.) Ordered Sets, pp. 445–470. Reidel, Dordrecht (1982)
15. Yan, S., Lee, D.: Toward Alternative Measures for Ranking Venues: A Case of Database Research Community. In: Proceedings of the 7th ACM/IEEE-CS Joint Conference on Digital Libraries, Vancouver, Canada, pp. 235–244 (2007)

Dynamic Tabu Search for Non Stationary Social Network Identification Based on Graph Coloring

Israel Rebollo Ruiz and Manuel Graña Romay

[1] Israel Rebollo Ruiz at Informática 68 Investigación y Desarrollo S.L.,
Computational Intelligence Group, University of the Basque Country
beca98@gmail.com
[2] Manuel Graña Romay at Computational Intelligence Group,
University of the Basque Country
ccpgrrom@gmail.com

Abstract. We introduce a new algorithm for the identification of Non Stationary Social Networks called Dynamic Tabu Search for Social Networks DTS-SN, that can analyze Social Networks by mapping them into a graph solving a Graph Coloring Problem (GCP). To map the Social Network into an unweighted undirected graph, to identify the users of the Social Networks, we construct a graph using the features that compound the Social Networks with a threshold that indicates if a pair of users have a relationship between them or not. We also take into account the dynamic behavior of the non stationary Social Network, where the relations between users change along time, adapting our algorithm in real time to the new structure of the Social Network.

Keywords: non stationary Social Networks, Dynamic Tabu Search, Graph Coloring.

1 Introduction

The Social Networks is a new phenomena that is grown very fast. Sometimes is necessary to control the social network by identifying users assigning them categories for different purposes. To make such categorization it is necessary to map the social network into a mathematical structure. The anthropologists have determined that a social network can be represented as a graph [1], using the graph theory to perform a clustering process to identify the actual social network. The use of Soft Computing to solve real life problems have been previously studied, like in building thermal insulation failures detection [2] or in identification of typical meteorological days [3].

The straightforward approach to model a social network as a graph is mapping it into a weighted graph, because the relation between users can depend on different factors that have different relevance. Opsahl [4] represent the social network as a graph, the vertices are users and the edges are the relationship between users, assigning different weight to the elements that link two users. For example, we can have two features linking people, the city and the friendship,

V. Snasel et al. (Eds.): SOCO Models in Industrial & Environmental Appl., AISC 188, pp. 495–503.
springerlink.com

if two people live in the same city they are related by this feature, but if two people live in different cities but are friends they also have a relation between them. We must assign weights to these relation to establish the weight of the edge between them in the graph representation.

The problem is to built a graph where all the relations between two users are represented with the proper weight. Zhao and Zang [5] introduce a new clustering method where the relationship between vertices are represented by an undirected and unweighted graph. They use hierarchical trees to establish an unweighted relation between users.

If we have a hierarchical relation between users we can apply specific method for clustering the users. Qiao [6] develop a Hierarchical Cluster Algorithm called HCUBE based on block modeling that can cluster Social Networks with difficult relations between users.

Another approach can be seen in Firat's [7]. In this work, we can see a new way of modeling the Social Network as a graph using a random-walk-based distance measure to find subgroups in the Social Network. Then, using a Genetic Algorithm, the authors make the group achieving a good result.

The application of clustering over a Social Network can be seen in Pekec's work [8]. In this work a role assignment is made for clustering the Social Network. The idea is not grouping users by their relationship, but their role in the Social Network. Lewis [9] studied the graph coloring problem, and used Social Network graphs to test various algorithms. The result is that the graph coloring can cluster these kind of Social Networks.

To identify a Social Network we can model it as a weighted or unweighted graphs. Applying a graph coloring algorithm we can solve the problem. However, a Social Network is a non stationary dynamic system so we need a dynamic approach to work with Social Networks. Dynamic Coloring of graph techniques appearing in the literature refer to a special type of graphs [10], not to be confused with the coloring of graphs whose structure of edges change along time.

The Graph Coloring Problem (GCP) is a classical NP-hard problem which has been widely studied [11,12,13,14,15]. There are a lot of algorithms dealing with it like Brelaz, Dutton or Corneil's [16,17,18], some of them using Ant Colony Optimization (ACO) [19] and Particle Swarm Optimization (PSO) [20]. The GCP consist in assigning a color to the vertices of a graph with the limitation that any pair of vertices linked by an edge cannot have the same color. This PSO approach have been used in other Soft Computing problem as appears in [21].

We present a new way of modeling a social Network as a Graph and a method based in the Tabu Search approach that works over a dynamically changing graph. The modeling allows to change the graph dynamically along time. The DTS-SN can return a clustering in every moment that correspond with the graph structure in that moment.

We map the social networks as undirected and unweighted graphs, in which the relation between vertices is defined as follows. We assign weights to different features of the relations between vertices. Then we are going to establish a threshold. If the relationship between two vertices has a weight greater than the threshold, then there is an edge between vertices. Therefore, we have a graph with edges establishing a relationship between vertices. Algorithms solving the GCP try to find collections of unconnected vertices, which is a dual problem of the vertex clustering problem, so we have to transform the social relation graph into its complementary graph. The complement (aka inverse) of a graph G is a graph H defined on the same vertices such that two vertices of H are adjacent if and only if they are not adjacent in G. Let $G = (V, E)$ be a simple graph and let K consist of all 2-element subsets of V. Then $H = (V, K \setminus E)$ is the complement of G.

On this graphs we are going to use a Tabu Search algorithm for graph coloring [22,23] which tries to find the minimum number of clusters for the network in a sequential and dynamic way. We use Tabu Search because it keeps a trace of the space visited by the algorithm and we can modify this trace when the relation graph change. As the solution can be in the visited space, without this trace we would need to revisit all the search space again. First, the algorithm is sequential, because we need to start assuming a big number of clusters, as far as we don't know the exact number of clusters . We reduce the number of clusters by one after each successful execution of the Tabu Search algorithm until we can solve the problem with that number of clusters in a given time or we reach to a given number of classes. Second, we introduce a variation into the Tabu Search algorithm that allows the relationship between vertices change. For that, we are going to remap the graph after each step of the algorithm. Then we are going to modify the Tabu tenure changing the Tabu value of any changing vertex relation with an unknown value. For that we need a special Tabu Search algorithm that can have unknown values in the Tabu tenure and also can have solutions repeated in the list.

The rest of the paper is organized as follow: section 2 presents our Dynamic Tabu Search Algorithm and the way we model the Social Network as a Graph. In Section 3 we show experimental results with random generated graphs that simulates a General Social Network, with a static behavior and also with a dynamic behavior. Finally, section 4 gives some conclusions and lines for future work.

2   Dynamic Tabu Search algorithm

In this section we are going to explain how we map a Social Network into a Graph. To apply our Dynamic Tabu Search (DTS-SN) algorithm for Social Networks we cluster the Social Network into the smallest number of user groups with the restriction of a given computational time. The aim of this algorithm is not to find the chromatic number , but to find a quick clustering of a dynamically changing network.

Let us have a Social Network called SN with I users and F features that establish a relation between users. $SN = \{I \mid F\}$. Let have a threshold U that is an scalar and represent the weight that must sum the features of a relationship between node to draw an edge between two users. We built a new graph G which have V vertex been $V = I$ and E edges such that

$$if \sum_F F(v_a, v_b) > U \ then \ \exists e(v_a, v_b) \ , \ e \in E, \ else \ not\exists e(v_a, v_b) \tag{1}$$

Where $e(v_a, v_b)$ is an edge between the vertex v_a and v_b. That way we have a graph $G = (V, E)$ that maps the Social Network SN. As SN can change dynamically, the graph G is also a Dynamic Graph so we need to recalculated E each time the SN change so we have a graph $G_t = (V, E_t)$ and the rule 1 is now depending of the time.

$$\exists e_{t_0}(v_a, v_b) \ if \sum_F F_{t_0}(v_a, v_b) > U \tag{2}$$

$$If \sum_F F_t(v_a, v_b) > U \ and \ \sum F_{t+1}(v_a, v_b) > U \ then \ \exists e_{t+1}(v_a, v_b) \ , \ e_{t+1} \in E \tag{3}$$

$$If \sum_F F_t(v_a, v_b) > U \ and \ \sum F_{t+1}(v_a, v_b) \le U \ then \ not\exists e_{t+1}(v_a, v_b) \tag{4}$$

We have to transform the Graph G or G_t into it's complementary graph $\overline{G}$ or $\overline{G_t}$. To follow a standard notation we call graph $\overline{G}$ graph G, and graph $\overline{G_t}$ graph G_t, but without forgetting that this graph is the complementary of the social relation graph.

We have developed a Tabu Search algorithm [24,22,23], and we are going to use it to solve the GCP. This algorithm works sequentially to reduce the number of colors used to solve the problem. The algorithm stops if it can't find a valid coloring after a given computational time has elapsed. In the Tabu tenure we keep a list of proper and improper solutions of the GCP. An improper solution is a solution in which not all the vertices have a color assigned, but if a vertex has a color assigned this is a valid color. The Tabu tenure has a special treatment because of the dynamic nature of the problem. Since the structure of the graph can change, the information keep in the Tabu tenure can become invalid, so we made a modification to the Tabu generation step. If the structure of the Graph change, we change the Tabu tenure deleting the assigned color to that pair of vertices in the list. Then me must go through the list to delete the repeated occurrences in the list.

It is important to say that the algorithm when detects a change in the social network, then it doesn't matter if the modification affects or not the validity of the solution. The algorithm change the Tabu tenure and continue searching the solution space.

Algorithm 1. Dynamic Tabu Search for Social Networks (DTS-SN)

Transform the Social Network into a Graph G
initialize maxiter,maxcolor,mincolor
let C = maxcolor
while C = mincolor and iter < maxiter
    let iter = 0
        while not solved(G) and iter < maxiter
            iter = iter + 1
            assign valid colors to G
            if in TabuList(G) then
                continue
            else
                Add to TabuList
            if G change then
                Change_TabuList()
        end while
    if solved(G) then
        Let C= C-1
end while

The algorithm returns a classification of the vertices $C = \{c_1, c_2, ..., c_k\} \mid \forall v \in V \ \exists c \to v \in C$, where C is the set of colors and k is the minimum number of colors find by the algorithm. The algorithm returns the first configuration that meets the GCP statement, but after that, an expert must supports this clustering. We can move the threshold up or down to get a more accurate clustering for the problem, but with this, we change the graph structure and the whole problem would have to be repeated.

3   Experimental Results

We have implemented our Dynamic Tabu Search algorithm for Social Networks DTS-SN in Microsoft Visual Studio 2005 Visual-Basic .Net. The graph generator is also implemented in the same platform. The experiments can be made in different machines so the computational time measured in second can change. The number of algorithm steps is an independent machine measurement that is useful to compare different experiments. The use of the same computer doesn't guarantee that the performance of the computer is the same for all the experiments (internal routines, disk memory swap, and other external events). Nevertheless, in this paper, we give also the computational time in seconds because the comparison between time measured in algorithm steps must be made with an important restriction: the number of clusters must be the same in all the experiments.

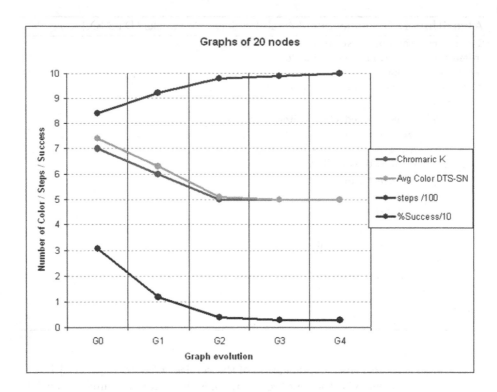

Fig. 1. Social Network of 20 users

We have built two simulated Social Networks (SN) with 20 and 50 users. We have use two features for all the SN: the city and the education degree. We have generated randomly the city of each user between 10 cities. We have assigned the education degree between 5 degrees. The influence of the city is 0 if the users don't live in the same city, 5 if the users live in the same city and 3 if the city of one user is near the city of the other user. The influence of the education degree is 5 if the users have the same degree, 3 if the degree of one user is one more degree or a one less degree, 1 if the degree of one user is two more degree of two less degree, and 0 is other case. The relation relevance threshold is set to 6. So there will be an edge between the nodes representing the users if two user have the same degree, and they belong to the same city or nearby cities, or the degree of one user is less or more one and they belong to the same city or nearby cities, or the users degree differ in two units and the belong to the same city.

We have change some relations aleatory generating 4 new graphs which chromatic number decreases. In the figure 1 we can see the results of the evolution of the social network clustering. In red we can see the chromatic number of each version of the associated graph obtained with a Backtracking algorithm. In light green we can see the average number of color obtained with our DTS-SN algorithm after 50 executions and a maximum of 5000 steps. In blue we can see the

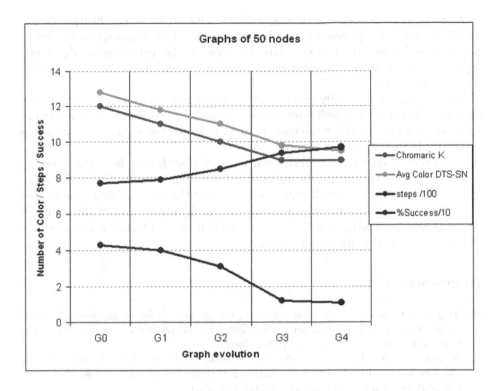

Fig. 2. Social Network of 50 users

average number of steps need to find the chromatic number divided by 100 to plot in the same figure. In Dark green we can see the average number of success divided by 10 for the same reason.

We can see that our algorithm evolves very quickly to the chromatic number. The number of steps is reducing from 3.100 steps in the initial graph to less than 400 if the last modified graph G4. The number of success also improves after each modification of the graph to a 100% of accuracy.

In figure 2 we can see that the average number of steps and the accuracy evolves in a similar way as in the 20 users social network. But in the Average number of colors we did not reach to the chromatic number as has happened in the previous experiment. But the evolution seems that we will find the correct solution after a big period of time. The results are an average after 50 executions, that means that we find the chromatic number in some cases.

4 Conclusions and Future Work

We have proposed a method to transform a Social Network into a Graph. This method helped by a threshold allows to built different graphs giving more emphasis to one feature or another. Then using the GCP to identify the social network using a Tabu Search algorithm.

As the social network can change, we have introduced a modification to our algorithm introducing into the graph structure the new relations in the social network, and delete the broken relations, allowing the Tabu Search algorithm, repair the Tabu tenure to adapt it to the new graphs structure and continue with the clustering.

We have show empirically that our algorithm can find a quick approach to the optimal solution of the graph representing the social network for small number of users social networks. When the number of users grows, the algorithm approaches very quickly to the optimal solution.

For future work, we want to test our algorithm with real data sets obtained from a specific environment, the users of a software system for Enterprise management, to enlarge the result and proof the algorithm over a more complicated scene. We also want to prepare a mathematical demonstration of the dynamic convergence of our algorithm. The approach can be also tested using more sophisticated algorithm for graph coloring.

References

1. Yegnanarayanan, V., UmaMaheswari, G.: Graph models for social relations. Electronic Notes in Discrete Mathematics 33(0), 101–108 (2009); International Conference on Graph Theory and its Applications.
2. Sedano, J., Curiel, L., Corchado, E., delacal, E., Villar, J.R.: A soft computing method for detecting lifetime building thermal insulation failures. Integrated Computer-Aided Engineering 17(2), 103–115 (2010)
3. Corchado, E., Angel Arroyo, V.T.: Soft computing models to identify typical meteorological days. Logic Journal of the IGPL 19(2), 373–383 (2011)
4. Opsahl, T., Panzarasa, P.: Clustering in weighted networks. Social Networks 31(2), 155–163 (2009)
5. Zhao, P., Zhang, C.Q.: A new clustering method and its application in social networks. Pattern Recognition Letters 32(15), 2109–2118 (2011)
6. Qiao, S., Li, T., Li, H., Peng, J., Chen, H.: A new blockmodeling based hierarchical clustering algorithm for web social networks. Engineering Applications of Artificial Intelligence 25(3), 640–647 (2012)
7. Firat, A., Chatterjee, S., Yilmaz, M.: Genetic clustering of social networks using random walks. Computational Statistics & Data Analysis 51(12), 6285–6294 (2007)
8. Pekec, A., Roberts, F.S.: The role assignment model nearly fits most social networks. Mathematical Social Sciences 41(3), 275–293 (2001)
9. Lewis, R., Thompson, J., Mumford, C., Gillard, J.: A wide-ranging computational comparison of high-performance graph colouring algorithms. Computers & Operations Research 39(9), 1933–1950 (2012)
10. Alishahi, M.: On the dynamic coloring of graphs. Discrete Applied Mathematics 159(2), 152–156 (2011)
11. Johnson, D.S., Mehrotra, A., Trick, M. (eds.) Proceedings of the Computational Symposium on Graph Coloring and its Generalizations, Ithaca, New York, USA (2002)
12. Johnson, D., Trick, M. (eds.): Proceedings of the 2nd DIMACS Implementation Challenge. DIMACS Series in Discrete Mathematics and Theoretical Computer Science, vol. 26. American Mathematical Society (1996)

13. Galinier, P., Hertz, A.: A survey of local search methods for graph coloring. Comput. Oper. Res. 33(9), 2547–2562 (2006)
14. Mehrotra, A., Trick, M.: A column generation approach for graph coloring. INFORMS Journal On Computing 8(4), 344–354 (1996)
15. Mizuno, K., Nishihara, S.: Toward ordered generation of exceptionally hard instances for graph 3-colorability. In: Discrete Applied Mathematics Archive, vol. 156, pp. 1–8 (January 2008)
16. Brèlaz, D.: New methods to color the vertices of a graph. Commun. ACM 22, 251–256 (1979)
17. Dutton, R.D., Brigham, R.C.: A new graph colouring algorithm. The Computer Journal 24(1), 85–86 (1981)
18. Corneil, D.G., Graham, B.: An algorithm for determining the chromatic number of a graph. SIAM J. Comput. 2(4), 311–318 (1973)
19. Ge, F., Wei, Z., Tian, Y., Huang, Z.: Chaotic ant swarm for graph coloring. In: 2010 IEEE International Conference on Intelligent Computing and Intelligent Systems (ICIS), vol. 1, pp. 512–516 (2010)
20. Hsu, L.Y., Horng, S.J., Fan, P., Khan, M.K., Wang, Y.R., Run, R.S., Lai, J.L., Chen, R.J.: Mtpso algorithm for solving planar graph coloring problem. Expert Syst. 38, 5525–5531 (2011)
21. Shi-Zheng Zhao, M., Willjuice Iruthayarajan, Zhao, S.-Z., Iruthayarajan, M.W., Baskar, S., Suganathan, P.N.: Multi-objective robust pid controller tuning using two lbests multi-objective particle swarm optimization. Information Science 181(16), 3323–3335 (2011)
22. Rebollo, I., Graña, M.: Gravitational Swarm Approach for Graph Coloring. In: Pelta, D.A., Krasnogor, N., Dumitrescu, D., Chira, C., Lung, R. (eds.) NICSO 2011. SCI, vol. 387, pp. 159–168. Springer, Heidelberg (2011)
23. Rebollo, I., Graña, M.: Further results of gravitational swarm intelligence for graph coloring. In: Nature and Biologically Inspired Computing (2011)
24. Rebollo, I., Graña, M., Hernandez, C.: Aplicacion de algoritmos estocosticos de optimizacion al problema de la disposicion de objetos no-convexos. Revista Investigacion Operacional 22(2), 184–191 (2001)

Immunity-Based Multi-Agent Coalition Formation for Elimination of Oil Spills

Martina Husáková

University of Hradec Králové, Faculty of Information Technologies,
Department of Information Technologies, Rokitanského 62,
Hradec Králové, Czech Republic
martina.husakova.2@uhk.cz

Abstract. Occurrence of oil spills is a serious ecological problem which negatively influences the environment, especially water ecosystems. It is necessary to use efficient approaches that can reduce this danger as fast as possible. Multi-agent coalition formation is investigated in conjunction with the immunity-based algorithm CLONALG-Opt for elimination of oil spills.

Keywords: Coalition, agent, immunity, lymphatic system, CLONALG-Opt.

1 Introduction

Instability of water ecosystems can be caused by the presence of oil spills. Oil spills contain dangerous chemicals which have negative influence on the development, physiology, immune system or reproduction of organisms living in (or near) the water ecosystem. It is often necessary to allocate a finance capital for elimination of oil spills. Costs are related to research or development of new approaches and technologies of reduction of this danger.

There are several methods typically used for cleaning of oil spills on the sea: self-cleaning and evaporation (natural processes), usage of booms or skimmers (for restriction the spreading of oil spills), usage of autonomous robots based on the methods of swarm intelligence [1], [2], [3].

It is obvious that presence of oil spills is a very serious problem. It is necessary to find efficient, ecological and economical solution. This paper investigates this problem with the usage of immunity-based multi-agent coalition formation, i. e. with the application of the clonal selection-based algorithm CLONALG-Opt [12], [13].

2 Multi-Agent Coalition Formation

The multi-agent coalition formation is one of the cooperation methods of artificial autonomous agents in the multi-agent system (MAS). The coalition is a goal-oriented and short-lived group of agents solving a specific pre-defined problem. Coalition can consist of cooperative (pro-social) agents that try to maximize the benefit of the group. The own benefit is not so important for them in contrast to self-interested (competitive) agents.

V. Snasel et al. (Eds.): SOCO Models in Industrial & Environmental Appl., AISC 188, pp. 505–514.
springerlink.com

Generation of an optimal coalition structure is one of the actual research challenges in the multi-agent coalition formation. The coalition structure is a group of different combinations of coalitions. Each coalition of agents solves a specific problem. The goal is to find efficient algorithms that are able to search the space of solutions with computational efficiency and minimal time consumption. Market-based approaches [5], graph-based techniques [6], [7], dynamic programming [8] or evolutionary techniques [9], [10] can be used for solution of this problem.

The problem of optimal coalition structure generation has not been sufficiently investigated by a different biology-inspired paradigm – artificial immune systems (AIS) resulting from mechanisms of biological immune systems (BIS).

3 Analogy between Lymphatic Nodes and Elimination of Oil Spills

BIS is a complex system which maintains the stability (homeostasis) of every living organism. Recognition of dangerous objects (antigens) is the main function of BIS. Different immune cells and organs are used for the elimination of antigens. The immune cell can be perceived as a pro-social biological agent that deals with problems for the sake of the BIS as a whole. Lymphatic nodes are powerful units responsible for elimination of antigens. Immune cells are able to form clusters in the lymphoid nodes for elimination of the danger, see Fig. 1 (a figure of a lymphoid node [15]). They are able to cooperatively solve a specific problem with others immune cells on the basis of specific stimuli. They use cellular signalling pathways which ensure transferring information from the source to the receiver. The similar clusters are necessary for elimination of oil spills by skimmers. The cooperation between them can lead to the faster problem solution. There is a potential for research of useful relations between multi-agent coalition formation, (artificial) immune systems and elimination of oil spills.

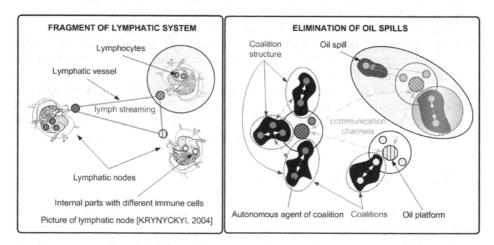

Fig. 1. Analogy between network of lymphatic nodes and elimination of oil spills

4 Artificial Immune System

Artificial immune systems (AIS) are perceived as a novel soft computing paradigm that is inspired by behaviour of BIS [17]. Different definitions of AIS exist. One of them is mentioned in [11]: *"Artificial immune systems (AIS) are adaptive systems, inspired by theoretical immunology and observed immune functions, principles and models, which are applied to problem solving."*

Four groups of immunity-based algorithms are used in present: gene library-based algorithms (bone marrow models), population-based algorithms, network-based algorithms, danger theory-based algorithm (dendritic cell algorithm).

Research is focused on the specification of the new immunity-based algorithms, improvement of these algorithms, usage of these algorithms in the new application areas, deeper theoretical study of AIS and development of hybridized AIS using "different" soft computing paradigms [4], [18], [19].

This paper investigates the population-based algorithm CLONALG-Opt. This algorithm is inspired by the clonal selection principle which explains the process of antibodies (B-lymphocytes) generation. Algorithm CLONALG was initially proposed for the pattern recognition [12], [13]. Optimized version of CLONALG (CLONALG-Opt) is mainly used for optimization tasks [13].

5 Proposed Solution

Elimination of oil spills is perceived as a multi-agent coalition formation problem in this paper. Algorithm CLONALG-Opt is proposed for the problem of generation coalition structures which can be applied for solving the above mentioned environmental problem.

5.1 Formal Model

Multi-agent system consists of a set of autonomous pro-social agents Cleaners $A = \{a_1, a_2, ..., a_m\}$. These ones are used for mitigation of the environmental danger. The agent a is evaluated by points b which are received thanks to the elimination of oil spills. The goal of the agent Cleaner is to collect maximum points for the elimination of oil spills. The agent has a collection of specific abilities that is represented as an ordered n-tuple of binary values $s = (s_1, s_2, ..., s_x)$ which are randomly generated. The agent uses an ordered n-tuple of filters for elimination of oil spills $f = (f_1, f_2, ..., f_y)$. Each filter f has a level of attrition o representing the amount of oil fragments that can be eliminated by the agent. This level is randomly generated for each agent. If the agent has a filter with high degree of attrition, the filter has to be cleaned or replaced.

Continuous oil spills (tasks) are represented in the set $U = \{OS_1, OS_2, ..., OS_n\}$. Each continuous oil spill is divided into the fragments (subtasks) os. Each oil fragment corresponds with the one type t of oil spill according to the level of its density. The types of oils spills are represented in the set $T = \{t_1, t_2, ..., t_n\}$. Each oil fragment is evaluated by points v. Agent receives more points for elimination of thicker oil spills because they are classified as more dangerous in comparison to the lighter oil spills (they spread more slowly and go down faster). The same property v is assigned for the type of subtask t.

Coalition c is used for the allocation of useful autonomous agents which eliminate specific oil spill of the only one type according their abilities and properties in this paper. This paper deals with only three non-overlapping coalitions, because three types of oil spills are distinguished (light, medium and heavy) in this paper. Coalition structure CS is a potential solution of the problem (antibody) that is divided into coalitions.

5.2 Optimization Model

Value of coalition structure $V(CS)$ depends on values of coalitions $v(c)$. Values of coalitions depend on the qualities of agents Cleaners. Quality of agent $v(a)$ depends on its ability to fulfil the task $a.s$, a detrition of a filter $f.o$ and points for the one type of subtask $t.v$. Fitness of an i-th agent performing a h-th type of task is calculated according to the equation (1).

$$v(a_i. t_h) = a_i. s_j \times f_r. o \times t_h. v \tag{1}$$

Stable coalition is defined as a set of agents that is able to accumulate minimal required amount of sources. Minimal required amount of sources is given by subtasks (fragments of oil spills) that have to be accomplished by agents Cleaners. Goal of the multi-agent coalition formation is to keep coalitions in a stable state as long as possible. It is necessary to ensure fast reaching of the goal without long time delays. Long-lasting stability of coalitions is supplied with the maximum amount of sources of agents Cleaners. Formula (2) should be valid for each stable coalition. Symbol $t_h. res$ denotes the minimal required amount of sources that is given by the h-th type of subtask t.

$$\forall\, v(c) \in N_0 : v(c_l) \geq t_h. res \tag{2}$$

Quality of coalition is specified as a sum of qualities of agents Cleaners that perform the specific subtask (e. g. an agent Cleaner eliminates an oil spill with the medium level of density). Quality of coalition structure can be represented as a tuple with coalition values $V(CS_k) = v(c_1), v(c_2), ..., v(c_n)$ or as a sum of coalition values, see equation (3).

$$V(CS_k) = \sum_{l=1}^{n} v(c_l) \tag{3}$$

Objective function is defined in the view of stable coalitions, see equation 4. Symbol x denotes the amount of oil spills of the specific type t (e. g. the amount of oil spills with medium density).

$$CS^* = \max \left(V(CS_k) - \sum_{h=1}^{n} x * t_h. v \right) \tag{4}$$

5.3 Formal and Conceptual Representation of Processes of CLONALG-Opt

Particular steps of the algorithm CLONALG-Opt are described e. g. in [14]. The following paragraphs are focused on the crucial steps of this algorithm, i. e.: representation and initialization of potential solutions, quality calculation of potential solutions, clonal selection process and affinity maturation.

5.3.1 Representation of Potential Solutions and Initialization

The coalition structure is represented as an antibody (a potential solution) and structured into the coalitions. Each coalition is created for fulfilment of the one specific type of subtask. Array of integers is used for the coalition structure representation. Each integer represents identification value of an agent Cleaner. The array of integers has a stable length. Each coalition has at least one agent Cleaner, because coalition without agents is not useful. This paper deals with the static multi-agent coalition formation. Number of agents Cleaners and types of subtasks are specified by the user and do not change during the calculation of the $V(CS)$. Length of antibody (L) is specified according to the equation (5) (a symbol m - the number of agents Cleaners involved in the coalition structure generation, a symbol n - the number of types of subtasks). Positions of coalitions are predefined in the coalition structure. The identity of coalition is defined for this purpose, see Fig. 2.

$$L = (m - (n - 1)) \times n \tag{5}$$

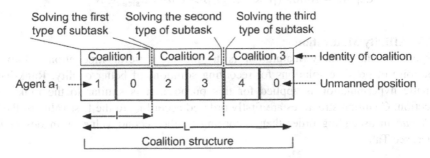

Fig. 2. Representation of potential solution with four agents

5.3.2 Quality of Potential Solution

Quality of coalition structure is represented as a list FCS of two values. The first value cuc represents number of unstable coalitions. The second one es represents the amount of sources allocated by agents Cleaners extra in comparison to the minimal requirements, see equation (6).

$$es = \sum_{l,h=1}^{n} (v(k_l) - t_h . res) \tag{6}$$

5.3.3 Clonal Selection

The operator of clonal selection is used for selection of high quality coalition structures. Rank-based selection is applied for this purpose. Coalition structures are firstly ordered according to the first value of the list *FCS cuc* in descending order, then according to the second value *es* in ascending order, see Tab. 1.

Table 1. Demonstration of rank-based selection for clonal selection

Coalition structure	Cuc	Es	Order	Probability of selection (%)
CS_1	2	6	1	10%
CS_2	1	10	2	20%
CS_3	0	17	3	30%
CS_4	0	20	4	40%

5.3.4 Clonal Expansion

The operator of clonal expansion is perceived as a reproduction operator with the goal of spreading of high quality solutions. Offsprings are generated for high quality potential solutions according to the equation (7) [14] (a symbol $CS_i.Nc$ - number of clones generated by the i-th coalition structure, a symbol β - clone factor, R_{size} – size of repertoire with potential solutions).

$$CS_i.Nc = round\ (\beta \times R_{size}), \beta > 0\ and\ R_{size} \in N \tag{7}$$

5.3.5 Affinity Maturation

Purpose of the affinity maturation is to include "useful innovations" into the repertoire of possible solutions for receiving solutions of better quality. Rank-based somatic hypermutation is applied for this purpose. It is similar to the rank-based selection. Coalition structures are firstly ordered according to the first value of the list *FCS cuc* in ascending order then according to the second value *es* in descending order, see Tab. 2.

Table 2. Demonstration of rank-based mutation for affinity maturation

Coalition structure	Cuc	Es	Order	Probability of mutation (%)
CS_1	0	20	1	10%
CS_2	0	17	2	20%
CS_3	1	10	3	30%
CS_4	2	6	4	40%

6 Experiments

NetLogo is a Java-based modelling and simulation tool for multi-agent systems. It is well suited for modelling complex problems related to biology, physics, chemistry, sociology, medicine or computer science. NetLogo is used for the implementation of the algorithm CLONALG-Opt [16]. BehaviourSpace is a build-in tool of the NetLogo which is applied for realization of experiments. This environment is utilized with the goal of observing the behaviour of the algorithm and verification of the application of algorithm for the multi-agent coalition formation and elimination of oil spills.

The goal of the experiments is to observe the behaviour of the CLONALG-Opt for different number of agents Cleaners and the size of repertoire with possible solutions at the usage of the constant seed value, see Tab. 3 with results. There are specific symbols in tables: $n_{cleaners}$ (a number of agents Cleaners), n_{rep} (a size of repertoire), n_{gen} (a number of generations), cf (a clonal factor), T_{min} (a duration of the partial experiment), WSF V(CS) (the worst possible solution founded so far), BSF V(CS) (the best possible solution founded so far).

The experiments use 6, 8 and 10 agents Cleaners. 50 and 40 generations are predefined by user with clonal factor 0.1 and a seed value 300. Size of repertoire is 10, 20 and 30 antibodies. More fluctuations appear in case of WSF value. BSF value is stabilized on: 66 units (6 agents Cleaners), 116 units (8 agents Cleaners) and 107 units (10 agents Cleaners) in two cases. Results of the experiments show that the algorithm CLONALG-Opt can be used for the multi-agent coalition formation. The algorithm is able to offer solution that can be used for elimination of oil spills. The algorithm CLONALG-Opt is not time-consuming, see Tab. 3. The following figures depict the best values of the coalition structures found in each generation. The Fig. 3 depicts the course of the algorithm CLONALG-Opt for different repertoires with 6 agents Cleaners. The Fig. 4 depicts the course of the algorithm for different repertoires with 8 agents Cleaners. The Fig. 5 depicts the course of the algorithm for different repertoires with 10 agents Cleaners.

Table 3. Results of the experiments

$n_{cleaners}$	Algorithm Clonalg-Opt			
	Parametres: n_{gen} = 50, 40*, cf = 0,1, seed = 300			
	n_{rep}	T_{min}	WSF V(CS)	BSF V(CS)
6	10	1	[0 51]	[0 66]
	20	2	[0 44]	[0 66]
	30	4*	[0 38]	[0 66]
8	10	1	[0 47]	[0 116]
	20	2	[0 45]	[0 116]
	30	5*	[0 72]	[0 116]
10	10	1	[0 44]	[0 107]
	20	1	[0 60]	[0 106]
	30	5*	[0 74]	[0 107]

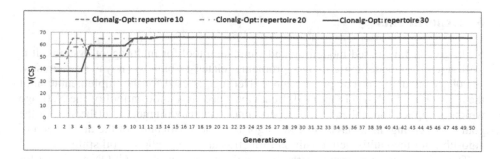

Fig. 3. Course of algorithm CLONALG-Opt for 6 agents Cleaners with different repertoires

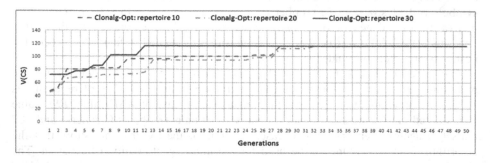

Fig. 4. Course of algorithm CLONALG-Opt for 8 agents Cleaners with different repertoires

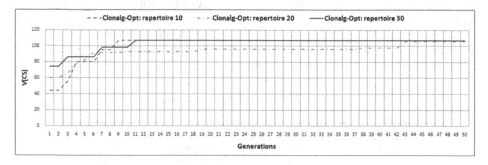

Fig. 5. Course of algorithm CLONALG-Opt for 10 agents Cleaners with different repertoires

7   Data Analysis

Statistical significance is investigated for uncovering possible differences in behaviour of the algorithm for different numbers of agents Cleaners. The IBM SPSS Statistics (ver. 20) is used for data analysis of experiments.

Kruskal-Wallis non-parametric test is used. Statistical significance is found out on the significance level α 5% for $n_{cleaners}$ 8 and 10, see Tab. 4. The Mann-Whitney post hoc test with Bonferroni correction is applied for discovering statistical significance between concrete groups of data. Statistical significance is find out between groups of n_{rep} 10 and n_{rep} 30 for $n_{cleaners}$ 8, between groups of n_{rep} 10 and n_{rep} 20 for $n_{cleaners}$ 10, between groups of n_{rep} 20 and n_{rep} 30 for $n_{cleaners}$ 10, see Tab. 5.

Table 4. Non-parametric test Kruskal-Wallis

	6 agents Cleaners	8 agents Cleaners	10 agents Cleaners
Chi-Square	0,259	17,206	63,964
Asymp. Sig.	0,878	0,000	0,000
p-value (Monte Carlo)	0,880	0,000	0,000
Decision	H_0	H_1	H_1

Table 5. Non-parametric test Mann-Whitney with Bonferroni correction

	8 agents Cleaners			10 agents Cleaners		
	10R/20R	10R/30R	20R/30R	10R/20R	10R/30R	20R/30R
Mann-Whitney	1013,5	868,5	1208	352	1222	316
Asymp. Sig.	0,090	0,003	0,964	0,000	0,773	0,000
Decision	H_0	H_1	H_0	H_1	H_0	H_1

8 Conclusion

The paper presents the usage of the immunity-based algorithm CLONALG-Opt for elimination of oil spills. This problem is perceived as a multi-agent coalition formation problem in this paper. CLONALG-Opt algorithm is analysed and designed for multi-agent coalition formation. NetLogo environment is used for the implementation of the algorithm for verification of its applicability for multi-agent coalition formation and elimination of oil spills. Experiments are realised with the aid of build-in NetLogo-based tool BehaviourSpace. Purpose of these experiments is not to optimize parameters of the algorithm, but observe the behaviour of the algorithm for different settings of the parameters. Experiments prove the applicability of the CLONALG-Opt algorithm for the multi-agent coalition formation and elimination of oil spills. Experiments demonstrate that the CLONALG-Opt algorithm is not time-consuming for used parameters. Two non-parametric tests are used for the data analysis – Kruskal-Wallis and Mann-Whitney with Bonferroni correction for analysis of statistical significance of data presented by experiments. It is find out that the algorithm CLONALG-Opt performs differently for various numbers of agents Cleaners, but it is necessary to do more experiments with different settings.

The paper presents initial study that connects immunity-based multi-agent coalition formation with serious environmental problem. The future research is firstly aimed at the investigation of the others immunity-based algorithms for the elimination of oil spills and their comparison with each other and with genetic algorithms. The second part of the future research is focused on the dynamic multi-agent coalition formation and application of relevant immunity-based algorithms for this purpose.

References

1. MIT: MIT researchers unveil autonomous oil-absorbing robot. In: MIT Media Relations (2010), http://web.mit.edu/press/2010/seaswarm.html
2. Fritsch, D., Wegener, K., Schraft, R.D.: Sensor concept for robotic swarms for the elimination of marine oil pollutions. In: Proceedings of the Joint Conference on Robotics. ISR 2006, The 37th International Symposium on Robotics, vol. 156(1-3), pp. 880–887 (2006)
3. Turan, O., et al.: Design and Operation of Small to Medium scale Oil-spill Cleaning Units. In: Proceedings of the International Conference on Towing and Salvage of Disabled Tankers Safetow, United Kingdom (2007)
4. Gan, Z., Li, G., Yang, Z., Jiang, M.: Automatic Modeling of Complex Functions with Clonal Selection-based Gene Expression Programming. In: The Third International Conference on Natural Computation (ICNC 2007), vol. 4, pp. 228–232 (2007)
5. Dias, M.B., et al.: Market-Based Multirobot Coordination: A Survey and Analysis. Robotics Institute, Carnegie Mellon University, Pittsburgh, Pennsylvania (2005)
6. Sandholm, T., et al.: Anytime Coalition Structure Generation with Worst Case Guarantees. In: AAAI 1998/IAAI 1998 Proceedings of the Fifteenth National Conference on Artificial Intelligence/Innovative Applications of Artificial Intelligence, pp. 46–53. American Association for Artificial Intelligence, Menlo Park (1998)
7. Sandholm, T., et al.: Coalition Structure Generation with Worst Case Guarantees. In: Artificial Intelligence, vol. 111(1-2), pp. 209–238 (1999)
8. Rahwan, T.: Algorithms for Coalition Formation in Multi-Agent Systems. University of Southhampton. Dissertation Thesis, p. 132 (2007)
9. Sen, S., Dutta, S.: Searching for optimal coalition structures. In: Proceedings of Fourth International Conference on Multi Agent Systems, pp. 287–292 (2000)
10. Ahmadi, M., Sayyadian, M., Rabiee, H.R.: Coalition Formation for Task Allocation via Genetic Algorithms. In: Advances in Information and Communication Technology. Springer, Germany (2002)
11. De Castro, L., Timmis, J.: Artificial Immune Systems: A New Computational Intelligence Approach, p. 398. Springer (2002)
12. Castro, L.N., Zuben, F.J.: Learning and Optimization Using the Clonal Selection Principle. IEEE Transaction on Evolutionary Computation 6(3), 239–251 (2002)
13. Castro, L.N., Zuben, F.J.: The Clonal Selection Algorithm With Engineering Application. In: Proceedings of the Genetic and Evolutionary Computation Conference (GECCO), pp. 36–37. Morgan Kaufmann, San Franciso (2002)
14. Castro, L.N.: Fundamentals of natural computing: basic concepts, algorithms, and applications, 1st edn., p. 696. Chapman and Hall/CRC (2006)
15. Krynyckyi, B.R., et al.: Clinical Breast Lymphoscintigraphy: Optimal Techniques for Performing Studies, Image Atlas, and Analysis of Images. Radiographics 24, 121–145 (2004), http://radiographics.rsna.org/content/24/1/121.full
16. Husáková, M.: Multi-Agent Coalition Formation – Algorithm Clonalg-Opt. (2012), http://edu.uhk.cz/~fshusam2/ClonalgOpt-final.html
17. De Castro, L.N., Timmis, J.I.: Artificial immune systems as a novel soft computing paradigm. In: Soft Computing, vol. 7, pp. 526–544. Springer (2003)
18. Hajela, P., Yoo, J., Lee, J.: GA based simulation of immune networks – applications in structural optimization. Journal of Engineering Optimization (29), 131–149 (1997)
19. Nasaroui, O., Gonzalez, F., Dasgupta, D.: The Fuzzy Artificial Immune System: Motivations, Basic Concepts, and Application to Clustering and Web Profiling. In: Proceedings of the 2002 IEEE International Conference on Fuzzy Systems (FUZZ-IEEE 2002), vol. 1, pp. 711–716 (2002)

Control Loop Model of Virtual Company in BPM Simulation

Roman Šperka[1], Marek Spišák[1], Kateřina Slaninová[1],
Jan Martinovič[2] and Pavla Dráždilová[2]

[1] Silesian University in Opava,
School of Business Administration in Karviná,
Univerzitní nám. 1934/3a,
733 40, Karvin, Czech Republic
{sperka,spisak,slaninova}@opf.slu.cz
[2] VŠB Technical University in Ostrava,
Faculty of Electrical Engineering,
17. listopadu 15/2172,
708 33 Ostrava, Czech Republic
{jan.martinovic,pavla.drazdilova}@vsb.cz

Abstract. The motivation of the paper is to introduce agent-based technology in the business process simulation. As in other cases, such simulation needs sufficient input data. However, in the case of business systems, real business data are not always available. Therefore, multi-agent systems often operate with randomly (resp. pseudo randomly) generated parameters. This method can also represent unpredictable phenomena. The core of the paper is to introduce the control loop model methodology in JADE business process simulation implementation. At the end of this paper analysis of agent-based simulation outputs through process mining methods and methods for analysis of agents' behavior in order to verify the correctness of used methodology is presented. The business process simulation inputs are randomly generated using the normal distribution. The results obtained show that using random number generation function with normal distribution can lead to the correct output data and therefore can be used to simulate real business processes.

1 Introduction

Simulations used in experiments in the paper could be described as agent-based simulations [13] of Business Process Management (BPM). Business process is an activity adding the value to the company. Usual business process simulation approaches are based on the statistical calculation (e.g. [14]). But only several problems can be identified while using the statistical methods. There are a lot of other influences those are not able to be captured by using any business process model (e.g. the effects of the collaboration of business process participants or their communication, experience level, cultural or social factors). This method has only limited capabilities of visual presentation while running the simulation. Finally, an observer does not actually see the participants of business process dealing with each other.

V. Snasel et al. (Eds.): SOCO Models in Industrial & Environmental Appl., AISC 188, pp. 515–524.
springerlink.com

Agent-based simulations dealing with a company simulation can bring several crucial advantages [4], [11]. They can overcome some of the problems identified above. It is possible to involve unpredictable disturbance of the environment into the simulation with the agents. All of the mentioned issues are the characteristics of a multi-agent system (MAS).

One of the problems the simulations of business processes tackle with is the lack of real business data. Many researchers [9], [20] use randomly generated data instead. On the basis of our previous research, we used the normal distribution in our simulation experiments. We reported on more issues dealing with the business process and financial market simulations [16], [17]. The simulation approach described in this paper uses bellow mentioned control loop model [17], [2] as a core principle. The influence of randomly generated parameters on the simulation outputs while using different kinds of distributions is presented in [18].

The novel methodology and workflow described in this paper are implemented in the form of MAS [19]. JADE [3] development platform was chosen for the realization. JADE provides robust running and simulation environment allowing the distribution of thousands of software agents. Multi-agent system is used as a BPM framework in this paper. When finished, it shall cover the whole company structure from supply of the material, through the production process, up to the selling and shipment. The overall idea of the proposed novel methodology is to simulate real business processes and to provide predictive results concerning the management impact. This should lead to improved and effective business process realization.

To achieve the suitable design of the proposed system a mechanism for verification is needed. As selected input data are generated randomly, or can be set for various types of simulations, the appropriate verification of agents' behavior is required. Process mining methods [1] were used for identification of agents' behavior as well as the methods for identification of behavioral patterns working with similarity of sequences [10], [8]. Used method for behavioral pattern identification was described more detailed e.g. in [15], where behavioral patterns of students in e-learning system were identified.

This paper is structured as follows. Section 2 briefly informs about the control loop model. Multi-agent system implementation and mathematical definition of production function are presented in Section 3. In Section 4 the simulation results are introduced. In Section 5 the process mining methods are used in order to verify the correctness of implemented model. In Section 6 the conclusion and future research steps find their places.

2 Control Loop Methodology

While analytical modeling approaches are based mostly on the mathematical theories [12] the approach followed in this paper is based on experimental simulations. The generic business company used for these simulations based on the control loop paradigm is presented in Figure 1.

The methodology is based on the idea that the company could be presented as a control loop, where the market conditions as well as customers behavior are seen as an external part of the modeled system while the internal company behavior is subject

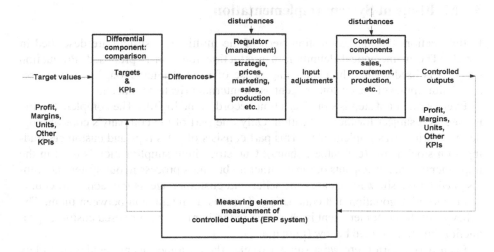

Fig. 1. Generic model of a business company (source: own)

to the simulation. We simulate one of the core business processes of a typical business company. The simulated business process is the selling of goods by sales representants (sales reps in the rest of the paper) to the customers. This process is only one part of the whole control loop. The subject presented in this paper consists of following types of agents: seller agents, customer agents, an informative, and a manager agent. The seller agent interacts with the customer agent according to the multi-agent approach. The interaction is based on the FIPA contract-net protocol [6]. This simplified system was extended by random disturbances influencing the agents' behavior. The number of customer agents is significantly higher than the number of seller agents. The behavior of agents is influenced by two randomly generated parameters using the normal distribution.

The control loop consists of controlled units like sales, purchase, production and others, managed by a regulator unit (the management of the company). The outputs of the controlled units are measured by the measuring unit and compared with the key performance indicators (KPIs). The differences found are sent to the regulator unit, which takes necessary measures in order to keep the system in the closeness with of the KPI values.

The overall workflow of the system proposed can be described as follows. The customer agents randomly generate the requests to buy random pieces of goods. The seller agents react to these requests according to their own internal decision functions and follow the contracting with the customers. The purpose of the manager agent is to manage the requests exchange. The contracting results into the sales events. More indicators of sale success like revenue, amount of sold goods, incomes, and costs are analyzed (more can be found in Section 4).

The motivation of this paper is to present a part of the whole system consisting of seller and customer agent types. In Section 3 the multi-agent system implementation is in detail described.

3 Multi-agent System Implementation

In this section, the implementation steps of the multi-agent system are described in details. The mathematical definition of production function is proposed. Production function is used during the contracting phase of agents' interaction. It serves to set up the limit price of the customer agent as an internal private parameter.

Every simulation step is stored as a time record in the log file. The complete log file is used as a subject for the verification. Only one part of the company's control loop, defined earlier, was implemented. This part consists of sales reps and customers trading with stock items (e.g. tables, chairs). One stock item simplification is used in the implementation. Participants of the contracting business process in our system are represented by the software agents - the seller and customer agents interacting in course of quotation, negotiation and contracting. There is an interaction between them. The behavior of the customer agent is characterized in our case by proposed customer production function defined bellow (Equation 1).

Each period turn (here we assume a week), the customer agent decides, if to buy something. His decision is defined randomly. If the customer agent decides not to buy anything, his turn is over; otherwise he creates a sales request and sends it to his seller agent. The seller agent answers with a proposal message (concrete quote starting with his maximal price − limit price ∗1.25). This quote can be accepted by the customer agent or not. The customer agents evaluate the quotes according to the production function. The production function was proposed to reflect the enterprise market share for the product quoted (market share parameter), sales reps' ability to negotiate, total market volume for the product quoted etc. (in e.g. [17]). If the price quoted is lower than the customer's price obtained as a result of the production function, the quote is accepted. In the opposite case, the customer rejects the quote and a negotiation is started. The seller agent decreases the price to the average of the minimum limit price and current price (in every iteration is getting effectively closer and closer to the minimum limit price), and resends the quote back to the customer. The message exchange repeats until there is an agreement or reserved time passes.

The sales production function for the m-th sales representative pertaining to i-th customer determines the price that i-th customer accepts [18].

$$c_n^m = \frac{\tau_n T_n \gamma \rho_m}{ZM \gamma_n^{mi}}, \tag{1}$$

where:

c_n^m - price of the n-th product quoted by m-th sales representative,
τ_n - company market share for the n-th product,
T_n - market volume for the n-th product in local currency,
γ - competition coefficient lowering the sales success,
ρ_m - quality parameter of the m-th sales representative,
Z - number of customers,
M - number of sales representatives in the company,
γ_n^{mi} - requested number of the n-th product by the i-th customer at m-th sales representative.

Customer agents are organized in groups and each group is being served by concrete seller agent. Their relationship is given; none of them can change the counterpart. Seller agent is responsible to the manager agent. Each turn, manager agent gathers data from all seller agents and makes state of the situation of the company. The data is the result of the simulation and serves to understand company behavior in a time depending on the agents' decisions and behavior. The customer agents need to know some information about the market. This information is given by informative agent. This agent is also responsible for the turn management and represents outside or controllable phenomena from the agents' perspective.

4 Simulation Results

In order to include randomly generated inputs, two important agents' attributes were chosen to be generated by pseudo random generator. Firstly, the seller agent's ability, and secondly the customer agent's decision about the quantity for purchase were used.

For generating random numbers from normal distribution (Gaussian) the Java library called Uncommon Maths written by [5] was used. For the values generation random `MerseneTwisterRNG` class was implemented. The class is a pure Java port of Makoto Matsumoto and Takuji Nishimura's proven and ultra-fast *Mersenne Twister Pseudo Random Number Generator for C*. The parametrization of MAS is listed in Table 1.

Table 1. List of agents' parameters

AGENT TYPE	AGENT COUNT	PARAMETER NAME	PARAMETER VALUE
Customer Agent	500	Maximum Discussion Turns	10
		Mean Quantity	50
		Quantity Standard Deviation	29
Seller Agent	50	Mean Ability	0.5
		Ability Standard Deviation	0.3
		Minimal Price	5
Manager Agent	0	Purchase Price	4
Market Info	1	Item Market Share	0.5
		Item Market Volume	5 000 000

One year of the selling/buying processes (52 turns weeks) was simulated (Figure 2). The output values are mostly closer to the mean than to the extremes. Obtained results demonstrate realistic KPIs of company simulated.

In Table 2 aggregated data can be seen, which reflects the most important characteristic of the normal distribution.

Both attributes the seller ability and also the selected quantity should be taken from the normal distribution for more realistic scope of the generation. The main significant data series from *Incomes* was used and the correlation analysis was made. The correlation coefficient -0.018 doesn't prove tight correlation binding between the results

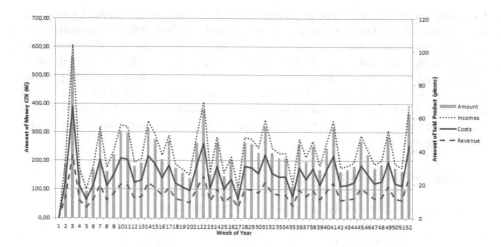

Fig. 2. The generation graph with normal distribution of attributes (source: own)

of simulation and random attributes generation with normal distribution. For the KPIs evaluation the amount, incomes, costs and revenue were counted. The values resulted from the 53 weeks of the company trading. Implemented MAS provides necessary results in the form of KPIs every week during one year of trading. Obtained KPIs could be compared from one simulation experiment to another. This could be used to analyze different simulation parameterizations and the impact on the company performance.

Table 2. Random Distribution Generation Output Results

AGGREGATION	AMOUNT (PIECES)	INCOMES (CZK)	COSTS (CZK)	REVENUE (CZK)
SUM	1 983.00	11 803.48	7 932.00	3 871.48
AVG	38.13	226.99	152.54	74.45
STD. DEV.	16.05	95.76	64.19	32.03

In Section 5 will be used the process mining method in order to verify the proposed model. If the verification confirms correct behavior of agents, the used model works also correct.

5   Verification of Simulation Implementation

The verification of implemented simulation of proposed multi-agent system was performed through the analysis of the agents' behavior. The system records all the actions performed by the agents into the log file. The log file structure corresponds to the following description: the rows of the log file represent events, which can be described by attributes like timestamp, performer (agent), type of action, and additional information (ability, stock items, price and other).

The agents' behavior was analyzed through the methods of process mining. Aalst in [1] defines a methodology for the analysis of business processes. This methodology was used for the extraction of sequences of actions performedx by the agents in the system. The sequences were extracted for each negotiation between the seller and customer agents; the other types of agents were analyzed as well. However, for the verification process, the behavior of the seller and customer agents was more important. More detailed description of sequence extraction and behavioral pattern identification using algorithms for the comparison of categorical sequences is described in our previous work, for example in [15], where behavioral patterns of students in e-learning system were analyzed. For the experiments in this paper, the method T-WLCS was used for finding the behavioral patterns of the agents.

After the sequence extraction phase, we have obtained 2 097 sequences. Using T-WLCS method for comparison of sequence similarity, we have constructed the similarity matrix, which can be represented using tools of graph theory. The finding of behavioral patterns of agents was performed through a weighted graph $G(V, E)$, where vertices V were the sequences and edges E represented the relations between the sequences on the basis of their similarity. The weight of edges w was defined by the sequence similarity (T-WLCS method). Detailed information of the original sequence graph is presented in Table 3.

Table 3. Description of Sequence Graph

INFORMATION	VALUE
Nodes	1975
Isolated Nodes	2
Edges	942 068
Connected Components	7

The sequence graph consisted of large amount of similar sequences. Moreover, it was dense and very large for further processing. Better interpretation of results was possible by finding the components, which can represent the behavioral patterns of agents. That was the reason, why we have used spectral clustering by Fiedler vector and algebraic connectivity [7]. The original sequence graph was divided into 7 components with similar sequences. Description of obtained components, their size and type of agents is presented in Table 4.

Visualization of obtained components using graph theory can be seen in Figure 3. Each component, determined by its color, consists of similar sequences and represents similar behavior. Each sequence is described by the information of its performer (agent). Therefore, we are able to identify groups of agents with similar behavior to verify the model.

The verification process of the implementation was based on finding behavioral patterns and groups of agents with similar behavior. Proposed method can be used successfully for better interpretation and description of real model working and of real behavior of the agents in the model. There are several input parameters, which can influence real behavior of the agents in the system. Proposed method can facilitate the analysis of

Table 4. Component Description

COMPONENT	SIZE	TYPE OF AGENT
C0	1 034	CustomerAgents
C1	498	CustomerAgents
C2	253	CustomerAgents
C3	136	SellerAgents
C4	51	SellerAgents
C5	1	ManagerAgent
C6	1	DisturbanceAgent

real behavior of the agents during the system simulation. As an example we can mention ability, which was set up randomly for the seller agents, where the values followed the normal distribution function. The ability is the parameter, which influents the seller agents' behavior. The higher the value is, the more skillful the agent is in the negotiation process. In the other words, the agent is able to finish the negotiation process in shorter time. This corresponds to shorter sequences extracted from the log file for this agent.

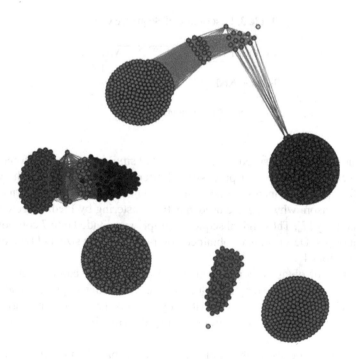

Fig. 3. Components with Similar Sequences

6   Conclusion

The BPM simulation experiment in the form of MAS was introduced in this paper. Proposed simulation model was implemented in order to simulate business process participants in virtual company. Overall methodology is based on the company's control loop. The simulation provides useful information about core business processes. Process mining methods used for the model verification confirmed the model correctness.

The verification process of the implementation was based on finding behavioral patterns and groups of agents with similar behavior. Proposed method can be used successfully for better interpretation and description of real model working and of real behavior of the agents in the model.

The next steps of our research will follow the rest parts of the whole control loop. The purchase side and the disturbance module will be implemented in order to analyze the performance of virtual company. The using of implemented MAS as a decision support tool for the management of company will be the leading idea in the future research.

Acknowledgment. This work was supported by the Bio-Inspired Methods: research, development and knowledge transfer project, reg. no. CZ.1.07/2.3.00/20.0073 funded by Operational Programme Education for Competitiveness, co-financed by ESF and state budget of the Czech Republic and by SGS, VSB – Technical University of Ostrava, Czech Republic, under the grant No. SP2012/151 Large graph analysis and processing.

References

1. van der Aalst, W.M.P.: Process Mining. Discovery, Conformance and Enhancement of Business Processes, p. 352. Springer, Heidelberg (2011) ISBN: 978-3-642-19344-6
2. Barnett, M.: Modeling & Simulation in Business Process Management. Gensym Corporation, pp. 6–7 (2003),
   http://bptrends.com/publicationfiles/
   11-03%20WP%20Mod%20Simulation%20of%20BPM%20-%20Barnett-1.pdf
   (accessed January 16, 2012)
3. Bellifemine, F., Caire, G., Trucco, T.: Jade Programmer's Guide. Java Agent Development Framework (2010), http://jade.tilab.com/doc/programmersguide.pdf
   (accessed January 16, 2012)
4. De Snoo, D.: Modelling planning processes with TALMOD. Master's thesis, University of Groningen (2005)
5. Dyer, D.W.: Uncommons Maths - Random number generators, probability distributions, combinatorics and statistics for Java (2010), http://maths.uncommons.org
   (accessed January 16, 2012)
6. Foundation for Intelligent Physical Agents (FIPA 2002) FIPA Contract Net Interaction Protocol. In Specification, FIPA,
   http://www.fipa.org/specs/fipa00029/SC00029H.pdf
7. Fiedler, M.: A property of eigenvectors of nonnegative symmetric matrices and its application to graph theory. Czechoslovak Mathematical Journal 25, 619–633 (1975)
8. Guo, A., Siegelmann, H.: Time-Warped Longest Common Subsequence Algorithm for Music Retrieval, pp. 258–261 (2004)

9. Hillston, J.: Random Variables and Simulation (2003),
   `http://www.inf.ed.ac.uk/teaching/courses/ms/notes/note13.pdf`
   (accessed January 16, 2012)
10. Hirschberg, D.S.: Algorithms for the longest common subsequence problem. J. ACM 24,
    664–675 (1977)
11. Jennings, N.R., Faratin, P., Norman, T.J., O'Brien, P., Odgers, B.: Autonomous agents for
    business process management. Int. Journal of Applied Artificial Intelligence 14, 145–189
    (2000)
12. Liu, Y., Trivedi, K.S.: Survivability Quantification: The Analytical Modeling Approach. De-
    partment of Electrical and Computer Engineering, Duke University, Durham, NC, U.S.A.
    (2011), `http://people.ee.duke.edu/~kst/surv/IoJP.pdf`
    (accessed January 16, 2012)
13. Macal, C.M., North, J.N.: Tutorial on Agent-based Modeling and Simulation. In: Proceed-
    ings: 2005 Winter Simulation Conference (2005)
14. Scheer, A.-W., Nüttgens, M.: ARIS Architecture and Reference Models for Business Process
    Management. In: van der Aalst, W.M.P., Desel, J., Oberweis, A. (eds.) Business Process
    Management. LNCS, vol. 1806, pp. 376–389. Springer, Heidelberg (2000)
15. Slaninova, K., Kocyan, T., Martinovic, J., Drazdilova, P., Snasel, V.: Dynamic Time Warping
    in Analysis of Student Behavioral Patterns. In: Proceedings of DATESO 2012, pp. 49–59
    (2012)
16. Spisak, M., Sperka, R.: Financial Market Simulation Based on Intelligent Agents - Case
    Study. Journal of Applied Economic Sciences VI(3(17)), 249–256 (2011) Print-ISSN 1843-
    6110
17. Vymetal, D., Sperka, R.: Agent-based Simulation in Decision Support Systems. In: Distance
    Learning, Simulation and Communication 2011. Proceedings, pp. 978–980 (2011) ISBN
    978-80-7231-695-3
18. Vymetal, D., Spisak, M., Sperka, R.: An Influence of Random Number Generation Function
    to Multiagent Systems. In: Proceedings: Agent and Multi-Agent Systems: Technology and
    Applications 2012, Dubrovnik, Croatia (2012)
19. Wooldridge, M.: Multi Agent Systems: An Introduction to, 2nd edn. John Wiley & Sons Ltd.,
    Chichester (2009)
20. Yuan, J., Madsen, O.S.: On the choice of random wave simulation in the surf zone processes.
    Coastal Engineering (2010),
    `http://censam.mit.edu/publications/ole3.pdf`

Intelligent Agents Design and Traffic Simulations

Michal Radecký and Petr Gajdoš

VŠB-TU Ostrava, IT4Innovations
17.listopadu 15, Ostrava, Czech Republic
{michal.radecky,petr.gajdos}@vsb.cz

Abstract. The development process of Multi-Agent Systems is mostly similar to development of standard information systems. However, there are a few special requests that have to be taken into account, e.g. multi-platform MAS architecture, internal architecture and implementation of particular agents, their autonomy and communication. This paper describes a development tool (**Agent**Studio) that covers main phases of agent development and simulation with emphasis to process modeling. These approaches support the creation of intelligent agents and reconfiguration of their behaviors to provide intelligence of them.

1 Motivation

A Multi-Agent System (MAS) attracts attention as an approach to complexity systems in recent years. Fundamental elements of MAS are represented by agents, individual software units in distributed systems. Efficiency of such systems depends on the quality of agents' internal architecture, features and their cooperation ability with other agents. Recent research deals with so called *intelligent agents* which dispose of additional skills that appear from logic, processes, information retrieval, etc. This paper describes a specification of agent behavior based on extended UML technique. Agents, which can modify their behaviors upon defined processes are main goal of our research.

Nowadays, there are standard approaches to develop whole information systems from the requirement specification to the deployment (UML, RUP, etc.). They can be used also for MAS, however, this is not proper way to build complex multi agent systems. Specific features of these systems (e.g. autonomy of element, communication among elements, logic) require some specific tools and/or approaches. The Agent UML (AUML) [1] and new features of UML 2.1 (and higher) and SysML [2] [3] can cover some phases of software process with respect to MAS. Nevertheless, the AUML is a quite old approach, that is no longer supported. On the other hand, new UML standards are too complicated for common users and they bring a lot of extensions which are not necessary for MAS development. In this case, we try to offer formal method to describe internal and external processes of the agents as well as user friedy approach. Selection of suitable implementation framework plays also an important role. A Java Agent Development framework (JADE) has been chosen for our purpose [4] [5] [6].

V. Snasel et al. (Eds.): SOCO Models in Industrial & Environmental Appl., AISC 188, pp. 525–535.
springerlink.com © Springer-Verlag Berlin Heidelberg 2013

2   Agent Behavior Modeling

Each agent is determined by its own objectives and the way to meet these objectives is founded on the internal behavior of a given agent. In the case of processional intelligent agents [7] [8], internal behavior is specified by an algorithm. Agent lives, behaves and reacts with respect to environment stimuli, its location and given algorithm [9]. Outer MAS behavior results from communication of particular agents and from interconnection of several internal agent behaviors. The behaviors can be changed dynamically with the aid of the *Reconfiguration approach* and communication is supported by *message passing*. These points are included in activity diagrams extension.

2.1   UML Activity Diagrams Extension

The UML Activity Diagrams represent a standard diagrammatic technique which describes the series of activities, processes and other control elements that together express an algorithm from the dynamical point of view [10] [11] [12]. Diagrams capture internal behavior of each agent, but they do not describe the interactions and communication between agents. Therefore, additional modifications and extensions are required to provide reconfiguration of behaviors and interaction among processes.

From now on, a *process* is a flow of atomic activities and sub-processes, that finally represents particular non-trivial action of the agents. An agent behavior consists of a set of such processes.

Our *Agent Behavior Diagrams* contain all the elements of the standard UML Activity Diagrams and new elements concerned with message passing, process hierarchy, input/output objects and resulting process scenarios (see the figure 1).

The simple, clear and formal specification of internal agent behaviors is a precondition for the following phases of multi-agent software process. Also other types of diagrams (e.g. sequential diagrams, maps of the agents communication) can be generated thanks to the new stereotypes and information (on objects, scenarios, activity scores, etc.), which were established in the *Agent Behavior Diagram*. The same holds for semi-automatic creation of source code templates of agents (e.g. agent interfaces, classes and methods).

2.2   ABD Rules

Several rules must be kept to build complex MAS model.

- Each process, as well as diagram, has just one *initial node* and just one *final node*. It is necessary for further joining of processes to obtain overall agent behavior.
- In the case of more final nodes within a process, the nodes are merged with appropriate process modification. Then, the previous final nodes are represented by *scenarios*, which are stored as an extra information in the merged node.

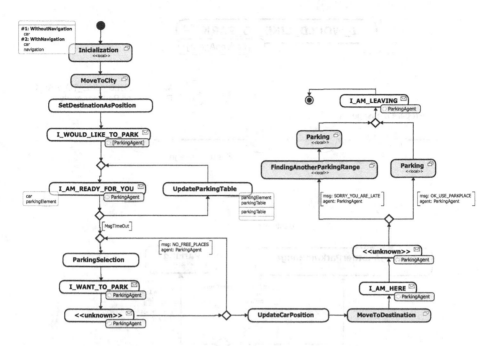

Fig. 1. Example of ABD (Agent Behavior Diagrams) with additional elements. It illustrates Car agent behavior during parking process.

- ABD are *well-formed* [13]. Van der Aalst defined a set of general structural rules for workflow nets (e.g. the level of split/join nodes preservation, no crossing of the levels of control flow). These rules were acquired and due this fact, the diagrams can be verified and formalized.

2.3   ABD and Workflow Nets

Thanks to mentioned rules it is possible to create such process specifications based on ABD, which will not only record the correct structure of the process, but will also further automated implemented or formalized. The transformation of ABD to the workflow network is one of possible step to formalization. Due to the above differences in the perception of objects within Petri nets/Workflow nets and activity diagrams, it is possible to speak only about formalization of the process control flow. If the transformation should include specific objects used as inputs or outputs of each activity, then the complete workflow network created from ABD can not longer meet the requirements of the definition and the result becomes the ordinary Petri Net (only one initial place, only one final place, tokens only in outplut place at the moment of cempletion of process, etc.). Figure 2 illustrates the transfer from parts of ABD to Workflow Net with respect to the formality of the process specifications, although generally the result is not complying with all rules for the Workflow nets.

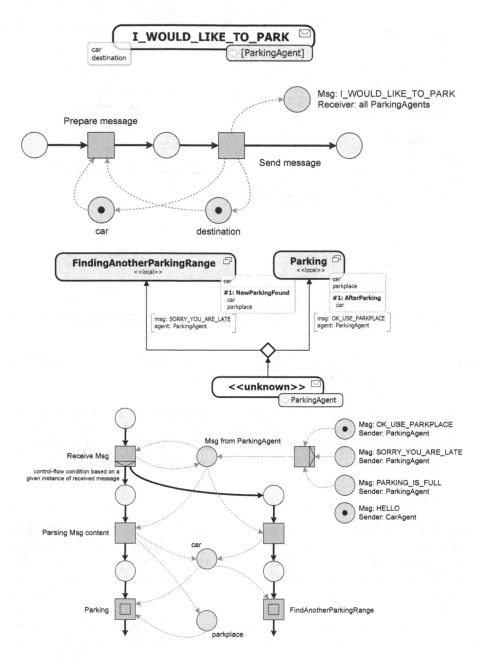

Fig. 2. Example of transformation of some ABD parts to corresponding Workfow Nets. Some parts of parking process are used there.

3   Reconfiguration of Behaviors

Every process, expect the *primary process*, can be specified by more than one diagram. Then, each diagram describes one *realization* of a given process. Realizations depend upon knowledge, experiences, environment and states of agents. The realizations can be stored within agent internal knowledge base or global MAS repository. Moreover, the agents can extend own sets of realizations thanks to communication and cooperation with other agents and/or platform facilities. This feature enables agents to learn.

The *behaviour reconfiguration approach* represents the way how to implement intelligent agents with respect to processes [14]. The idea behind the reconfiguration comes from the hypothesis, that each process (*reconfiguration point*) can be realized by different ways - realizations. Reconfiguration algorithm is applied in time of process firing. Each process requires a set of input objects and can produce the outputs. The same holds for the realizations.

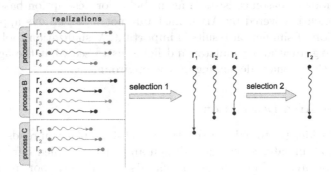

Fig. 3. The reconfiguration algorithm scheme

The figure 3 depicts the basic scheme of mentioned reconfiguration method. At the beginning, the set of all processes and their realizations is defined. Next, the selection phase is initiated. Depicted selection consists of two steps. The first one represents a simple selection of applicable realizations, based on input objects occurrences. The second one chooses the most suitable realization according to input objects properties, scores, etc. Methods of multicriterial analysis or logical tools can be used during the selection phase.

4   Traffic Simulations

Mentioned ideas and approaches are projected in an implementation of MAS in the area of traffic and transport. Developed **Agent**Studio applications are suitable for this purpose. [15]

Proposed approaches and methods have been proven in the area of traffic management. These are the reasons, that led us to choose this application area. First, the computational simulations are becoming increasingly important because, in some cases, it is the only way to study and interpret designed processes. These simulations may require very large computational power and the calculations have to be distributed on several computers. The MAS technology supports such kind of computation because of its independence from platforms, operation systems, etc.[16] [17] However, just selected traffic situations were taken into account to demonstrate the power of MAS technology, logic and **Agent**Studio. Next, several commercial systems pick up actual traffic data, create digital models of traffic infrastructure (roads, crossroads, traffic signs, etc.) and provide such data sets to use them within other projects. Then it is quite easy to use provided data in simplified form for **Agent**Studio Simulator and test agents' behaviors on real data. Logic and intelligent decision-making process play an important part in our project. Also this point is inherent with traffic simulations, e.g. if a traffic light is red, cars should stop before a crossroad. In other words, the car has to change its behavior. Most of such rules are well described in Highway Code and can be rewritten into Prolog and/or TIL (Transparent Intensional Logic) formulas. The fourth reason consists in agent behavior description based on UML modeling which is covered by ABD. Last but not least, an eye appealing way of presentation of simulation results is important. Visualization tool makes one part of the **Agent**Studio Simulator and helps us to see what the agents really do and how they behave depending on the environment.

4.1   Target Area Description

Traffic simulations try to reflect real situations taking place on roads. Nowadays, **Agent**Studio Simulator allows us to design and edit simplified infrastructure to test agents' behaviors. In the future, it will also enable to import real GIS data. These are important situations which we focused on:

- Cars overtake each other and they will recognize traffic obstacles.
- They safety pass through crossroads.
- They keep safety distance from other agents (cars).
- They keep basic rules defined in Highway Code.

Particular situation is solved during the internal agent's life with respect to agent's ability to make decisions. Finally, overall MAS development and simulation can be generally divided into steps as real world requirement analysis, Agents' begahirs modeling, source code generation, simulation, visualization, etc. So, the development of particular agents is only part of this overall process.

4.2   Traffic MAS Architecture

The figure 4 illustrates mentioned MAS architecture. Platforms **1** and **3** represent two parts of the real world, e.g. a town district with a separated parking lot. The data of such platform consists of a traffic infrastructure map. Next, each platform has a description of traffic elements located on the traffic infrastructure.

Finally, the data holds the information on mobile agents obtained from *proxy agents*. Generally, particular platform data reflects the state of the real world. Second parts of these platforms make environments for system agents which are responsible for communication with other agents (Proxy agents), map services (MapDispatcher agent) and for agent registration (WorldRegister agent). The platform **2** consists of mobile agents that represent cars moving in the real world. The platform **2** can be distributed on many hardware nodes according to FIPA standard.

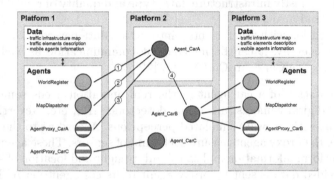

Fig. 4. MAS architecture for traffic simulation

The main relationships between platforms/agents are described in the following points (see the figure 4).

- A single car (Agent_CarA) registers itself into a given part of the world (Platform 1). It is done through the communication with WorldRegister agent that also creates a proxy agent for Agent_CarA (st. like proxy in the Object Oriented Programming).
- This connection provides an access to map services (road finding, infrastructure description, etc.) ensured by MapDispatcher agent (see the Yellow Pages in [18]).
- This communication realizes a synchronization of mobile agent data. It runs during whole car agent's life.
- Mobile agents can communicate between each other to negotiate emergency situations and/or to get some new knowledge, addition information on surrounding world, etc.

5   Car as an Intelligent Agent

Previously mentioned approach of reconfiguration is applied within every car agent. Moreover, a car agent needs certain form of inner structure for intelligent determination. The basic features of an intelligent mobile agent are perception, decision making and acting. [19] [20]

Perception. It is a natural feature of each live organism and the same holds for the intelligent agent. The perceptions are based on technical facilities. There are several information sources, e.g:

- sensors – speed information, accelerometer data, distance measurement, camera data, temperature, GPS or another positioning system, etc.
- computed knowledge – the information based on some approaches (logic, soft-computing, experting, etc.) that bring added values to the raw data from sensors or data sources, e.g. meteorological characteristics [21].
- GIS data – mostly infrastructure data, type and quality of roads, traffic signs locations and areas, etc.
- communication with other agents – interaction data passed with surrounding agents. Generally, it can be multi-agent communication as well as special approaches like Car2Car, Car2X, etc. [22] [23]

An Implementation of a car agent in the real operation needs some hardware sensors, e.g. digital cameras, ultrasonic detectors, GPS, etc. Nevertheless, this research deals with software simulated perceptions. According to the mentioned architecture, the proxy agents ensure such kind of sensors. These agents have full access to platform information and can simulate sight and location of substituted agents. Provided data is sent by proxy agent to its car agent via ACL message.

The last information source appears from communication among agents. Agents can interchange some knowledge (traffic jam, accident location, etc.) or they use services (parking payment, call for help, etc.).

Decision Making. A rational agent in a multi-agent world is able to reason about the world (what holds true and what does not), about its own cognitive state, and about that of other agents [24]. A theory formalizing reasoning of autonomous intelligent agents has thus to be able to "talk about" and quantify over the objects of agents' attitudes, iterate attitudes of distinct agents, express self-referential statements and respect different inferential abilities of agents. Since agents have to communicate, react to particular events in the outer world, learn by experience and be less or more intelligent, a powerful logical tool is of a critical importance.

Acting. Acting is a natural consequence of two previous features. It means the concrete process firing which leads to pass agent's objectives. At the end of this, agent's state, as well as the state of whole MAS, has to be updated. Then, the process compound of perception, decision making, and acting is repeated.

6   Complex Architecture Example

The described architecture of whole simulation and operation model is ready to be released as a robust framework for agents, management elements and infrastructure model. The simulation time shot is depicted (Fig. 5) as a screenshot

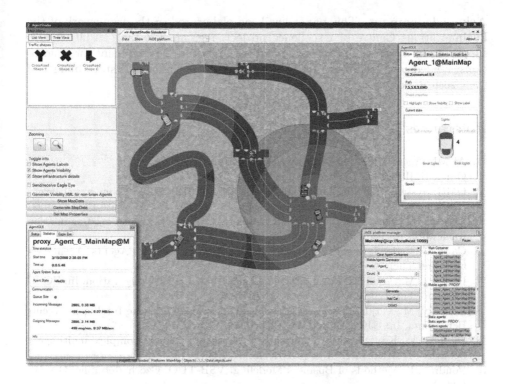

Fig. 5. Screenshot of AgensStudio with more complex simulation model

of **Agent**Studio application. There are more cars, more crossroads and roads. Also the JADE platform manager is presented there. it is upon on implemnetation if the cars (agents) are implemented as real object with real world data or as pure simulation object with generated data.

7 Conclusion and Future Work

The result of our research on Agent bahevior specification was described in this paper. The extended UML diagrams (ABD) were used to model agents' behaviors. The reconfiguration principles were developed and implemented in the **Agent**Studio application. So, in previous papers we prosented the vasic ideas of behavior specification and reconfiguration with. In this paper, we introduced the relation of ABD with formalization based on Petri Nets/Workflow Nets that is ground for the automated transformation, verification and implemetation. The future work attention has to be paid to imperfect and vague information, specially according to automated transformation and Workflow Nets interpretation as well as formal specification of some parts of our approach.

Acknowledgement. This paper has been elaborated in the framework of the IT4Innovations Centre of Excellence project, reg. no. CZ.1.05/1.1.00/02.0070 supported by Operational Programme 'Research and Development for Innovations' funded by Structural Funds of the European Union and state budget of the Czech Republic.

References

1. AUML: Agent UML (2005), http://www.auml.org
2. OMG: UML 2.4 OMG Specification (2011), http://www.uml.org
3. OMG: System Modeling Language Specification (2012), http://www.sysml.org
4. Bellifemine, F., Caire, G., Poggi, A., Rimassa, G.: JADE - A White Paper (1999)
5. Derksen, C., Branki, C., Unland, R.: Agent.gui: A multi-agent based simulation framework. In: FedCSIS, pp. 623–630 (2011)
6. Caire, G.: Jade: Java agents development framework. Software Agents, Agent Systems and Their Applications, 28–56 (2012)
7. Radecký, M., Vondrák, I.: Agents and Their Behavior's Reconfiguration. In: ECEC 2006, pp. 113–120 (2006)
8. Radecký, M., Gajdoš, P.: Process and Logic Approaches in the Intelligent Agents Behaviours. In: EJC 2006, pp. 264–270 (2006)
9. Sadilek, A., Kautz, H.A.: Location-based reasoning about complex multi-agent behavior. J. Artif. Intell. Res (JAIR) 43, 87–133 (2012)
10. Vondrák, I.: Methods of Business Modeling. VŠB-TUO, Czech Republic, Ostrava (2004)
11. Pilone, D., Pitman, N.: UML 2.0 in a Nutshell. O'Reilly Media, Sebastopol (2004)
12. Fortino, G., Rango, F., Russo, W.: Statecharts-Based JADE Agents and Tools for Engineering Multi-Agent Systems. In: Setchi, R., Jordanov, I., Howlett, R.J., Jain, L.C. (eds.) KES 2010, Part I. LNCS, vol. 6276, pp. 240–250. Springer, Heidelberg (2010)
13. van der Aalst, W.: Verification of Workflow Nets. Application and Theory of Petri Nets (1997)
14. Radecký, M., Gajdoš, P.: Process and Logic Approaches in the Intelligent Agents Behaviours. In: Information Modelling and Knowledge Bases XVIII. IOS Press, Amsterdam (2007)
15. Radecký, M., Gajdos, P.: Traffic simulation and intelligent agents. In: ICITST, pp. 1–6 (2009)
16. Navarro, L., Flacher, F., Corruble, V.: Dynamic level of detail for large scale agent-based urban simulations. In: AAMAS, pp. 701–708 (2011)
17. Dignum, F.: Agents for games and simulations. Autonomous Agents and Multi-Agent Systems 24(2), 217–220 (2012)
18. Odell, J.: The Foundation for Intelligent Physical Agents (2006), http://www.fipa.org/
19. Liang, J., Chen, L., Cheng, X.Y., Bo Chen, X.: Multi-agent and driving behavior based rear-end collision alarm modeling and simulating. Simulation Modelling Practice and Theory 18(8), 1092–1103 (2010)
20. Ehlert, P., Rothkrantz, L.: Microscopic traffic simulation with reactive driving agents. In: Proceedings. 2001 IEEE Intelligent Transportation Systems, pp. 860–865 (2001)

21. Corchado, E., Arroyo, N., Tricio, V.: Soft computing models to identify typical meteorological days
22. Car2Car: CAR 2 CAR Communication Consortium (2012), http://www.car-to-car.org/
23. Festag, A., Noecker, G., Strassberger, M., Lübke, A., Bochow, B., Torrent-moreno, M., Schnaufer, S., Eigner, R., Catrinescu, C., Kunisch, J.: Now Network on Wheels: Project Objectives, Technology and Achievements (2008)
24. Schelfthout, K., Holvoet, T.: An environment for coordination of situated multiagent systems. In: 1st Int. Workshop on Environments for MAS, E4MAS (2004)

Soft Computing Techniques Applied to a Case Study of Air Quality in Industrial Areas in the Czech Republic

Ángel Arroyo[1], Emilio Corchado[2,5], Verónica Tricio[3],
Laura García-Hernández[4], and Václav Snášel[5]

[1] Department of Civil Engineering, University of Burgos, Burgos, Spain
aarroyop@ubu.es
[2] Departmento de Informática y Automática, University of Salamanca, Salamanca, Spain
escorchado@usal.es
[3] Department of Physics, University of Burgos, Burgos, Spain
vtricio@ubu.es
[4] Area of Project Engineering, University of Cordoba, Spain
ir1gahel@uco.es
[5] Department of Computer Science,
VSB-Technical University of Ostrava, Czech Republic
IT4Innovations, Ostrava, Czech Republic
vaclav.snasel@vsb.cz

Abstract. This multidisciplinary research analyzes the atmospheric pollution conditions of two different places in Czech Republic. The case study is based on real data provided by the Czech Hydrometeorological Institute along the period between 2006 and 2010. Seven variables with atmospheric pollution information are considered. Different Soft Computing models are applied to reduce the dimensionality of this data set and show the variability of the atmospheric pollution conditions among the two places selected, as well as the significant variability of the air quality along the time.

Keywords: Artificial neural networks, soft computing, meteorology, statistical models, environmental conditions.

1 Introduction

Soft computing [1] [2] [3] consists of various techniques which are used to solve inexact and complex problems. It is used to investigate, simulate, and analyze complex issues and phenomena in an attempt to solve real-world problems.

There are many studies based on the application of different soft computing paradigms [4] [5] [6] to the field of air quality and environmental conditions. Some of them are based in the Czech Republic [7] where an air quality model is approached. Other studies visualize high dimensionality data sets with environmental information in order to find patterns of behavior in the climatology and pollution in local areas [8] or global areas [9].

In this study it is tested the validity of soft computing models to analyze the atmospheric pollution in two different places in Czech Republic. The data are provided by the ISKO (Czech Hydrometeorological Institute) [10].

V. Snasel et al. (Eds.): SOCO Models in Industrial & Environmental Appl., AISC 188, pp. 537–546.
springerlink.com   © Springer-Verlag Berlin Heidelberg 2013

The Czech Republic is the object of this study for its important tradition in the field of environmental conditions analysis [11]. This is a contribution to the study of environmental pollution in areas of high industrial activity.

The rest of this study is organized as follows. Section 2 presents the soft computing paradigms applied throughout this research. Section 3 details the real case study and Section 4 describes the experiments and results. Finally, Section 5 sets out the conclusions and future work.

2 Soft Computing Techniques

In order to analyze data sets with atmospheric pollution information, several dimensionality reduction techniques are applied, although the results are only shown for those that achieve the best performance.

Principal Components Analysis (PCA). PCA [12] gives the best linear data compression in terms of least mean square error and can be implemented by several artificial neural networks [13] [14].

Isometric Featured Mapping (ISOMAP). ISOMAP [15] is a nonlinear dimensionality reduction method which preserves the global properties of the data. Methods for nonlinear dimensionality reduction have proven successful in many applications, although the weakness of a method such as Multidimensional Scaling (MDS) [16] is that they are based on Euclidean distances and do not take the distribution of the neighboring data points into account. ISOMAP nonlinear dimensionality reduction [17] resolves this problem by attempting to preserve pairwise geodesic (or curvilinear) distance between data points. Geodesic distance is the distance between two points measured over the manifold. ISOMAP defines the geodesic distance as the sum of edge weights along the shortest path between two nodes (computed using Dijkstra's algorithm [18], for example). The doubly-centered geodesic distance matrix K in ISOMAP is presented by "Eq. 1":

$$K = \frac{1}{2} H D^2 H \tag{1}$$

Where $D^2 = D^2_{ij}$ means the element wise square of the geodesic distance matrix $D=[D_{ij}]$, and H is the centring matrix, given by "Eq. 2":

$$H = I_n - \frac{1}{N} e_N e_N^T \tag{2}$$

In which $e_N = [1...1]^T \in R^N$.

The top N eigenvectors of the geodesic distance matrix represent the coordinates in the new n-dimensional Euclidean space.

Local Linear Embedding (LLE). LLE [19] is an unsupervised learning algorithm that computes low-dimensional, neighborhood-preserving embeddings of high-dimensional inputs [20]. LLE attempts to discover nonlinear structure in high dimensional data by exploiting the local symmetries of linear reconstructions. Notably, LLE maps its inputs into a single global coordinate system of lower dimensionality, and its optimizations — though capable of generating highly nonlinear embeddings — do not involve local minima.

Suppose the data consist of N real-valued vectors x_i, each of dimensionality D, sampled from some smooth underlying manifold. Provided there is sufficient data (such that the manifold is well-sampled), it is expected that each data point and its respective neighbors will lie on or close to a locally linear patch of the manifold. The method can be defined as follows:

1. Compute the neighbors of each vector, x_i.
2. Compute the weights W_{ij} that best reconstruct each vector x_i from its neighbors minimizing the cost in by constrained linear fits, "Eq. 3".

$$\varepsilon(W) = \sum_i \left| x_i - \sum_j W_{ij} x_j \right|^2 \tag{3}$$

3. Finally, find point y_i in a lower dimensional space to minimize:, "Eq. 4":

$$\Phi(Y) = \sum_i \left| y_i - \sum_j W_{ij} y_j \right|^2 \tag{4}$$

This cost function in "Eq. 4" like the previous one in "Eq. 3" is based on locally linear reconstruction errors, but here the weights W_{ij} are fixed while optimizing the coordinates y_i. The embedding cost in "Eq. 4" defines a quadratic form in the vectors y_i.

3 An Atmospheric Pollution Real Case Study

This multidisciplinary study is focused on the analysis of real pollution data recorded in the Czech Republic [21]. It is based on the information collected by the national network stations in the Czeh Republic. The National air pollution network (SIS) is an open network [22], member of the International Air Quality network AIRNOW [23]. This network classifies the stations as traffic, urban, suburban, rural and industrial stations. In this study two industrial data acquisition stations are selected in order to analyze its air pollution.

1. Station coded as TOPRA, located in Ostrava-Privoz, Region Moravskoslezsky. Coordinates: 49° 51´ North latitude 18° 16´ East longitude, 207 masl (meters above sea level), the official estimated population of Ostrava was 310,464 inhabitants in 2011. The main industrial activity is focused on steel works, concentrates on metallurgy and machine engineering.
2. Station coded as SVELA, located in Veltrusy, Region Stredocesky. Coordinates: 50° 16´ North latitude 14° 19´ East longitude, 174 masl, 1,600 inhabitants. The main industrial activity is focused on engineering, chemical industry, food industry, glass industry, ceramics manufacture and printing industry.

The years under study are the following: 2006, 2008, 2009, 2010. Year 2007 is omitted because there are multiple data corrupt or missing, (the monthly summarized database is used). The study is focused on the twelve months of 2006, 2008, 2009 and 2010 and the 2 stations analyzed, 12 samples for each station and year, one sample per month. There is a total of 96 samples. In this research, the following seven variables were analyzed:

- Nitrogen Oxide (NO) - ug/m^3, Nitrogen Dioxide (NO$_2$)-ug/m³, Nitrogen Oxides (NO$_x$) - ug/m^3, Particulate Matter (PM10) - ug/m³, Sulphur Dioxide (SO$_2$) - ug/m^3. The choice of these contaminants is because they represent the majority of air emissions in areas of industrial activity.
- Benzene (BZN) - ug/m^3, Toluene (TLN) - ug/m^3. These contaminants belong to the Volatile Organic Compounds (VOCs). VOCs are linked to automobile emissions as a result of incomplete gas combustion. VOCs are also used in the production of paints, varnishes, lacquers, thinners, adhesives, plastics, resins and synthetic fibers.

This study examines the performance of several statistical and soft computing methods, in order to analyze the behaviour of the pollutants cited above to compare the air quality of an important industrial area in Czech Republic as Ostrava with other industrial point located in a very different location in Czech Republic as Veltrusy, along a wide period of time.

4   Results and Discussions

The results shown below correspond to the mean monthly values of the parameters described in the previous section, after applying the soft computing methods described above for clustering and dimensionality reduction:

Table 1. Label interpretation for (Figure 1, Figure 2 and Figure 3)

First Parameter - Month	Second Parameter - Year	Third Parameter - Station
(1 to 12)	2006, 2008, 2009, 2010	O – Ostrava - Privoz
		V – Veltrusy

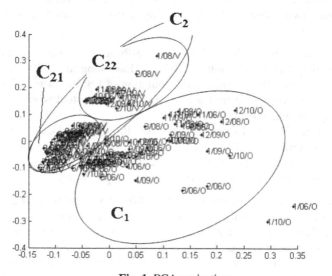

Fig. 1. PCA projections

Fig. 1. PCA identifies two main clusters, C_1 corresponds with the samples related to Ostrava station and C_2 corresponds with the samples of Veltrusy station. In cluster C_1 there is an important concentration of samples corresponding to the seasons of spring and summer, (see Table 2). The samples in C_1 corresponding to the seasons of winter and autumn are not so concentrated and present high levels of air pollution. Table 2 indicates the high values of PM10 and NO_x. The points labeled with '1/10/O' and '1/06/O' represents the days when the levels of SO_2 and specially PM10 reach values which exceed the health protection fixed in 40 ug/m³ [24], in more than 80 ug/m³ in both cases. In [24], the report about stations with annual average concentration corroborates this fact. Cluster C_2 (see Fig. 1) contains the samples corresponding to the seasons of spring, summer and autumn in Veltrusy, where the levels of air pollution are lower than winter in Veltrusy and lower than in Ostrava. In the season of winter, the concentration of NO_x is higher than in the rest of the year in both stations.

Table 2. Samples of each cluster identified by PCA and the range of values of each variable and units

Cluster #	Samples						
C_1	2/10/O 3/10/O 4/10/O 5/10/O 6/10/O 7/10/O 8/10/O 9/10/O 10/10/O 11/10/O 12/10/O 1/09/O 2/09/O 3/09/O 4/09/O 5/09/O 6/09/O 7/09/O 8/09/O 9/09/O 10/09/O 11/09O/ 12/09/O 1/08/O 2/08/O 3/08/O 4/08/O 5/08/O 6/08/O 7/08/O 8/08/O 9/08/O 10/08/O 11/08/O 12/08/O 2/06/O 3/06/O 4/06/O 5/06/O 6/06/O 7/06/O 8/06/O 9/06/O 10/06/O 11/06/O 12/06/O. Total number of samples: 46						
Var.	BZN (ug/m³)	TLN (ug/m³)	SO₂ (ug/m³)	PM10 (ug/m³)	NO₂(ug/m³)	NO (ug/m³)	NO$_X$ (ug/m³)
Range of values	2.2-21.9	4.3 -26.4	4.3 – 20.3	30.7 – 87.2	3.0-28.9	3.0-26.1	25.0-76.4
Description	Groups all the samples from the station based in Ostrava, except '1/10/O' and '1/06/O'. High values of PM10 and NO$_x$ in the season of winter						
C_{21}	3/10/V 4/10/V 5/10/V 6/10/V 7/10/V 8/10/V 9/10/V 10/10/V 3/09/V 4/09/V 5/09/V 6/09/V 7/09/V 8/09/V 9/09/V 10/09/V 11/09/V 3/08/V 4/08/V 5/08/V 6/08/V 7/08/V 8/08/V 9/08/V 10/08/V 11/08/V 12/08/V 1/06/V 11/06/V 12/06/V. Total number of samples: 30						
Range of Values	0.4 – 10.3	0.2 – 22.9	2.6 – 6.4	9.7 – 24.0	11.4 – 27.8	0.9 – 15.8	12.0 – 46.5
Description	Groups most of the samples from the station based in Veltrusy, samples corresponding to the seasons of spring, summer and autumn. High values only in NO$_x$.						
C_{22}	1/10/V 2/10/V 11/10/V 12/10/V 1/09/V 2/09/V 12/09/V 1/08/V 2/08/V 11/08/V 12/08/V 1/06/V 11/06/V 12/06/V. Total number of samples: 16						
Range of Values	1.3 – 10.5	1.2 – 14.6	2.6 – 11.8	11.5 – 28.1	23.9 – 40	3.4 – 17.9	30.0 – 75.0
Description	Groups most of the samples from the station based in Veltrusy corresponding to the season of winter. The samples with very high levels of NOx are located out of the cluster.						

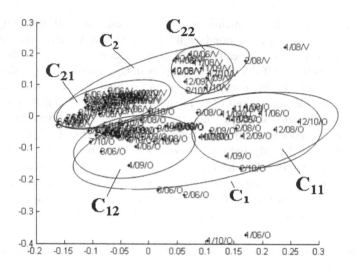

Fig. 2. LLE projections. Number of dimensions: 2, Neighbors: 20. Applies Euclidian distance function.

Fig. 2. LLE identifies the same two clusters of data than PCA (see Fig. 1) but in a more clear way. LLE is also capable of differentiating two subclusters of data (C_{11} and C_{12}), corresponding to the samples of data in C_1 (see Fig. 1). C_{11} corresponds to the most samples of winter in Ostrava (see Table 3). These samples offer the highest values of air pollution presenting in the data set, and C_{12} corresponds to the samples of the rest of the year. LLE in this case is more sensitive with samples with very high value in any of its parameters; i.e. ('2/06/O', '3/06/O'), where an important level of PM10 is detected and '1/08/V' and '2/08/V' where the concentrations of NO, NO_2, NO_X and PM10 are higher than in the rest of the year.

Table 3. Samples of each cluster identified by LLE and the range of values of each variable and units

Cluster #	Samples						
C_{11}	11/10/O 12/10/O 1/09/O 2/09/O 11/09/O 12/09/O 1/08/O 2/08/O 11/08/O 12/08/O 11/06/O 12/06/O. Total number of samples: 12						
Var.	BZN (ug/m³)	TLN(ug/m³)	SO_2 (ug/m³)	PM10(ug/m³)	NO_2 (ug/m³)	NO (ug/m³)	NO_X (ug/m³)
Range of Values	5.0 – 10.9	1.6 – 6.9	6.7 – 20.3	48.2 – 76.5	30.2 – 44.2	16.4 - 31	49.9 – 76.4
Description	Groups most of the samples of winter based in Ostrava. These samples offer the highest values of air pollution presenting in the data set. The samples with the maximum levels of NO_x and PM10 are located out of the cluster.						
C_{12}	3/10/O 4/10/O 5/10/O 6/10/O 7/10/O 8/10/O 9/10/O 10/10/O 3/09/O 4/09/O 5/09/O 6/09/O 7/09/O 8/09/O 9/09/O 10/09/O 3/08/O 4/08/O 5/08/O 6/08/O 7/08/O 8/08/O 9/08/O 10/08/O 3/06/O 4/06/O 5/06/O 6/06/O 7/06/O 8/06/O 9/06/O 10/06/O. Total number of samples: 32						

Table 3. (*Continued*)

Range of Values	3.1 – 21.9	1.6 – 8.0	4.1 – 16.7	26.4 – 75.0	21.0 – 42.3	2.8 – 21.9	25.0 – 68.1
Description	Groups most of the samples of summer, spring and autumn based in Ostrava. These samples offer high values of pollution, but not so high as in C_{11}.						
C_{21}	3/10/V 4/10/V 5/10/V 6/10/V 7/10/V 8/10/V 9/10/V 10/10/V 3/09/V 4/09/V 5/09/V 6/09/V 7/09/V 8/09/V 9/09/V 10/09/V 11/09/V 3/08/V 4/08/V 5/08/V 6/08/V 7/08/V 8/08/V 9/08/V 10/08/V 11/08/V 12/08/V 1/06/V 11/06/V 12/06/V. Total number of samples: 30						
Range of Values	0.4 – 10.3	0.2 – 22.9	2.6 – 6.4	9.7 - 24	11.4 – 27.8	0.9 – 15.8	12.0 – 46.5
Description	Groups most of the samples from the station located in Veltrusy, samples corresponding to the seasons of spring, summer and autumn. High values only of NO_x.						
C_{22}	1/10/V 2/10/V 11/10/V 12/10/V 1/09/V 2/09/V 2/09/V 11/08/V 12/08/V 1/06/V 12/06/V. Total number of samples: 11						
Range of Values	1.3 – 10.5	1.2 – 14.6	2.6 – 11.8	11.0 – 28.1	22.3 – 38.4	3.4 – 12.1	30.0 – 54.4
Description	Groups most of the samples from the station located in Veltrusy corresponding to the season of winter. Very high values of NO_x.						

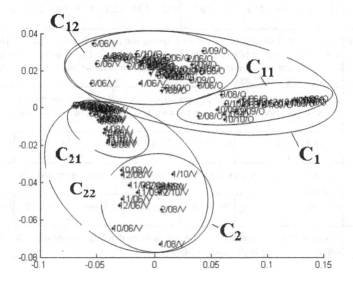

Fig. 3. ISOMAP projections. Number of dimensions: 2, Number of neighbors: 12. Applies Geodesic distance function, using Dijkstra's algorithm.

Fig. 3. ISOMAP offers the most clear results. Identifies four clusters as LLE does (LLE), but in a more sparse way. The clusters contain the same samples of data than applying LLE, except C_{11} (see Table 4). The drawback is that ISOMAP can not identify isolate samples with unusually high values in any of its parameters, as LLE does (see Fig. 2), which is very useful to identify exceptional situations. This property of ISOMAP is due to the fact that this method tries to preserve the global properties of the data, which may be an advantage or a disadvantage depending on the situation.

Table 4. Samples of each cluster identified by ISOMAP and the range of values of each variable and units

Cluster #	Samples						
C_{11}	11/10/O 12/10/O 1/09/O 2/09/O 11/09/O 12/09/O 1/08/O 2/08/O 11/08/O 12/08/O 11/06/O 12/06/O 9/08/O 9/10/O 10/10/O 10/09/O 6/06/O. Total number of samples: 17						
Var.	BZN (ug/m$^{3)}$	TLN (ug/m^3)	SO$_2$ (ug/m³)	PM10(ug/m³)	NO$_2$(ug/m³)	NO (ug/m³)	NO$_X$ (ug/m³)
Range of Values	2.2 – 16.6	1.6 – 6.9	6.2 – 33.2	28.9 – 123.7	21.1 – 48.5	2.8 – 27.7	25.0 – 76.4
Description	Groups most of the samples of winter based in Ostrava. These samples offer the highest values of air pollution presenting in the data set. ISOMAP is able to group samples than in LLE are located out of the cluster (see Fig. 2).						
C_{12}	3/10/O 4/10/O 5/10/O 6/10/O 7/10/O 8/10/O 9/10/O 10/10/O 3/09/O 4/09/O 5/09/O 6/09/O 7/09/O 8/09/O 9/09/O 10/09/O 3/08/O 4/08/O 5/08/O 6/08/O 7/08/O 8/08/O 9/08/O 10/08/O 1/06/O 2/06/O 3/06/O 4/06/O 5/06/O 7/06/O 8/06/O 9/06/O 10/06/O 1/06/V 5/06/V 6/06/V 7/06/V 8/06/V. Total number of samples: 38						
Range of Values	3.1 – 21.9	1.6 – 29.9	2.8 – 26.4	10.1 – 130.1	12.8 – 51.5	3.0 – 21.9	17.7 – 88.6
Description	Groups most of the samples of summer, spring and autumn based in Ostrava. These samples offer high values of air pollution but not so high as in C_{11}. This time some samples based in Veltrusy are presented, those with a similar pollution to Ostrava, belong to the season of summer of 2006.						
C_{21}	3/10/V 4/10/V 5/10/V 6/10/V 7/10/V 8/10/V 9/10/V 10/10/V 3/09/V 4/09/V 5/09/V 6/09/V 7/09/V 8/09/V 9/09/V 10/09/V 11/09/V 3/08/V 4/08/V 5/08/V 6/08/V 7/08/V 8/08/V 9/08/V 10/08/ 11/06/V 12/06/V. Total number of samples: 27						
Range of Values	0.5 – 3.1	0.2 - 3.5	0.6 – 4.5	9.7 – 19.9	10.3 – 31.5	0.9 – 7.6	12.0 – 29.6
C_{22}	1/10/V 2/10/V 11/10/V 12/10/V 1/09/V 2/09/V 1/08/V 2/08/V 11/08/V 12/08/V 10/06/V 11/06V 12/06/V. Total number of samples: 13						
Range of Values	0.4 – 10.5	0.2 – 14.6	2.6 – 11.8	11.0 – 28.1	22.3 – 40.0	3.4 – 25.0	30.0 – 75.0
Description	Groups most of the samples from the station located in Veltrusy corresponding to the season of winter. Very high values of NO$_x$.						

5 Conclusions

This study shows the different values of air pollution over several years in two Czech Republic localities with high industrial activity, Ostrava - Privoz and Veltrusy, appreciating higher levels in Ostrava. The study also reflects a similar behavior in both of the localities along the year, showing a significant increase of air pollution in the season of winter, especially in January. The pollutant with a higher increase of level in winter is PM10 and in a less important way NO$_X$. Ostrava is an important core of industrial production, especially production of steel and metal industries. These metal industries emit large amounts of PM10 and NO_x, which justifies these results, also the heavy traffic, specially the diesel engine type, represent another major source of PM10 emissions to the atmosphere. The high industrial production in winter and the low levels of precipitations in the first months of the year lead to more air pollution in winter. In Veltrusy the type of industry is different and more diversified than in Ostrava, resulting in a slight increase in all contaminants in winter when the weather is drier, but not so marked an increase of PM10 as in Ostrava.

Finally, the study demonstrates the behavior of the methods applied. PCA is the first method used in the data analysis process. It identifies the internal structure of the data. The rest of the techniques applied reaffirms and improves these graphical results. ISOMAP is able to group the data in compacted clusters. LLE is able to identify these same clusters and also detect unusual or interesting situations, that could be due to abnormal climatological situations in those days of high pollution, identifying these significant data points.

Acknowledgments. This research is partially supported through a projects of the Spanish Ministry of Economy and Competitiveness [ref: TIN2010-21272-C02-01] (funded by the European Regional Development Fund). This work was also supported in the framework of the IT4Innovations Centre of Excellence project, reg. no. CZ.1.05/1.1.00/02.0070 by operational programme \Research and Development for Innovations\ funded by the Structural Funds of the European Union and state budget of the Czech Republic.

References

[1] Corchado, E., Herrero, E.: Neural visualization of network traffic data for intrusion detection. Appl. Soft Comput. 11(2), 2042–2056 (2011)

[2] Corchado, E., Arroyo, A., Tricio, V.: Soft computing models to identify typical meteorological days. Logic Journal of the IGPL 19(2), 373–383 (2011)

[3] Vaidya, V., Park, J.H., Arabnia, H.R., Pedrycz, W., Peng, S.: Bio-inspired computing for hybrid information technology. Soft. Comput. 16(3), 367–368 (2012)

[4] Maqsood, I., Abraham, A.: Weather analysis using ensemble of connectionist learning paradigms. Applied Soft Computing 7(3), 995–1004 (2007)

[5] Wang, W., Men, C., Lu, W.: Online prediction model based on support vector machine. Neurocomputing 71(4–6), 550–558 (2008)

[6] Chattopadhyay, G., Chattopadhyay, S., Chakraborthy, P.: Principal component analysis and neurocomputing-based models for total ozone concentration over different urban regions of India. Theoretical and Applied Climatology, 1–11 (December 2011)

[7] Glezakos, T.J., Tsiligiridis, T.A., Iliadis, L.S., Yialouris, C.P., Maris, F.P., Ferentinos, K.P.: Feature extraction for time-series data: An artificial neural network evolutionary training model for the management of mountainous watersheds. Neurocomputing 73(1–3), 49–59 (2009)

[8] Arroyo, A., Corchado, E., Tricio, V.: Soft computing models to analyze atmospheric pollution issues. Logic Jnl. IGPL (February 2011)

[9] Arroyo, A., Corchado, E., Tricio, V.: A climatologycal analysis by means of soft computing models. Advances in Intelligent and Soft Computing 87, 551–559 (2011)

[10] CHMI Portal: Home,
    http://www.chmi.cz/portal/
    dt?portal_lang=en&menu=JSPTabContainer/P1_0_Home
    (accessed: April 19, 2012)

[11] Dvorska, A., Lammel, G., Klanova, J., Holoubek, I.: Kosetice, Czech Republic-ten years
    of air pollution monitoring and four years of evaluating the origin of persistent organic
    pollutants. Environmental Pollution 156(2), 403–408 (2008)

[12] Pearson, K.: On lines and planes of closest fit to systems of points in space. Philosophical
    Magazine 2(2), 559–572 (1901)

[13] Oja, J., Ogawa, E., Wangviwattana, H.: Principal component analysis by homogeneous
    neural networks, Part I: The weighted subspace criterion. IEICE Trans. Inf. Syst. E75-
    D(3), 366–375 (1992)

[14] Oja, E.: Neural networks, principal components, and subspaces. International Journal of
    Neural Systems 1(1), 61–68 (1989)

[15] Balasubramanian, M., Schwartz, E.L.: The isomap algorithm and topological stability.
    Science 295(5552), 7 (2002)

[16] Cox, A.A., Cox, T.F.: Multidimensional Scaling. In: Handbook of data visualization.
    Springer Handbooks of Computational Statistics, vol. III, pp. 315–347 (2008)

[17] Huang, S., Cai, C., Zhang, Y.: Dimensionality Reduction by Using Sparse Reconstruction
    Embedding. In: Qiu, G., Lam, K.M., Kiya, H., Xue, X.-Y., Kuo, C.-C.J., Lew, M.S. (eds.)
    PCM 2010, Part II. LNCS, vol. 6298, pp. 167–178. Springer, Heidelberg (2010)

[18] Shankar, N.R., Sireesha, V.: Using Modified Dijkstra's Algorithm for Critical Path Me-
    thod in a Project Network. International Journal of Computational and Applied Mathe-
    matics 5(2), 217–225 (2010)

[19] Roweis, S.T., Saul, L.K.: Nonlinear dimensionality reduction by locally linear embed-
    ding. Saul. Science 290(5500), 2323–2326 (2000)

[20] Smola, A.J., Schölkopf, B.: A tutorial on support vector regression. Statistics and Compu-
    ting 14(3), 199–222 (2004)

[21] Annual Tabular Overview,
    http://portal.chmi.cz/files/portal/docs/uoco/isko/
    tab_roc/tab_roc_EN.html (accessed: April 19, 2012)

[22] Networks of Ambient Air Quality Monitoring Stations,
    http://www.chmi.cz/files/portal/docs/uoco/isko/grafroc/
    groce/gr10e/akap21.html (accessed: May 08, 2012)

[23] International Air Quality,
    http://airnow.gov/index.cfm?action=topics.world
    (accessed: May 09, 2012)

[24] Stations with annual average conc.,
    http://portal.chmi.cz/files/portal/docs/uoco/isko/tab_roc/
    2010_enh/eng/pdf/MaximaH-RAP.pdf (accessed: April 19, 2012)

Author Index